D0053951

Numerical Modeling
in Science and Engineering

Numerical Modeling in Science and Engineering

Myron B. Allen III
Department of Mathematics
University of Wyoming
Laramie, Wyoming

Ismael Herrera
Institute of Geophysics
National University of Mexico
Mexico City, Mexico

George F. Pinder
Department of Civil Engineering
Princeton University
Princeton, New Jersey

A WILEY-INTERSCIENCE PUBLICATION

JOHN WILEY & SONS

New York • Chichester • Brisbane • Toronto • Singapore

Copyright © 1988 by John Wiley & Sons, Inc.

All rights reserved. Published simultaneously in Canada.

Reproduction or translation of any part of this work
beyond that permitted by Section 107 or 108 of the
1976 United States Copyright Act without the permission
of the copyright owner is unlawful. Requests for
permission or further information should be addressed to
the Permissions Department, John Wiley & Sons, Inc.

Library of Congress Cataloging in Publication Data:

Allen, Myron B., 1954–
    Numerical modeling in science and engineering.

    "A Wiley-Interscience publication."
    Includes bibliographies and index.
    1. Science—Mathematical models.   2. Engineering—
Mathematical models.   I. Pinder, George Francis,
1942–   . II. Herrera, Ismael.   III. Title.

Q158.5.A45   1988   507.2′4   87–28047
ISBN 0–471–80635–8

Printed in the United States of America

10 9 8 7 6 5 4 3 2 1

We dedicate this book to

Vernelle I. Baird
Socorro de Herrera
Ivan and Marjorie Charlton

PREFACE

Numerical modeling has become an essential tool in most branches of science and engineering. As a methodology, numerical modeling encompasses at least three phases. First, the modeler must formulate a mathematical description of the phenomena of interest. In many fields of engineering and applied science, these phenomena, and hence their governing equations, belong to the realm of continua. Second, the modeler must devise effective techniques for solving the governing equations. Since many physical systems give rise to complicated sets of partial differential equations together with auxiliary conditions, numerical approximation amenable to implementation on a computer often offers the only practical approach to this task. Finally, these first two phases cannot remain independent. The modeler must marshal a thorough qualitative understanding of the physical system, the mathematical equations, and their numerical analogs to ensure that the numbers coming out of the machine reflect the physics being modeled.

It seems to us that academia has traditionally separated these three phases into distinct disciplines such as continuum mechanics, differential equations, and numerical analysis. Each of these fields is indeed deep in its own right, and yet in combination the three viewpoints form a powerful approach to understanding and predicting natural phenomena.

This book is an effort to meld all three aspects of the numerical modeler's task into a unified treatment. We intend for the book to serve as a text for a year-long graduate course aimed at engineers, scientists, and applied mathematicians. Most of the material appearing here has already been used in various courses at the University of Wyoming, Princeton University, and the National University of Mexico (UNAM). In our experience students, perhaps more than their professors, seem to appreciate the insights gained in crossing traditional boundaries between disciplines.

The book is by no means exhaustive in scope. We treat only enough continuum mechanics in Chapter One to serve our purpose, which is to provide a consistent framework for the formulation of governing equations. Chapter Two, were it to stand alone, would be a whirlwind tour of numerical methods for partial differential equations. Our intent here is to present the basics, leaving many subtleties to applications in later chapters (and occasionally to the references cited). Chapters Three, Four, and Five all discuss physical, mathematical, and numerical aspects of the three major classes of partial differential equations: elliptic, parabolic, and hyperbolic. In keeping with our emphasis on physical applications, we have identified these classes of equations with steady-state systems, dissipative systems, and nondissipative systems, respectively, deliberately ignoring exceptions to this pigeonholing. Chapter Six examines several systems that are in some sense exotic: higher-order equations, nonlinear equations, and coupled systems of equations. Each topic covered here deserves its own book, and there exist equally important topics under this rubric that receive no treatment at all. Given limited space, we chose a few favorite aspects of a few favorite topics.

We owe debts of gratitude to many people for their help and intellectual guidance. Among these, of course, are many teachers and colleagues. Quite a few of these people may have no inkling of their contribution. Not wishing to fail at attempting an all-inclusive list, we shall be content to mention explicitly only a few individuals. Certainly deserving our thanks are Gonzalo Alduncin, Larry Bentley, Paul Imhoff, Joe Guarnaccia, Stuart Stothoff, and Tullio Tucciarelli, each of whom patiently read sections of the book before its publication. We also appreciate the diligence of our TEX-nical typists, Paula Sircin, Joanne Wyrick, and Mitzi Stephens, and our technical artist, Carolina Herrera. They deserve credit for a beautiful manuscript; we deserve blame for any remaining mistakes.

Laramie, Wyoming
Mexico City, Mexico
Princeton, New Jersey
January, 1988.

CONTENTS

Numerical Modeling
in Science and Engineering

CHAPTER ONE
BASIC EQUATIONS
OF MACROSCOPIC SYSTEMS

1.1. The Concept of a Continuum.

In our everyday experience we typically perceive events as continuum phenomena. By this we mean, intuitively, that matter appears to be a collection of bodies, each of which is infinitely divisible as is the three-dimensional space of Euclidean geometry. This description is deeply ingrained in our verbal descriptions of macroscopic events. For example, if we climb to a high altitude we feel a decrease in pressure that is apparently continuous as a function of distance from sea level. When we place a drop of ink into still water we see a small region of highly concentrated ink spread in a continuous fashion until the process appears to equilibrate at a lower, uniform concentration. If we lay the end of a cool metal bar against a hot stove, we expect the temperature at any point along the bar to increase continuously with time. As far as our senses are concerned, the air, water, and metal in these examples are continua. Concomitantly, the notions of pressure, concentration, and temperature that we use to describe these material responses are native to the continuum point of view. When made precise, such macroscopic descriptions commonly give rise to partial differential equations, and mathematicians, engineers, and applied scientists often face the task of solving such equations. This book presents methods for the systematic analysis and prediction of continuum phenomena using numerical solutions of partial differential equations.

It may appear that modern theories of matter and energy do not support the classical continuum picture. The quantum physics developed in the twentieth century treats the phenomena of everyday experience as ultimately corpuscular in nature. Thus matter consists of molecules, which consist of atoms, which consist of more elementary particles, and so on. Energy too is quantized, occurring, for example, as electron orbitals and photons. The corpuscular theories of the world about us seem to contradict the continuum picture to which we are so accustomed.

Appearances notwithstanding, the continuum picture that has descended from classical physics yields theories that are indispensible to modern engineering and applied science. From a pragmatic point of view, the question whether to adopt the continuum picture or the corpuscular pic-

1

ture reduces to the question of which set of theories proves most useful for the purposes we have in mind. The usefulness of continuum theories in describing many macroscopic phenomena contrasts with the impracticality (or even the impossibility) of deriving accurate descriptions of these phenomena from corpuscular theories. Thus, while statistical mechanics may have succeeded in yielding simple, qualitatively correct models of gas dynamics, diffusion, and heat flow, classical continuum mechanics furnishes better predictive tools in each case. No competent engineer would resort to the purely corpuscular theories to develop usable descriptions of the three examples mentioned here, much less to model the behaviors of airplane wings, petroleum reservoirs, or solar heating panels. In the words of Truesdell and Toupin (1960),

> It is classical physics by which we grasp the world about us: the heavenly motions, the winds and the tides, the terrestrial spin and the subterraneous tremors, prime movers and mechanisms, sound and flying, heat and light.

The field of continuum mechanics is a rich one, and we cannot hope to treat it properly in just one chapter. Nevertheless, the fundamentals of continuum mechanics provide a rigorous language in which to analyze problems in engineering and science. In particular it is through the kinematics, balance laws, and constitutive assumptions of continua that we arrive at most of the partial differential equations of practical interest. We shall therefore find it worthwhile to review these principles briefly.

1.2. Kinematics.

The goal of kinematics is to describe the movements of matter in space, without reference to forces. In the general case we consider some **body** B, being a set of **material points**, having labels such as X, and ask ourselves what point $\mathbf{x}$ in three-dimensional Euclidean space each material point X occupies at any time t. Let us assume that the collection B of material points takes the shape of some connected region of three-space. For example, B can be a body of fluid or a chunk of rigid solid, and our object is to describe the succession of spatial locations $\mathbf{x}$ that any material point $X \in B$ will occupy during a given time interval.

It is convenient to label material points X in the body B by their coordinates $\mathbf{X}$ measured in some reference configuration of B. We might, for example, identify the material point X by referring to the specific spatial point $\mathbf{X}$ that it occupies at time $t = 0$. This done, we need no longer use the somewhat abstract label X; instead, we can refer to X concretely by citing its spatial position $\mathbf{X}$ in the reference configuration. The coordinates $\mathbf{X}$

of X are called **material coordinates**, while the coordinates $\mathbf{x}$ are called **spatial coordinates**.

Motion and mass.

So far, we have not been very specific about the manner in which material points may move about in space. Let us now postulate that matter is indestructible, in other words, that volumes of matter that are initially nonzero and finite never vanish or become infinite. In addition, let us assume that matter is impenetrable, so that distinct material points in a single body never occupy the same point in space. Mathematically, this **axiom of continuity** implies the existence of a one-to-one correspondence between material points, identified by their reference coordinates $\mathbf{X}$, and spatial points $\mathbf{x}$ that they occupy. Hence we have a functional relationship

$$(1.2\text{-}1) \qquad \begin{aligned} \mathbf{x} &= \chi(\mathbf{X}, t) \\ &= (\chi_1(\mathbf{X}, t), \chi_2(\mathbf{X}, t), \chi_3(\mathbf{X}, t)) \end{aligned}$$

giving the place $\mathbf{x}$ where the material point labeled $\mathbf{X}$ is at time t. What is more, we require that this function be continuously differentiable and that the determinant of its Jacobian matrix, $\mathbf{J}$, whose (i,j)-th entry is $\partial \chi_i(\mathbf{X}, t)/\partial X_j = \partial x_i/\partial X_j$, never vanish. Then, by the inverse function theorem (see Williamson et al., 1972, Chapter Four), there exists an inverse mapping giving the material coordinates in terms of the spatial coordinates:

$$\chi^{-1}(\mathbf{x}, t) = \mathbf{X}.$$

The forward mapping (1.2-1) is called the **motion** of the body and is drawn schematically in Figure 1-1.

For convenience we shall allow χ to be differentiable as many times as we need, except perhaps at certain sets of singular points such as shocks, which will require special attention. Strictly speaking, χ is a function, while $\mathbf{x}$ denotes a position in space. Nevertheless, we shall frequently lapse into a mild abuse of notation, writing $\mathbf{x}(\mathbf{X}, t)$ for the functional relationship (1.2-1).

The motion provides all we really need to give a complete kinematic description of the body. However, there are two important modes of description available to us now. One approach is to describe the motion by specifying what happens to material points, labeled by their reference or material coordinates $\mathbf{X}$. This is the **Lagrangian** picture. It serves as the continuum analog of the mode of description most common in classical particle mechanics, where one commonly considers the motions of discrete sets of particles labeled, for example, 1, 2, ... As an alternative, we can

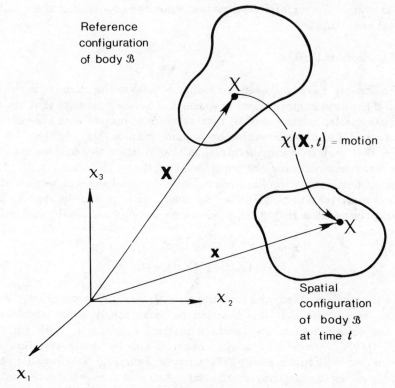

Figure 1-1. Schematic of the relationship between the reference configuration of a body $\mathcal{B}$ and the material coordinates $\mathbf{X}$ and spatial
coordinates $\mathbf{x}$ of a material point X in $\mathcal{B}$.

focus on what is happening at specified points $\mathbf{x}$ in space as the material
points move about. This second mode yields the **Eulerian** picture, and it
is often useful for describing the motions of bodies in which the trajectories
of material points may be difficult to describe in practice. Fluid flows serve
as examples of complicated motions in which the Eulerian picture is often
appropriate.

Given the motion $\mathbf{x} = \mathbf{x}(\mathbf{X}, t) = \chi(\mathbf{X}, t)$ of a material point $\mathbf{X}$, its
velocity is

$$\mathbf{V}(\mathbf{X}, t) = \frac{\partial \mathbf{x}}{\partial t}(\mathbf{X}, t),$$

the partial derivative of the motion with respect to time. Since this definition refers to a function of material coordinates, it belongs to the Lagrangian picture; its Eulerian counterpart is the spatial velocity defined by
$\mathbf{v}(\mathbf{x}, t) = \mathbf{V}(\chi^{-1}(\mathbf{x}, t), t)$. Less formally, we can also write $\mathbf{v}(\mathbf{x}(\mathbf{X}, t), t) =$
$\mathbf{V}(\mathbf{X}, t)$. The quantity $\mathbf{V}(\mathbf{X}, t)$ is the velocity of a given material point $\mathbf{X}$;

the quantity $v(x, t)$ is the velocity of whichever material point may be passing through the spatial point x. Thus, for a given motion, v and V have the same values for any given point (x, t). They are, however, functions of different variables. The point behind these different ways of relating V and v is that the Eulerian and Lagrangian descriptions are completely interchangeable owing to the invertibility of the motion. Similar definitions stand for the **acceleration**: in the Lagrangian picture

$$A(X, t) = \frac{\partial^2 x}{\partial t^2}(X, t),$$

and in the Eulerian picture $a(x, t) = A(X(x, t), t)$.

Also associated with the body B is a nonnegative scalar function M defined on subsets of B such that, for a subset or part $P \subset B$, $M(P, t)$ is the **mass** of P at time t. Given certain technical assumptions about B, M, and the subsets P on which M is defined (see Truesdell, 1966, Part 1), we can assert the existence of a nonnegative scalar density $\rho(x, t)$ defined on spatial points. This function ρ is the **mass density** of the body. The mass of the subset P is just the integral of ρ over the spatial region $x(P, t)$ that P occupies:

$$M(P, t) = \int_{x(P, t)} \rho(x, t) dx.$$

Here $dx = dx_1 dx_2 dx_3$ denotes the differential element of volume in spatial coordinates.

The material derivative.

We shall often need to consider the rates of change of various quantities with respect to time. Since we have available several frames of reference, defining temporal rates requires some care. We denote the time rate of change of a scalar, vector, or tensor quantity f *following a specified material point* X as follows:

(1.2-2)
$$\frac{Df}{Dt} = \frac{\partial f}{\partial t}\bigg|_X.$$

The notation on the right indicates that we hold the material point X constant during differentiation. This equation defines the **material derivative** of the quantity f.

The method for evaluating Df/Dt depends on whether f belongs to the Lagrangian point of view or to the Eulerian point of view. If f is a Lagrangian quantity, so that $f = f(X, t)$, then Equation (1.2-2) reduces to

$$\frac{Df}{Dt} = \frac{\partial f}{\partial t}(X, t).$$

On the other hand, if f is an Eulerian quantity, then $f = f(\mathbf{x}(\mathbf{X}, t), t)$, and by the chain rule we have

$$\frac{Df}{Dt}(\mathbf{x}, t) = \frac{\partial f}{\partial t}(\mathbf{x}, t) + \sum_{i=1}^{3} \frac{\partial f}{\partial x_i}(\mathbf{x}, t) \frac{\partial x_i}{\partial t}(\mathbf{X}, t).$$

If we denote $(\partial f/\partial x_1, \partial f/\partial x_2, \partial f/\partial x_3) = \nabla f$ and observe that the quantities $\partial x_i/\partial t$ are just components of the spatial velocity $\mathbf{v}$, we obtain

$$(1.2\text{-}3) \qquad \frac{Df}{Dt} = \frac{\partial f}{\partial t}(\mathbf{x}, t) + \mathbf{v}(\mathbf{x}, t) \cdot \nabla f(\mathbf{x}, t).$$

The first term on the right side of this equation is the **accumulation** of f; the second term on the right is the **advective rate** of f. It is an easy exercise to show that, regardless of whether one's viewpoint is Lagrangian or Eulerian, the operator D/Dt obeys familiar rules of differentiation:

$$\frac{D}{Dt}(f + g) = \frac{Df}{Dt} + \frac{Dg}{Dt},$$
$$\frac{D}{Dt}(fg) = f\frac{Dg}{Dt} + \frac{Df}{Dt}g.$$

A reader encountering D/Dt for the first time may feel apprehensive toward the distinction between D/Dt and $\partial/\partial t$, not to mention the total time derivative d/dt. To get a clearer picture of these operators, imagine following several smooth paths $\mathbf{x} = \boldsymbol{\xi}(t)$ in space-time coordinates, all the while measuring the rate of change of some quantity ϑ. For arbitrary smooth paths, the rate measured will be $d\vartheta/dt$, and obviously the actual function recorded for this rate may depend on the path $\mathbf{x} = \boldsymbol{\xi}(t)$ followed. The quantities $\partial\vartheta/\partial t$ and $D\vartheta/Dt$ just correspond to special choices of $\boldsymbol{\xi}(t)$. When $\mathbf{x} = \boldsymbol{\xi}(t)$ is a constant path, meaning $\boldsymbol{\xi}(t) = \mathbf{x}_0$ for some fixed spatial location $\mathbf{x}_0$, then the rate reflects the derivative of $\vartheta(\mathbf{x}, t)$ with respect to t when $\mathbf{x}$ remains fixed. Thus the rate measured along such a path is $\partial\vartheta/\partial t$. When $\mathbf{x} = \boldsymbol{\xi}(t)$ coincides with the trajectory of a fixed material point labeled $\mathbf{X}$, then the rate reflects the derivative of ϑ with respect to t with $\mathbf{X}$ held constant, that is, $D\vartheta/Dt$. For an amusing and vivid illustration of these concepts, we recommend a classic passage in Bird, Stewart, and Lightfoot's *Transport Phenomena* (1960, pp. 73–74) analyzing the fish concentration in the Kickapoo River.

Finally, recall that changing variables in a volume integral requires knowledge of the Jacobian determinant of the transformation. For our purposes, we shall find it useful to be able to change variables in integrals over **material volumes**, in other words, volumes defined as sets of specified material points. Since these volumes ordinarily change under the motion, the Jacobian determinant of interest will be det $\mathbf{J}$, which as a function of $(\mathbf{X}, t)$ may also vary in time. Let us examine how.

Because det $\mathbf{J}(\mathbf{X}, t)$ is a Lagrangian quantity,

$$\frac{D}{Dt}[\det \mathbf{J}(\mathbf{X}, t)] = \frac{\partial}{\partial t}[\det \mathbf{J}(\mathbf{X}, t)].$$

By an elementary rule for differentiating determinants,

(1.2-4)

$$\frac{\partial}{\partial t}(\det \mathbf{J}) = \det \begin{bmatrix} \partial J_{11}/\partial t & \partial J_{12}/\partial t & \partial J_{13}/\partial t \\ J_{21} & J_{22} & J_{23} \\ J_{31} & J_{32} & J_{33} \end{bmatrix}$$

$$+ \det \begin{bmatrix} J_{11} & J_{12} & J_{13} \\ \partial J_{21}/\partial t & \partial J_{22}/\partial t & \partial J_{23}/\partial t \\ J_{31} & J_{32} & J_{33} \end{bmatrix}$$

$$+ \det \begin{bmatrix} J_{11} & J_{12} & J_{13} \\ J_{21} & J_{22} & J_{23} \\ \partial J_{31}/\partial t & \partial J_{32}/\partial t & \partial J_{33}/\partial t \end{bmatrix}.$$

But notice

$$\frac{\partial J_{ij}}{\partial t} = \frac{\partial}{\partial t}\frac{\partial x_i}{\partial X_j} = \frac{\partial}{\partial X_j}\frac{\partial x_i}{\partial t} = \frac{\partial v_i}{\partial X_j}$$

$$= \sum_{k=1}^{3}\frac{\partial v_i}{\partial x_k}\frac{\partial x_k}{\partial X_j} = \sum_{i=1}^{3}\frac{\partial v_i}{\partial x_k}J_{kj}.$$

So, for example, we can rewrite the first determinant on the right side of Equation (1.2-4) as follows:

$$\sum_{k=1}^{3}\det \begin{bmatrix} (\partial v_1/\partial x_k)J_{k1} & (\partial v_1/\partial x_k)J_{k2} & (\partial v_1/\partial x_k)J_{k3} \\ J_{21} & J_{22} & J_{23} \\ J_{31} & J_{32} & J_{33} \end{bmatrix}$$

$$= \det \begin{bmatrix} (\partial v_1/\partial x_1)J_{11} & (\partial v_1/\partial x_1)J_{12} & (\partial v_1/\partial x_1)J_{13} \\ J_{21} & J_{22} & J_{23} \\ J_{31} & J_{32} & J_{33} \end{bmatrix}$$

$$= \frac{\partial v_1}{\partial x_1}\det \mathbf{J}.$$

(The other determinants in the sum both have two rows differing only by a factor of $\partial v_1/\partial x_k$ and so vanish.) If we reason similarly for the second and third terms on the right side of Equation (1.2-4), we find

$$\frac{\partial}{\partial t}(\det \mathbf{J}) = (\frac{\partial v_1}{\partial x_1} + \frac{\partial v_2}{\partial x_2} + \frac{\partial v_3}{\partial x_3})\det \mathbf{J} = (\nabla \cdot \mathbf{v})\det \mathbf{J}$$

or, following our observation that det **J** is a Lagrangian quantity,

$$(1.2\text{-}5) \qquad\qquad \frac{D}{Dt}(\det \mathbf{J}) = (\nabla \cdot \mathbf{v}) \det \mathbf{J}.$$

1.3. Balance Laws.

The axioms of mass, momentum, angular momentum, and energy balance and the entropy inequality furnish the basic laws from which we derive field equations modeling the behavior of continua. These laws arise as fundamental postulates, their primitive forms not being deduced from other mechanical propositions but rather serving as primary assumptions to which the equations governing specific materials are subordinate. To introduce the balance laws, we shall state the general form of a global balance law for a generic conserved quantity, which we provisionally denote as Ψ. Then we shall demonstrate how the global or integral form of the general law reduces to a differential equation, valid at points in the body where the field is continuous, with a jump condition valid at loci of discontinuities. This accomplished, we shall establish the particular balance laws for mass, momentum, angular momentum, and energy, and we shall review the entropy inequality.

General form of balance laws.

The general form of a **global balance law** for a conserved quantity Ψ is as follows:

$$(1.3\text{-}1) \qquad \frac{d}{dt}\int_{\mathcal{V}-\Sigma} \rho\Psi\, d\mathbf{x} - \oint_{\partial\mathcal{V}-\Sigma} \boldsymbol{\tau}\cdot\mathbf{n}\, d\mathbf{x} - \int_{\mathcal{V}-\Sigma} \rho g\, d\mathbf{x} = 0.$$

In this equation, Ψ can be a scalar, vector, or tensor whose value changes in time and space. The tensorial characters of $\boldsymbol{\tau}$ and g must preserve consistency among all terms in Equation (1.3-1). The quantity $\rho\Psi$ must be a density; in physical terms this means $\rho\Psi$ must be a quantity defined per unit volume occupied by the body under investigation. The quantity $\boldsymbol{\tau}$ is the **flux** of $\rho\Psi$ across mathematical surfaces in space, and the quantity g representes the **external supply** of Ψ. The physical interpretations of Ψ, $\boldsymbol{\tau}$, and g will become clearer when we consider particular balance laws, but for now let us allow them to remain abstract.

The integrals appearing in Equation (1.3-1) are volume and surface integrals associated with a material volume $\mathcal{V}$ that is a subset of the body's configuration. (We shall not use any special notation for the element of surface integration, even though this is a fairly common practice elsewhere.) Σ denotes an orientable surface containing all points in space at which the variables appearing in the integrands may be discontinuous, and $\mathcal{V} - \Sigma$ signifies the set of all points in $\mathcal{V}$ that do not belong to Σ. The symbol $\partial\mathcal{V}$ denotes the surface bounding the volume $\mathcal{V}$, and $\mathbf{n}$ is a unit vector normal to $\partial\mathcal{V}$ and pointing outward from $\mathcal{V}$. Also, for future use, let us denote by $\mathbf{v}_\Sigma$ the velocity of the surface Σ and by $\mathbf{n}_\Sigma$ the unit vector normal to Σ. Provided Σ is orientable, we can unambiguously choose a direction "outward" from Σ and stipulate that $\mathbf{n}_\Sigma$ point in that direction. Figure 1-2 depicts this geometry.

Being a global balance law, Equation (1.3-1) is an integral equation. Let us review an argument that reduces this integral form to a local or differential balance equation along with a jump condition governing loci of discontinuities. The first task in the reduction is to convert the left side of Equation (1.3-1) to a single volume integral accompanied by an integral over the surface Σ of discontinuity. To achieve this objective, let us first state the **generalized Gauss Theorem**. Suppose $\mathcal{V}$ is a connected region in three-dimensional Euclidean space with orientable boundary $\partial\mathcal{V}$ having outward unit normal vector $\mathbf{n}$. Also, let $\mathbf{f}(\mathbf{x})$ be a vector-valued function that is continuously differentiable on $\mathcal{V}$ with the possible exception of jump discontinuities on an orientable, smooth surface Σ intersecting $\mathcal{V}$ and having unit normal $\mathbf{n}_\Sigma$. Then

$$(1.3\text{-}2) \qquad \oint_{\partial\mathcal{V}-\Sigma} \mathbf{f} \cdot \mathbf{n} \, d\mathbf{x} = \int_{\mathcal{V}-\Sigma} \nabla \cdot \mathbf{f} \, d\mathbf{x} + \int_{\Sigma} \lfloor \mathbf{f} \rceil \cdot \mathbf{n}_\Sigma \, d\mathbf{x},$$

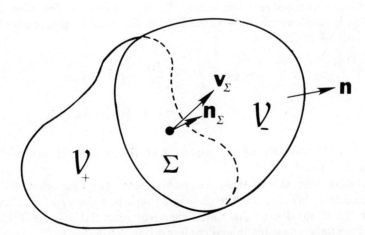

Figure 1-2. Schematic of a material volume $\mathcal{V}$ with boundary $\partial\mathcal{V}$ and unit outward normal vector $\mathbf{n}$. The surface Σ of discontinuity has normal vector $\mathbf{n}_\Sigma$ and velocity $\mathbf{v}_\Sigma$. Σ divides $\mathcal{V}$ into disjoint parts, $\mathcal{V}_+$ and $\mathcal{V}_-$.

where the notation $[f]$ denotes the **jump** in $\mathbf{f}$ across Σ:

$$[f] = \lim_{x \to \Sigma_+} f(x) - \lim_{x \to \Sigma_-} f(x) = f_+ - f_-.$$

We shall show how the generalized theorem (1.3-2) follows from the standard Gauss (or divergence) theorem, whose proof may be found in Williamson et al. (1972, Chapter Seven). Consider a volume $\mathcal{V}$ split by a surface Σ on which $\mathbf{f}$ has a jump discontinuity, as shown in Figure 1-2. Applying the standard Gauss theorem to the subvolumes $\mathcal{V}_+$, $\mathcal{V}_-$ formed by Σ yields

$$\int_{\mathcal{V}_+} \nabla \cdot f \, dx = \int_{\partial \mathcal{V}_+} f \cdot n \, dx + \int_{\Sigma} [\lim_{x \to \Sigma_+} f(x)] \cdot n_{\Sigma} \, dx,$$

$$\int_{\mathcal{V}_-} \nabla \cdot f \, dx = \int_{\partial \mathcal{V}_-} f \cdot n \, dx + \int_{\Sigma} [\lim_{x \to \Sigma_-} f(x)] \cdot (-n_{\Sigma}) \, dx.$$

Adding these two equations gives the desired result,

$$\int_{\mathcal{V}-\Sigma} \nabla \cdot f \, dx = \int_{\partial \mathcal{V}-\Sigma} f \cdot n \, dx + \int_{\Sigma} [f] \cdot n_{\Sigma} \, dx.$$

Next, we state the **extended Reynolds transport theorem**, which tells us how volume integrals change over time. Let $\mathcal{V}$ be a connected material region, and suppose Σ is an orientable smooth surface intersecting $\mathcal{V}$ and having unit normal vector n_{Σ}. If f is a real-valued function defined on $\mathcal{V}$ that is continuously differentiable with the possible exception of jump discontinuities at Σ, then

(1.3-3)
$$\frac{d}{dt} \int_{\mathcal{V}-\Sigma} f \, dx = \int_{\mathcal{V}-\Sigma} \left(\frac{Df}{Dt} + f \nabla \cdot v \right) dx$$
$$+ \int_{\Sigma} [f(v - v_{\Sigma})] \cdot n_{\Sigma} \, dx,$$

where v is the velocity of the material at the surface Σ and v_{Σ} is the velocity of Σ itself.

To show why this is true, let us first treat the case when f has no discontinuities. We can use the change-of-variables theorem of calculus to convert the integral over the time-dependent material volume $\mathcal{V}(t)$ to an integral over the constant initial configuration $\mathcal{V}(0)$:

$$\frac{d}{dt} \int_{\mathcal{V}(t)} f \, dx = \frac{d}{dt} \int_{\mathcal{V}(0)} f \det \mathbf{J} \, dx,$$

where det **J** is the Jacobian determinant of the motion. Since $\mathcal{V}(0)$ is independent of t,

$$\frac{d}{dt} \int_{\mathcal{V}(0)} f \det \mathbf{J} \, d\mathbf{x} = \int_{\mathcal{V}(0)} \frac{D}{Dt} (f \det \mathbf{J}) \, d\mathbf{x}$$

$$= \int_{\mathcal{V}(0)} \left[\frac{Df}{Dt} \det \mathbf{J} + f \frac{D}{Dt} (\det \mathbf{J}) \right] d\mathbf{x}.$$

Now we apply Equation (1.2-5) and then change variables back to an integral over $\mathcal{V}(t)$:

$$\int_{\mathcal{V}(0)} \left[\frac{Df}{Dt} \det \mathbf{J} + f \frac{D}{Dt} (\det \mathbf{J}) \right] d\mathbf{x}$$

$$= \int_{\mathcal{V}(0)} \left(\frac{Df}{Dt} + f \nabla \cdot \mathbf{v} \right) \det \mathbf{J} \, d\mathbf{x}$$

$$= \int_{\mathcal{V}(t)} \left(\frac{Df}{Dt} + f \nabla \cdot \mathbf{v} \right) d\mathbf{x}.$$

This result is the standard Reynolds transport theorem. By the Gauss theorem we can also write it as follows:

$$\frac{d}{dt} \int_{\mathcal{V}(t)} f \, d\mathbf{x} = \int_{\mathcal{V}(t)} \frac{Df}{Dt} \, d\mathbf{x} + \oint_{\partial\mathcal{V}(t)} f \mathbf{v} \cdot \mathbf{n} \, d\mathbf{x}.$$

Now suppose a surface Σ of discontinuity is present as drawn in Figure 1-2. Applying the result just proved to the regions $\mathcal{V}_+$ and $\mathcal{V}_-$ where f is continuously differentiable, we get

$$\frac{d}{dt} \int_{\mathcal{V}_+} f \, d\mathbf{x} = \int_{\mathcal{V}_+} \frac{Df}{Dt} \, d\mathbf{x} + \int_{\partial\mathcal{V}_+} f \mathbf{v} \cdot \mathbf{n} \, d\mathbf{x}$$

$$+ \int_{\Sigma} \left(\lim_{\mathbf{x} \to \Sigma_+} f \right) \mathbf{v}_\Sigma \cdot (-\mathbf{n}_\Sigma) \, d\mathbf{x},$$

$$\frac{d}{dt} \int_{\mathcal{V}_-} f \, d\mathbf{x} = \int_{\mathcal{V}_-} \frac{Df}{Dt} \, d\mathbf{x} + \int_{\partial\mathcal{V}_-} f \mathbf{v} \cdot \mathbf{n} \, d\mathbf{x}$$

$$+ \int_{\Sigma} \left(\lim_{\mathbf{x} \to \Sigma_-} f \right) \mathbf{v}_\Sigma \cdot \mathbf{n}_\Sigma \, d\mathbf{x}.$$

Adding these two equations gives the equation

$$\frac{d}{dt} \int_{\mathcal{V}-\Sigma} f \, d\mathbf{x} = \int_{\mathcal{V}-\Sigma} \frac{Df}{Dt} \, d\mathbf{x} + \int_{\partial\mathcal{V}-\Sigma} f \mathbf{v} \cdot \mathbf{n} \, d\mathbf{x}$$

$$+ \int_{\Sigma} \lfloor f \rfloor \mathbf{v}_\Sigma \cdot \mathbf{n}_\Sigma \, d\mathbf{x}.$$

Now applying the generalized Gauss theorem (Equation 1.3-2) to the second term on the right, we obtain the desired result, Equation (1.3-3).

With these theorems in hand, we can resume work on the global balance law. Applying the generalized Gauss theorem to the second term in Equation (1.3-1) and using the extended Reynolds transport theorem (1.3-3) on the first term allows us to rewrite the general global balance law as the sum of a volume integral and a surface integral over Σ:

(1.3-4) $$\int_{\mathcal{V}-\Sigma} \left[\frac{D}{Dt}(\rho\Psi) + \rho\Psi\nabla\cdot\mathbf{v} - \nabla\cdot\boldsymbol{\tau} - \rho g \right] d\mathbf{x}$$

$$+ \int_{\Sigma} \left[\rho\Psi(\mathbf{v} - \mathbf{v}_\Sigma) - \boldsymbol{\tau} \right] \cdot \mathbf{n}_\Sigma \, d\mathbf{x} = 0.$$

The second task in reducing the global balance law (1.3-1) to differential form is to impose the **principle of localization**. Mathematically, this principle consists in the **extended duBois-Reymond lemma**. In its usual form, the duBois-Reymond lemma is as follows: suppose $f(\mathbf{x})$ is continuous on a region $\mathcal{U}$ and $\int_{\mathcal{V}} f(\mathbf{x}) \, d\mathbf{x} = 0$ for every subregion $\mathcal{V} \subset \mathcal{U}$. Then $f(\mathbf{x}) = 0$ for all $\mathbf{x} \in \mathcal{U}$. The extended version of the lemma accommodates the case when a surface Σ of discontinuity is present. Specifically, let $f(\mathbf{x})$ be continuous except for jump discontinuities along an orientable smooth surface $\Sigma \subset \mathcal{U}$, let $g(\mathbf{x})$ be continuous on Σ, and suppose

$$\int_{\mathcal{V}-\Sigma} f(\mathbf{x}) \, d\mathbf{x} + \int_{\Sigma\cap\mathcal{V}} g(\mathbf{x}) \, d\mathbf{x} = 0$$

for arbitrary subregions $\mathcal{V} \subset \mathcal{U}$. Then $f(\mathbf{x}) = 0$ for every $\mathbf{x} \in \mathcal{U} - \Sigma$, and $g(\mathbf{x}) = 0$ for every $\mathbf{x} \in \Sigma$.

To see why the usual version of the lemma holds, we can assume the contrary and thereby deduce a contradiction. Thus, assume f is continuous on $\mathcal{U}$ and suppose $f(\mathbf{x}_0) \neq 0$ for some $\mathbf{x}_0 \in \mathcal{U}$, even though $\int_{\mathcal{V}} f(\mathbf{x}) \, d\mathbf{x} = 0$ for every subregion $\mathcal{V} \subset \mathcal{U}$. Without loss of generality, let us assume $f(\mathbf{x}_0) > 0$. (Otherwise consider $-f$.) Since f is continuous by hypothesis, there must be some small neighborhood $\mathcal{V}_0 \subset \mathcal{U}$ containing $\mathbf{x}_0$ such that $f(\mathbf{x}) > 0$ for every $\mathbf{x} \in \mathcal{V}_0$. But if this is so then $\int_{\mathcal{V}_0} f \, d\mathbf{x} > 0$, the desired contradiction. Now let us demonstrate the extended form of the lemma. Assume that f has jump discontinuities along an orientable smooth surface $\Sigma \subset \mathcal{U}$, that g is continuous on Σ, and that the integral equation stated in the hypothesis of the lemma holds. Observe $\int_S g \, d\mathbf{x} = 0$ for any connected piece S of the surface Σ, since on any subregion $\mathcal{V} \subset \mathcal{U}$ with $\mathcal{V} - \Sigma = S$,

$$\int_{\mathcal{V}-S} f \, d\mathbf{x} + \int_S g \, d\mathbf{x} = 0$$

by hypothesis, and $\int_{\mathcal{V}-s} f \, dx = 0$ by the usual form of the lemma. Since g is continuous on Σ, we can mimic the argument proving the usual duBois-Reymond lemma to complete the proof of the extended form.

To apply this lemma, let $\mathbf{x}(\mathcal{B})$ denote the spatial configuration of the body $\mathcal{B}$. We must assume that the quantity $D(\rho\Psi)/Dt + \rho\Psi\nabla \cdot \mathbf{v} - \nabla \cdot \boldsymbol{\tau} - \rho g$ is a continuous function of spatial position over $\mathbf{x}(\mathcal{B}) - \Sigma$ with the possible exception of jump discontinuities on the surface Σ. Furthermore, the jump $\lfloor \rho\Psi(\mathbf{v} - \mathbf{v}_\Sigma) - \boldsymbol{\tau} \rfloor \cdot \mathbf{n}_\Sigma$ must be continuous on Σ. Then, if the integral equation (1.3-4) holds for arbitrary subregions $\mathcal{V}$ of $\mathbf{x}(\mathcal{B})$, no matter how small, we can conclude

$$(1.3\text{-}5) \qquad \frac{D}{Dt}(\rho\Psi) + \rho\Psi\nabla \cdot \mathbf{v} - \nabla \cdot \boldsymbol{\tau} - \rho g = 0 \qquad \text{on} \quad \mathbf{x}(\mathcal{B}) - \Sigma,$$

$$(1.3\text{-}6) \qquad \lfloor \rho\Psi(\mathbf{v} - \mathbf{v}_\Sigma) - \boldsymbol{\tau} \rfloor \cdot \mathbf{n}_\Sigma = 0 \qquad \text{on} \quad \Sigma.$$

Equations (1.3-5) and (1.3-6) are, respectively, the general **local** or **differential balance law** and the general **jump condition**.

Although the deduction of these local laws from the global balance (1.3-4) is relatively uncontroversial from a mathematical standpoint, the assumption that the global balance law (1.3-4) holds for arbitrarily small volumes $\mathcal{V}$ may arguably be unwarranted on physical grounds in some cases. If this is the case, the breakdown occurs not because of corpuscular phenomena, which we know prevail on a very fine scale but have logically disregarded in formulating the continuum theory. Rather, the local forms (1.3-5) and (1.3-6) can conceivably fail because of interactions at the continuum level whose effects disrupt the local balance but preserve the integral law (1.3-4) for any subregion $\mathcal{V}$ whose diameter exceeds some small number. Theories admitting phenomena of this sort are called **nonlocal theories**. We shall not consider them further.

Mass balance.

To make the general balance law (1.3-5) and the jump condition (1.3-6) concrete, we must assign physical meanings to the expressions Ψ, $\boldsymbol{\tau}$, and g. Table 1-1 does this for four conserved quantities. Thus, for example, the local **mass balance** assumes the form

$$\frac{D\rho}{Dt} + \rho\nabla \cdot \mathbf{v} = 0 \qquad \text{on} \quad \mathbf{x}(\mathcal{B}) - \Sigma$$

or, by the identity (1.2-4) and the fact that $\nabla \cdot (\rho\mathbf{v}) = \rho\nabla \cdot \mathbf{v} + \mathbf{v} \cdot \nabla\rho$,

$$(1.3\text{-}7) \qquad \frac{\partial \rho}{\partial t} + \nabla \cdot (\rho\mathbf{v}) = 0 \qquad \text{on} \quad \mathbf{x}(\mathcal{B}) - \Sigma.$$

TABLE 1-1. Assignments of physical meaning to symbols in the general local balance law.

Balance Law	Ψ	τ	g
Mass	1	0	0
Momentum	$\mathbf{v}$	$\mathbf{t}$	$\mathbf{b}$
Angular momentum	$\mathbf{x} \times \mathbf{v}$	$\mathbf{x} \times \mathbf{t}$	$\mathbf{x} \times \mathbf{b}$
Energy	$E + \frac{1}{2}\mathbf{v} \cdot \mathbf{v}$	$\mathbf{q} + \mathbf{t} \cdot \mathbf{v}$	$h + \mathbf{b} \cdot \mathbf{v}$

Interpretations and physical dimensions*

$\mathbf{t} =$ stress tensor $[ML^{-1}T^{-2}]$

$\mathbf{b} =$ body force $[LT^{-1}]$

$E =$ internal energy $[L^2T^{-2}]$

$\mathbf{q} =$ heat flux $[MT^{-3}]$

$h =$ heat source $[L^2T^{-3}]$

---

* $L =$ length; $M =$ mass; $T =$ time.

The corresponding jump condition then becomes

$$\lfloor \rho(\mathbf{v} - \mathbf{v}_\Sigma)\rfloor = 0 \quad \text{on} \quad \mathbf{x}(\mathcal{B}) \cap \Sigma.$$

The differential equation of mass balance plays a basic role in most models of continua, and as such it is in a sense the most fundamental of the balance laws. Its analogs arise in a wide range of applications of continuum mechanics, including models of such seemingly exotic "continua" as automobile traffic and biological species populations. Equation (1.3-7) sometimes appears under the somewhat misleading alias, "equation of continuity."

Momentum balance.

Again, referring to Table 1-1, the local **momentum balance** is as follows:

(1.3-8) $$\frac{D}{Dt}(\rho\mathbf{v}) + \rho\mathbf{v}\nabla \cdot \mathbf{v} - \nabla \cdot \mathbf{t} - \rho\mathbf{b} = 0.$$

Notice that this equation is actually vector notation for three equations in which Ψ successively takes the values of the velocity components v_1, v_2, v_3. To preserve tensorial consistency, then, the flux $\mathbf{r}$ must be a second-order tensor, called the **stress tensor t**, and the external supply g must be a vector, namely, the **body force b** per unit mass. In many cases $\mathbf{b}$ is simply the gravitational acceleration $\mathbf{g}$. Also, notice that if we view Equation (1.3-8) as a differential equation governing the velocity field $\mathbf{v}$, then the term $\rho \mathbf{v} \cdot \nabla \mathbf{v}$ makes the equation nonlinear. This term, called the **inertial force**, gives rise to some of the most important (and troublesome) nonlinearities in fluid mechanics. The jump condition corresponding to the differential equation (1.3-8) is

$$\lfloor \rho \mathbf{v}(\mathbf{v} - \mathbf{v}_\Sigma) - \mathbf{t} \rfloor \cdot \mathbf{n}_\Sigma = 0.$$

We can reduce the local momentum balance (1.3-8) to a slightly simpler form by using the local mass balance (1.3-7). To do this, let us expand the material derivative in Equation (1.3-8) to get

$$\rho \frac{D\mathbf{v}}{Dt} + \mathbf{v} \frac{D\rho}{Dt} + \rho \mathbf{v}(\nabla \cdot \mathbf{v}) - \nabla \cdot \mathbf{t} - \rho \mathbf{b} = 0.$$

The second and third terms in this equation combine to give

$$\mathbf{v}\left(\frac{D\rho}{Dt} + \rho \nabla \cdot \mathbf{v}\right),$$

which vanishes by the mass balance (1.3-7). Therefore we are left with

(1.3-9) $$\rho \frac{D\mathbf{v}}{Dt} - \nabla \cdot \mathbf{t} - \rho \mathbf{b} = 0.$$

This form of the momentum balance is called **Cauchy's first law** of continuum mechanics.

The stress tensor $\mathbf{t}$ appearing in the momentum balance deserves some explanation. It is clear from the general balance that $\mathbf{t}$ is the flux of momentum, that is, the time rate of change of momentum (which is force) across a unit area of surface in the body. To see what this means, consider a mathematical plane surface in the body oriented perpendicular to the x_i-axis. The force per unit area acting across this plane has three components: two tangent to the plane surface and the other normal to it, as shown in Figure 1-3.

The component t_{ij} of $\mathbf{t}$ in Cartesian coordinates is the j-th component of the force per unit area, or momentum flux, across a plane surface perpendicular to the i-th coordinate axis. Thus the diagonal elements t_{ii} of the stress tensor are momentum fluxes normal to coordinate planes or **normal stresses**, while the off-diagonal elements t_{ij}, $i \neq j$, are momen-

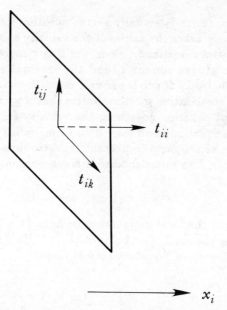

Figure 1-3. Components of the stress tensor relative to a plane perpendicular to the x_i-axis. The components t_{ij}, t_{ik}, $j, k \neq i$, are shear stresses; t_{ii} is a normal stress.

tum fluxes parallel to coordinate planes, or **shear stresses**. In Newtonian fluid mechanics, for example, the normal stresses account for the pressure in a fluid, while the shear stresses are proportional to the fluid's viscosity.

Angular momentum balance.

Next, let us examine the balance of angular momentum. Straightforward substitution from Table 1-1 into the general local balance equations (1.3-5) and (1.3-6) produces the differential equation

$$\frac{D}{Dt}(\rho \mathbf{x} \times \mathbf{v}) + \rho(\mathbf{x} \times \mathbf{v})\nabla \cdot \mathbf{v} - \nabla \cdot (\mathbf{x} \times \mathbf{t}) - \rho(\mathbf{x} \times \mathbf{b}) = 0$$

and the jump condition

$$\lfloor (\rho \mathbf{x} \times \mathbf{v})(\mathbf{v} - \mathbf{v}_\Sigma) - \mathbf{x} \times \mathbf{t} \rfloor \cdot \mathbf{n}_\Sigma = 0.$$

Observe that each term in the row of Table 1-1 labeled "angular momentum" is a moment of the corresponding term in the row labeled "momentum." Hence the terms in the differential momentum balance (1.3-8) have the dimensions of force per unit volume, while those in the differential angular momentum balance have the dimensions of torque per unit volume.

An argument similar to that given for the momentum balance reduces the differential angular momentum balance to the simpler form,

$$(1.3\text{-}10) \qquad \rho \mathbf{x} \times \frac{D\mathbf{v}}{Dt} - \nabla \cdot (\mathbf{x} \times \mathbf{t}) - \rho \mathbf{x} \times \mathbf{b} = \mathbf{0}.$$

The principal use of the balance of angular momentum is to prove that the stress tensor $\mathbf{t}$ is symmetric, that is, $\mathbf{t} = \mathbf{t}^\mathsf{T}$, or $t_{ij} = t_{ji}$ for $i, j = 1, 2, 3$. This is known as **Cauchy's second law of continuum mechanics.** The argument is as follows: We can rewrite Equation (1.3-10) to get the the equation

$$\rho \mathbf{x} \times \frac{D\mathbf{v}}{Dt} - \mathbf{x} \times (\nabla \cdot \mathbf{t}) - (\nabla \mathbf{x}) \times \mathbf{t} - \rho \mathbf{x} \times \mathbf{b} = 0$$

or, rearranging,

$$\mathbf{x} \times \left(\rho \frac{D\mathbf{v}}{Dt} - \nabla \cdot \mathbf{t} - \rho \mathbf{b} \right) - (\nabla \mathbf{x}) \times \mathbf{t} = \mathbf{0}.$$

By Cauchy's first law (1.3-9) the first term vanishes, and there remains $(\nabla \mathbf{x}) \times \mathbf{t} = \mathbf{0}$. To see what this means, observe that $(\nabla \mathbf{x})_{j\ell} = \partial x_j / \partial x_\ell = \delta_{j\ell}$, the Kronecker symbol (see Appendix). By using the Levi-Civita symbol ϵ_{ijk} (also described in the Appendix) we can thus rewrite the i-th component of $(\nabla \mathbf{x}) \times \mathbf{t}$ in indicial form as follows:

$$\sum_{j,k,\ell=1}^{3} \epsilon_{ijk} \delta_{j\ell} t_{\ell k} = \sum_{j,k=1}^{3} \epsilon_{ijk} t_{jk}.$$

This quantity is a vector with components $(t_{23} - t_{32}, t_{31} - t_{13}, t_{12} - t_{21})$, and since we have deduced that this vector vanishes, we have established that $\mathbf{t} = \mathbf{t}^\mathsf{T}$.

Energy balance.

The last true balance law we shall consider is the energy balance. From Table 1-1 and Equations (1.3-5) and (1.3-6), the differential energy balance takes the form

$$(1.3\text{-}11) \qquad \frac{D}{Dt} [\rho(E + \tfrac{1}{2}\mathbf{v} \cdot \mathbf{v})] + \rho(E + \tfrac{1}{2}\mathbf{v} \cdot \mathbf{v})\nabla \cdot \mathbf{v}$$
$$- \nabla \cdot (\mathbf{q} + \mathbf{t} \cdot \mathbf{v}) - \rho(h + \mathbf{b} \cdot \mathbf{v}) = 0,$$

while the jump condition is the following:

$$\lfloor (E + \tfrac{1}{2}\mathbf{v} \cdot \mathbf{v})(\mathbf{v} - \mathbf{v}_\Sigma) - \mathbf{q} - \mathbf{t} \cdot \mathbf{v} \rfloor \cdot \mathbf{n}_\Sigma = 0.$$

In these equations E is the internal energy per unit mass; q is the heat flow per unit area or heat flux, and h represents the contribution to the energy gain per unit volume from heat sources in the body. The quantity $\rho \mathbf{v} \cdot \mathbf{v}/2$ appearing in the first and second terms of Equation (1.3-11) accounts for the body's kinetic energy; $\mathbf{t} \cdot \mathbf{v}$ and $\mathbf{b} \cdot \mathbf{v}$ are the rates of work done per unit volume by stresses and body forces, respectively.

We can simplify the energy balance equation by recognizing that certain of its terms cancel each other in accordance with the laws of mass and momentum balance. To see this, let us first expand the material derivative in Equation (1.3-11) and rewrite the flux of work done by stresses using the product rule, $\nabla \cdot (\mathbf{t} \cdot \mathbf{v}) = \mathbf{v} \cdot (\nabla \cdot \mathbf{t}) + \mathbf{t} : \nabla \mathbf{v}$. We get

$$\rho \frac{DE}{Dt} + \rho \frac{D}{Dt} \left(\tfrac{1}{2}\mathbf{v} \cdot \mathbf{v}\right) + \left(E + \tfrac{1}{2}\mathbf{v} \cdot \mathbf{v}\right)\left(\frac{D\rho}{Dt} + \rho \nabla \cdot \mathbf{v}\right) - \nabla \cdot \mathbf{q}$$
$$- \mathbf{v} \cdot (\nabla \cdot \mathbf{t}) - \mathbf{t} : \nabla \mathbf{v} - \rho h - \rho \mathbf{b} \cdot \mathbf{v} = 0.$$

Notice that the third term in this equation vanishes, since one of its factors is zero by the mass balance law. Therefore, we are left with

$$(1.3\text{-}12) \qquad \rho \frac{DE}{Dt} + \rho \frac{D}{Dt} \left(\tfrac{1}{2}\mathbf{v} \cdot \mathbf{v}\right) - \nabla \cdot \mathbf{q} - \mathbf{v} \cdot (\nabla \cdot \mathbf{t}) - \mathbf{t} : \nabla \mathbf{v}$$
$$- \rho h - \rho \mathbf{b} \cdot \mathbf{v} = 0.$$

Now we are in a position to observe that several terms in the energy balance look like terms in the momentum balance dotted with the velocity $\mathbf{v}$. This is indeed the case. If we form the dot product of Equation (1.3-9) with $\mathbf{v}$ and recognize that $\mathbf{v} \cdot D\mathbf{v}/Dt = D(\tfrac{1}{2}\mathbf{v} \cdot \mathbf{v})/Dt$, we arrive at a **mechanical energy balance**:

$$(1.3\text{-}13) \qquad \underset{\substack{\text{rate of change of}\\\text{kinetic energy}}}{\rho \frac{D}{Dt} \left(\tfrac{1}{2}\mathbf{v} \cdot \mathbf{v}\right)} \;-\; \underset{\substack{\text{rate of work}\\\text{by stress}}}{\mathbf{v} \cdot (\nabla \cdot \mathbf{t})} \;-\; \underset{\substack{\text{rate of work}\\\text{by body forces}}}{\rho \mathbf{b} \cdot \mathbf{v}} \;=\; 0.$$

This equation is really just a scalar form of the momentum balance. We can subtract it from Equation (1.3-12) to obtain a reduced form of the energy balance:

$$(1.3\text{-}14)$$
$$\underset{\substack{\text{rate of change of}\\\text{internal energy}}}{\rho \frac{DE}{Dt}} \;-\; \underset{\substack{\text{heat flow}}}{\nabla \cdot \mathbf{q}} \;-\; \underset{\substack{\text{rate of heating by com-}\\\text{pression and dissipation}}}{\mathbf{t} : \nabla \mathbf{v}} \;-\; \underset{\substack{\text{heat sources}}}{\rho h} \;=\; 0.$$

Since this equation is an energy balance with the mechanical work terms eliminated, it is natural to call it the **thermal energy balance**.

In elementary mechanics one frequently considers the total energy to be the sum of kinetic energy and potential energy. The question then arises, where does potential energy appear in our formulation of the energy balance? The answer is that, under certain special assumptions, the mechanical energy balance (1.3-13) assumes a form in which a potential energy appears explicitly. In particular, if the body force $\mathbf{b}$ is a conservative vector field, then it must be the gradient of some scalar potential function, say $\mathbf{b} = -\nabla\Phi$. For example, if $\mathbf{b} = \mathbf{g}$, the gravitational acceleration, and the x_3-axis points vertically upward, then $\Phi = gx_3$. If in addition, $\partial\Phi/\partial t = 0$, then the body-force term in the mechanical energy balance (1.3-13) becomes

$$-\rho(\mathbf{v} \cdot \mathbf{b}) = \rho(\mathbf{v} \cdot \nabla\Phi) = \rho\frac{D\Phi}{Dt} - \rho\frac{\partial\Phi}{\partial t} = \rho\frac{D\Phi}{Dt}.$$

Since Φ has the dimensions of work, it is natural to associate it with the potential energy arising from the body force field. Therefore the mechanical energy balance reduces to an equation governing the material derivative of kinetic energy plus potential energy:

$$\frac{D}{Dt}\left(\tfrac{1}{2}\mathbf{v} \cdot \mathbf{v} + \Phi\right) - \mathbf{v} \cdot (\nabla \cdot \mathbf{t}) = 0.$$

The Clausius-Duhem inequality.

Finally, in modern continuum mechanics it is important to recognize constraints imposed by the second law of thermodynamics. Clausius stated this law as,

Die Entropie der Welt strebt einem Maximum zu.

"The entropy of the world tends to a maximum." In somewhat more precise terms, the local form of the second law of thermodynamics takes the form of a differential entropy inequality:

(1.3-15) $$\frac{D}{Dt}(\rho S) + \rho S\nabla \cdot \mathbf{v} - \nabla \cdot (\mathbf{q}/T) - \rho h/T \geq 0.$$

This assertion is sometimes called the **Clausius-Duhem inequality**. The jump condition corresponding to this differential law is $\lfloor \rho S(\mathbf{v} - \mathbf{v}_\Sigma) - \mathbf{q}/T \rfloor \cdot \mathbf{n}_\Sigma \geq 0$. Here S stands for entropy per unit mass, and T is the absolute temperature, with $T > 0$. Observe that the inequality (1.3-15) is similar in form to the general differential balance law (1.3-5) with $\Psi = S$, $\boldsymbol{\tau} = \mathbf{q}/T$, and $g = h/T$. The important difference is the appearance of the sign $\geq$ where equality stands in the balance laws. As we shall see, while the entropy inequality does not generally provide a differential equation to solve for the behavior of continuum properties, it does impose certain limitations on the types of functional dependencies we can assume to hold between various properties.

As with the momentum and energy balance laws, it is useful to simplify the Clausius-Duhem inequality by observing that the balance laws lead to some cancellation of terms. First, expanding the material derivative and the flux $\nabla \cdot (\mathbf{q}/T)$ in Equation (1.3-14), we see that

$$\rho \frac{DS}{Dt} + S\left(\frac{D\rho}{Dt} + \rho \nabla \cdot \mathbf{v}\right) - \frac{1}{T}\nabla \cdot \mathbf{q} - \mathbf{q} \cdot \nabla\left(\frac{1}{T}\right) - \frac{\rho h}{T} \geq 0.$$

The second term vanishes by mass balance. Now from the thermal energy balance (1.3-13) we can deduce

$$\frac{\rho h}{T} = \frac{\rho}{T}\frac{DE}{Dt} - \frac{1}{T}\nabla \cdot \mathbf{q} - \frac{1}{T}\mathbf{t} : \nabla\mathbf{v}.$$

Subtracting this from the previous equation and recognizing that $\nabla(1/T) = T^{-2}\nabla T$ gives

(1.3-16) $$\rho\left(\frac{DS}{Dt} - \frac{1}{T}\frac{DE}{Dt}\right) + \frac{1}{T}\mathbf{t} : \nabla\mathbf{v} + \frac{1}{T^2}\mathbf{q} \cdot \nabla T \geq 0.$$

It is common practice in thermodynamics to write energetic relationships in terms of various energy functions (also known somewhat misleadingly as "thermodynamic potentials"), depending on which material properties one considers to be independent variables. While we do not wish to delve deeply into this arcana, it will be convenient later in this chapter to rewrite the Clausius-Duhem inequality by eliminating the internal energy E in favor of the **Helmholtz free energy** $A = E - ST$. The reason for this stems from the thermodynamicists' convention that at thermostatic equilibrium E is a function of S and ρ^{-1} while A is a function of T and ρ^{-1}. For reasons discussed in the next section, a function of the form $A(\rho^{-1}, T)$ fits more naturally into the formalism of constitutive theory than one of the form $E(S, T)$.

To effect the transformation, notice that

$$\frac{DA}{Dt} = \frac{DE}{Dt} - S\frac{DT}{Dt} - T\frac{DS}{Dt},$$

and so

$$\frac{DS}{Dt} - \frac{1}{T}\frac{DE}{Dt} = -\frac{1}{T}\left(\frac{DA}{Dt} + S\frac{DT}{Dt}\right).$$

Therefore we can rewrite the Clausius-Duhem inequality (1.3-16) as follows:

(1.3-17) $$-\frac{\rho}{T}\left(\frac{DA}{Dt} + S\frac{DT}{Dt}\right) + \frac{1}{T}\mathbf{t} : \nabla\mathbf{v} + \frac{1}{T^2}\mathbf{q} \cdot \nabla T \geq 0.$$

1.4. Constitutive Laws.

So far we have established a set of balance laws governing the variation of any body's material properties in space and time. We have seen that the differential forms of these balance laws reduce to the following equations:

$$(1.4\text{-}1) \qquad \frac{D\rho}{Dt} + \rho \nabla \cdot \mathbf{v} = 0, \qquad \qquad \text{(mass balance)}$$

$$(1.4\text{-}2) \qquad \rho \frac{D\mathbf{v}}{Dt} - \nabla \cdot \mathbf{t} - \rho \mathbf{b} = \mathbf{0}, \qquad \text{(momentum balance)}$$

$$(1.4\text{-}3) \qquad \mathbf{t} = \mathbf{t}^{\mathsf{T}}, \qquad \qquad \text{(angular momentum balance)}$$

$$(1.4\text{-}4) \qquad \rho \frac{DE}{Dt} - \nabla \cdot \mathbf{q} - \mathbf{t} : \nabla \mathbf{v} - \rho h = 0. \qquad \text{(energy balance)}$$

One way to assess the sufficiency of this set of equations is simply to count the number of equations and compare it with the number of unknowns. The mass and energy balances each furnish one equation that we can use to solve for the body's behavior. The momentum balance, being a vector equation, gives three equations. The angular momentum balance nominally provides nine equations, but only three of them—namely the identities $t_{23} = t_{32}$, $t_{31} = t_{13}$, and $t_{12} = t_{21}$—are independent. Thus the balance laws offer eight independent equations governing the material properties of a body.

Now let us count the number of variables appearing in these equations. If we consider the spatial position $\mathbf{x}$ and time t to be independent variables, we are left with ρ, $\mathbf{v}$, $\mathbf{t}$, $\mathbf{b}$, E, $\mathbf{q}$, and h. Observe that $\mathbf{v}$, $\mathbf{b}$, and $\mathbf{q}$, being vectors, are really arrays containing three variables each, while the stress tensor $\mathbf{t}$ is an array of nine variables. Therefore, excluding the independent variables $\mathbf{x}$ and t, there are 21 different parameters describing the material attributes of the body. This rather naive accounting shows that the number of equations developed so far falls short of the number of variables to be evaluated. To solve the balance equations for the behavior of a body, then, we must have more information in the form, for example, of independent equations giving mathematical relationships among the variables. **Constitutive theory** is a systematic approach to selecting such relationships.

A more physically motivated way of viewing constitutive theory is as the collection of principles by which we distinguish the different materials observed as continua in nature. The kinematic principles and balance laws are universal in the sense that we intend for them to apply to all continua. It is therefore impossible to distinguish between, say, the air flowing over a jet wing and the water flowing in a river on the basis of the mechanical theory presented so far. Constitutive laws describe in mathematical terms our empirical observations concerning the responses of particular materials to applied loads. Such laws are nearly ubiquitous in the applied sciences: for inviscid fluids, shear stresses are absent; for Newtonian fluids, shear stresses are proportional to the deformation rate; for elastic solids in the "linear" range, stresses are proportional to the strain; for simple heat-conducting solids the heat flux is proportional to the temperature gradient. From an engineer's or physicist's viewpoint, the precise, quantitative forms of these laws specify what sort of material we are modeling and therefore allow us to distinguish the behaviors of different continua.

Although materials in nature exhibit a spectacular variety of behaviors, there are restrictions on the forms of constitutive laws that can legitimately be used to describe a continuum. Constitutive theory, in essence, is a systematically developed set of rules for specifying constitutive laws that do not conflict with fundamental principles of physics. We shall review these rules and look at some examples of constitutive laws that often arise in mathematical modeling of continua.

Internal constraints.

Before beginning this review, however, let us identify a class of special assumptions about continua, analogous to constitutive laws, called **internal constraints**. There are two noteworthy examples of internal constraints, one of which is the constraint that a body be **rigid**. This concept is similar to but distinct from the concept of **rigid motion**. A body's motion is rigid (although the body itself need not be) when all distances and angles formed by sets of material points in the body remain unchanged under the motion. A useful mathematical equivalent to this condition is the assertion that any differential element of arclength $ds = \sqrt{dx \cdot dx}$ remains invariant under the motion; in particular, $dx \cdot dx = dX \cdot dX$, the square of the arclength element in the initial configuration (see Figure 1-4).

But by the chain rule, $dx = J\,dX$, where J is the Jacobian matrix of the motion, having entries $J_{ij} = \partial x_i / \partial X_j$. Therefore, for a rigid motion we require

$$dx \cdot dx = J\,dX \cdot J\,dX = dX \cdot (J^\top J)\,dX = dX \cdot dX,$$

where $J^\top$ signifies the transpose of the Jacobian matrix. From this consideration we can see that a body's motion will be rigid if and only if

$$J^\top J = I,$$

where I is the identity matrix.

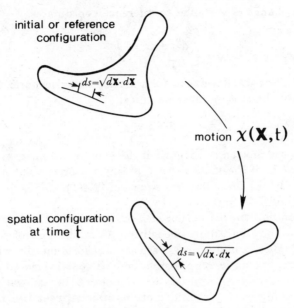

initial or reference
configuration

$ds = \sqrt{d\mathbf{X} \cdot d\mathbf{X}}$

motion $\chi(\mathbf{X}, t)$

spatial configuration
at time t

$ds = \sqrt{d\mathbf{x} \cdot d\mathbf{x}}$

Figure 1-4. Relationships between material and spatial arclength elements under rigid motion.

Notice that to say a motion is rigid only describes what actually happens to the body as it responds to a particular set of applied loads. It does not place any restriction on conceivable responses of the body under different applied loads. Thus the rigidity of motions is a kinematic notion with no necessary implications regarding the specific materials involved. Rigidity becomes an internal constraint, however, when we stipulate a priori that, whatever the applied loads may be, the body can only undergo rigid motions. In this case we say the **body** is rigid. Mathematically, a body undergoing rigid motion and a rigid body both satisfy the condition $\mathbf{J}^T\mathbf{J} = \mathbf{I}$. Conceptually, though, the body undergoing rigid motion may in fact be quite vulnerable to stretching and shearing motions under certain external influences, while the internal constraint on the material constituting the rigid body prohibits stretch and shear.

The second internal constraint arises if we prohibit volume changes under the motion. Here again we confront the distinction between kinematic description and internal constraint. Kinematically, we say that a motion is **isochoric** if the volume of the spatial configuration $\mathcal{V}(t)$ occupied by any part of the body remains unchanged from its initial configuration:

$$\int_{\mathcal{V}(t)} d\mathbf{x} = \int_{\mathcal{V}(0)} d\mathbf{x}$$

or, using the change of variables to reference coordinates,

$$\int_{\mathcal{V}(0)} \det \mathbf{J} \, d\mathbf{x} = \int_{\mathcal{V}(0)} d\mathbf{x}.$$

From this equation it follows that, for a motion to be isochoric, the Jacobian determinant of the motion must be unity:

$$\det \mathbf{J} = 1.$$

Equivalently, the motion is isochoric if the volume occupied by any mass is invariant, that is, if the mass density at any material point is a constant, $D\rho/Dt = 0$. By the mass balance equation (1.4-1), this is also equivalent to the condition $\nabla \cdot \mathbf{v} = 0$.

Corresponding to this notion of isochoric motion is the internal constraint that a body be **incompressible**. As in the case of rigid motions versus rigid bodies, a body can experience isochoric motion without being incompressible, but an incompressible body is constrained to undergo isochoric motion. Fluid dynamicists have adopted the somewhat confusing practice of calling isochoric fluid motions **incompressible flows**. Such flows occur in such clearly compressible bodies as volumes of air in steady motion at speeds that are small compared with the speed of sound (Batchelor, 1967, Section 3.6). In the fluid dynamicists' argot, "incompressible flow" refers to a kinematic notion while "incompressible fluid" expresses an internal constraint.

Principles of constitutive theory.

Now let us return to constitutive theory proper. In formulating constitutive laws, we must first decide which variables are to be expressed in terms of which others. As hinted somewhat earlier, we naturally think of position $\mathbf{x}$ and time t as independent variables, since in classical mechanics, at least, we regard space and time as the ambient fields in which material bodies move and against which their responses may be measured. Also, we consider the material points $\mathbf{X}$ to be primitive quantities having no need of functional description. Thus we do not expect to seek constitutive relations for these variables; rather, $\mathbf{X}$, $\mathbf{x}$, and t serve as suitable arguments for constitutive functionals. Certain other quantities appearing in the balance laws are directly derivable from $\mathbf{x}$ and t. Examples include the velocity $\mathbf{v} = \partial\mathbf{x}/\partial t$, the acceleration $\mathbf{a} = \partial^2\mathbf{x}/\partial t^2$, and the velocity gradient $\nabla\mathbf{v}$, a tensor whose (i,j)-th entry in Cartesian coordinates is $\partial v_j/\partial x_i$. Given the motion $\mathbf{x}(\mathbf{X}, t)$ we can readily compute these quantities, so we have no need of constitutive laws for these variables.

We shall further agree that the temperature T, along with quantities such as ∇T that are directly derivable from T, are independent constitutive variables also. Formally, this agreement is nothing more than a fiat. Intuitively, though, we might regard temperature as an indicator of motion at the corpuscular scale, so that the inclusion of T in the list of independent constitutive variables has some appeal on physical grounds. We should caution, however, that temperature is not a consistent measure of microscopic motion even in the corpuscular theories. Besides, an appeal to corpuscular physics is logically out of place in continuum mechanics.

These considerations lead to a formal rule governing the formulation of constitutive laws.

Principle of Causality. *The independent variables in any constitutive law are the material points* $\mathbf{X}$*, the variables* $\mathbf{x}$*,* t*, and* T*, and quantities directly derivable from these. The only quantities for which constitutive laws are necessary are other quantities not included among the independent constitutive variables.*

As we have observed, $\mathbf{x}$, t, $\mathbf{v}$, $\mathbf{a}$, $\nabla\mathbf{v}$, T, and ∇T are independent constitutive variables. So is ρ, since it is directly derivable from $\mathbf{v}$ and the mass balance (1.4-1), and therefore $\nabla\rho$, $\partial\rho/\partial t$, etc., are independent constitutive variables, too. The external supplies $\mathbf{b}$ and h are not derivable from $\mathbf{X}$, $\mathbf{x}$, t, or T, so they cannot be considered independent constitutive variables. However, in practice $\mathbf{b}$ and h often appear as quantities prescribed by the physics of the problem; for example, $\mathbf{b}$ commonly reduces to the gravitational acceleration. As a result, the body force and heat supply frequently do not warrant much concern as constitutive variables.

The remaining quantities—the stress $\mathbf{t}$, the heat flux $\mathbf{q}$, and the internal energy E—are intimately connected to the properties of specific materials. Consequently, these quantities are the constitutive variables that are of special concern to constitutive theory. If we wish to accommodate the Clausius-Duhem inequality, we must also consider a constitutive law for the entropy S, and instead of a constitutive law for E we might as well examine constitutive laws for the Helmholtz free energy $A = E - ST$. Let us adopt the general symbol $\mathcal{F}$ to denote any dependent constitutive variable. Then mathematically what we seek are functional relationships of the form

$$\mathcal{F} = \mathcal{F}(\mathbf{x}, T, \ldots, \mathbf{X}, t),$$

where we have allowed for explicit dependence of the constitutive variable $\mathcal{F}$ on material points $\mathbf{X}$ as well as on their spatial positions $\mathbf{x}$. The ellipsis in this general form stands for the collection of quantities directly derivable from the variables displayed explicitly inside the parentheses.

Having decided which quantities are dependent constitutive variables and which are independent constitutive variables, let us review four axioms that delimit the physically allowable forms of constitutive laws.

Axiom 1. Determinism. *The value of a dependent constitutive variable $\mathcal{F}$ at time t_0 depends only on the values of independent constitutive variables at times $t \leq t_0$.*

A somewhat clumsy way of writing this idea occasionally appears in the literature:

$$\mathcal{F} = \mathop{\mathcal{F}}_{t'=0}^{\infty} [\mathbf{x}(t_0 - t'), T(t_0 - t'), \ldots, \mathbf{X}, t_0 - t'].$$

This formidable notation expresses nothing more than the intuitive idea that the future does not determine the present.

The second axiom of constitutive theory prohibits the dependence of a constitutive variable $\mathcal{F}$ at a material point $\mathbf{X}$ on values of the independent constitutive variables at material points distant from $\mathbf{X}$.

Axiom 2. Local Action. *The value of a dependent constitutive variable $\mathcal{F}$ at a material point $\mathbf{X}$ is determined by the values of the independent constitutive variables in an arbitrarily small neighborhood of $\mathbf{X}$.*

We shall often assume that the independent variables $\mathbf{x}(\mathbf{X}, t)$ and $T(\mathbf{x}, t)$ are analytic functions of the material coordinates $\mathbf{X}$, so that it is possible to represent their variations over any neighborhood of a material point $\mathbf{X}_0$ by Taylor series:

$$x_i(\mathbf{X}, t) = x_i(\mathbf{X}_0, t) + (\mathbf{X} - \mathbf{X}_0) \cdot \nabla_{\mathbf{X}} x_i(\mathbf{X}_0, t) + \cdots,$$

$$T(\mathbf{X}, t) = T(\mathbf{X}_0, t) + (\mathbf{X} - \mathbf{X}_0) \cdot \nabla_{\mathbf{X}} T(\mathbf{X}_0, t) + \cdots.$$

Here, $\nabla_{\mathbf{X}} = (\partial/\partial X_1, \partial/\partial X_2, \partial/\partial X_3)$ stands for the gradient with respect to material coordinates. Thus by the axiom of local action a constitutive variable $\mathcal{F}$ at $\mathbf{X}$ may depend on the values of $\mathbf{x}$, T, $\nabla_{\mathbf{X}}\mathbf{x}$, $\nabla_{\mathbf{X}}T$, and higher-order derivatives of $\mathbf{x}$ and T at $\mathbf{X}$, but not on the values of these fields at other material points. In the special case that the constitutive equations depend only on derivatives through the first order, as in

$$\mathcal{F} = \mathcal{F}(\mathbf{x}, T, \nabla_{\mathbf{X}}\mathbf{x}, \nabla_{\mathbf{X}}T, \mathbf{X}, t),$$

we say that the material constituting the body is **simple**.

The third axiom expresses the requirement that the mathematical forms of constitutive laws be independent of the coordinate system, or frame, that we use to measure them.

Axiom 3. Objectivity. *A constitutive law $\mathcal{F} = \mathcal{F}(\mathbf{x}, T, \ldots, \mathbf{X}, t)$ must be invariant under changes of frame.*

To see the significance of this axiom, we must first clarify what we mean by a change of frame. **A change of frame** is a function $\mathcal{Q}$ mapping each point $(\mathbf{x}, t)$ of space-time onto another point $(\mathbf{x}^*, t^*)$ in space-time. The mapping must satisfy three requirements.

(i) $\mathcal{Q}$ must be a one-to-one correspondence, that is, $\mathcal{Q}(\mathbf{x}_A, t_A) \neq \mathcal{Q}(\mathbf{x}_B, t_B)$ whenever $(\mathbf{x}_A, t_A) \neq (\mathbf{x}_B, t_B)$, and moreover every $(\mathbf{x}^*, t^*)$ in space-time can be expressed as $\mathcal{Q}(\mathbf{x}, t)$ for some point $(\mathbf{x}, t)$ in space-time.

(ii) $\mathcal{Q}$ must preserve distances: If $\mathcal{Q}(\mathbf{x}_A, t) = (\mathbf{x}_A^*, t^*)$ and $\mathcal{Q}(\mathbf{x}_B, t) = (\mathbf{x}_B^*, t^*)$, then $\|\mathbf{x}_A^* - \mathbf{x}_B^*\|_2 = \|\mathbf{x}_A - \mathbf{x}_B\|_2$, where $\|\mathbf{x}\|_2$ denotes the Euclidean norm, or length, of the vector $\mathbf{x}$.

(iii) $\mathcal{Q}$ must preserve time intervals: If $\mathcal{Q}(\mathbf{x}_A, t_A) = (\mathbf{x}_A^*, t_A^*)$ and $\mathcal{Q}(\mathbf{x}_B, t_B) = (\mathbf{x}_B^*, t_B^*)$, then $t_A - t_B = t_A^* - t_B^*$.

These requirements capture the idea that a change of frame is a change in the position, orientation, time, and rigid motion of an observer. Thus the axiom of objectivity essentially requires that the form of a constitutive law be independent of the coordinate systems used by the observers.

The general change of frame $\mathcal{Q} : (\mathbf{x}, t) \mapsto (\mathbf{x}^*, t^*)$ has the following form:

(1.4-5)
$$\mathbf{x}^* = \mathbf{Q}(t)(\mathbf{x} - \mathbf{x}_0) + \mathbf{x}_1(t),$$
$$t^* = t - t_0.$$

Here, $\mathbf{x}_0$ is a fixed spatial point; $\mathbf{x}_1(t)$ is a spatial locus that may be moving in the original coordinate system, and t_0 is a constant. $\mathbf{Q}$ is a time-dependent linear transformation; in particular it must be a proper rotation. Such transformations are also known as **orthogonal matrices**; their defining properties are the following:

$$\mathbf{Q}\mathbf{Q}^\top = \mathbf{I},$$
$$\det \mathbf{Q} = 1.$$

Notice that these properties are exactly those enjoyed by the Jacobian matrices of rigid motion.

Now we are in a position to state exactly what we mean by invariance under changes of frame. In what follows, let the superscript * signify the value of a quantity after a change of frame. A scalar function s is invariant under changes of frame if its functional form is preserved under any change of frame: $s^* = s$. A vector-valued function $\mathbf{V}$ is invariant if its functional form remains unchanged under transformations of the form $\mathbf{V}^* = \mathbf{Q}\mathbf{V}$. Finally, a tensor-valued function $\mathbf{T}$ is invariant if its functional form is unchanged under transformations of the form $\mathbf{T}^* = \mathbf{QTQ}^\top$. The axiom of objectivity states that all scalar, vector, and tensor constitutive laws must be invariant according to these criteria.

One of the simplest examples of an quantity that fails to be objective is the spatial velocity $\mathbf{v}$. For, by definition, the velocity in a transformed frame is $\mathbf{v}^* = \partial\mathbf{x}^*/\partial t$. Since the transformation to the "starred" frame has the form $\mathbf{x}^* = \mathbf{Q}(\mathbf{x} - \mathbf{x}_0) + \mathbf{x}_1$, we have

$$\mathbf{v}^* = \frac{\partial}{\partial t}\left[\mathbf{Q}(\mathbf{x} - \mathbf{x}_0) + \mathbf{x}_1\right]$$

$$= \mathbf{Q}\,\frac{\partial}{\partial t}\,(\mathbf{x} - \mathbf{x}_0) + (\mathbf{x} - \mathbf{x}_0)\,\frac{\partial\mathbf{Q}}{\partial t} + \frac{\partial\mathbf{x}_1}{\partial t}$$

$$= \mathbf{Q}\mathbf{v} + (\mathbf{x} - \mathbf{x}_0)\,\frac{\partial\mathbf{Q}}{\partial t} + \frac{\partial\mathbf{x}_1}{\partial t}.$$

Thus only for special cases of $\mathbf{Q}$ and $\mathbf{x}_1$ can we conclude that $\mathbf{v}^* = \mathbf{Q}\mathbf{v}$, and hence the velocity is not objective. However, the reader should check that the **relative velocity** $\mathbf{v}_2 - \mathbf{v}_1$ between any two material points $\mathbf{X}_2$ and $\mathbf{X}_1$ is indeed an objective quantity. Therefore, while a constitutive law depending on absolute velocities will not generally be objective, constitutive laws may depend on relative velocities without violating the axiom of objectivity.

The last axiom of constitutive theory that we shall discuss essentially forbids constitutive laws that violate the second law of thermodynamics.

Axiom 4. Thermodynamic Admissibility. *Constitutive laws must not entail violations of the Clausius-Duhem inequality.*

We shall examine this axiom more closely in the following example.

Example: inviscid fluids.

So far our discussion of constitutive theory has remained rather abstract. In fact, concrete examples of constitutive laws appear quite frequently in applied sciences, although in most applications the distinction between the balance laws, which are universal, and the constitutive laws, which are really special cases, seldom arises explicitly. Let us therefore consider a simple example showing how constitutive laws reduce the balance laws to usable field equations.

Consider a body on which body forces $\mathbf{b}$ and heat sources h have no effect and in which we can neglect heat fluxes $\mathbf{q}$. (These assumptions amount to trivial constitutive laws.) Assume also that the body suffers no shear stresses, so that the off-diagonal elements of $\mathbf{t}$ vanish in any coordinate system. Intuitively, this assumption suggests a fluid that does not transport momentum in directions perpendicular to the direction in which an applied load acts. The assumption models molasses poorly, for example, as may be seen by the following argument. If we push a thin knife through molasses we expect to see motion at points in the fluid at a distance perpendicular to the direction in which the knife moves, as shown in Figure 1-5.

This momentum transport orthogonal to the direction of mean flow over the knife occurs via shear stresses in the fluid. On the other hand, the assumption that shear forces are negligible might be a good model for a "thinner" fluid such as air. As for the normal stresses, we shall assume that they are independent of direction, or isotropic. In particular, we shall assume that there is some scalar function p such that $\mathbf{t} = -p\mathbf{1}$, where $\mathbf{1}$ is the unit tensor. This scalar parameter is the **mechanical pressure** of the fluid.

These constitutive laws leave the mass balance equation intact:

(1.4-6a)
$$\frac{D\rho}{Dt} + \rho\nabla \cdot \mathbf{v} = 0.$$

However, the constitutive law for the stress tensor reduces the flux term $\nabla \cdot \mathbf{t}$ in the momentum balance to $-\nabla \cdot (p\mathbf{1}) = -\nabla p$, so the momentum balance becomes simply

(1.4-6b)
$$\frac{D\mathbf{v}}{Dt} + \frac{1}{\rho}\nabla p = 0.$$

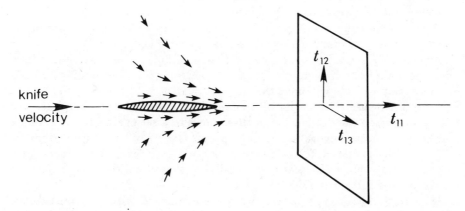

Figure 1-5. Profile of a knife moving through molasses along with part of the direction field of the fluid's motion. Observe that the motion of material points at the knife's surface induces motion at points located away from the surface and perpendicular to the knife's velocity. The accompanying diagram shows schematically the normal and shear components of the momentum flux or stress tensor.

Notice that the stress tensor $-p\mathbf{1}$ is automatically symmetric since the unit tensor $\mathbf{1}$ is, so there is no need to consider the angular momentum balance in this example. Now let us see what happens to the energy balance. Accounting for the facts that $h = 0$ and $\mathbf{q} = \mathbf{0}$ and observing that $-\mathbf{t} : \nabla\mathbf{v} = p\mathbf{1} : \nabla\mathbf{v} = p\nabla\cdot\mathbf{v}$, we have

$$(1.4\text{-}6c) \qquad\qquad \frac{DE}{Dt} + \frac{p}{\rho}\nabla\cdot\mathbf{v} = 0.$$

Equations (1.4-6a), (1.4-6b), and (1.4-6c) are known collectively as **Euler's equations** for an inviscid fluid. They amount to five partial differential equations in the unknowns ρ, v_1, v_2, v_3, p, and E. To close the system we need a sixth equation. This equation already exists implicitly in our constitutive assumption that the stress tensor reduces to the mechanical pressure. To make the equation explicit, we must specify some functional relationship, called an **equation of state**, giving the mechanical pressure in terms of independent constitutive variables. Empirical evidence suggests an equation of state of the form

$$(1.4\text{-}6d) \qquad\qquad p = p(\rho^{-1}, T),$$

where we have cast the density dependence in terms of the specific volume ρ^{-1} merely to promote accord between our mechanics and the conventions of classical thermodynamics. In practice such an equation of state may be nothing more than a table of measured values, or it may take an analytic form whose parameters satisfy some criterion of "best fit" with experimental data. One equation of state that is interesting because of its close connections with corpuscular theories is the ideal gas law, $p = \rho RT/\hat{M}$, where $R = 8.315$ J/K and $\hat{M}$ is the mass of 1 mole of fluid. The constitutive laws for the Helmholtz free energy A and the entropy S corresponding to such an equation of state will have the forms $A = A(\rho^{-1}, T)$ and $S = S(\rho^{-1}, T)$.

In one sense the physics of inviscid fluid flows are now complete. We have converted the description of these flows to a set of six equations in six unknowns. Assuming we can find an appropriate set of boundary and initial data for the partial differential equations, we have reduced the mechanics to the purely mathematical task of solving the equations–a task that we shall pursue further in Chapter Five.

Strictly speaking, however, we are not finished with the constitutive theory. There remains the job of checking that our constitutive laws satisfy the four axioms stated previously. Actually, we need to do this only for the stress tensor and entropy, since our assumptions for the other constitutive variables reduce to trivial cases in which the variables vanish or, in the case of internal energy (or Helmholtz free energy) to the deduction of a partial differential equation governing the evolution of E.

Let us attack the constitutive assumption for the stress tensor. For the simple law at hand the axioms of determinism and local action pose no real difficulty: We simply require that the value of p at any time t and material point $\mathbf{X}$ depend only on the values of ρ^{-1} and T at t and $\mathbf{X}$. The axiom of objectivity is equally easy to check in this case. Under any change of frame $\mathbf{x}^* = \mathbf{Q}(\mathbf{x} - \mathbf{x}_0) + \mathbf{x}_1$, the stress tensor $\mathbf{t} = -p\mathbf{1}$ transforms as follows:

$$\mathbf{t}^* = \mathbf{Q}\mathbf{t}\mathbf{Q}^\top = -p\mathbf{Q}\mathbf{1}\mathbf{Q}^\top = -p\mathbf{Q}\mathbf{Q}^\top = -p\mathbf{1},$$

since $\mathbf{Q}$ must be an orthogonal tensor. Thus the functional relationship is invariant under changes of frame.

Checking the axiom of thermodynamic admissibility is a less trivial matter. Let us begin by writing the Clausius-Duhem inequality (1.3-17) and incorporating our assumptions that $\mathbf{q} = 0$, $\mathbf{t} = -p(\rho^{-1}, T)\mathbf{1}$, $A = A(\rho^{-1}, T)$, $S = S(\rho^{-1}, T)$:

$$-\frac{\rho}{T}\left[\frac{D}{Dt}A(\rho^{-1}, T) + S(\rho^{-1}, T)\frac{DT}{Dt}\right] - \frac{1}{T}p(\rho^{-1}, T)\mathbf{1} : \nabla\mathbf{v} \geq 0.$$

Using the chain rule, we can convert this inequality to one in which material derivatives of only ρ^{-1} and T appear:

$$-\frac{\rho}{T}\frac{\partial A}{\partial \rho^{-1}}\frac{D\rho^{-1}}{Dt} - \frac{\rho}{T}\left(\frac{\partial A}{\partial T} + S\right)\frac{DT}{Dt} - \frac{\rho}{T}\mathbf{1} : \nabla\mathbf{v} \geq 0.$$

Now we can observe that $D\rho^{-1}/Dt = -\rho^{-2}D\rho/Dt$ and $\mathbf{1} : \nabla\mathbf{v} = \nabla \cdot \mathbf{v}$ and use the fact that $T > 0$ to simplify:

$$-\frac{\partial A}{\partial \rho}\frac{1}{\rho}\frac{D\rho}{Dt} - \rho\left(\frac{\partial A}{\partial T} + S\right)\frac{DT}{Dt} - p\nabla \cdot \mathbf{v} \geq 0.$$

By mass balance, $\nabla \cdot \mathbf{v} = -\rho^{-1}D\rho/Dt$, so we have

$$(1.4\text{-}7) \qquad -\left(\frac{\partial A}{\partial \rho^{-1}} - p\right)\frac{1}{\rho}\frac{D\rho}{Dt} - \rho\left(\frac{\partial A}{\partial T} + S\right)\frac{DT}{Dt} \geq 0.$$

This inequality states that a linear combination of the quantities $D\rho/Dt$ and DT/Dt can never assume negative values. There is nothing, however, to prevent $D\rho/Dt$ or DT/Dt from being either positive or negative independently of each other. Indeed, we must allow for all combinations of the signs of these two derivatives, including the possibility that either may vanish. Given this observation, it is clear that if the coefficients of $D\rho/Dt$ or DT/Dt have definite signs, then one can pick values of $D\rho/Dt$ and DT/Dt that force the linear combination to be negative. Therefore, for our constitutive laws to be thermodynamically admissible, the coefficients of $D\rho/Dt$

and DT/Dt in the inequality (1.4-7) must vanish. Since the mass density will typically be nonzero, we must conclude

$$(1.4\text{-}8a) \qquad\qquad S = -\frac{\partial A}{\partial T},$$

$$(1.4\text{-}8b) \qquad\qquad p = -\frac{\partial A}{\partial \rho^{-1}}.$$

These equations will look familiar to readers versed in classical thermodynamics, where they are sometimes called "Maxwell relations." In our context, Equation (1.4-8b) is the more interesting of the identities. The quantity $-\partial A/\partial \rho^{-1}$ is called the **thermodynamic pressure**, since it appears as the coefficient of isothermal volume changes in Gibbs' equation for changes in Helmholtz free energy at thermostatic equilibrium. Equation (1.4-8b) tells us that in inviscid flows the mechanical pressure, defined for moving bodies in terms of the stresses, is identical to the thermodynamic pressure defined by the energetics of motionless bodies at thermostatic equilibrium.

For the time being, we shall close our discussion of inviscid fluids with two remarks. First, the approach presented here for checking the Clausius-Duhem inequality differs somewhat from the customary one. The usual approach considers quite general functional relationships as candidates for constitutive laws, allowing for dependence of the constitutive variables on many quantities at the outset and then following a procedure akin to the one just presented to derive restrictions on the assumed relationships. In the arguments leading to Equations (1.4-8) we considered rather specific candidates for constitutive laws, chosen on the basis of fairly common empirical results, and derived restrictions on these particular relationships. This approach offers the advantage of greater simplicity in the chain-rule expansions of the Clausius-Duhem inequality. Nevertheless, it suffers the disadvantage that it fails to yield general constitutive relationships and therefore cannot reveal new and untested functional dependencies that may exist in nature. For an introduction to the more general approach, we refer the reader to Eringen (1980, Chapter Five).

Second, as one might imagine, checking the Clausius-Duhem inequality can involve large and tedious calculations when the bodies under investigation exhibit richer functional dependencies than the inviscid fluid just studied. A thorough study of these calculations for all of the constitutive laws used in this book would carry us far afield. Therefore we shall leave further formal exploitation of the Clausius-Duhem inequality to more advanced works on continuum mechanics.

Example: viscous, incompressible fluids.

Let us now consider a more general class of fluids, namely, those in which shear stresses are present. We shall simplify matters by restricting attention to fluids that are incompressible, so that $D\rho/Dt = 0$ or, by the mass balance, $\nabla \cdot \mathbf{v} = 0$. Assume further that body forces and heat sources are absent, so that $\mathbf{b} = \mathbf{0}$ and $h = 0$, and that heat fluxes are negligible, so that $\mathbf{q} = \mathbf{0}$.

To accommodate the effects of shear stresses, we must introduce a more complicated constitutive law for the stress tensor $\mathbf{t}$ than the law used to model inviscid fluids. Intuitively, one may expect that shear stresses will be strongest in regions of space where the fluid momentum varies rapidly within small distances. Thus, over a small region where density is nearly uniform, fluid momentum will exhibit a flux from points having high fluid velocity to nearby points where the fluid velocity is low. The momentum fluxes in such a flow will, roughly speaking, proceed down the velocity gradient $\nabla \mathbf{v}$. While it is tempting to postulate that the stress $\mathbf{t}$ is therefore proportional to $\nabla \mathbf{v}$, such a law would contradict the symmetry of the stress tensor, since $\nabla \mathbf{v}$ is not symmetric. However, the **deformation rate tensor** $\mathbf{D} = \frac{1}{2}[\nabla \mathbf{v} + (\nabla \mathbf{v})^\top]$ is symmetric, so it is reasonable to expect a direct relationship between $\mathbf{t}$ and $\mathbf{D}$.

Landau and Lifschitz (1959, Section 15) show that the most general law giving a linear dependence of $\mathbf{t}$ on the deformation rate has the form

$$\mathbf{t} = -p\mathbf{1} + \lambda \, \text{tr}(\mathbf{D}) + 2\mu\mathbf{D},$$

where p is, as before, the mechanical pressure, a positive parameter accounting for normal stresses in the fluid. The coefficients λ and μ are also positive, being called the **coefficients of viscosity** of the fluid. In the particular case when the fluid is incompressible, it is relatively easy to see that $\text{tr}(\mathbf{D}) = \nabla \cdot \mathbf{v} = 0$, so that the constitutive law for $\mathbf{t}$ reduces to the following equation:

(1.4-9) $\mathbf{t} = -p\mathbf{1} + 2\mu\mathbf{D}.$

Now we are in a position to write the laws of mass, momentum, and energy balance for this class of fluids. The mass balance is quite simple: by incompressibility,

$$\frac{D\rho}{Dt} = \frac{\partial \rho}{\partial t} + \mathbf{v} \cdot \nabla \rho = 0.$$

For the momentum balance, straightforward substitution of the constitutive law (1.4-9) into Cauchy's first law (1.3-9), neglecting body forces, produces

$$\rho \frac{D\mathbf{v}}{Dt} = -\nabla p + \mu \nabla \cdot [\nabla \mathbf{v} + (\nabla \mathbf{v})^\top],$$

assuming μ is uniform. The reader should check that, for an incompressible fluid, $\nabla \cdot [\nabla \mathbf{v} + (\nabla \mathbf{v})^\top] = \nabla^2 \mathbf{v}$, so we get

(1.4-10)
$$\rho \frac{\partial \mathbf{v}}{\partial t} + \rho \mathbf{v} \cdot \nabla \mathbf{v} = -\nabla p + \mu \nabla^2 \mathbf{v}.$$

This is the celebrated **Navier-Stokes equation** modeling linearly viscous, incompressible fluid flows. We shall discuss this equation further in Chapter Five. Finally, under our assumptions, the thermal energy equation reduces to the following law:

$$\rho \frac{DE}{Dt} + p\nabla \cdot \mathbf{v} - 2\mu \mathbf{D} : \nabla \mathbf{v} = 0,$$

in which the last term on the left represents viscous dissipation of the fluid's kinetic energy into internal energy.

Example: heat flow.

The examples cited so far have ignored the effects of nonuniform temperatures and heat transfer. To accommodate these phenomena, we must take a closer look at the energy balance or, more specifically, at the thermal energy balance (1.3-13):

$$\rho \frac{DE}{Dt} - \nabla \cdot \mathbf{q} - \mathbf{t} : \nabla \mathbf{v} - \rho h = 0.$$

If heat sources are absent and if the material under consideration exhibits no viscous dissipation of kinetic energy into heat, then $h = 0$ and the rate of heating reduces to the compression term, $\mathbf{t} : \nabla \mathbf{v} = -p\nabla \cdot \mathbf{v}$. Thus,

$$\frac{DE}{Dt} - \frac{1}{\rho}\nabla \cdot \mathbf{q} + \frac{p}{\rho}\nabla \cdot \mathbf{v} = 0.$$

By mass balance, the last term on the left side of this equation is equal to $-(p/\rho^2)D\rho/Dt$, or $p\,D(1/\rho)/Dt$. Now we can use the chain rule to express the material derivatives DE/Dt and $D\rho/Dt$ in terms of the rate of change DT/Dt of temperature:

$$\frac{\partial E}{\partial T}\frac{DT}{Dt} + p\frac{\partial}{\partial T}\left(\frac{1}{\rho}\right)\frac{DT}{Dt} - \frac{1}{\rho}\nabla \cdot \mathbf{q} = 0.$$

The combination

$$c_p = \frac{\partial E}{\partial T} + p\frac{\partial}{\partial T}\left(\frac{1}{\rho}\right)$$

is the **heat capacity** of the material, measured under conditions of constant pressure. Using this new quantity allows us to rewrite the thermal energy balance as follows:

$$c_p \frac{DT}{Dt} - \frac{1}{\rho}\nabla \cdot \mathbf{q} = 0.$$

There remains the problem of establishing a constitutive law for $\mathbf{q}$. The most common law is that advanced by Fourier in his *Theorie Analytique de la Chaleur* in 1822. Fourier's law states that heat flux is proportional to the temperature gradient, $\mathbf{q} = k_H \nabla T$. (One often sees this law written with the opposite sign from that shown here. There is no inconsistency, however; the sign appropriate for Fourier's law depends on the sign convention adopted for $\mathbf{q}$ in the energy balance.) In simple cases the positive coefficient k_H may be constant and uniform; more generally it can vary in space or time or even change with the temperature of the material. Fourier's law further transforms the thermal energy equation to

$$c_p \frac{DT}{Dt} = \frac{1}{\rho} \nabla \cdot (k_H \nabla T)$$

or, assuming k_H is uniform

$$\frac{\partial T}{\partial t} + \mathbf{v} \cdot \nabla T = \kappa_H \nabla^2 T$$

where $\kappa_H = k_H / \rho c_p$ is called the **thermal diffusivity**. When convection is absent $(\mathbf{v} = \mathbf{0})$, we get the standard constant-coefficient heat equation,

(1.4-11)
$$\frac{\partial T}{\partial t} = \kappa_H \nabla^2 T.$$

We shall examine this equation in Chapter Four; Chapter Three discusses its steady-state counterpart.

Example: longitudinal waves in an elastic solid.

The mechanics of elastic solids offers fertile ground for the study of a wide variety of continuum-mechanical phenomena. We shall focus here on the development of the wave equation governing the propagation of compressive or longitudinal waves in one dimension along a bar composed of an elastic solid. Since these waves arise through the transfer of momentum along the bar, the balance equation of interest is the momentum balance, which in the absence of body forces is

$$\rho \frac{D\mathbf{v}}{Dt} - \nabla \cdot \mathbf{t} = \mathbf{0}.$$

In the theory of elasticity the actual spatial coordinates of material points are often of less concern than the **displacements** of material points from their initial configurations. If we take the initial configuration as the reference configuration for the motion, then the displacement will be $\mathbf{U} = \mathbf{x} - \mathbf{X}$, which, when regarded as a function of $\mathbf{x}$ and t, is an Eulerian quantity. Since $D\mathbf{X}/Dt = 0$, the Eulerian velocity $\mathbf{v}$ is the following:

$$\mathbf{v} = \frac{D\mathbf{x}}{Dt} = \frac{D\mathbf{U}}{Dt} = \frac{\partial\mathbf{U}}{\partial t} + \mathbf{v}\cdot\nabla\mathbf{U}.$$

For motions of solids in which the displacement and its derivatives are small, the nonlinear term $\mathbf{v}\cdot\nabla\mathbf{U}$ is the product of two small quantities, and the linearizing approximation $\mathbf{v} \simeq \partial\mathbf{U}/\partial t$ may be acceptable. Under similar hypotheses, the acceleration

$$\frac{D\mathbf{v}}{Dt} = \frac{\partial\mathbf{v}}{\partial t} + \mathbf{v}\cdot\nabla\mathbf{v}$$

has a nonlinear term that is the product of small quantities, and we can therefore argue that

$$\frac{D\mathbf{v}}{Dt} \simeq \frac{\partial\mathbf{v}}{\partial t} = \frac{\partial^2\mathbf{U}}{\partial t^2}.$$

This approximation reduces the momentum balance to

$$\rho\frac{\partial^2\mathbf{U}}{\partial t^2} - \nabla\cdot\mathbf{t} = \mathbf{0}.$$

Now we need to make some constitutive assumption about the stress $\mathbf{t}$. The most common constitutive law in the theory of elasticity is **Hooke's law**: Stress is proportional to strain. Actually, the general tensor form of this law gives a relationship between $\mathbf{t}$ and the **infinitesimal Eulerian strain**, defined as the symmetric part of the displacement gradient:

$$\mathbf{e} = \tfrac{1}{2}[\nabla\mathbf{U} + (\nabla\mathbf{U})^\top].$$

Hooke's law states that

(1.4-12) $$\mathbf{t} = 2\mu\mathbf{e} + \lambda\operatorname{tr}(\mathbf{e})\mathbf{1},$$

where λ and μ are positive parameters, known as **Lamé constants**, characteristic of the particular solid under investigation.

Substituting the definition of $\mathbf{e}$ into Hooke's law and taking the divergence yields

$$\nabla\cdot\mathbf{t} = \mu\nabla^2\mathbf{U} + (\mu + \lambda)\nabla\nabla\cdot\mathbf{U},$$

which reduces the momentum balance to a vector equation in $\mathbf{U}$:

$$\rho\frac{\partial^2\mathbf{U}}{\partial t^2} - \mu\nabla^2\mathbf{U} - (\mu + \lambda)\nabla\nabla\cdot\mathbf{U} = \mathbf{0}.$$

For purely longitudinal motions along, say, the x_1-axis, only the first component U_1 of the displacement vector $\mathbf{U}$ is nonzero. Furthermore, this component varies only in the x_1-direction, so $U_1 = U_1(x_1, t)$. This fact implies that

$$-\mu \nabla^2 \mathbf{U} - (\mu + \lambda)\nabla \nabla \cdot \mathbf{U} = -(\lambda + 2\mu) \begin{bmatrix} \partial^2 U_1/\partial x_1^2 \\ 0 \\ 0 \end{bmatrix},$$

and so the momentum balance reduces even further to a single equation,

$$\rho \frac{\partial^2 U_1}{\partial t^2} - (\lambda + 2\mu) \frac{\partial^2 U_1}{\partial x_1^2} = 0.$$

To simplify notation, let us denote $x_1 = x$ and $U_1 = U$ and call $k = (\lambda + 2\mu)/\rho$, which is positive. This allows us to write the equation for the displacement as follows:

$$(1.4\text{-}13) \qquad \frac{\partial^2 U}{\partial t^2} = k \frac{\partial^2 U}{\partial x^2}.$$

This is the classic wave equation of mathematical physics. Chapter Five discusses the solution of this equation.

Example: plane strain in an elastic solid.

Another, more complicated example of motion in an elastic material is plane strain. This type of motion occurs, for example, in a cylindrical body whose axis lies along the x_3-direction, provided (i) there is no displacement in the x_3-direction,

$$\mathbf{U} = \begin{bmatrix} U_1 \\ U_2 \\ 0 \end{bmatrix},$$

and (ii) there is no variation of U_1, U_2 or the stress components t_{ij} in the x_3-direction, $\partial U_1/\partial x_3 = \partial U_2/\partial x_3 = \partial t_{ij}/\partial x_3 = 0$. Let us assume further that the body is stationary, that its material properties are spatially uniform, and that body forces are negligible. We shall derive a fourth-order partial differential equation governing the configuration of such a body.

Owing to stationarity and the absence of body forces, the momentum balance simplifies to $\nabla \cdot \mathbf{t} = \mathbf{0}$. What is more, $\partial t_{i3}/\partial x_3 = 0$ for $i = 1, 2, 3$, and, by virtue of the fact that $e_{3j} = (\partial U_j/\partial x_3 + \partial U_3/\partial x_j)/2 = 0$ for $j = 1$ and 2, we have $\partial t_{31}/\partial x_1 = \partial t_{32}/\partial x_2 = 0$. Therefore the only nontrivial components of the momentum balance are the following:

$$\frac{\partial t_{11}}{\partial x_1} + \frac{\partial t_{12}}{\partial x_2} = 0,$$

(1.4-14)

$$\frac{\partial t_{21}}{\partial x_1} + \frac{\partial t_{22}}{\partial x_2} = 0,$$

or, more suggestively,

$$\nabla \times \begin{bmatrix} -t_{12} \\ t_{11} \\ 0 \end{bmatrix} = \mathbf{0}, \qquad \nabla \times \begin{bmatrix} -t_{22} \\ t_{21} \\ 0 \end{bmatrix} = \mathbf{0}.$$

These last two equations state that the two vector fields defined by $(-t_{12}, t_{11}, 0)^\top$, $(-t_{22}, t_{21}, 0)^\top$ are irrotational, which implies that they are gradients of some scalar fields $V_1(x_1, x_2)$ and $V_2(x_1, x_2)$. Therefore we can write

$$-t_{12} = \frac{\partial V_1}{\partial x_1}, \qquad t_{11} = \frac{\partial V_1}{\partial x_2},$$

$$-t_{22} = \frac{\partial V_2}{\partial x_1}, \qquad t_{21} = \frac{\partial V_2}{\partial x_2}.$$

Now the stress **t** is symmetric, so $-\partial V_1/\partial x_1 = \partial V_2/\partial x_2$, or, what is equivalent,

$$\nabla \times \begin{bmatrix} -V_2 \\ V_1 \\ 0 \end{bmatrix} = \mathbf{0}.$$

Hence, the vector field $(-V_2, V_1, 0)^\top$ must also be the gradient of some scalar field $\psi(x_1, x_2)$, so $-V_2 = \partial\psi/\partial x_1$ and $V_1 = \partial\psi/\partial x_2$. Substituting these identities into the stress relationships gives

(1.4-15) $$t_{11} = \frac{\partial^2 \psi}{\partial x_2^2}; \qquad t_{12} = t_{21} = -\frac{\partial^2 \psi}{\partial x_1 \, \partial x_2}; \qquad t_{22} = \frac{\partial^2 \psi}{\partial x_1^2}.$$

The function $\psi(x_1, x_2)$ is called **Airy's stress function.**

The stress **t** computed in this way determines the strain **e** according to Hooke's law. To be physically consistent, the strain must in turn be the symmetric part of the gradient of some displacement **U**, and nothing in the stress equations (1.4-15) guarantees that such a function **U** exists. We must, therefore, impose the additional condition that **e** arise from a mathematically realizable displacement **U**, and for the case of plane strain this condition requires that there exist functions $U_1(x_1, x_2)$ and $U_2(x_1, x_2)$ such that

$$e_{11} = \frac{\partial U_1}{\partial x_1}, \qquad e_{22} = \frac{\partial U_2}{\partial x_2}, \qquad e_{12} + e_{21} = \frac{\partial U_1}{\partial x_2} + \frac{\partial U_2}{\partial x_1}.$$

These equations are called **compatibility conditions**. We can compress them into a single equation by applying $\partial^2/\partial x_2^2$ to the first, $\partial^2/\partial x_1^2$ to the second, and $\partial^2/\partial x_1\,\partial x_2$ to the third to deduce

$$(1.4\text{-}16) \qquad \frac{\partial^2 e_{11}}{\partial x_2^2} + \frac{\partial^2 e_{22}}{\partial x_1^2} = \frac{\partial^2}{\partial x_1\,\partial x_2}(e_{12} + e_{21}).$$

We wish to convert the compatibility condition (1.4-15) to a condition on the stress components t_{ij} and, ultimately, on Airy's stress function ψ. The first step in this conversion is to solve Hooke's law (1.4-12) for the strain components e_{ij} as functions of the stress components t_{ij}. We find

$$e_{11} = \frac{1}{E}[t_{11} - \nu(t_{22} + t_{33})],$$

$$e_{22} = \frac{1}{E}[t_{22} - \nu(t_{11} + t_{33})],$$

$$e_{12} = t_{12}/2\mu,$$

$$e_{21} = t_{21}/2\mu,$$

where the combination

$$E = \frac{\mu(2\mu + 3\lambda)}{\lambda + \mu}$$

is called **Young's modulus**, and

$$\nu = \frac{\lambda}{2(\lambda + \mu)}$$

is **Poisson's ratio**. The equations for e_{11} and e_{22} admit further simplification if we observe that

$$t_{11} + t_{22} + t_{33} = 2\mu e_{11} + \lambda(e_{11} + e_{22}) + 2\mu e_{22}$$
$$+ \lambda(e_{11} + e_{22}) + \lambda(e_{11} + e_{22})$$
$$= \frac{2\mu + 3\lambda}{\lambda}\lambda(t_{11} + t_{22} + t_{33})$$
$$= \frac{2\mu + 3\lambda}{\lambda}t_{33},$$

so that $t_{33} = \nu(t_{11} + t_{22})$. Using this fact, we can write

$$e_{11} = \frac{(1+\nu)}{E}(t_{11} - \nu\Theta),$$

$$e_{22} = \frac{(1+\nu)}{E}(t_{22} - \nu\Theta),$$

where $\Theta = t_{11} + t_{22}$.

The second step in converting the compatibility condition is to substitute for the strain components in Equation (1.4-16) to get equations in the stress components. This tactic yields

$$(1.4\text{-}17) \qquad \frac{(1+\nu)}{E} \left[\frac{\partial^2}{\partial x_2^2} (t_{11} - \nu\Theta) + \frac{\partial^2}{\partial x_1^2} (t_{22} - \nu\Theta) \right]$$

$$= \frac{1}{2\mu} \frac{\partial^2}{\partial x_1 \, \partial x_2} (t_{12} + t_{21}).$$

The third step is to differentiate the momentum balance equations (1.4-14) to observe that

$$\frac{\partial^2 t_{12}}{\partial x_1 \, \partial x_2} = -\frac{\partial^2 t_{11}}{\partial x_1^2}, \qquad \frac{\partial^2 t_{21}}{\partial x_2 \, \partial x_1} = -\frac{\partial^2 t_{22}}{\partial x_2^2}.$$

These relationships, together with the fact that $2\mu(1+\nu)/E = 1$, reduce Equation (1.4-17) to

$$\frac{\partial^2}{\partial x_2^2} (t_{11} - \nu\Theta) + \frac{\partial^2}{\partial x_1^2} (t_{22} - \nu\Theta) = -\frac{\partial^2 t_{11}}{\partial x_1^2} - \frac{\partial^2 t_{22}}{\partial x_2^2}.$$

This equation is equivalent to the simpler equation,

$$(1 - \nu)\nabla^2\Theta = 0,$$

where, in this case, $\nabla^2 = \partial^2/\partial x_1^2 + \partial^2/\partial x_2^2$ denotes the two-dimensional Laplace operator.

This equation expresses the compatibility condition that the stress components must obey if they are to give rise to realizable displacements. If we recall that $\Theta = t_{11} + t_{22}$ and make use of the momentum balance in the form (1.4-15), we discover that

$$\nabla^2 \left(\frac{\partial^2\psi}{\partial x_2^2} + \frac{\partial^2\psi}{\partial x_1^2} \right) = \nabla^2(\nabla^2\psi) = 0,$$

or, more compactly,

$$\nabla^4\psi = 0.$$

This **biharmonic equation** for Airy's stress function $\psi(x_1, x_2)$ combines the compatibility condition with the balance of momentum in plane-strain problems. We shall discuss the biharmonic equation further in Chapter Six.

1.5. Mixtures.

The continuum theory that we have discussed so far is not sufficient to describe phenomena in which chemical, mechanical, or thermal interactions among several bodies may be important. Such phenomena are indeed quite common in our experience. They include, for example, the behaviors of solutions in which diffusion occurs, flows of fluids through permeable solids, and the mechanics of chemically reacting mixtures. To model these phenomena, we need to augment the standard formulation of continuum mechanics to account for the effects of interaction among various constituents. **Mixture theory**, introduced by Eringen and Ingram (1965), provides a rigorous approach to such problems.

In a sense, mixture theory is nothing more than an extension of standard continuum physics to cases in which more than one body is present. One should recognize, however, that the extension is not altogether straightforward and that there are subtleties that make mixture theory an active area of current inquiry. We do not propose to treat these matters here; interested readers may consult Atkin and Craine (1976) for a review. Rather, this section focuses on the practical fundamentals of mixture mechanics with the aim of showing how one may derive governing equations from sound basic principles of mixtures of continua.

Basic mixture mechanics.

By definition, a mixture is a collection of some number (say, N) of bodies, called **constituents**, structured so that they form overlapping continua. To gain a physical sense of what this means, let us consider two examples. In a chemical solution such as salt water the constituents are molecular **species**, namely, water and the ionic species Na^+ and Cl^- of salt. While these constituents may occupy different spatial positions at any instant on the corpuscular scale, to macroscopic observers it appears that salt and water are present at each spatial point in the mixture. Thus the constituents of salt water form overlapping continua at the macroscopic scale.

As the second example, consider a chunk of porous and permeable rock, such as sandstone, whose interstices are completely filled with water. This mixture is different from a chemical solution in that the spatial segregation of the constituents, or **phases**, is observable at a microscopic scale much larger than the corpuscular scale. Thus there is a scale of observation, albeit a very fine one, at which the constituents appear as continua but do not appear to overlap. Nevertheless, at the macroscopic scales of observation familiar, for example, to groundwater hydrologists, rock and water truly appear to be present at every "point" in space, and the constituents may fairly be said to overlap. Since the microscopic distribution of water and rock has macroscopic implications (such as the total volume of water stored in an aquifer), it is common to introduce the notion of porosity or volume fraction when a mixture such as this exhibits continuum structure on a fine scale.

It should be clear from these examples that the notion of overlapping continua is a hypothesis about materials in much the same way as is the notion of a continuum itself. Models developed using mixture theory will therefore be useful provided their predictions are interpreted at scales consistent with the mixture hypothesis.

To formalize the notion of overlapping continua, we associate with each constituent $\alpha = 1, \ldots, N$ a **motion** giving the spatial position $\mathbf{x}$ of any material point of α as a function of its coordinates $\mathbf{X}^\alpha$ in a reference configuration and time t: $\mathbf{x} = \mathbf{x}^\alpha(\mathbf{X}^\alpha, t)$. As with single-body continua, we assume that the function $\mathbf{x}^\alpha$ is a one-to-one correspondence having nonzero Jacobian determinant, so there exists an inverse motion $\mathbf{X}^\alpha = \mathbf{X}^\alpha(\mathbf{x}, t)$. Using these functions we can compute, for example, the Lagrangian velocity

$$\mathbf{V}(\mathbf{X}^\alpha, t) = \frac{\partial \mathbf{x}^\alpha}{\partial t}(\mathbf{X}^\alpha, t),$$

the Eulerian velocity $\mathbf{v}^\alpha(\mathbf{x}^\alpha, t) = \mathbf{V}^\alpha(\mathbf{X}^\alpha(\mathbf{x}^\alpha, t), t)$, and other derivatives of the motion for each constituent α.

Also paralleling the single-body theory, we assign to each constituent α a density function $\rho^\alpha(\mathbf{x}^\alpha, t)$ giving the mass of α per unit volume of α, so that the mass of α residing in any part $\mathcal{P}$ of the mixture is the following:

$$M^\alpha(\mathcal{P}) = \int_{\mathbf{x}^\alpha(\mathcal{P})} \rho^\alpha \, d\mathbf{x}.$$

As mentioned previously, for certain mixtures such as water-saturated porous media we also wish to account at the macroscopic level for some of the microscopic (but not corpuscular) structure of the constituents. Specifically, for these mixtures we assign to each constituent α a volume fraction $\phi^\alpha(\mathbf{x}^\alpha, t)$ such that, for any part $\mathcal{P}$ of the mixture, the total volume within $\mathcal{P}$ occupied by material from α is the following:

$$F^\alpha(\mathcal{P}) = \int_{\mathbf{x}^\alpha(\mathcal{P})} \phi^\alpha \, d\mathbf{x}.$$

It should be clear from this definition that volume fractions obey the inequality $0 \le \phi^\alpha \le 1$. Since the spatial volume occupied by $\mathcal{P}$ must be exactly accounted for by material from all constituents $\alpha = 1, \ldots, N$, the volume fractions must also obey the restriction

$$\sum_{\alpha=1}^{N} \phi^\alpha = 1.$$

In what follows we shall distinguish between **multiphase** mixtures, in which volume fractions play an important role, and **multispecies** mixtures, such as chemical solutions, in which volume fractions play no role.

Using the basic constituent variables introduced so far, we can define several quantities describing the mixture as a whole. For example, the following expression gives the **overall mass density** of the mixture:

$$\rho = \begin{cases} \displaystyle\sum_{\alpha=1}^{N} \rho^{\alpha}, & \text{multispecies mixtures;} \\[2em] \displaystyle\sum_{\alpha=1}^{N} \phi^{\alpha}\rho^{\alpha}, & \text{multiphase mixtures.} \end{cases}$$

In terms of this new quantity, the **mass fraction** of constituent α is the mass of α per unit mass of mixture, that is,

$$\omega^{\alpha} = \begin{cases} \rho^{\alpha}/\rho, & \text{multispecies mixtures;} \\ \phi^{\alpha}\rho^{\alpha}/\rho, & \text{multiphase mixtures.} \end{cases}$$

It is clear that the mass fractions must satisfy the restriction $\sum_{\alpha=1}^{N} \omega^{\alpha} = 1$. The **barycentric velocity** of the mixture is the mass-weighted mean of the constituent velocities:

$$\mathbf{v} = \begin{cases} \dfrac{1}{\rho}\displaystyle\sum_{\alpha=1}^{N} \rho^{\alpha}\mathbf{v}^{\alpha}, & \text{multispecies mixtures;} \\[2em] \dfrac{1}{\rho}\displaystyle\sum_{\alpha=1}^{N} \phi^{\alpha}\rho^{\alpha}\mathbf{v}^{\alpha}, & \text{multiphase mixtures.} \end{cases}$$

The deviation of a constituent's velocity from the barycentric velocity of the mixture is the **diffusion velocity**, $\boldsymbol{\nu}^{\alpha} = \mathbf{v}^{\alpha} - \mathbf{v}$. One can easily show that $\sum_{\alpha=1}^{N} \omega^{\alpha}\boldsymbol{\nu}^{\alpha} = \mathbf{0}$.

Mixtures obey balance laws that are derivable from a general form in a fashion similar to single-body continua. The general global balance law for mixtures is analogous to that for single-body continua: for multispecies mixtures,

$$\sum_{\alpha=1}^{N} \left(\frac{d}{dt}\int_{\mathcal{V}_{\alpha}} \rho^{\alpha}\Psi^{\alpha}\, d\mathbf{x} - \oint_{\partial\mathcal{V}_{\alpha}} \boldsymbol{\tau}^{\alpha}\cdot\mathbf{n}\, d\mathbf{x} - \int_{\mathcal{V}_{\alpha}} \rho^{\alpha}g^{\alpha}\, d\mathbf{x} \right) = 0,$$

while for multiphase mixtures

$$\sum_{\alpha=1}^{N} \left(\frac{d}{dt}\int_{\mathcal{V}_{\alpha}} \phi^{\alpha}\rho^{\alpha}\Psi^{\alpha}\, d\mathbf{x} - \oint_{\partial\mathcal{V}_{\alpha}} \boldsymbol{\tau}^{\alpha}\cdot\mathbf{n}\, d\mathbf{x} - \int_{\mathcal{V}_{\alpha}} \phi^{\alpha}\rho^{\alpha}g^{\alpha}\, d\mathbf{x} \right) = 0.$$

Here, Ψ^α stands for a generic conserved quantity associated with each constituent; $\boldsymbol{\tau}^\alpha$ and g^α are generalized fluxes and supplies, respectively, and $\mathcal{V}_\alpha$ signifies a material volume for constituent α with orientable boundary $\partial\mathcal{V}_\alpha$. These equations ignore the possibility that surfaces of discontinuity may intersect $\mathcal{V}_\alpha$; however, the extension of the balance laws to accommodate such surfaces is straightforward by analogy with single-body continuum mechanics.

A sequence of arguments paralleling those invoked in the single-body case reduces the general global balances to **local mixture balances**. We get

$$\sum_{\alpha=1}^{N}\left[\overset{\alpha}{\frac{D}{Dt}}\left(\rho^\alpha\Psi^\alpha\right) + \rho^\alpha\Psi^\alpha\nabla\cdot\mathbf{v}^\alpha - \nabla\cdot\boldsymbol{\tau}^\alpha - \rho^\alpha g^\alpha\right] = 0$$

for multispecies mixtures and

$$\sum_{\alpha=1}^{N}\left[\overset{\alpha}{\frac{D}{Dt}}\left(\phi^\alpha\rho^\alpha\Psi^\alpha\right) + \phi^\alpha\rho^\alpha\Psi^\alpha\nabla\cdot\mathbf{v}^\alpha - \nabla\cdot\boldsymbol{\tau}^\alpha - \phi^\alpha\rho^\alpha g^\alpha\right] = 0$$

for multiphase mixtures. Observe that the material derivative operators appearing in these equations are indexed by constituents, being defined by

$$\overset{\alpha}{\frac{D}{Dt}} = \begin{cases} \dfrac{\partial}{\partial t} & \text{for functions of material coordinates } (\mathbf{X}^\alpha, t); \\[2mm] \dfrac{\partial}{\partial t} + \mathbf{v}^\alpha\cdot\nabla & \text{for functions of spatial coordinates } (\mathbf{x}^\alpha, t). \end{cases}$$

These two definitions correspond, respectively, to the Lagrangian and Eulerian points of view in mixture theory.

In contrast to the single-body theory, the local mixture balances govern sums over all constituents. These laws are equivalent to sets of individual constituent balances provided we take careful account of interactions among constituents. We then obtain the general **local constituent balances**

(1.5-1a) $$\overset{\alpha}{\frac{D}{Dt}}\left(\rho^\alpha\Psi^\alpha\right) + \rho^\alpha\Psi^\alpha\nabla\cdot\mathbf{v}^\alpha - \nabla\cdot\boldsymbol{\tau}^\alpha - \rho^\alpha g^\alpha = e^\alpha,$$

$$\alpha = 1,\ldots,N,$$

for multispecies mixtures and

(1.5-1b) $$\overset{\alpha}{\frac{D}{Dt}}\left(\phi^\alpha\rho^\alpha\Psi^\alpha\right) + \phi^\alpha\rho^\alpha\Psi^\alpha\nabla\cdot\mathbf{v}^\alpha - \nabla\cdot\boldsymbol{\tau}^\alpha - \phi^\alpha\rho^\alpha g^\alpha = e^\alpha,$$

$$\alpha = 1,\ldots,N,$$

for multiphase mixtures. The new variable e^α in these laws is a quantity measuring the exchange of the conserved quantity Ψ into constituent α from other constituents. If the constituent balances (1.5-1) are to be consistent with the local mixture balances, these exchange terms must obey the restriction $\sum_{\alpha=1}^{N} e^\alpha = 0$.

To derive particular balance laws from the general forms (1.5-1), one simply assigns to Ψ^α, τ^α, and g^α the physical interpretations set forth in Table 1-1, taking care to preserve the tensorial consistency of the resulting equations and noting the corresponding interpretations of the exchange term e^α. We shall do this for the mass and momentum balances using two examples.

Example: passive, isothermal transport of solute in a fluid.

First, let us examine a multispecies mixture in which a single solute moves isothermally in a fluid, as for instance when a soluble but otherwise inert contaminant undergoes transport in a river. Thus we have two constituents: the solute S and the fluid F. Let us assume that the transport is **passive**, meaning that the solute is chemically unreactive and has negligible effect on the mechanical properties of the mixture. We shall develop a mass balance for the solute S from the general multispecies constituent balance (1.5-1a).

Table 1-1 indicates that for the mass balance $\Psi^\alpha = 1$, $\tau^\alpha = 0$, and $g^\alpha = 0$. Furthermore, the exchange term e^α in this case represents the production of species α via chemical reactions, which we can denote by r^α. The species mass balance therefore becomes

$$(1.5\text{-}2) \qquad \frac{\overset{\alpha}{D} \rho^\alpha}{Dt} + \rho^\alpha \nabla \cdot \mathbf{v}^\alpha = \frac{\partial \rho^\alpha}{\partial t} + \nabla \cdot (\rho^\alpha \mathbf{v}^\alpha) = r^\alpha, \qquad \alpha = \text{S,F},$$

or, using the mass fractions and barycentric velocity,

$$\frac{\partial}{\partial t} (\rho^\alpha) + \nabla \cdot (\rho^\alpha \mathbf{v}) + \nabla \cdot \mathbf{j}^\alpha = r^\alpha, \qquad \alpha = \text{S,F}$$

in which $\mathbf{j}^\alpha = \rho^\alpha \boldsymbol{\nu}^\alpha$, the **diffusive flux** of constituent α. When chemical reactions are absent, we find

$$(1.5\text{-}3) \qquad \frac{\partial \rho^\alpha}{\partial t} + \nabla \cdot (\rho^\alpha \mathbf{v}) + \nabla \cdot \mathbf{j}^\alpha = 0, \qquad \alpha = \text{S,F}.$$

Now consider the solute S. To make sense out of the balance equation just derived, we need a constitutive law for the diffusive flux $\mathbf{j}^S$. The simplest nontrivial constitutive law in common use is **Fick's law**, stating that diffusive flux is proportional to the gradient in concentration:

$$\mathbf{j}^{S} = -K^{S}\nabla\rho^{S}.$$

Here K^{S} is a positive scalar called the **diffusion coefficient**. Substituting this simple assumption into the balance law (1.5-3) yields

$$(1.5\text{-}4) \qquad \frac{\partial\rho^{S}}{\partial t} + \nabla\cdot(\rho^{S}\mathbf{v}) - \nabla\cdot(K^{S}\nabla\rho^{S}) = 0.$$

This second-order differential equation is known as the **advection-diffusion transport equation**. We shall say more about this equation in Chapters Four and Five.

Example: flow of a fluid in porous rock.

As a second example of mixture mechanics, let us examine the flow of a homogeneous fluid F through a rock matrix R. For simplicity, assume that the two constituents F and R in this multiphase mixture are chemically inert, so that the mass exchanges vanish: $r^{R} = r^{F} = 0$. Assume further that the rock matrix is immobile, so that $\mathbf{v}^{R} = 0$ and the volume fraction ϕ^{R} is constant in time. We shall derive a field equation for the fluid velocity.

We begin with the local momentum balance for any constituent α in a multiphase mixture. To obtain this balance from the general form (1.5-1b), we set $\Psi^{\alpha} = \mathbf{v}^{\alpha}$, $\boldsymbol{r}^{\alpha} = \mathbf{t}^{\alpha}$, and $g^{\alpha} = \mathbf{b}^{\alpha}$, where $\mathbf{t}^{\alpha}$ represents the stress tensor in phase α and $\mathbf{b}^{\alpha}$ stands for the body force per unit mass in phase α. The exchange term in this case must be the momentum transfer $\mathbf{m}^{\alpha}$ into phase α from other phases. Thus we have the constituent momentum balance

$$\overset{\alpha}{\frac{D}{Dt}}(\phi^{\alpha}\rho^{\alpha}\mathbf{v}^{\alpha}) + \phi^{\alpha}\rho^{\alpha}\mathbf{v}^{\alpha}\nabla\cdot\mathbf{v}^{\alpha} - \nabla\cdot\mathbf{t}^{\alpha} - \phi^{\alpha}\rho^{\alpha}\mathbf{b}^{\alpha} = \mathbf{m}^{\alpha}, \qquad \alpha = \text{F,R.}$$

By expanding the material derivative in this equation we find

$$\phi^{\alpha}\rho^{\alpha}\overset{\alpha}{\frac{D\mathbf{v}^{\alpha}}{Dt}} + \mathbf{v}^{\alpha}\left[\overset{\alpha}{\frac{D}{Dt}}(\phi^{\alpha}\rho^{\alpha}) + \phi^{\alpha}\rho^{\alpha}\nabla\cdot\mathbf{v}^{\alpha}\right] - \nabla\cdot\mathbf{t}^{\alpha}$$
$$- \phi^{\alpha}\rho^{\alpha}\mathbf{b}^{\alpha} = \mathbf{m}^{\alpha}, \qquad \alpha = \text{F,R.}$$

However, the sum appearing in square brackets here equals the mass exchange r^{α} according to the constituent mass balance, and we have stipulated that, for the rock-fluid mixture, each $r^{\alpha} = 0$. Therefore we have

$$(1.5\text{-}5) \qquad \phi^{\alpha}\rho^{\alpha}\overset{\alpha}{\frac{D\mathbf{v}^{\alpha}}{Dt}} - \nabla\cdot\mathbf{t}^{\alpha} - \phi^{\alpha}\rho^{\alpha}\mathbf{b}^{\alpha} = \mathbf{m}^{\alpha}.$$

To develop the velocity field equation for the fluid F we must introduce some constitutive laws for the stress $\mathbf{t}^F$, the body force $\mathbf{b}^F$, and the exchange term $\mathbf{m}^F$. Let us assume that the fluid is Newtonian and that momentum transfers attributable to shear stresses in the fluid are negligible compared with momentum transfers between the fluid and the rock. Hence, we need only consider normal stresses in the fluid, so $\mathbf{t}^F = -p^F \mathbf{1}$, p^F being the mechanical pressure in the fluid. Further, assume that gravity is the only body force, so that we can write $\phi^F \mathbf{b}^F = g\nabla Z$, where g is the gravitational acceleration and Z signifies depth below some datum. Finally, let us model the interphase momentum transfer as a **Stokes drag**, meaning $\mathbf{m}^F$ is proportional to the difference between rock and fluid velocities. In simple cases this relationship will be isotropic, so that the constant of proportionality will be a scalar. It is also reasonable to expect that $\mathbf{m}^F$ will be proportional to the local volume fraction of the fluid. Since $\mathbf{v}^R = \mathbf{0}$, we thus get a constitutive law of the form $\mathbf{m}^F = -\phi^F \mathbf{v}^F / \Lambda$. The constant Λ is called the **mobility**.

Substituting these constitutive assumptions into Equation (1.5-5), we arrive at the fluid momentum balance

$$\phi^F \rho^F \frac{\overset{F}{D}\mathbf{v}^F}{Dt} - \nabla p^F - \rho^F g\nabla Z = -\frac{\phi^F}{\Lambda}\mathbf{v}^F.$$

Most theories of flow in porous media assume that the inertia of the fluid has a negligible influence on the flow compared with the pressure, gravity, and momentum exchange terms. Mathematically, this assumption leads to the approximation $\overset{\alpha}{D}\mathbf{v}^\alpha / Dt = \mathbf{0}$, which, after some rearranging, reduces our fluid momentum balance to

$$\mathbf{v}^F = -\frac{\Lambda}{\phi^F}\left(\nabla p^F - \rho^F g\nabla Z\right).$$

Finally, experimentalists usually interpret the mobility Λ by decomposing it into a "rock" property and a "fluid" property: $\Lambda = k/\mu^F$. Here μ^F is the fluid property, namely, the fluid's dynamic viscosity, and k is regarded as a property of the rock matrix, called the **permeability** and having the dimensions of area. With this decomposition, the fluid velocity field equation becomes

$$(1.5\text{-}6) \qquad \mathbf{v}^F = -\frac{k}{\mu^F \phi^F}\left(\nabla p^F - \rho^F g\nabla Z\right).$$

This equation is familiar to groundwater hydrologists and petroleum engineers as **Darcy's law** for a homogeneous fluid flowing in a solid porous medium. The application of Darcy's law to systems of one or more fluids flowing in porous media leads to a huge variety of systems of partial differential equations. We shall examine some important special cases in Chapters Three and Six.

This rather brief overview of mixture theory omits consideration of many of the intricacies of the subject. We have not discussed any applications of the mixture energy balance, nor, for example, have we discussed implications of momentum exchanges for the symmetry of constituent stress tensors. Finally, we have neglected altogether the mixture entropy inequality and its applications to constitutive theory. While these matters would carry us far afield of our present purposes, readers should be aware that there is much more to the mechanics of mixtures—and indeed to continuum mechanics as a whole—than we have discussed in this chapter.

1.6. Problems for Chapter One.

1. Show that the Eulerian velocity and acceleration are related by

$$\mathbf{a}(\mathbf{x}, t) = \frac{D\mathbf{v}}{Dt}(\mathbf{x}, t).$$

Use this fact to verify the decomposition

$$\mathbf{a} = \frac{\partial \mathbf{v}}{\partial t} - \mathbf{v} \times (\nabla \times \mathbf{v}) + \nabla(\tfrac{1}{2}\mathbf{v} \cdot \mathbf{v}).$$

The quantity $\boldsymbol{\omega} = \nabla \times \mathbf{v}$ is called the **vorticity**; $\tfrac{1}{2}\mathbf{v} \cdot \mathbf{v}$ represents the kinetic energy per unit mass.

2. Consider the motion $\mathbf{x} = \mathbf{x}(\mathbf{X}, t)$ whose component functions are the following:

$$x_1 = X_1 + X_2 + 2X_3,$$
$$x_2 = X_2 + 3t,$$
$$x_3 = X_3 e^{-t}.$$

Compute the inverse motion $\mathbf{X}(\mathbf{x}, t)$, the Eulerian velocity, and the Jacobian determinant $\det \mathbf{J}$.

3. If the velocity field $\mathbf{v}$ is sufficiently smooth, Taylor's theorem gives its variation in a neighborhood of a spatial point $\mathbf{x}$ as

$$\mathbf{v}(\mathbf{x} + \mathbf{h}, t) = \mathbf{v}(\mathbf{x}, t) + \nabla \mathbf{v}(\mathbf{x}, t) \cdot \mathbf{h} + \mathcal{O}(\|\mathbf{h}\|^2).$$

One can decompose $\nabla \mathbf{v}$ into symmetric and antisymmetric tensors, $\nabla \mathbf{v} = \mathbf{D} + \boldsymbol{\Omega}$, where $\mathbf{D} = \tfrac{1}{2}[\nabla \mathbf{v} + (\nabla \mathbf{v})^\top]$ is the deformation rate tensor and the tensor $\boldsymbol{\Omega} = \tfrac{1}{2}[\nabla \mathbf{v} - (\nabla \mathbf{v})^\top]$ is the **spin**. Write the entries of $\boldsymbol{\Omega}$ in terms of the vorticity $\boldsymbol{\omega} = \nabla \times \mathbf{v}$, and show that

$$\mathbf{v}(\mathbf{x} + \mathbf{h}, t) = \mathbf{v}(\mathbf{x}, t) + \mathbf{D}(\mathbf{x}, t) \cdot \mathbf{h} + \tfrac{1}{2}\mathbf{h} \times \boldsymbol{\omega}(\mathbf{x}, t) + \mathcal{O}(\|\mathbf{h}\|^2).$$

4. In **plane motion** there is no variation in one spatial direction, say x_3. In this case the mass balance equation reduces to

$$\frac{\partial \rho}{\partial t} + \nabla \cdot (\rho \mathbf{v}) = \frac{\partial \rho}{\partial t} + \frac{\partial}{\partial x_1}(\rho v_1) + \frac{\partial}{\partial t}(\rho v_2) = 0.$$

If, in addition, the motion is steady $(\partial \rho / \partial t = 0)$ and ρ is uniform $(\nabla \rho = \mathbf{0})$, then $\mathbf{v} = (v_1, v_2, 0)$ is **solenoidal**, meaning $\nabla \cdot \mathbf{v} = 0$. When this condition holds, we can introduce a **stream function** $\boldsymbol{\psi} = (0, 0, \psi)$ such that $\mathbf{v} = \nabla \times \boldsymbol{\psi}$. Verify that this representation of the velocity forces $\nabla \cdot \mathbf{v} = 0$ automatically. When $\mathbf{v}$ is **irrotational**, meaning

$$\boldsymbol{\omega} \equiv \nabla \times \mathbf{v} = \left(0, 0, \frac{\partial v_2}{\partial x_1} - \frac{\partial v_1}{\partial x_2}\right) = (0, 0, \omega) = \mathbf{0},$$

then we can introduce a **velocity potential** φ such that $\mathbf{v} = \nabla \varphi$. Show that, for velocity fields $\mathbf{v}$ that are both solenoidal and irrotational, curves where φ is constant are orthogonal to curves where ψ is constant. Thus φ and ψ form a "natural" orthogonal coordinate system for such flows. (Hint: Recall that, provided f is differentiable, curves where f is constant are orthogonal to ∇f.)

5. By examining the scalar vorticity ω defined in Problem 4, show that in irrotational, solenoidal plane motion the scalar stream function ψ satisfies Laplace's equation, $\nabla^2 \psi = 0$.

6. Consider a motion in which no discontinuities are present. Show that the global mass balance

$$\frac{d}{dt} \int_{\mathcal{V}} \rho \, d\mathbf{x} = 0$$

can be written in local form as $\rho_0 = \rho \det \mathbf{J}$, where ρ_0 signifies the mass density in any reference configuration and $\mathbf{J}$ denotes the Jacobian matrix of the motion.

7. Show that, for an incompressible fluid, $\nabla \cdot [\nabla \mathbf{v} + (\nabla \mathbf{v})^\top] = \nabla^2 \mathbf{v} \equiv \nabla \cdot \nabla \mathbf{v}$.

8. Recall Darcy's law, giving the velocity of a fluid in a porous medium in the form $\mathbf{v} = -\lambda(\nabla p - \rho g \nabla Z)$ [see Equation (1.5-6)]. The aim of this problem is to identify conditions under which the quantity $(\rho g)^{-1} \nabla p - \nabla Z$ can be written as the gradient of a **potential** h, so that $\mathbf{v} = -\rho g \lambda \nabla h$. From Problem 4, we know that $(\rho g)^{-1} \nabla p - \nabla Z = \nabla h$ provided $\nabla \times [(\rho g)^{-1} \nabla p - \nabla Z] = \mathbf{0}$. Show that a sufficient condition for this provision to hold is that the fluid be **barotropic**, that is, ρ depends only on p.

9. Referring to Problem 8, show that for a barotropic fluid the function h defined by the equation

$$h = \int_{p_{\text{ref}}}^{p} \frac{dq}{g\rho(q)},$$

where p_{ref} signifies any fixed reference value of pressure, serves as a potential for the Darcy velocity. That is, show that $\nabla h = (\rho g)^{-1} \nabla p - \nabla Z$. Hint: Use the Leibnitz formula,

$$\frac{\partial}{\partial u} \int_{a(u)}^{b(u)} f(u, v) \, dv = \int_{a(u)}^{b(u)} \frac{\partial f}{\partial u}(u, v) \, dv + f(u, b(u))\frac{\partial b}{\partial u} - f(u, a(u))\frac{\partial a}{\partial u}.$$

10. We may specify certain deformations of a body by giving a functional relationship $\mathbf{x} = \mathbf{f}(\mathbf{X})$ between the position $\mathbf{x}$ after deformation and the coordinates $\mathbf{X}$ in the initial (reference) configuration, before deformation. Compute the infinitesimal Eulerian strains for each of the following deformations:

   (a) **uniaxial strain**

$$\begin{bmatrix} x_1 \\ x_2 \\ x_3 \end{bmatrix} = \begin{bmatrix} \lambda & 0 & 0 \\ 0 & 1 & 0 \\ 0 & 0 & 1 \end{bmatrix} \begin{bmatrix} X_1 \\ X_2 \\ X_3 \end{bmatrix},$$

   (b) **uniform dilatation**

$$\begin{bmatrix} x_1 \\ x_2 \\ x_3 \end{bmatrix} = \begin{bmatrix} \lambda & 0 & 0 \\ 0 & \lambda & 0 \\ 0 & 0 & \lambda \end{bmatrix} \begin{bmatrix} X_1 \\ X_2 \\ X_3 \end{bmatrix},$$

   (c) **simple shear**

$$\begin{bmatrix} x_1 \\ x_2 \\ x_3 \end{bmatrix} = \begin{bmatrix} 1 & s & 0 \\ 0 & 1 & 0 \\ 0 & 0 & 1 \end{bmatrix} \begin{bmatrix} X_1 \\ X_2 \\ X_3 \end{bmatrix}.$$

Sketch the effects of these deformations on a cube whose edges have unit length.

11. The **principal axes of strain** are lines collinear with the eigenvectors of the infinitesimal Eulerian strain, and the **principal strains** are its eigenvalues. Show that simple shear, defined in Problem 10(c), is isochoric, and determine its principal axes of strain and the principal strains.

1.7. References.

Atkin, R.J. and Craine,R.E., "Continuum theories of mixtures: basic theory and historical development," *Q. J. Mech. Appl. Math., 29:2* (1976), 209-244.

Batchelor, G.K., *An Introduction to Fluid Dynamics*, Cambridge, U.K.: Cambridge University Press, 1967.

Bird, R.B., Stewart, W.E., and Lightfoot, E.N., *Transport Phenomena,* New York: Wiley, 1960.

Eringen, A.C., *Mechanics of Continua,* 2nd ed., Huntington, New York: Krieger, 1980.

Eringen, A.C. and Ingram, J.D., "A continuum theory of chemically reacting media–I," *Int. J. Eng. Sci., 3* (1965), 197-212.

Landau, L.D. and Lifschitz, E.M., *Fluid Mechanics,* New York: Pergamon Press, 1959.

Truesdell, C.A., *Elements of Continuum Mechanics,* New York: Springer-Verlag, 1966.

Truesdell, C.A. and Toupin, R.A., "The classical field theories," in S. Flügge, Ed., *Handbuch der Physik III/1,* Berlin: Springer-Verlag, 1960, pp. 226-793.

Williamson, R.E., Crowell, R.H., and Trotter, H.F., *Calculus of Vector Functions,* 3rd ed., Englewood Cliffs, New Jersey: Prentice-Hall, 1972.

CHAPTER TWO
INTRODUCTION TO NUMERICAL METHODS

2.1. Introduction.

In Chapter One we illustrated a methodology for the development of equations normally encountered in engineering and scientific applications. Now we turn our attention to the task of solving these equations numerically. After a brief introduction to the classification of partial differential equations (PDEs), we describe the initial and boundary conditions required to assure that a given problem is mathematically well posed. Next, we introduce the topic of interpolation theory, which provides the mathematical foundation upon which we develop our approximation methods. Subsequent sections of this chapter describe the finite-difference method, the Galerkin finite-element method, and the collocation method. Having developed these numerical tools, we shall finally be in a position to examine the solution of the equations developed in Chapter One. The remainder of the book is devoted to this task.

2.2. Partial Differential Equations.

A partial differential equation in two space dimensions can be written, albeit rather abstractly, as

$$(2.2\text{-}1) \qquad \mathcal{F}\left(x, y, u, \frac{\partial u}{\partial x}, \frac{\partial u}{\partial y}, \frac{\partial^2 u}{\partial x^2}, \frac{\partial^2 u}{\partial y^2}, \frac{\partial^2 u}{\partial x \, \partial y}, \cdots\right) = 0,$$

where at least one partial derivative must exist. The PDE holds for all points (x, y) inside some open region Ω of two-dimensional space, called the **domain** of the PDE. A well-known example of a PDE is Laplace's equation,

$$(2.2\text{-}2) \qquad \frac{\partial^2 u}{\partial x^2} + \frac{\partial^2 u}{\partial y^2} = 0,$$

and we have encountered others in Chapter One.

The **order** of a PDE is defined to be the order of the highest-order derivative appearing in the equation. Thus Laplace's equation (2.2-2) is of second order, as is Burgers' equation

$$(2.2\text{-}3) \qquad \frac{\partial u}{\partial t} + u \frac{\partial u}{\partial x} - \epsilon \frac{\partial^2 u}{\partial x^2} = 0,$$

while the equation

$$(2.2\text{-}4) \qquad \frac{\partial u}{\partial t} + b \frac{\partial u}{\partial y} = 0$$

is of first order.

Another important concept is that of linearity. The PDE (2.2-1) is **linear** if $\mathcal{F}$ can be expressed as a linear combination of u and its derivatives. The general linear second-order PDE in two independent variables x and y has the form

$$a(x,y) \frac{\partial^2 u}{\partial x^2} + b(x,y) \frac{\partial^2 u}{\partial x \, \partial y} + c(x,y) \frac{\partial^2 u}{\partial y^2} + e(x,y) \frac{\partial u}{\partial x}$$

$$+ f(x,y) \frac{\partial u}{\partial y} + g(x,y)u + h(x,y) = 0.$$

Observe that each of the functions $a, \ldots, h$ varies with x and y only and does not depend on the unknown u or any of its derivatives. One observes, then, that Laplace's equation (2.2-2) is linear, while Burgers' equation (2.2-3) is nonlinear. As we shall see in later chapters, the property of linearity plays an important role in the selection of a solution methodology.

The physical phenomenon captured in a PDE is reflected in the behavior of its solutions. Solution behavior, in turn, plays a major role in the selection of a numerical strategy. One approach to the description of solution behavior, prior to solving an equation, is through the classification of PDEs. Let us consider a PDE that is at most second order, having the form

$$(2.2\text{-}5) \qquad a \frac{\partial^2 u}{\partial x^2} + b \frac{\partial^2 u}{\partial x \, \partial y} + c \frac{\partial^2 u}{\partial y^2} + e = 0,$$

where the coefficients a, b, c, and e in Equation (2.2-5) can be functions of the independent variables x and y and possibly the dependent variables u, $\partial u/\partial x$, and $\partial u/\partial y$. It is convenient at this point to introduce the change of variables

$$p = \frac{\partial u}{\partial x}, \qquad q = \frac{\partial u}{\partial y}, \qquad r = \frac{\partial^2 u}{\partial x^2}, \qquad s = \frac{\partial^2 u}{\partial x \, \partial y}, \qquad t = \frac{\partial^2 u}{\partial y^2}.$$

Consider now a curve C in the (x, y)-plane, parametrized as $(x(\sigma), y(\sigma))$, along which u, p, q, r, s, and t satisfy the model equation (2.2-5). One can employ the chain rule for partial differentiation along C to yield

(2.2-6a)
$$\frac{dp}{d\sigma} = \frac{\partial p}{\partial x}\frac{dx}{d\sigma} + \frac{\partial p}{\partial y}\frac{dy}{d\sigma} = r\frac{dx}{d\sigma} + s\frac{dy}{d\sigma},$$

(2.2-6b)
$$\frac{dq}{d\sigma} = \frac{\partial q}{\partial x}\frac{dx}{d\sigma} + \frac{\partial q}{\partial y}\frac{dy}{d\sigma} = s\frac{dx}{d\sigma} + t\frac{dy}{d\sigma}.$$

Combination of Equations (2.2-5) and (2.2-6) yields, after some manipulation,

(2.2-7)
$$s\left[a\left(\frac{dy}{dx}\right)^2 - b\left(\frac{dy}{dx}\right) + c\right] - \left[a\frac{dp}{dx}\frac{dy}{dx} + c\frac{dq}{dx} + e\frac{dy}{dx}\right] = 0.$$

Let us now select the curve C so that its slope dy/dx at every point satisfies

(2.2-8)
$$a\left(\frac{dy}{dx}\right)^2 - b\left(\frac{dy}{dx}\right) + c = 0.$$

Under this constraint Equation (2.2-7) reduces to a set of ordinary differential equations (ODEs) in either x or y. While an entire computational strategy can be built around the solution of these ODEs along C, for now we are interested only in the functional form of C as defined through Equation (2.2-8).

Equation (2.2-8) is a quadratic equation in dy/dx. When $b^2 - 4ac < 0$, the solutions are complex; when $b^2 - 4ac = 0$, the solutions are real and equal, and when $b^2 - 4ac > 0$, the solutions are real and distinct. The number of real solutions to Equation (2.2-8) is just the number of distinct, real differential equations defining curves C passing through the point (x, y) where the coefficients a, b, c are evaluated. Thus the nature of the discriminant $b^2 - 4ac$ indicates the number of distinct families of real curves C along which the given PDE reduces to ODEs.

Consider now the equation of a conic section,

(2.2-9)
$$ax^2 + bxy + cy^2 + dx + ey + f = 0.$$

This equation describes an ellipse when $b^2 - 4ac < 0$, a parabola when $b^2 - 4ac = 0$, and a hyperbola when $b^2 - 4ac > 0$. Borrowing from this nomenclature, we define PDEs with $b^2 - 4ac < 0$ as **elliptic**, those with $b^2 - 4ac = 0$ as **parabolic**, and those with $b^2 - 4ac > 0$ as **hyperbolic**. We employ this nomenclature through the remainder of this book. This classification scheme can be extended to higher-order equations, equations in more independent variables, and systems of PDEs; the interested reader should consult Zauderer (1983, Chapter Three).

2.3. Boundary and Initial Conditions.

A given PDE posed on a domain Ω generally has an infinite number of solutions if it has any at all. If solutions exist, they will be unique only when we impose appropriate auxiliary conditions. For example, to specify a unique solution to Laplace's equation

$$(2.3\text{-}1) \qquad \frac{\partial^2 u}{\partial x^2} + \frac{\partial^2 u}{\partial y^2} = 0$$

on a domain Ω we may prescribe the values of $u(x, y)$ along the boundary $\partial\Omega$ of Ω. On the other hand, to guarantee a unique solution to the heat equation

$$(2.3\text{-}2) \qquad \frac{\partial^2 u}{\partial x^2} + \frac{\partial^2 u}{\partial y^2} - \frac{\partial u}{\partial t} = 0$$

on a space-time domain Ω, we would need to specify the values of $u(x, y, t)$ at some initial time, say $t = 0$, in addition to prescribing values along the spatial boundary. Prescribed values of $u(x, y, t)$ along the spatial boundary are **boundary conditions**, while prescribed values of $u(x, y, t)$ or any of its derivatives for a specific time $t = t_0$ are **initial conditions**.

Boundary and initial conditions can be written in the general form

$$(2.3\text{-}3) \qquad \alpha(x, y, t)u(x, y, t) + \beta(x, y, t)\frac{\partial u}{\partial n}(x, y, t) = \gamma(x, y, t),$$

$$(x, y, t) \in \partial\Omega,$$

where α, β, and γ are known functions, $\partial\Omega$ is the boundary of the space-time domain Ω, and $\partial u/\partial n$ denotes the derivative normal to a boundary in either space or time. **Dirichlet** conditions are boundary conditions wherein u is specified, that is, when $\beta = 0$. **Neumann** conditions describe the case when the normal derivative of the function is specified, that is, when $\alpha = 0$. When $\alpha \neq 0$ and $\beta \neq 0$, one has specified a **Robin** boundary condition. Dirichlet, Neumann and Robin boundary conditions are also known by the aliases Type 1, Type 2, and Type 3 boundary conditions, respectively. Whenever $\gamma = 0$, the boundary condition is said to be **homogeneous**.

The issue of what constitutes an appropriate set of initial and boundary conditions for a given PDE can be quite complex. In fact, there exists little in the way of general mathematical theory to guide us here outside the realm of fairly simple, thoroughly studied PDEs. In practice, auxiliary conditions usually arise from physical considerations, and their mathematical justification, if any, occurs a posteriori. The interested reader should

consult Garabedian (1964, Chapter Two) for a more detailed discussion
than we shall pursue here.

2.4. Polynomial Approximations.

Polynomial approximation theory constitutes the foundation upon which
we shall build the various numerical methods. In developing this foundation
we follow closely the approach of Botha and Pinder (1983). We omit several
proofs of error estimates; however, problems at the end of the chapter
outline some of the arguments. Let us begin by discretizing a closed interval
$[a, b]$ by a **grid** Δ, which we define with the assistance of Figure 2-1 as
follows:

$$(2.4\text{-}1) \qquad \Delta : (a =)\ x_0 < x_1 < x_2 < \cdots < x_n\ (= b).$$

Equation (2.4-1) states that our grid will consist of $n + 1$ **nodes** x_0, x_1,
$\ldots, x_n$ and n discrete segments or **finite elements** $[x_{i-1}, x_i]$, such that
$x_0 = a$ and $x_n = b$. We call the largest interval length, $\max_{1 \le i \le n}\{x_i - x_{i-1}\}$, the **mesh** of the grid.

Lagrange interpolating polynomials.

We now seek a polynomial $P_n(x)$ of degree n that **interpolates** a
given function $f(x)$ between the nodes x_i of the grid. That is, we want

$$(2.4\text{-}2) \qquad P_n(x_i) = f(x_i),$$

where $\{f(x_i)\}_{i=0}^n$ is a set of values of the prescribed function $f(x) \in C^m[a, b]$ at the nodal points $\{x_i\}_{i=0}^n$, as illustrated in Figure 2-1. The
notation $C^m[a, b]$ designates the class of functions that are continuously
differentiable m times over the closed interval $[a, b]$. The conditions (2.4-2)
suffice to specify $P_n(x)$ as

$$(2.4\text{-}3) \qquad P_n(x) = \sum_{i=0}^n \ell_i(x) f(x_i),$$

where the polynomials $\ell_i(x)$ have the form

$$(2.4\text{-}4) \qquad \ell_i(x) = \prod_{\substack{j=0 \\ j \ne i}}^n \frac{x - x_j}{x_i - x_j}, \qquad x \in [a, b].$$

The approximation $P_n(x)$ to $f(x)$ is known as a **Lagrange interpolation
polynomial**, and the function $\ell_i(x)$ is called a **Lagrange basis polynomial**. The simplest Lagrange basis polynomial is the linear $(n = 1)$
member, which is illustrated in Figure 2-2a.

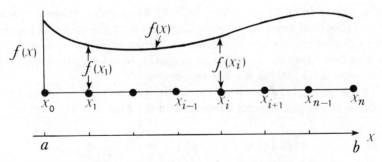

Figure 2-1. Specification of the grid Δ defined over the interval $[a, b]$. The nodal locations are designated by the symbol •. Specific values of the function $f(x)$ are designated $f(x_i)$.

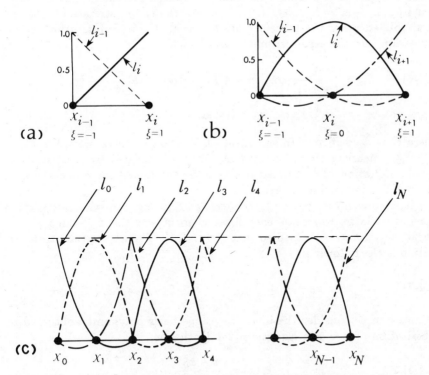

Figure 2-2. (a) Linear Lagrange "chapeau" basis functions ℓ_i. ξ is the local coordinate $\xi \in [-1, 1]$. (b) Quadratic Lagrange basis functions ℓ_i. (c) Global arrangement of local Lagrange quadratic basis functions to give piecewise quadratic interpolation.

Quadratic $(n = 2)$ and cubic $(n = 3)$ basis polynomials are also employed in practical applications, but Lagrange polynomial bases higher than cubic in degree are seldom encountered. When working with grids having large numbers of nodes x_i, we shall typically resort to **piecewise** polynomial interpolation, which we discuss shortly.

The error $E_n(x)$ associated with the interpolation of $f(x)$ by $P_n(x)$ over the interval $[x_0, x_n]$ can be estimated for $f(x) \in C^{n+1}[a, b]$ as

$$(2.4\text{-}5a) \qquad E_n(x) \equiv f(x) - P_n(x) = \frac{w_n(x)}{(n + 1)!} \frac{d^{n+1} f}{dx^{n+1}}(\varsigma),$$

where ς is some number lying in the open interval (x_0, x_n), and

$$(2.4\text{-}5b) \qquad w_n(x) = (x - x_0)(x - x_1) \cdots (x - x_n).$$

When the grid Δ is defined such that the spatial increments are uniform, that is, $x_{i+1} - x_i = h$, $i = 0, 1, 2, \ldots, n - 1$, the error estimate (2.4-5) can be simplified using the transformation $x = x_0 + \alpha h$. Substitution of this expression into the error estimate (2.4-5) yields

$$(2.4\text{-}6) \qquad E_n = \frac{1}{(n + 1)!}(\alpha h)[(\alpha - 1)h] \cdots [(\alpha - n)h] \frac{d^{n+1} f}{dx^{n+1}}(\varsigma)$$

$$= Ch^{n+1}, \qquad x_0 < \varsigma < x_n,$$

where C is a coefficient independent of h. We can therefore write $E_n = \mathcal{O}(h^{n+1})$, meaning that the ratio E_n/h^{n+1} is bounded by a constant as $h \to 0$. Equation (2.4-6) illustrates that, as the increment h decreases, so also will the interpolation error E_n.

It is convenient, particularly when performing numerical integration as required later, to rescale each finite element, say $[x_0, x_1]$, using a new, "local" coordinate ξ defined so that $\xi \in [-1, 1]$. The appropriate transformation is the following:

$$(2.4\text{-}7) \qquad x = x_0 + \frac{h}{2} + \frac{h}{2}\, \xi, \qquad -1 \le \xi \le 1, \; x_0 \le x \le x_1.$$

As an example consider the linear Lagrange basis polynomial obtained by substitution of $n = 1$ in Equation (2.4-4), that is,

$$(2.4\text{-}8) \qquad \ell_1(x) = \frac{x - x_0}{x_1 - x_0}.$$

Making the change of variables (2.4-7) in Equation (2.4-8) yields

$$(2.4\text{-}9) \qquad \hat{\ell}_1(\xi) \equiv \frac{h(1 + \xi)/2}{h} = \tfrac{1}{2}(1 + \xi), \qquad -1 \le \xi \le 1.$$

TABLE 2-1. Lagrange polynomial basis in the
ξ coordinate system.

DEGREE	ASSOCIATED NODE ξ_i	POLYNOMIAL	VALUE OF ξ_i
Linear (1)		$\frac{1}{2}(1 + \xi\xi_i)$	± 1
Quadratic (2)	End Node	$\frac{1}{2}\xi\xi_i(1 + \xi\xi_i)$	± 1
	Interior Node	$(1 - \xi^2)$	$= 0$
Cubic (3)	End Node	$(9\xi^2 - 1)(\xi_i\xi + 1)/16$	± 1
	Interior Node	$9(1 - \xi^2)(1 + 3\xi\xi_i)/16$	$\pm\frac{1}{3}$

Observe that we have adopted the notation $\hat{\ell}_i(\xi)$ to represent the function that takes the same values as $\ell_i(x)$ for values of ξ corresponding to values of x through the transformation (2.4-7). Thus, while $\hat{\ell}_i$ and ℓ_i are logically different functions, they take the same values over a given finite element. The basis function $\hat{\ell}_1(\xi)$ is illustrated in Figure 2-2a. Other common members of the Lagrange family of basis polynomials are given in Table 2-1. The general form for $\hat{\ell}_i(\xi)$ is

$$(2.4\text{-}10) \qquad \hat{\ell}_i(\xi) = \begin{cases} \displaystyle\prod_{\substack{j=0 \\ i \neq j}}^{n} \frac{\xi - \xi_j}{\xi_i - \xi_j}, & (-1 \leq \xi \leq 1) \\ \\ 0, & \text{otherwise.} \end{cases}$$

So far we have examined the Lagrange interpolating polynomials only over grids in which the number of elements equals the degree of the polynomial. When working with grids having large numbers of intervals $[x_i, x_{i+1}]$, one typically assigns a set of low-degree ($n = 1$, 2, or 3) basis functions of the form (2.4-4) to each adjacent set of $n+1 = 2$, 3, or 4 nodes. Figure 2-2c illustrates this arrangement for the case $n = 2$. Observe that one collection of three quadratic basis functions "lives" over the interval $[x_0, x_2]$, another over $[x_2, x_4]$, and so forth. Globally, we still associate one basis function to each node x_i. The nodes x_{ni} are common to two adjacent intervals and mark the boundaries where the form of the interpolation changes from one polynomial expression to another. In contrast to global interpolation by one polynomial of the form (2.4-3), this **piecewise** Lagrange polynomial interpolation is preferred in most numerical applications involving large grids. Problem 2 at the end of the chapter compares global and piecewise polynomial interpolation.

Hermite interpolation polynomials.

In general, given a interval $[a, b]$ partitioned by a grid Δ, a piecewise Lagrange interpolation polynomial belongs to $C^0([a, b])$. The interpolating function itself is continuous, but its derivative is discontinuous over domains composed of several intervals. The derivative discontinuity occurs at the boundary nodes x_{ni}, a fact easily established through an examination of Figure 2-2. There are occasions, however, when it is desirable to have higher-order continuity. A class of functions that satisfy this constraint are the **Hermite interpolation polynomials**. Let us assume the same grid Δ employed earlier, and let us further assume that, at each node, the values of both the function and its derivative are specified. That is, we are given $\{f(x_i), f'(x_i)\}_{i=0}^n$. Then there exists a unique polynomial

$$(2.4\text{-}11) \qquad H_n(x) = \sum_{i=1}^n \left[h_i^0(x) f(x_i) + h_i^1(x) f'(x_i) \right],$$

of degree at most $2n + 1$, such that $H_n \in C^1([a, b])$ and

$$(2.4\text{-}12) \qquad \frac{d^q H_n}{dx^q}(x_i) = \frac{d^q f}{dx^q}(x_i), \qquad q = 0, 1; \ i = 0, 1, \ldots, n.$$

Thus through a judicious choice of the functions h_i^0 and h_i^1 one can generate an interpolation polynomial that, when used in the piecewise sense described above, everywhere exhibits continuous first derivatives.

The Hermite polynomial basis functions h_i^0 and h_i^1 can be written, using the notation introduced in Equation (2.4-4), as follows:

$$(2.4\text{-}13\text{a}) \quad h_i^0(x) = \begin{cases} \dfrac{[\ell_i(x)]^2}{[\ell_i(x_i)]^2} \left\{ 1 - \dfrac{[\ell_i'(x_i)]^2}{[\ell_i(x_i)]^2}(x - x_i) \right\}, & x \in [x_0, x_n], \\ \\ 0, & \text{otherwise,} \end{cases}$$

$$(2.4\text{-}13\text{b}) \quad h_i^1(x) = \begin{cases} \dfrac{[\ell_i(x)]^2}{[\ell_i(x_i)]^2}(x - x_i), & x \in [x_0, x_n], \\ \\ 0, & \text{otherwise.} \end{cases}$$

These functions appear in Figure 2-3.

Examination of Equations (2.4-13) reveals that a cubic is the lowest degree polynomial that can be used successfully in Hermite interpolation; such functions are often nicknamed **Hermite cubics**. One can show that the Hermite interpolation error for a function $f \in C^{2(n+1)}[a, b]$ is given by

$$(2.4\text{-}14) \qquad E_n(x) = f(x) - H_n(x) = \frac{[w_n(x)]^2}{(2n + 1)!} \frac{d^{2(n+1)} f}{dx^{2(n+1)}}(\varsigma),$$

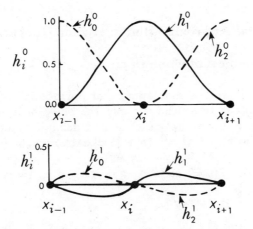

Figure 2-3. Hermite cubic basis functions.

where $w_n(x)$ is defined in (2.4-5b) and ς is some number in the interval (a, b).

It is possible to write the interpolation error for either Lagrange or Hermite polynomial approximations in another form that will be useful later. The new representation is achieved by combining expressions (2.4-5) and (2.4-14) with a Taylor series expansion, using the integral form of the remainder. The resulting expression gives the q-th derivative of the interpolation error as

$$(2.4\text{-}15\text{a}) \qquad \frac{d^q E_n}{dx^q}(x) = \frac{1}{N!} \int_a^b \frac{d^q K}{dx^q}(x, t) \frac{d^{N+1} f}{dt^{N+1}}(t)\, dt,$$

where

$$(2.4\text{-}15\text{b}) \qquad K(x, t) = (x - t)_+^N - \sum_{i=0}^n \sum_{j=0}^{m-1} h_i^j(x) \frac{d^j}{dt^j}(x_i - t)_+^N$$

and $N = m(n + 1) - 1$. In Equation (2.4-15b) the term $(x - t)_+^N$ is defined as follows:

$$(2.4\text{-}15\text{c}) \qquad (x - t)_+^N = \begin{cases} (x - t)^N & x \geq t, \\ 0 & x < t. \end{cases}$$

In these equations, n is the degree of the interpolating polynomial, $m = 1$ for Lagrange polynomials and $m = 2$ for Hermite polynomials, and $K(x, t)$ is known as the **Peano kernel**. The relationship (2.4-15) is the **Peano kernel theorem**. Rather than proving this theorem here, we refer the reader to Botha and Pinder (1983, Chapter Two) for further explanation.

2.5. Polynomial Approximation in Higher Dimensions.

Tensor-product interpolation.

One approach to the development of polynomial approximations in higher dimensions is to extend directly the one-dimensional concepts of Section 2.4. Using this approach let us first define a two-dimensional grid over a rectangular region $\Omega = [a, b] \times [c, d]$ as $\Delta = \Delta_x \times \Delta_y$, where

$$(2.5\text{-}1a) \qquad \Delta_x : (a =) \; x_0 < x_1 < x_2 < \cdots < x_n \; (= b)$$

and

$$(2.5\text{-}1b). \qquad \Delta_y : (c =) \; y_0 < y_1 < y_2 < \cdots < y_m \; (= d)$$

Figure 2-4 illustrates a typical grid of this kind; note that n and m have different meanings here than in the preceding section.

To obtain a suitable polynomial approximation over Ω, we proceed in two steps. First, we hold one independent variable, say x, fixed while we use our one-dimensional theory [see Equation (2.4-3)] to interpolate the function $f(x, y)$ along the y-axis. Then we interpolate the resulting functions $\{f(x, y_j)\}_{j=0}^m$ along the x-axis. The procedure can be stated algebraically as

$$(2.5\text{-}2a) \qquad f(x, y) = \sum_{j=0}^{m} \ell_j(y) f(x, y_j) + E_m$$

for the interpolation in y, and

$$(2.5\text{-}2b) \qquad f(x, y) = \sum_{i=0}^{n} \sum_{j=0}^{m} \ell_i(x) \ell_j(y) f(x_i, y_j) + E_{mn}(x, y)$$

for the interpolation over x and y. The error term $E_{mn}(x, y)$ is given by

$$
\begin{aligned}
E_{mn}(x, y) = {} & \frac{w_n(x)}{(n+1)!} \frac{\partial^{n+1} f}{\partial x^{n+1}} (\varsigma_1, y) \\
& + \frac{w_m(y)}{(m+1)!} \frac{\partial^{m+1} f}{\partial y^{m+1}} (x, \varsigma_3) \\
& - \frac{w_n(x) w_m(y)}{(m+n+2)!} \frac{\partial^{n+1}}{\partial x^{n+1}} \frac{\partial^{m+1} f}{\partial y^{m+1}} (\varsigma_2, \varsigma_4),
\end{aligned}
$$

for some numbers ς_1, ς_2 lying in the open interval (a, b) and some ς_3, ς_4 lying in (c, d). The same conceptual model can be used to extend our

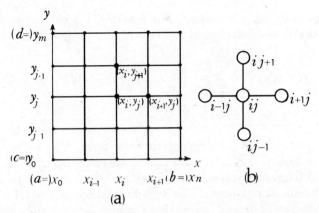

(a)

(b)

Figure 2-4. Specification of the grid Δ defined over the domain $\Omega = [a, b] \times [c, d]$. The nodal locations are designated by the symbol $\bullet$. Figure 2-4b is a computational molecule that can be used to represent the nomenclature at a typical node (x_i, y_j).

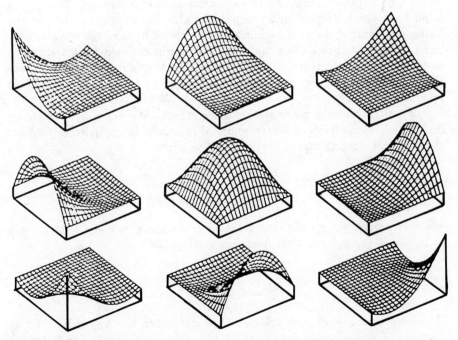

Figure 2-5. Tensor-product quadratic basis functions (after Botha and Pinder, 1983).

formulation to higher dimensions. For example, the extension of Equation (2.5-2) to the three-dimensional cell $\Omega = [a, b] \times [c, d] \times [e, g]$ would read

$$(2.5\text{-}3) \qquad f(x, y, z) = \sum_{i=0}^{n} \sum_{j=0}^{m} \sum_{k=0}^{p} \ell_i(x) \ell_j(y) \ell_k(z) f(x_i, y_j, z_k)$$
$$+ E_{mnp}(x, y, z),$$

where E_{mnp} stands for the appropriate error term. The polynomial bases appearing in Equations (2.5-2) and (2.5-3) are made up of products of one-dimensional Lagrange polynomials. They are often referred to as **tensor-product basis functions**. The two-dimensional forms of the tensor-product basis for the case of quadratic polynomials (see Figure 2-2b) appear in Figure 2-5. Figure 2-6a shows a typical Lagrange biquadratic element.

While the tensor-product formulation of basis functions is a straight-forward approach applicable to many problems encountered in mathematical physics, there are occasions when another strategy may be more appropriate. Consider, for example, a formulation wherein the center node of a two-dimensional biquadratic element is considered unnecessary, or where we prefer not to include the four internal nodes of a bicubic element. One approach might be to use a trial-and-error strategy. Keeping in mind both the necessary and computationally desirable properties of the bases, one could determine by ingenuity and good fortune the sought functions. It is thus appropriate that the elements introduced next are known as **Serendipity** elements, after the heroes in the fairy tale, "The Three Princes of Serendip," who possessed the faculty of finding valuable things by chance.

We shall follow a more systematic strategy. We begin by writing an incomplete biquadratic basis polynomial for node (x_i, y_j), neglecting for sheer convenience the term proportional to $x^2 y^2$:

$$(2.5\text{-}4) \qquad g_{ij}(x, y) = a_0 + a_1 x + a_2 y + a_3 xy$$
$$+ a_4 x^2 + a_5 y^2 + a_6 x^2 y + a_7 xy^2.$$

We now impose on g_{ij} the interpolation constraints, namely, that $g_{ij} = 1$ at the node (x_i, y_j) and that it vanish at all other nodes:

$$(2.5\text{-}5) \qquad g_{ij}(x_k, y_\ell) = \delta_{ik} \delta_{j\ell}, \qquad 0 \le i, j, k, \ell \le 2; \text{ except } i = j = 1.$$

(Here, δ_{ik} signifies the Kronecker symbol, defined in the Appendix.) Combination of Equations (2.5-4) and (2.5-5) yields, for example, the following matrix equation for the coefficients a_m of g_{00}:

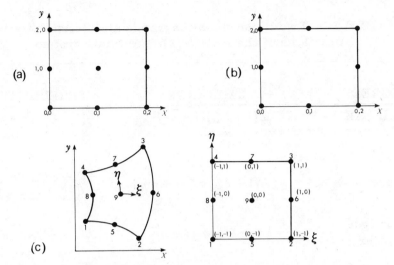

Figure 2-6. Biquadratic finite elements. (a) A Lagrange biquadratic element, (b) a Serendipity biquadratic element, and (c) an isoparametric biquadratic element.

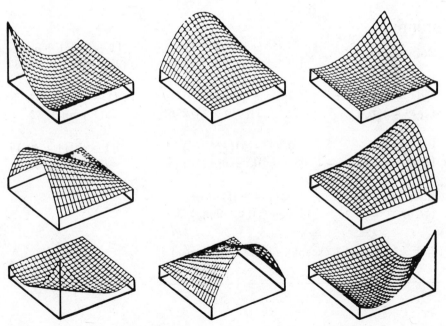

Figure 2-7. Serendipity quadratic basis functions (after Botha and Pinder, 1983).

TABLE 2-2. Two-dimensional Lagrange and Serendipity basis functions in the local (ξ, η) coordinate system.

DEGREE	LAGRANGE	SERENDIPITY
BILINEAR	$\frac{1}{4}(1 + \xi\xi_i)(1 + \eta\eta_j)$	$\frac{1}{4}(1 + \xi\xi_i)(1 + \eta\eta_j)$
BIQUADRATIC		
Corner node	$\frac{1}{4}\xi\xi_i(1 + \xi\xi_i)\eta\eta_j(1 + \eta\eta_j)$	$\frac{1}{4}(1 + \xi\xi_i)(1 + \eta\eta_j)$ $\cdot(\xi\xi_i + \eta\eta_j - 1)$
Side node $(0, \eta_j)$	$\frac{1}{2}(1 - \xi^2)\eta\eta_j(1 + \eta\eta_j)$	$\frac{1}{2}(1 - \xi^2)(1 + \eta\eta_j)$
Side node $(\xi_i, 0)$	$\frac{1}{2}\xi\xi_i(1 + \xi\xi_i)(1 - \eta^2)$	$\frac{1}{2}(1 + \xi\xi_i)(1 - \eta^2)$
Interior node	$(1 - \xi^2)(1 - \eta^2)$	–
BICUBIC		
Corner node	$(9\xi^2 - 1)(\xi_i\xi + 1)$ $\cdot(9\eta^2 - 1)(\eta_j\eta + 1)/256$	$(1 + \xi\xi_i)(1 + \eta\eta_j)$ $\cdot[9(\xi^2 + \eta^2) - 10]/32$
Side nodes $(\xi_i = \pm\frac{1}{3}, \eta_j)$	$9(1 - \xi^2)(1 + 9\xi\xi_i)$ $\cdot(9\eta^2 - 1)(\eta_j\eta + 1)/256$	$9(1 - \xi^2)(1 + 9\xi\xi_i)$ $\cdot(1 + \eta\eta_j)/32$
Side nodes $(\xi_i, \eta_j = \pm\frac{1}{3})$	$9(9\xi^2 - 1)(\xi_i\xi + 1)$ $\cdot(1 - \eta^2)(1 + 9\eta\eta_j)/256$	$9(1 - \eta^2)(1 + \xi\xi_i)$ $\cdot(1 + 9\eta\eta_j)/32$
Interior nodes $(\xi_i, \eta_j = \pm\frac{1}{3})$	$81(1 - \xi^2)(1 - \eta^2)$ $\cdot(1 + 9\xi\xi_i)(1 + 9\eta\eta_j)/256$	–

$(\xi_i = \pm1; \eta_j = \pm1$ except as stated otherwise$)$

(2.5-6)

$$
\begin{bmatrix}
1 & x_0 & y_0 & x_0y_0 & x_0^2 & y_0^2 & x_0^2y_0 & x_0y_0^2 \\
1 & x_0 & y_1 & x_0y_1 & x_0^2 & y_1^2 & x_0^2y_1 & x_0y_1^2 \\
1 & x_0 & y_2 & x_0y_2 & x_0^2 & y_2^2 & x_0^2y_2 & x_0y_2^2 \\
1 & x_1 & y_0 & x_1y_0 & x_1^2 & y_0^2 & x_1^2y_0 & x_1y_0^2 \\
1 & x_1 & y_2 & x_1y_2 & x_1^2 & y_2^2 & x_1^2y_2 & x_1y_2^2 \\
1 & x_2 & y_0 & x_2y_0 & x_2^2 & y_0^2 & x_2^2y_0 & x_2y_0^2 \\
1 & x_2 & y_1 & x_2y_1 & x_2^2 & y_1^2 & x_2^2y_1 & x_2y_1^2 \\
1 & x_2 & y_2 & x_2y_2 & x_2^2 & y_2^2 & x_2^2y_2 & x_2y_2^2
\end{bmatrix}
\begin{bmatrix}
a_0 \\ a_1 \\ a_2 \\ a_3 \\ a_4 \\ a_5 \\ a_6 \\ a_7
\end{bmatrix}
=
\begin{bmatrix}
1 \\ 0 \\ 0 \\ 0 \\ 0 \\ 0 \\ 0 \\ 0
\end{bmatrix}.
$$

Equation (2.5-6) can be solved formally once we have selected an appropriate coordinate system and have evaluated the elements of the coefficient matrix. Having thus obtained the coefficients a_m, one obtains the basis function g_{ij} via Equation (2.5-4). A plot of the two-dimensional biquadratic Serendipity basis functions appears in Figure 2-7.

A comparison of Figures 2-5 and 2-7 illustrates that the difference between the biquadratic Lagrange and Serendipity elements is subtle but significant. Note particularly the relatively large region over which the Serendipity corner basis function is negative. This gives rise to a negative contribution to the volume under this surface as opposed to a strictly positive volume under the Lagrange surface. This property poses some numerical difficulties in selected circumstances. Table 2-2 gives explicit formulas for the Lagrange and Serendipity bilinear, biquadratic, and bicubic bases.

Interpolation on triangles.

While the rectangular quadrilateral elements just discussed are applicable to a wide range of practical problems, the triangular element has been a mainstay of the finite-element procedure. The triangular element has the advantage of maintaining mathematical simplicity while permitting an accurate representation of an irregular region. One reason that this element is so effective in applications is the triangular or area coordinate system on which it is defined. The concept of area coordinates is easily understood with the aid of Figure 2-8.

Consider first Figure 2-8a, defining a typical triangular element in Cartesian (x, y)-coordinates. A linear polynomial approximation of a function $f(x, y)$ defined over this element can be written

$$
f(x, y) \simeq P_1(x, y) = ax + by + c,
$$

where a, b, and c are undetermined coefficients. Let us impose the standard

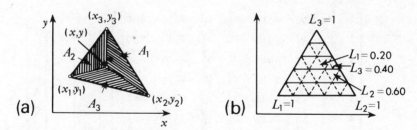

Figure 2-8. Triangular finite element in Cartesian and natural coordinates (from Botha and Pinder, 1983).

interpolation constraints on $P_1(x, y)$. Thus, at the node i, $P_1(x_i, y_i) = f(x_i, y_i)$. One can show in a straightforward manner that

$$(2.5\text{-}7) \qquad P_1(x, y) = \sum_{i=1}^{3} \phi_i(x, y) f(x_i, y_i),$$

where, assuming (i, j, k) represents a cyclic permutation of the node numbers $(1, 2, 3)$, the basis function associated with the i-th node is as follows:

$$(2.5\text{-}8) \qquad \phi_i(x, y) = \frac{[x(y_j - y_k) + y(x_k - x_j) + (x_j y_k - x_k y_j)]}{\det(\mathbf{P})}.$$

Here, the matrix $\mathbf{P}$ is given in terms of the nodal coordinates as

$$(2.5\text{-}9) \qquad \mathbf{P} = \begin{bmatrix} x_i & y_i & 1 \\ x_j & y_j & 1 \\ x_k & y_k & 1 \end{bmatrix},$$

where, again, we assume that (i, j, k) is a cyclic permutation of $(1, 2, 3)$. We now observe that the numerator in Equation (2.5-8) is the determinant of the matrix

$$(2.5\text{-}10) \qquad \mathbf{P}_i = \begin{bmatrix} x, & y, & 1 \\ x_j & y_j & 1 \\ x_k & y_k & 1 \end{bmatrix}.$$

Thus the function $\phi_i(x, y)$ defined in (2.5-8) can be written in the equivalent form

$$(2.5\text{-}11) \qquad \phi_i(x, y) = \frac{\det(\mathbf{P}_i)}{\det(\mathbf{P})}.$$

To see how these basis functions lead to an area coordinate system, observe that the area of a triangle with vertices at the nodes (x_i, y_i), $i = 1, 2, 3$, is given by

$$(2.5\text{-}12) \qquad A = \tfrac{1}{2} \det \begin{bmatrix} x_1 & y_1 & 1 \\ x_2 & y_2 & 1 \\ x_3 & y_3 & 1 \end{bmatrix}.$$

Therefore the areas A and A_i shown in Figure 2-8a are given by

$$(2.5\text{-}13) \qquad A = \tfrac{1}{2} \det(\mathbf{P})$$

$$(2.5\text{-}14) \qquad A_i = \tfrac{1}{2} \det(\mathbf{P}_i).$$

Thus one can rewrite $\phi_i(x, y)$ as a ratio of areas:

$$(2.5\text{-}15) \qquad \phi_i(x, y) = \frac{A_i}{A} = L_i(x, y),$$

where L_i is called an **area basis function**. The **area coordinate system** over which L_i is defined is illustrated in Figure 2-8b.

It is important to recognize that the three basis functions L_i, $i = 1, 2, 3$, are not independent. This is because of the relationship

$$(2.5\text{-}16) \qquad \sum_{i=1}^{3} L_i = 1,$$

which implies, for example, $L_3 = L_3(L_1, L_2)$. This dependence has implications in partial differentiation. Indeed, by the chain rule,

$$(2.5\text{-}17) \qquad \frac{\partial L_i}{\partial x} = \frac{\partial L_i}{\partial L_1} \frac{\partial L_1}{\partial x} + \frac{\partial L_i}{\partial L_2} \frac{\partial L_2}{\partial x}.$$

For the specific case of $L_3 = 1 - L_1 - L_2$ we have

$$(2.5\text{-}18) \qquad \frac{\partial L_3}{\partial x} = -\frac{\partial L_1}{\partial x} - \frac{\partial L_2}{\partial x}.$$

It is important to keep this relationship in mind when we differentiate interpolatory representations such as Equation (2.5-7).

The computational efficiency achieved using triangular elements is in no small measure due to the very simple formulas that can be used to integrate the terms arising in the finite-element method. One such formula for products of the functions L_i is

$$(2.5\text{-}19) \qquad \int_A L_1^{m_1} L_2^{m_2} L_3^{m_3} \, dx \, dy = 2A \frac{m_1! \, m_2! \, m_3!}{(m_1 + m_2 + m_3 + 2)!},$$

where the exponents m_i are nonnegative integers and A denotes the element area. The significance of this relationship will become apparent in our later discussion of the finite-element formulation.

Let us now examine the interpolation error associated with the area basis functions. While there are several approaches to this problem (see, for example, Prenter, 1975, and Mitchell and Wait, 1977), we follow that of Botha and Pinder (1983). Consider the polynomial approximation $P_1(x, y)$ that interpolates a function $f \in C^2(\Omega)$ over an arbitrary triangle Ω_e, that is,

$$(2.5\text{-}20) \qquad P_1(x,y) = \sum_{i=1}^{3} L_i(x,y)\, f(x_i,y_i), \quad (x,y) \in \Omega_e.$$

The interpolation error committed at a point $(x, y) \in \Omega_e$ is, by definition,

$$(2.5\text{-}21) \qquad g(x,y) = f(x,y) - P_1(x,y).$$

Because the interpolation error at a node is zero, $g(x_i, y_i) = 0$. For future reference we define the maximum side length of the triangle Ω_e as

$$(2.5\text{-}22) \qquad h = \max_{1 \le i,j \le 3} \sqrt{(x_i - x_j)^2 + (y_i - y_j)^2}$$

and observe that $|x - x_i| \le h$ and $|y - y_i| \le h$ whenever $(x, y) \in \Omega_e$.

Now let us observe that, since $f \in C^2(\Omega_e)$ and $P \in C^2(\Omega_e)$, the interpolation error $g \in C^2(\Omega_e)$ also. Thus $g(x, y)$ can be written using Taylor's theorem with remainder as follows. For some point ς along the line segment connecting the point $(x, y) = \mathbf{x}$ to the node $\mathbf{x}_i$ (see Figure 2-9),

$$(2.5\text{-}23) \qquad \begin{aligned} g(\mathbf{x}) &= g(\mathbf{x}_i) + \nabla g(\mathbf{x}_i) \cdot (\mathbf{x} - \mathbf{x}_i) \\ &+ \tfrac{1}{2}(\mathbf{x} - \mathbf{x}_i) \cdot \text{Hess}\, g(\varsigma) \cdot (\mathbf{x} - \mathbf{x}_i). \end{aligned}$$

Here, $\text{Hess}\, g(\varsigma)$ denotes the **Hessian** of g, defined as

$$(2.5\text{-}24) \qquad \text{Hess}\, g \equiv \begin{bmatrix} \dfrac{\partial^2 g}{\partial x^2} & \dfrac{\partial^2 g}{\partial x\, \partial y} \\[2ex] \dfrac{\partial^2 g}{\partial y\, \partial x} & \dfrac{\partial^2 g}{\partial y^2} \end{bmatrix},$$

evaluated at the point ς.

The first term on the right side of Equation (2.5-23) is zero because, as noted earlier, the interpolation error vanishes at the nodes. To estimate the second term it is necessary to define the unit vector $\mathbf{e}_{ij} = (\mathbf{x}_j - \mathbf{x}_i)/\|\mathbf{x}_j - \mathbf{x}_i\|_2$, which is collinear with the side of the triangle defined by $\mathbf{x}_j - \mathbf{x}_i$ (see

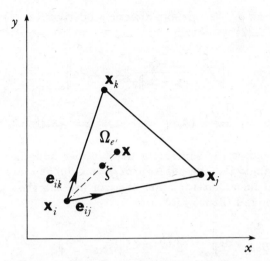

Figure 2-9. Triangle used to determine interpolation error estimates. The locations x_i, x_j, and x_k denote nodes, x is an arbitrary point, ς is a point along the line segment connecting x_i and x, and e_{ij} is a unit vector defined along the line segment connecting x_i and x_j, which is a side of triangle Ω_e.

again Figure 2-9). Similarly, let $e_{ik} = (x_k - x_i)/\|x_k - x_i\|_2$. Let us now rewrite the second term of the expansion (2.5-23) as follows:

$$(2.5 - 25) \qquad \nabla g(x_i) \cdot \frac{x - x_i}{\|x - x_i\|_2} \|x - x_i\|_2.$$

Notice that $(x - x_i)/\|x - x_i\|_2$ is a unit vector in the direction defined by the line connecting x and x_i. These unit vector definitions therefore imply that $(x - x_i)/\|x - x_i\|_2$ can be written as a linear combination of the unit vectors e_{ij} and e_{ik}:

$$\frac{x - x_i}{\|x - x_i\|_2} = \alpha_1 e_{ij} + \alpha_2 e_{ik}, \qquad 0 \le |\alpha_1|, |\alpha_2| \le 1.$$

Hence, we can rewrite the expression (2.5-25) as follows:

$$
\begin{aligned}
(2.5\text{-}26) \qquad & \nabla g(x_i) \cdot \frac{x - x_i}{\|x - x_i\|_2} \|x - x_i\|_2 \\
& = [\alpha_1 \nabla g(x_i) \cdot e_{ij} + \alpha_2 \nabla g(x_i) \cdot e_{ik}] \|x - x_i\|_2 \\
& = \left[\alpha_1 \frac{\partial g}{\partial e_{ij}}(x_i) + \alpha_2 \frac{\partial g}{\partial e_{ik}}(x_i) \right] \|x - x_i\|_2,
\end{aligned}
$$

where $\partial g/\partial e_{ij}$ and $\partial g/\partial e_{ik}$ denote directional derivatives. From Equations (2.5-26) and (2.5-22) we obtain

$$\left\| \nabla g \cdot \frac{\mathbf{x} - \mathbf{x}_i}{\|\mathbf{x} - \mathbf{x}_i\|_2} \|\mathbf{x} - \mathbf{x}_i\|_2 \right\|_\infty$$

$$\leq |\alpha_1| \left\| \frac{\partial g}{\partial e_{ij}} (\mathbf{x}_i) \right\|_\infty h + |\alpha_2| \left\| \frac{\partial g}{\partial e_{ik}} (\mathbf{x}_i) \right\|_\infty h,$$

where $\| \cdot \|_\infty$ denotes the supremum (least upper bound) of the function $(\cdot)$ over the triangle Ω_e. Because $\partial g/\partial e_{ij}$ and $\partial g/\partial e_{ik}$ are derivatives of one-dimensional linear interpolants, one can use the theory presented in Section 2.4 to bound them as follows:

$$\left\| \frac{\partial g}{\partial e_{ij}} (\mathbf{x}_i) \right\|_\infty \leq Mh; \qquad \left\| \frac{\partial g}{\partial e_{ik}} (\mathbf{x}_i) \right\|_\infty \leq Mh,$$

where

$$M = \max_{(x,y) \in \Omega_e} \left\{ \|\partial^2 f/\partial x^2\|_\infty, \|\partial^2 f/\partial x\, \partial y\|_\infty, \|\partial^2 f/\partial y^2\|_\infty \right\}.$$

Thus we can estimate the second term in the Taylor expansion (2.5-23) as follows:

$$(2.5\text{-}27) \quad \|\nabla g(\mathbf{x}_i) \cdot (\mathbf{x} - \mathbf{x}_i)\|_\infty \leq \left\| \nabla g(\mathbf{x}_i) \cdot \frac{\mathbf{x} - \mathbf{x}_i}{\|\mathbf{x} - \mathbf{x}_i\|_2} \|\mathbf{x} - \mathbf{x}_i\|_2 \right\|_\infty$$

$$\leq 2Mh^2.$$

We now turn our attention to the last term in the Taylor expansion (2.5-23). Because $P_1(x, y)$ is linear along any line in the (x, y)-plane, differentiating Equation (2.5-21) twice yields

$$(2.5\text{-}28) \qquad \frac{\partial^2 g}{\partial x^2} = \frac{\partial^2 f}{\partial x^2} - \frac{\partial^2 P_1}{\partial x^2} = \frac{\partial^2 f}{\partial x^2},$$

and similarly for other entries in the Hessian. From the definition of M and Equation (2.5-28), each entry of the Hessian can now be bounded in magnitude by M, so that the third term in Equation (2.5-23) can also be bounded as follows:

$$(2.5\text{-}29) \qquad \left\| \tfrac{1}{2}(\mathbf{x} - \mathbf{x}_i) \cdot \text{Hess } g(\varsigma) \cdot (\mathbf{x} - \mathbf{x}_i) \right\|_\infty$$

$$\leq \tfrac{1}{2} \begin{bmatrix} h \\ h \end{bmatrix}^\top \begin{bmatrix} M & M \\ M & M \end{bmatrix} \begin{bmatrix} h \\ h \end{bmatrix} = 2Mh^2.$$

Combination of the estimates (2.5-27) and (2.5-29) yields the final result,

$$(2.5\text{-}30) \qquad \|g(\mathbf{x})\|_\infty \leq 2Mh^2 + 2Mh^2 = 4Mh^2, \qquad \mathbf{x} \in \Omega_e.$$

Equation (2.5-30) tells us that the triangular interpolation on Ω_e converges as the square of its longest side. An equally useful result is obtained by restating the convergence properties in terms of the minimum interior angle of Ω_e. This result is achieved by employing the trigonometric identity

$$(2.5\text{-}31) \qquad h = h_0 \frac{\sin \theta}{\sin \theta_0},$$

where θ and θ_0 are the maximum and minimum interior angles in triangle Ω_e and h_0 is the minimum side length. It is always possible to find a constant H independent of the maximum side length h such that

$$(2.5\text{-}32) \qquad h \leq \frac{H}{\sin \theta_0}.$$

Combination of equations (2.5-30) through (2.5-32) yields

$$(2.5\text{-}33) \qquad \|g(\mathbf{x})\|_\infty \leq 4Mh^2 = 4M\left(\frac{H}{\sin \theta_0}\right)^2.$$

This very important result states that, as the smallest interior angle of Ω_e decreases, the interpolation error increases. Thus one must take great care in designing a finite-element triangulation so that very small angles are avoided.

It is perhaps worth mentioning that one can also derive rather general relationships for the interpolation error (2.5-21) using the **Sard kernel theorem**, which is a multidimensional extension of the Peano kernel theorem. The use of this approach on triangles, however, is less transparent than in our earlier applications. The reader interested in this approach is referred to the work Barnhill and Mansfield (1974).

Interpolation on isoparametric elements.

Occasionally it may be advantageous to employ elements that are neither rectangles nor straight-sided triangles. Consider, for example, the element illustrated in Figure 2-6c. This element can be used effectively to accommodate curved boundaries. It is useful to describe this element, of very general geometry in the global (x, y)-coordinate system, by transforming to a coordinate system that is appropriate for numerical integration. This transformation is based upon the interpolating polynomials introduced in Section 2.5. Using the coordinates of the element corners, we can specify any location (x, y) in the element Ω_e in terms of the local (ξ, η) co-ordinate system $(-1 \leq \xi, \eta \leq 1)$ using an interpolation basis $\{\psi_i(x, y)\}_{i=1}^m$ as follows:

$$(2.5\text{-}34a) \qquad\qquad x = \sum_{i=1}^m x_i \, \psi_i(\xi, \eta),$$

$$(2.5\text{-}34b) \qquad\qquad y = \sum_{i=1}^m y_i \, \psi_i(\xi, \eta),$$

where m is the number of nodes in the element Ω_e. When the functions $\psi_i(\xi, \eta)$ defining the coordinate transformation are the same as the local basis functions $\phi_i(\xi, \eta)$ for the interpolation scheme, tabulated in Table 2-2, the transformation is called **isoparametric**.

To illustrate the use of the transformation (2.5-34), consider a typical integral that we shall encounter in the finite-element method (see Section 2.12):

$$(2.5\text{-}35) \qquad\qquad I_1 \equiv \int_{\Omega_e} \frac{\partial \phi_j}{\partial x} \frac{\partial \phi_i}{\partial x} \, dx \, dy.$$

For concreteness, let us take Ω_e to be a quadrilateral with straight sides. Let us assume that $\phi_j(\xi, \eta) = \psi_j(\xi, \eta)$ and that the basis functions $\phi_i(\xi, \eta)$ are the Lagrange polynomials of degree one. Employing the chain rule, we obtain

$$(2.5\text{-}36a) \qquad\qquad \frac{\partial \phi_k}{\partial \xi} = \frac{\partial \phi_k}{\partial x} \frac{\partial x}{\partial \xi} + \frac{\partial \phi_k}{\partial y} \frac{\partial y}{\partial \xi},$$

$$(2.5\text{-}36b) \qquad\qquad \frac{\partial \phi_k}{\partial \eta} = \frac{\partial \phi_k}{\partial x} \frac{\partial x}{\partial \eta} + \frac{\partial \phi_k}{\partial y} \frac{\partial y}{\partial \eta}.$$

In matrix form Equations (2.5-36) become

$$(2.5\text{-}37) \qquad\qquad \begin{bmatrix} \dfrac{\partial \phi_k}{\partial \xi} \\[2ex] \dfrac{\partial \phi_k}{\partial \eta} \end{bmatrix} = \begin{bmatrix} \dfrac{\partial x}{\partial \xi} & \dfrac{\partial y}{\partial \xi} \\[2ex] \dfrac{\partial x}{\partial \eta} & \dfrac{\partial y}{\partial \eta} \end{bmatrix} \begin{bmatrix} \dfrac{\partial \phi_k}{\partial x} \\[2ex] \dfrac{\partial \phi_k}{\partial y} \end{bmatrix},$$

Where the matrix

$$\mathbf{J} = \begin{bmatrix} \dfrac{\partial x}{\partial \xi} & \dfrac{\partial y}{\partial \xi} \\[2ex] \dfrac{\partial x}{\partial \eta} & \dfrac{\partial y}{\partial \eta} \end{bmatrix}$$

is the Jacobian matrix of the coordinate transformation from the (x, y)-plane to the (ξ, η)-plane. When $\mathbf{J}$ is nonsingular we can invert it to yield

$$(2.5\text{-}38) \quad \begin{bmatrix} \dfrac{\partial \phi_k}{\partial x} \\[2mm] \dfrac{\partial \phi_k}{\partial y} \end{bmatrix} = \mathbf{J}^{-1} \begin{bmatrix} \dfrac{\partial \phi_k}{\partial \xi} \\[2mm] \dfrac{\partial \phi_k}{\partial \eta} \end{bmatrix}$$

$$= \left(\frac{\partial x}{\partial \xi} \frac{\partial y}{\partial \eta} - \frac{\partial x}{\partial \eta} \frac{\partial y}{\partial \xi} \right)^{-1} \begin{bmatrix} \dfrac{\partial y}{\partial \eta} & -\dfrac{\partial y}{\partial \xi} \\[2mm] -\dfrac{\partial x}{\partial \eta} & \dfrac{\partial x}{\partial \xi} \end{bmatrix} \begin{bmatrix} \dfrac{\partial \phi_k}{\partial \xi} \\[2mm] \dfrac{\partial \phi_k}{\partial \eta} \end{bmatrix}.$$

We can now use Equation (2.5-38) to write the integral (2.5-35) in terms of ξ and η:

$$(2.5\text{-}39) \quad \int_{-1}^{1} \int_{-1}^{1} \left[\left(\frac{\partial x}{\partial \xi} \frac{\partial y}{\partial \eta} - \frac{\partial x}{\partial \eta} \frac{\partial y}{\partial \xi} \right)^{-2} \left(\frac{\partial y}{\partial \eta} \frac{\partial \phi_j}{\partial \xi} - \frac{\partial y}{\partial \xi} \frac{\partial \phi_j}{\partial \eta} \right) \right.$$
$$\left. \cdot \left(\frac{\partial y}{\partial \eta} \frac{\partial \phi_i}{\partial \xi} - \frac{\partial y}{\partial \xi} \frac{\partial \phi_i}{\partial \eta} \right) \left(\frac{\partial x}{\partial \xi} \frac{\partial y}{\partial \eta} - \frac{\partial x}{\partial \eta} \frac{\partial y}{\partial \xi} \right) \right] d\xi \, d\eta.$$

Here we have used the change-of-variables theorem to equate

$$dx \, dy = \det(\mathbf{J}) \, d\xi \, d\eta = \left(\frac{\partial x}{\partial \xi} \frac{\partial y}{\partial \eta} - \frac{\partial x}{\partial \eta} \frac{\partial y}{\partial \xi} \right) d\xi \, d\eta.$$

In actual computations the elements of the Jacobian matrix are readily obtained from Equations (2.5-34). For example, for an element with four nodes (x_i, y_i),

$$(2.5\text{-}40) \quad \mathbf{J} = \begin{bmatrix} \dfrac{\partial \phi_1}{\partial \xi} & \dfrac{\partial \phi_2}{\partial \xi} & \dfrac{\partial \phi_3}{\partial \xi} & \dfrac{\partial \phi_4}{\partial \xi} \\[3mm] \dfrac{\partial \phi_1}{\partial \eta} & \dfrac{\partial \phi_2}{\partial \eta} & \dfrac{\partial \phi_3}{\partial \eta} & \dfrac{\partial \phi_4}{\partial \eta} \end{bmatrix} \begin{bmatrix} x_1 & y_1 \\ x_2 & y_2 \\ x_3 & y_3 \\ x_4 & y_4 \end{bmatrix}$$

$$= \begin{bmatrix} \displaystyle\sum_{i=1}^{4} x_i \frac{\partial \phi_i}{\partial \xi} & \displaystyle\sum_{i=1}^{4} y_i \frac{\partial \phi_i}{\partial \xi} \\[4mm] \displaystyle\sum_{i=1}^{4} x_i \frac{\partial \phi_i}{\partial \eta} & \displaystyle\sum_{i=1}^{4} y_i \frac{\partial \phi_i}{\partial \eta} \end{bmatrix}.$$

However, even though $\mathbf{J}$ is relatively easy to compute, the integrand that results from the change of variables may be quite complicated. While the exact integration of Equation (2.5-39) appears awesome, we shall see in Section 2.13 how to accomplish the task, at least approximately, with the aid of numerical quadrature.

So far we have established some of the more common schemes for approximating functions using piecewise polynomials. In the context of numerical solutions to PDEs the function to be approximated is unknown. Thus at the outset of solving a PDE numerically we can specify the functional form of the approximate solution (chapeau, Serendipity, or triangular elements, for example), but the specific nodal values remain to be computed. We now proceed to examine methods for determining the nodal values in seeking approximate solutions to PDEs.

2.6. Finite-Difference Approximations.

In the **finite-difference method**, the differential equation is approximated at discrete points using interpolating polynomials. A difference equation at each point results directly, and, given the theory presented above, the associated approximation error is readily determined. Of the various numerical techniques for solving differential equations, the finite-difference approach is by far the simplest to understand and implement.

One-dimensional approximations.

Let us begin with one-dimensional approximations. The first step is the specification of the interval $[a, b]$ spanned by the interpolating polynomial. Consider first the case $(a =\ )\ x_0 < x_1\ (\ = b)$ illustrated in Figure 2-10a.

The appropriate Lagrange interpolating polynomial is

$$
\begin{aligned}
(2.6\text{-}1) \qquad P_1(x) &= \ell_0(x)f(x_0) + \ell_1(x)f(x_1) \\
&= \frac{x - x_1}{x_0 - x_1}\,f(x_0) + \frac{x - x_0}{x_1 - x_0}\,f(x_1) \\
&= \frac{1}{h}\left[(x_1 - x)f(x_0) + (x - x_0)f(x_1)\right],
\end{aligned}
$$

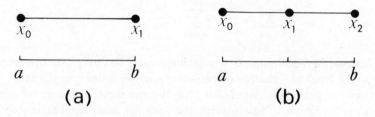

Figure 2-10. (a) A two-point finite-difference template; (b) a three-point finite-difference template.

where $h = x_1 - x_0$. Thus, given a function $f \in C^2[a, b]$, we use the approximation $f(x) = P_1(x) + E_1(x)$, where E_1 is the error. Differentiating this approximation at x_0 gives

(2.6-2)
$$\frac{df}{dx}(x_0) = \frac{dP_1}{dx}(x_0) + \frac{dE_1}{dx}(x_0)$$
$$= \frac{1}{h}[f(x_1) - f(x_0)] + \frac{dE_1}{dx}(x_0).$$

This is a **forward difference approximation**, often written in the abbreviated form

(2.6-3)
$$\frac{df}{dx}(x_0) = \frac{\Delta f(x_0)}{h} + \frac{dE_1}{dx}(x_0).$$

We define the **forward difference operator** Δ in this expression by the identity $\Delta f(x_i) = f(x_{i+1} - f(x_i)$. The error term in Equation (2.6-3) is, from Equation (2.4-5a),

(2.6-4)
$$\frac{dE_1}{dx}(x_0) = \frac{d}{dx}\left[\frac{1}{2!}(x - x_0)(x - x_1)\frac{d^2 f}{dx^2}(\varsigma)\right]_{x_0}$$
$$= -\frac{h}{2!}\frac{d^2 f}{dx^2}(\varsigma) = \mathcal{O}(h),$$

where ς is some point in the open interval (a, b).

An analogous **backward difference approximation** is obtained by evaluating the interpolating polynomial at x_1:

(2.6-5)
$$\frac{df}{dx}(x_1) = \frac{dP_1}{dx}(x_1) + \frac{dE_1}{dx}(x_1)$$
$$= \frac{1}{h}[f(x_1) - f(x_0)] + \frac{dE_1}{dx}(x_1).$$

Here, the error has the form

(2.6-6)
$$\frac{dE_1}{dx}(x_1) = \frac{h}{2!}\frac{d^2 f}{dx^2}(\varsigma) = \mathcal{O}(h),$$

for some $\varsigma \in (a, b)$. While the formula (2.6-5) appears identical to that found in Equation (2.6-2), it is different because it approximates df/dx at x_1 rather than x_0. Thus the difference is taken in a backward sense. The error in Equation (2.6-5), while having the same form as that in Equation (2.6-4), has a different sign. The compact form of Equation (2.6-5) analogous to Equation (2.6-3) is

(2.6-7)
$$\frac{df}{dx}(x_1) = \frac{\nabla f(x_1)}{h} + \mathcal{O}(h),$$

where $\nabla f(x_i) = f(x_i) - f(x_{i-1})$ defines the **backward difference operator**. To avoid possible confusion with the gradient operator, also denoted ∇, we shall not use this notation very often.

Now consider the interval $(a =\ ) \; x_0 < x_1 < x_2 \; (\, = b)$ illustrated in Figure 2-10(b). Assuming $x_1 - x_0 = x_2 - x_1 = h$, the appropriate three-point Lagrange interpolating polynomial over this interval is as follows:

$$
\begin{aligned}
\text{2.6-8)} \quad P_2(x) &= \ell_0(x)f(x_0) + \ell_1(x)f(x_1) + \ell_2(x)f(x_2) \\
&= \frac{1}{2h^2}[(x - x_1)(x - x_2)f(x_0) \\
&\quad - 2(x - x_0)(x - x_2)f(x_1) + (x - x_0)(x - x_1)f(x_2)].
\end{aligned}
$$

We obtain an approximation for df/dx at x_1 by differentiating the approximation $f = P_2 + E_2$:

$$
\begin{aligned}
\text{(2.6-9)} \qquad \frac{df}{dx}(x_1) &= \frac{dP_2}{dx}(x_1) + \frac{dE_2}{dx}(x_1) \\
&= \frac{1}{2h}[f(x_2) - f(x_0)] + \frac{dE_2}{dx}(x_1).
\end{aligned}
$$

The error term in this difference approximation is

$$
\begin{aligned}
\text{(2.6-10)} \qquad \frac{dE_2}{dx}(x_1) &= \frac{d}{dx}\left[\frac{1}{3!}(x - x_0)(x - x_1)(x - x_2)\frac{d^3 f}{dx^3}(\varsigma)\right]_{x_1} \\
&= -\frac{h^2}{3!}\frac{d^3 f}{dx^3}(\varsigma) = \mathcal{O}(h^2),
\end{aligned}
$$

ς representing some point in (x_0, x_2). Observe that the three-point approximation is accurate to $\mathcal{O}(h^2)$, whereas the two-point, one-sided difference is accurate only to $\mathcal{O}(h)$. Thus the error in the three-point approximation diminishes more rapidly as we shrink h. The **central difference approximation** defined in Equation (2.6-9) can be conveniently written as

$$
\text{(2.6-11)} \qquad \frac{df}{dx}(x_1) = \frac{1}{2h}\left[\delta f\left(x_1 + \frac{h}{2}\right) + \delta f\left(x_1 - \frac{h}{2}\right)\right] + \mathcal{O}(h^2),
$$

where the **central difference operator** δ is defined by

$$
\text{(2.6-12)} \qquad \delta f(x_i) = f\left(x_i + \frac{h}{2}\right) - f\left(x_i - \frac{h}{2}\right).
$$

Finite-difference approximations of higher-order derivatives are obtained in a similar manner. Consider, for example, the approximation of a second-order derivative at x_1. Differentiation of $f = P_2 + E_2$ twice yields

$$
\text{(2.6-13)} \qquad \frac{d^2 f}{dx^2} = \frac{d^2 P_2}{dx^2} + \frac{d^2 E_2}{dx^2}.
$$

Evaluation of this equation at $x = x_1$ provides the desired relationship directly:

(2.6-14) $$\frac{d^2 f}{dx^2}(x_1) = \frac{1}{h^2}\left[f(x_0) - 2f(x_1) + f(x_2)\right] + \frac{d^2 E_2}{dx^2}(x_1),$$

where, for some $\varsigma \in (x_0, x_2)$,

(2.6-15) $$\frac{d^2 E_2}{dx^2}(x_1) = -\frac{2h^2}{3!}\frac{d^4 f}{dx^4}(\varsigma) = \mathcal{O}(h^2).$$

Using central difference notation, we may write Equation (2.6-15) as follows:

(2.6-16) $$\frac{d^2 f}{dx^2}(x_1) = \frac{\delta^2 f}{h^2}(x_1) + \mathcal{O}(h^2).$$

It is evident at this point that a vast array of difference approximations can be generated with relative ease using the interpolating polynomial technique. Several commonly encountered formulas are given in Table 2-3.

An alternative and commonly used strategy for the development of finite-difference formulas employs Taylor series expansions. Because this approach is well documented elsewhere we consider here only an illustrative example. Let us write Taylor series expansions about the point $x = ih$ for the values of f at $x = (i - 1)h$ and $x = (i + 1)h$:

(2.6-17a) $$f\big((i-1)h\big) = f(ih) - \frac{h}{1}\frac{df}{dx}(ih) + \frac{h^2}{2!}\frac{d^2 f}{dx^2}(ih)$$
$$- \frac{h^3}{3!}\frac{d^3 f}{dx^3}(ih) + \mathcal{O}(h^4),$$

(2.6-17b) $$f\big((i+1)h\big) = f(ih) + \frac{h}{1}\frac{df}{dx}(ih) + \frac{h^2}{2!}\frac{d^2 f}{dx^2}(ih)$$
$$+ \frac{h^3}{3!}\frac{d^3 f}{dx^3}(ih) + \mathcal{O}(h^4).$$

Finite-difference approximations can be obtained directly from the expansions (2.6-17). From Equation (2.6-17a) we obtain a backward difference formula by rearranging the equation to give

(2.6-18a) $$\frac{df}{dx}(ih) = \frac{1}{h}\left[f(ih) - f\big((i-1)h\big)\right] + \mathcal{O}(h).$$

We obtain a forward difference formula from Equation (2.6-17b) similarly:

(2.6-18b) $$\frac{df}{dx}(ih) = \frac{1}{h}\left[f\big((i+1)h\big) - f(ih)\right] + \mathcal{O}(h).$$

TABLE 2-3. One-dimensional finite-difference formulas.

LAGRANGE POLYNOMIAL	DERIVATIVE	APPROXIMATION+ERROR
$\displaystyle\prod_{\substack{j=0 \\ j\neq i}}^{1}\left(\frac{x-x_j}{x_i-x_0}\right)$	$\dfrac{df}{dx}(x_0)$	$[f(x_1)-f(x_0)]/h+\mathcal{O}(h)$
	$\dfrac{df}{dx}(x_1)$	$[f(x_1)-f(x_0)]/h+\mathcal{O}(h)$
$\displaystyle\prod_{\substack{j=0 \\ j\neq i}}^{2}\left(\frac{x-x_j}{x_i-x_j}\right)$	$\dfrac{df}{dx}(x_0)$	$[-3f(x_0)+4f(x_1)-f(x_2)]/2h$ $+\mathcal{O}(h^2)$
	$\dfrac{df}{dx}(x_1)$	$[f(x_2)-f(x_0)]/2h+\mathcal{O}(h^2)$
	$\dfrac{df}{dx}(x_2)$	$[f(x_0)-4f(x_1)+3f(x_2)]/2h$ $+\mathcal{O}(h^2)$
	$\dfrac{d^2f}{dx^2}(x_0)$	$[f(x_0)-2f(x_1)+f(x_2)]/h^2+\mathcal{O}(h)$
	$\dfrac{d^2f}{dx^2}(x_1)$	$[f(x_0)-2f(x_1)+f(x_2)]/h^2$ $+\mathcal{O}(h^2)$
$\displaystyle\prod_{\substack{j=0 \\ j\neq 1}}^{4}\left(\frac{x-x_j}{x_i-x_j}\right)$	$\dfrac{df}{dx}(x_2)$	$[f(x_0)-8f(x_1)+8f(x_3)$ $-f(x_4)]/12h+\mathcal{O}(h^4)$
	$\dfrac{d^2f}{dx^2}(x_2)$	$[-f(x_0)+16f(x_1)-30f(x_2)$ $+16f(x_3)-f(x_4)]/12h^2+\mathcal{O}(h^4)$
	$\dfrac{d^3f}{dx^3}(x_2)$	$[-f(x_0)+2f(x_1)-2f(x_3)$ $+f(x_4)]/2h^3+\mathcal{O}(h^2)$

Finally, we derive the more accurate central difference approximation by subtracting (2.6-17a) from (2.6-17b) and rearranging to get

$$(2.6\text{-}18c)\qquad \frac{df}{dx}(ih)=\frac{1}{2h}\big[f\big((i+1)h\big)-f\big((i-1)h\big)\big]+\mathcal{O}(h^2).$$

Formulas for higher derivative approximations can be obtained in an analogous manner.

Two-dimensional approximations.

The extension of the interpolatory approach to accommodate higher dimensions is straightforward. If the finite-difference formula is to be obtained directly from a Lagrange interpolating polynomial, one begins by writing the tensor product of one-dimensional Lagrange polynomials. If $\ell_i(x)$ and $\ell_i(y)$ denote second-degree Lagrange polynomials, one can write, using Equation (2.5-2b),

$$(2.6\text{-}19) \qquad f(x,y) = \sum_{i=0}^{2} \sum_{j=0}^{2} \ell_i(x)\ell_j(y)f(x_i,y_j) + E_{22}(x,y).$$

This yields the first-order derivative approximation

$$(2.6\text{-}20) \quad \frac{\partial f}{\partial x}(x_1,y_1) = \sum_{i=0}^{2} \sum_{j=0}^{2} \frac{d\ell_i}{dx}(x_1)\ell_j(y_1)f(x_i,y_j) + \frac{\partial E_{22}}{\partial x}(x,y)$$

$$= -\frac{1}{h^2}[2x_1 - x_0 - x_2]\left[-\frac{1}{h^2}(y - y_0)(y - y_1)\right] + \frac{\partial E_{22}}{\partial x}(x,y)$$

$$= \frac{1}{h^2}[x_0 - 2x_1 + x_2] + \mathcal{O}(h^2).$$

Higher-order derivatives are obtained using an analogous procedure.

The Taylor-series approach also generalizes to two dimensions. If $f(x,y)$ is twice continuously differentiable around the point (ih, jh) in the (x,y)-plane, Taylor's theorem with remainder says that, for some number $\theta \in (0,1)$,

$$(2.6\text{-}21) \qquad f\big((i+1)h, (j+1)h\big)$$

$$= f(ih,jh) + \left(h\frac{\partial}{\partial x} + h\frac{\partial}{\partial y}\right)f(ih,jh) + \cdots$$

$$+ \frac{1}{n!}\left(h\frac{\partial}{\partial x} + h\frac{\partial}{\partial y}\right)^n f\big((i+\theta)h, (j+\theta)h\big).$$

Equation (2.6-21) may now be manipulated algebraically to obtain finite-difference approximations in a manner similar to that employed in generating the one-dimensional forms such as (2.6-18).

We have observed how Taylor series and interpolating polynomials provide mathematical vehicles for creating discrete approximations and their associated errors. However, these errors are "local" in the sense that they describe the interpolation property only over a specified discretization interval, for example, $(i-1)h \leq x \leq (i+1)h$. In the next section we address the more difficult question of "global" discretization errors.

2.7. Error Estimates for Finite Differences.

Let us now examine the global error associated with the finite-difference method. We begin by selecting a model equation of the form

$$(2.7\text{-}1) \qquad \frac{d^2 u}{dx^2} - p(x) \frac{du}{dx} - q(x)u = f(x),$$

$$q(x) \geq K > 0, \quad x \in (a, b),$$

with boundary conditions

$$(2.7\text{-}2\text{a}) \qquad\qquad u(a) = \alpha,$$

$$(2.7\text{-}2\text{b}) \qquad\qquad u(b) = \beta.$$

We shall seek an approximation $\hat{u}(x)$ to the true solution $u(x)$ by determining nodal values $\hat{u}(x_i)$ on a uniform grid $\Delta : a = x_0 < x_1 < \cdots < x_N = b$ with mesh h. Employing second-order accurate difference approximations in Equation (2.7-1), one obtains the difference analog

$$(2.7\text{-}3) \qquad \frac{1}{h^2}[\hat{u}(x_{i-1}) - 2\hat{u}(x_i) + \hat{u}(x_{i+1})]$$

$$- \frac{p(x_i)}{2h}[\hat{u}(x_{i+1}) - \hat{u}(x_{i-1})] - q(x_i)\hat{u}(x_i) = f(x_i).$$

Define the **solution error** as $e(x_i) \equiv \hat{u}(x_i) - u(ih)$, where $\hat{u}(x_i)$ denotes the difference approximation at the node x_i and $u(ih)$ is the corresponding value of the exact solution. Substitution of $u(ih)$ for $\hat{u}(x_i)$ in Equation (2.7-3) yields

$$(2.7\text{-}4) \qquad \frac{1}{h^2}[u((i-1)h) - 2u(ih) + u((i+1)h)]$$

$$- \frac{p(ih)}{2h}[u((i+1)h) - u((i-1)h)] - q(ih)u(ih)$$

$$= f(ih) + \mathcal{O}(h^2),$$

where the term $\mathcal{O}(h^2)$ is the interpolation error derived in Section 2.6. Subtracting Equation (2.7-4) from Equation (2.7-3) and recognizing that $x_i = ih$, we find

$$(2.7\text{-}5) \qquad \frac{1}{h^2}[e(x_{i-1}) - 2e(x_i) + e(x_{i+1})]$$

$$- \frac{p(x_i)}{2h}[e(x_{i+1}) - e(x_{i-1})] - q(x_i)e(x_i) = \mathcal{O}(h^2).$$

Since we can represent the Dirichlet boundary conditions (2.7-2) exactly, $e_0 = e_N = 0$, and the set of difference equations (2.7-5) can be written

$$
(2.7\text{-}6) \quad
\begin{bmatrix}
a_1 & -c_1 & & & \\
-b_2 & a_2 & -c_2 & & \\
& \ddots & \ddots & \ddots & \\
& & & -b_{N-1} & a_{N-1}
\end{bmatrix}
\begin{bmatrix}
e_1 \\ e_2 \\ e_3 \\ \vdots \\ e_{N-1}
\end{bmatrix}
= -
\begin{bmatrix}
f_1 \\ f_2 \\ f_3 \\ \vdots \\ f_{N-1}
\end{bmatrix},
$$

where

$$
a_i = \left[\frac{2}{h^2} + q(x_i) \right],
$$

$$
b_i = \frac{1}{2}\left[\frac{2}{h^2} + \frac{1}{h}\, p(x_i) \right],
$$

$$
c_i = \frac{1}{2}\left[\frac{2}{h^2} - \frac{1}{h}\, p(x_i) \right],
$$

$$
f_i = \mathcal{O}(h^2).
$$

We rewrite the matrix equation (2.7-6) as

$$(2.7\text{-}7) \qquad\qquad\qquad \mathbf{Ae} = \mathbf{f}$$

and call the vector $\mathbf{e}$ the **discretization** or **global error** of the difference analog.

One can use the eigenvalue $\lambda_{\min}(\mathbf{A})$ of $\mathbf{A}$ having smallest magnitude to bound this error. Using the Euclidean norm, we find

$$(2.7\text{-}8) \qquad\qquad\qquad \|\mathbf{Ae}\|_2 = \|\mathbf{f}\|_2.$$

But since $\|\mathbf{Ae}\|_2 \geq |\lambda_{\min}(\mathbf{A})|\,\|\mathbf{e}\|_2$, we have

$$(2.7\text{-}9) \qquad\qquad\qquad \|\mathbf{f}\|_2 \geq |\lambda_{\min}(\mathbf{A})|\,\|\mathbf{e}\|_2.$$

Whenever $\mathbf{A}$ is nonsingular, $\lambda_{\min}(\mathbf{A}) \neq 0$, and we have the following bound on the global error:

$$(2.7\text{-}10) \qquad\qquad\qquad \|\mathbf{e}\|_2 \leq \frac{\|\mathbf{f}\|_2}{|\lambda_{\min}(\mathbf{A})|}.$$

If $\lambda_{\min}(\mathbf{A})$ is bounded away from zero, that is, $\lambda_{\min}(\mathbf{A})$ approaches a nonzero value as $h \to 0$, then the norm of the global error $\|\mathbf{e}\|_2 \to 0$ at least as fast as the truncation error $\|\mathbf{f}\|_2$. In the case of our example problem (2.7-6), the eigenvalues of $\mathbf{A}$ are the following:

$$(2.7\text{-}11) \qquad \lambda_s(\mathbf{A}) = a + 2\sqrt{bc}\,\frac{\cos s\pi}{N+2}, \qquad s = 1, 2, \ldots, N+1.$$

Since for our problem $q(x) \geq K > 0$, we can conclude that $\lambda_{\min}(\mathbf{A})$ is bounded away from zero as h approaches zero. Therefore the global error norm $\|e\|_2 \rightarrow 0$ at least as fast as the truncation error $\|f\|_2$, which we found in Section 2.6 to be $\mathcal{O}(h^2)$.

2.8. Consistency of Finite-Difference Approximations.

The **consistency** of a finite-difference approximation refers to the convergence of the approximate algebraic equation to the desired differential equation as the grid mesh $h \rightarrow 0$. While the majority of difference approximations exhibit this property, it is possible to encounter approximations that, in the limit as $h \rightarrow 0$, may not converge to the desired differential equation. To illustrate the concept of consistency let us examine two finite-difference approximations to the heat equation

$$\frac{\partial^2 u}{\partial x^2} = \frac{\partial u}{\partial t}$$

that have been formulated using Taylor series expansions. This PDE is a simplified version of Equation (1.4-11), derived from the energy balance.

Consider first the centered-in-space, backward-in-time difference formula at the node $(x_i, t_n) = (ih, nk)$:

$$(2.8\text{-}1) \qquad u\big(ih, (n+1)k\big) - u(ih, nk) - \frac{k}{h^2}\big[u\big((i+1)h, (n+1)k\big)$$

$$- 2u\big(ih, (n+1)k\big) + u\big((i-1)h, (n+1)k\big)\big]$$

$$= \left[-\frac{1}{2!}\, k^2\, \frac{\partial^2 u}{\partial t^2} + \frac{1}{3!}\, k^3\, \frac{\partial^3 u}{\partial t^3} + \cdots\right]_{(ih,(n+1)k)}$$

$$- \frac{k}{h^2}\left[+\frac{2}{4!}\, h^4\, \frac{\partial^4 u}{\partial x^4} + \frac{2}{6!}\, h^6\, \frac{\partial^6 u}{\partial x^6} + \cdots\right]_{(ih,(n+1)k)}$$

The right side of Equation (2.8-1) represents k times the truncation error $E(x, t)$ of the finite-difference formula. We can rewrite this error as follows

$$(2.8\text{-}2) \qquad E(x, t) = \left[-\frac{1}{2}\, k\, \frac{\partial^2 u}{\partial t^2} + \frac{1}{3!}\, k^2\, \frac{\partial^3 u}{\partial t^3} + \cdots\right.$$

$$\left. - \frac{2}{4!}\, h^2\, \frac{\partial^4 u}{\partial x^4} - \frac{2}{6!}\, h^4\, \frac{\partial^6 u}{\partial x^6} - \cdots\right]_{(ih,(n+1)k)}.$$

Examination of (2.8-2) reveals that, as h and k approach zero, the truncation error E must vanish, no matter how h and k behave relative to one another. Thus this difference approximation is **unconditionally consistent**.

To illustrate that this is not always the case, we use an alternative approximation to the same model equation namely,

(2.8-3) $u\big(ih, (n+1)k\big) - u\big(ih, (n-1)k\big)$

$$- \frac{2k}{h^2}\big[u\big((i+1)h, k\big) - u\big(ih, (n+1)k\big)$$

$$- u\big(ih, (n-1)k\big) + u\big((i-1)h, k\big)\big]$$

$$= -2k\left[\left(\frac{k}{h}\right)^2 \frac{\partial^2 u}{\partial t^2} + \frac{k^2}{6}\frac{\partial^3 u}{\partial t^3} - \frac{h^2}{12}\frac{\partial^4 u}{\partial y^4} + \cdots\right]_{(ih,nk)}.$$

This difference formula is referred to as the **DuFort-Frankel approximation**. Consider the asymptotic behavior of the error on the right side of Equation (2.8-3) in the limit as h and k approach zero. Whereas in the previous example the truncation error described by Equation (2.8-2) vanished in the limit as h and k approached zero, irrespective of the relative behavior of h and k, such is not the case with the DuFort-Frankel approximation. If $k/h \to 0$ as $k \to 0$, then the error term in Equation (2.8-3) vanishes, and the difference formula approximates the model equation. However, if k/h tends to a nonzero constant as $k \to 0$, then the difference formula is an approximation to

$$\frac{\partial^2 u}{\partial x^2} = \frac{\partial u}{\partial t} + \left[\lim_{k,h\to 0}\left(\frac{k}{h}\right)^2\right]\frac{\partial^2 u}{\partial t^2},$$

which is hyperbolic rather than parabolic. Thus one concludes that the difference expression appearing in Equation (2.8-3) is consistent with the model equation if and only if k goes to zero faster than h, and therefore the DuFort-Frankel approximation is **conditionally consistent** with the heat equation.

2.9. Stability of Finite-Difference Approximations.

The notion of **stability** addresses the computational behavior of the algebraic equations arising from the numerical approximation of differential equations. There are several ways of approaching the concept of stability. Here we define a **stable** algorithm to be one for which every component of an initial function, possibly containing numerical errors, is limited in the degree to which it is amplified by the numerical procedure. Unstable algorithms are virtually useless in digital calculations, since they tend to amplify unavoidable roundoff errors without bound. Let us now consider the question of how to establish the stability of a finite-difference equation.

One approach consists of programming the algorithm for the digital computer and then conducting a series of numerical experiments. Through examination of the behavior of the method over a broad spectrum of mesh geometries and coefficient values, one can come to certain conclusions regarding the stability bounds of the method. The approach is often referred to as **heuristic stability analysis**.

A more structured approach was proposed by von Neumann (Charney et al., 1950). It furnishes a necessary and, in the case of certain pure initial value problems, sufficient condition for stability. The procedure is based on Fourier analysis. A reasonably well behaved function $u(x,t)$ on a spatial interval $[0, X]$ can be represented by an infinite series

$$(2.9\text{-}1) \qquad u(x,t) = \sum_{m=-\infty}^{\infty} g_m(t)\, e^{\hat{i}2\pi m(x/X)},$$

where $\hat{i} = \sqrt{-1}$ and

$$(2.9\text{-}2) \qquad g_m(t) = \frac{1}{X} \int_0^X u(x,t)\, e^{-\hat{i}2\pi m(x/X)}\, dx.$$

The quantity g_m is the amplitude of the m-th harmonic or Fourier mode; this mode has a wavelength X/m.

Assume we know $u(x,t)$ at $+1L$ distinct points or nodes (ih, t), $i = 0, 1, 2, \ldots, L$. Then the values $u(ih, t)$, $i = 0, 1, 2, \ldots, L$, can be approximated by a set of L Fourier modes, where x is replaced by ih and X by Lh:

$$(2.9\text{-}3) \qquad u(ih, t) \simeq \sum_{m=0}^{L-1} \hat{g}_m(t)\, e^{\hat{i}2\pi m(i/L)}.$$

Here the amplitude of the m-th mode is given by a discrete version of the integral (2.9-2):

$$(2.9\text{-}4) \qquad \hat{g}_m(t) = \frac{1}{L} \sum_{i=0}^{L-1} u(ih, t)\, e^{-\hat{i}2\pi m(i/L)}.$$

Note that this representation constitutes a "long wavelength" approximation to $u(x,t)$, which is appropriate since no wavelength smaller than h can be represented on the finite-difference grid.

Our task now is to examine the propagation of errors in finite-difference expressions as the simulation advances through time. Consider once again a difference approximation to the one-dimensional heat equation. Employing a forward-in-time, centered-in-space approximation, we obtain the following algebraic equations for the approximate solution $\hat{u}(x,t)$:

$$(2.9\text{-}5) \quad \hat{u}\big(ih, (n+1)k\big) - \hat{u}(ih, nk)$$

$$- \frac{k}{h^2} \big[\hat{u}\big((i+1)h, nk\big) - 2\hat{u}(ih, nk) + \hat{u}\big((i-1)h, nk\big)\big] = 0.$$

Let us assume that, at $t = 0$, machine limitations introduce an error at the spatial nodes. We denote this error as $\epsilon(ih, 0)$. The machine will force the function $\hat{u} + \epsilon$ to satisfy the difference equation, while in principle we want $\hat{u}$ to satisfy it. Moreover, let us represent the error ϵ by a Fourier expansion of the form presented in (2.9-3):

$$(2.9\text{-}6) \qquad \epsilon(ih, 0) = \sum_{m=0}^{L-1} \hat{g}_m(0) e^{i2\pi m(i/L)},$$

where

$$(2.9\text{-}7) \qquad \hat{g}_m(0) = \frac{1}{L} \sum_{i=0}^{L-1} \epsilon(ih, 0) e^{-i2\pi m(i/L)}.$$

Because, by definition, the true values of $\hat{u}$ satisfy the difference equation, which is linear, the error also satisfies the difference equation. We can thus write

$$(2.9\text{-}8) \qquad \epsilon\big(ih, (n+1)k\big) - \epsilon(ih, nk)$$
$$= \frac{k}{h^2} \big[\epsilon\big((i+1)h, nk\big) - 2\epsilon(ih, nk) + \epsilon\big((i-1)h, nk\big)\big].$$

From the principle of superposition we know that we can establish the behavior of the scheme (2.9-8) by examining the propagation of a typical term in the series (2.9-6) for ϵ. Such a term has the form

$$(2.9\text{-}9) \qquad \epsilon_m(ih, nk) = \hat{g}_m(nk) e^{i\beta_m ih},$$

where $\beta_m = 2\pi m/Lh$. To simplify notation we define $\xi^n = \hat{g}_m(nk)$ and designate ξ the **amplification factor**. Substitution of the term (2.9-9) into the difference equation (2.9-8), suppression of the mode number m, and replacement of k/h^2 by the symbol $\bar{p}$ yield

$$(2.9\text{-}10) \qquad \xi^{n+1} e^{i\beta ih} - \xi^n e^{i\beta ih}$$
$$= \bar{p}\big[\xi^n e^{i\beta(i+1)h} - 2\xi^n e^{i\beta ih} + \xi^n e^{i\beta(i-1)h}\big].$$

This expression can be solved for the amplification factor, giving

$$(2.9\text{-}11) \qquad \begin{aligned} \xi &= 1 + \bar{p}(e^{i\beta h} - 2 + e^{-i\beta h}) \\ &= (1 - 2\bar{p}) + \bar{p}(e^{i\beta h} + e^{-i\beta h}) \\ &= 1 - 2\bar{p}(1 - \cos \beta h) \\ &= 1 - 4\bar{p}\sin^2\left(\frac{\beta h}{2}\right). \end{aligned}$$

To interpret ξ, observe from Equation (2.9-9) that

$$(2.9\text{-}12) \qquad \frac{\epsilon(ih, (n+1)k)}{\epsilon(ih, nk)} = \frac{\xi^{n+1}}{\xi^n}$$

or, upon rearranging,

$$(2.9\text{-}13) \qquad \epsilon(ih, (n+1)k) = \xi\epsilon(ih, nk).$$

Thus the error component ϵ will not grow provided $|\xi| \leq 1$. This is the **von Neumann necessary condition for stability**.

Returning to Equation (2.9-11) we observe that von Neumann stability for this example requires

$$(2.9\text{-}14) \qquad \left| 1 - 4\bar{\rho}\sin^2\left(\frac{\beta h}{2}\right) \right| \leq 1.$$

For this inequality to hold for all values of β, it is necessary that

$$(2.9\text{-}15) \qquad \bar{\rho} \leq \tfrac{1}{2}$$

or, recalling the definition of $\bar{\rho}$,

$$(2.9\text{-}16) \qquad k \leq \tfrac{1}{2}h^2.$$

Thus for a given spatial increment h the scheme (2.9-8) will be stable only for small enough time steps k. Similar relationships can be developed for other difference approximations, albeit with possibly more tedious algebraic manipulations.

A third approach to stability, and one particularly suited to approximation methods exclusive of finite differences, is **matrix stability analysis**. We illustrate the methodology using the same example presented in the preceding section on von Neumann stability.

The finite-difference expression (2.9-8), in combination with Dirichlet boundary conditions, can be written in matrix form as

$$(2.9\text{-}17) \qquad \mathbf{A}\boldsymbol{\epsilon}^{(n)} + \mathbf{B}\big(\boldsymbol{\epsilon}^{(n+1)} - \boldsymbol{\epsilon}^{(n)}\big) = \mathbf{0},$$

where the nonzero elements of $\mathbf{A}$ are given by

$$\begin{aligned}
a_{i,i-1} &= -\bar{\rho}, & i \neq 1, \\
a_{i,i} &= 2\bar{\rho}, & \\
a_{i,i+1} &= -\bar{\rho}, & i \neq L.
\end{aligned}$$

The matrix $\mathbf{B}$ is diagonal, having nonzero entries $b_{i,i} = 1$, while the error vectors $\boldsymbol{\epsilon}^{(n)}$ and $\boldsymbol{\epsilon}^{(n+1)}$ have entries

$$\epsilon_i^{(n)} = \epsilon(ih, nk),$$
$$\epsilon_i^{(n+1)} = \epsilon(ih, (n+1)k).$$

Equation (2.9-17) can be rearranged to yield $\mathbf{B}\epsilon^{(n+1)} = \mathbf{B}\epsilon^{(n)} - \mathbf{A}\epsilon^{(n)} \equiv \mathbf{C}\epsilon^{(n)}$, where $\mathbf{C} = \mathbf{B} - \mathbf{A}$. Because $\mathbf{B}$ is the identity matrix $\mathbf{I}$ in our example, this expression can be written

$$\epsilon^{(n+1)} = \mathbf{B}^{-1}\mathbf{C}\epsilon^{(n)} = \mathbf{C}\epsilon^{(n)} = \mathbf{C}^n\epsilon^{(0)},$$

where the nonzero elements of $\mathbf{C}$ are as follows

$$
\begin{aligned}
c_{i,i-1} &= \bar{p}, & i \neq 1, \\
c_{i,i} &= 1 - 2\bar{p}, & \\
c_{i,i+1} &= \bar{p}, & i \neq L.
\end{aligned}
$$

The error growth will be bounded provided the norm of the error vector is bounded with respect to the initial error as time progresses. Thus, for any norm $\| \cdot \|$, stability requires

$$(2.9\text{-}18) \qquad \|\epsilon^{(n+1)}\| \leq K\|\epsilon^{(0)}\| \qquad \text{as } n \to \infty,$$

for some constant K independent of the discretization intervals h and k. Employing the Schwarz inequality and assuming $\mathbf{C}$ is independent of time, we get

$$(2.9\text{-}19) \qquad \|\epsilon^{(n+1)}\| = \|\mathbf{C}\epsilon^{(n)}\| = \|\mathbf{C}^n\epsilon^{(0)}\| \leq \|\mathbf{C}^n\| \, \|\epsilon^{(0)}\|$$
$$\leq \|\mathbf{C}\|^n\|\epsilon^{(0)}\|.$$

Here the norm of the matrix $\mathbf{C}$ is defined relative to the vector norm $\| \cdot \|$ via the equation

$$\|\mathbf{C}\| = \sup\frac{\|\mathbf{C}\mathbf{x}\|}{\|\mathbf{x}\|},$$

where "sup" denotes the least upper bound taken over all nonzero vectors $\mathbf{x}$. Thus a *sufficient* condition for stability is as follows:

$$(2.9\text{-}20) \qquad \|\mathbf{C}\|^n \leq K \qquad \text{as } n \to \infty.$$

A more useful relationship can be derived using the relationship between the Euclidean matrix norm and the spectral radius or maximum-modulus eigenvalue, $\rho(\mathbf{C}) = \max_i |\lambda_i(\mathbf{C})|$:

$$(2.9\text{-}21) \qquad \rho^n(\mathbf{C}) \leq \|\mathbf{C}\|_2^n.$$

Therefore a *necessary* condition for (2.9-20) to hold is

$$(2.9\text{-}22) \qquad \rho^n(\mathbf{C}) \leq K \qquad \text{as } n \to \infty,$$

which implies the requirement

(2.9-23) $\rho(\mathbf{C}) \leq 1.$

In general the Euclidean norm is related to the spectral radius through the equation

(2.9-24) $\|\mathbf{C}\|_2 = \sqrt{\rho(\mathbf{C}\mathbf{C}^\top)}.$

Thus for symmetric matrices we can replace the inequality (2.9-21) by

(2.9-25) $\|\mathbf{C}\|_2 = \rho(\mathbf{C}),$

and (2.9-23) becomes both necessary and sufficient.

   Let us now return to our example problem for the heat equation. Because of the particular structure of $\mathbf{C}$, the eigenvalues can be computed analytically in this problem. They are as follows:

(2.9-26) $\lambda_i(\mathbf{C}) = 1 - 4\bar{\rho}\sin^2\left(\dfrac{i\pi}{2L}\right), \qquad i = 1, 2, \ldots, L - 1.$

Hence the criterion (2.9-23) becomes

(2.9-27) $\left| 1 - 4\bar{\rho}\sin^2\left(\dfrac{i\pi}{2L}\right) \right| \leq 1, \qquad i = 1, 2, \ldots, L - 1,$

which implies the same stability criterion derived using the von Neumann method.

2.10. The Method of Weighted Residuals.

The **method of weighted residuals** provides a conceptual foundation upon which to construct Galerkin finite-element, collocation, and boundary element methods, all to be considered in this chapter. While we shall become more specific later, at this point let us consider a differential equation written in operator form as

(2.10-1) $\mathcal{L}\big(u(\mathbf{x})\big) - f(\mathbf{x}) = 0, \qquad \mathbf{x} \in \Omega,$

where $\mathcal{L}(\cdot)$ is the differential operator and $f(\mathbf{x})$ is a known **forcing function**, both defined over a region Ω with boundary $\partial\Omega$. Let $\phi_i(\mathbf{x})$, $i = 1, 2, \ldots, N$ represent N functions selected from a set of known, linearly independent **basis functions**. In general these functions will satisfy homogeneous boundary conditions on $\partial\Omega$, but to see precisely what these conditions should be we need to carry the development a little further. Let

$\phi_0(\mathbf{x})$ be a function whose values on $\partial\Omega$ are those essential to the construction of an approximate solution satisfying the boundary conditions imposed on $u(\mathbf{x})$. Again, we must carry the development somewhat further to see what the conditions on ϕ_0 must be.

We define a **trial function** $\hat{u}(\mathbf{x})$ approximating $u(\mathbf{x})$ as a linear superposition of the functions ϕ_i:

$$(2.10\text{-}2) \qquad \hat{u}(\mathbf{x}) = \phi_0(\mathbf{x}) + \sum_{i=1}^{N} u_i\, \phi_i(\mathbf{x}),$$

where the constants $\{u_i\}_{i=1}^{N}$ are yet to be determined. Because N is finite, $\hat{u}$ is generally different from the true solution u, whose specification may require an infinite number of degrees of freedom and hence an infinite series of basis functions chosen from a complete set. Therefore substitution of $\hat{u}$ into (2.10-1) will typically result in a nonzero **residual** $\mathcal{R}(\mathbf{x}, \mathbf{U})$, where $\mathbf{U}$ stands for the vector of N unknown coefficients u_i. In symbols,

$$(2.10\text{-}3) \qquad \mathcal{L}\big(\hat{u}(\mathbf{x})\big) - f(\mathbf{x}) = \mathcal{R}(\mathbf{x}, \mathbf{U}).$$

The objective of the method of weighted residuals is to select $\mathbf{U}$ so as to minimize $\mathcal{R}(\mathbf{x}, \mathbf{U})$ in some sense. This can be accomplished by first multiplying $\mathcal{R}(\mathbf{x}, \mathbf{U})$ by a set of **weighting functions** $w_j(\mathbf{x})$, $j = 1, 2, \ldots, N$. The resulting product is then integrated over Ω and the resulting integral set to zero. These two procedures force the residual to vanish in a weighted-average sense:

$$(2.10\text{-}4) \qquad \int_{\Omega} \big[\mathcal{L}\big(\hat{u}(\mathbf{x})\big) - f(\mathbf{x})\big]\, w_j(\mathbf{x})\, d\mathbf{x} = 0, \qquad j = 1, 2, \ldots, N.$$

If we regard functions as vectors, then the operation defined by $\langle f, g \rangle = \int_{\Omega} f(\mathbf{x})\, g(\mathbf{x})\, d\mathbf{x}$ acts as an **inner product** on suitably restricted sets of functions. Employing this inner product notation, the weighted-residual conditions (2.10-4) can be rewritten

$$(2.10\text{-}5) \qquad \big\langle \mathcal{L}\big(\hat{u}(\mathbf{x})\big) - f(\mathbf{x}), w_j(\mathbf{x}) \big\rangle = 0, \qquad j = 1, 2, \ldots, N,$$

which forces the residual to be orthogonal to each weighting function w_j. Substitution of the trial function (2.10-2) into this expression yields

$$(2.10\text{-}6) \qquad \sum_{i=1}^{N} u_i \big\langle \mathcal{L}\big(\phi_i(\mathbf{x})\big), w_j(\mathbf{x}) \big\rangle = -\big\langle \mathcal{L}\big(\phi_0(\mathbf{x})\big), w_j(\mathbf{x}) \big\rangle$$

$$+ \langle f(\mathbf{x}), w_j(\mathbf{x}) \rangle, \qquad j = 1, 2, \ldots, N.$$

Now we are in a position to see what boundary values we must impose on the functions $\phi_0, \ldots, \phi_N$ to guarantee that $\hat{u}$ satisfy given boundary conditions. Let us treat the typical operator defined by

$$\mathcal{L}(u(\mathbf{x})) = -\nabla \cdot [p(\mathbf{x})\nabla u(\mathbf{x})] + q(\mathbf{x})u(\mathbf{x})$$

for concreteness. With this choice, Equation (2.10-6) becomes

$$(2.10\text{-}7) \quad \sum_{i=1}^{N} u_i \int_{\Omega} \{-\nabla \cdot [p(\mathbf{x})\nabla\phi_i(\mathbf{x})] + q(\mathbf{x})\phi_i(\mathbf{x})\}w_j(\mathbf{x})d\mathbf{x}$$

$$= \int_{\Omega} \{\nabla \cdot [p(\mathbf{x})\nabla\phi_0(\mathbf{x})] - q(\mathbf{x})\phi_0(\mathbf{x})\}w_j(\mathbf{x})d\mathbf{x}$$

$$+ \int_{\Omega} f(\mathbf{x})w_j(\mathbf{x})d\mathbf{x}.$$

For the time being, let us assume that the weighting functions w_j are piecewise differentiable. (This will be the case in the Galerkin-based versions of the method of weighted residuals introduced in the next section.) Then we can integrate the terms involving the functions $\phi_0, \ldots, \phi_N$ using Green's theorem (or integration by parts, in the one-dimensional case) to get

$$\sum_{i=1}^{N} u_i \int_{\Omega} [p(\mathbf{x})\nabla\phi_i(\mathbf{x}) \cdot \nabla w_j(\mathbf{x}) + q(\mathbf{x})\phi_i(\mathbf{x})w_j(\mathbf{x})]d\mathbf{x}$$

$$- \sum_{i=1}^{N} u_i \oint_{\partial\Omega} w_j(\mathbf{x})p(\mathbf{x})\nabla\phi_i(\mathbf{x}) \cdot \mathbf{n}(\mathbf{x})d\mathbf{x}$$

$$= - \int_{\Omega} [p(\mathbf{x})\nabla\phi_0(\mathbf{x}) \cdot \nabla w_j(\mathbf{x}) + q(\mathbf{x})\phi_0(\mathbf{x})w_j(\mathbf{x})]d\mathbf{x}$$

$$+ \oint_{\partial\Omega} w_j(\mathbf{x})p(\mathbf{x})\nabla\phi_0(\mathbf{x}) \cdot \mathbf{n}(\mathbf{x})d\mathbf{x} + \int_{\Omega} f(\mathbf{x})w_j(\mathbf{x})d\mathbf{x}.$$

This is equivalent to

$$(2.10\text{-}8) \quad \sum_{i=1}^{N} u_i \int_{\Omega} [p(\mathbf{x})\nabla\phi_i(\mathbf{x}) \cdot \nabla w_j(\mathbf{x}) + q(\mathbf{x})\phi_i(\mathbf{x})w_j(\mathbf{x})]d\mathbf{x}$$

$$= - \int_{\Omega} [p(\mathbf{x})\nabla\phi_0(\mathbf{x}) \cdot \nabla w_j(\mathbf{x}) + q(\mathbf{x})\phi_0(\mathbf{x})w_j(\mathbf{x})]d\mathbf{x}$$

$$+ \int_{\Omega} f(\mathbf{x})w_j(\mathbf{x})d\mathbf{x}$$

$$+ \oint_{\partial\Omega} w_j(\mathbf{x})p(\mathbf{x})\nabla\Big[\phi_0(\mathbf{x}) + \sum_{i=1}^{N} u_i\phi_i(\mathbf{x})\Big] \cdot \mathbf{n}(\mathbf{x})d\mathbf{x}.$$

The integrals in the sum on the left side of Equation (2.10-8) are directly calculable, since we know the basis functions $\phi_1, \ldots, \phi_N$ and the weighting functions $w_1, \ldots, w_N$ in advance of solving the problem. Similarly, the two volume integrals on the right side of Equation (2.10-8) are also directly calculable if we know what ϕ_0 is. This leaves the boundary integral, whose calculation will depend on the type of boundary conditions imposed.

Suppose first that we are solving a boundary-value problem of the Dirichlet type, where $u(\mathbf{x}) = \gamma(\mathbf{x})$ on $\partial\Omega$ for some prescribed data $\gamma(\mathbf{x})$. Then we can calculate the boundary integral in Equation (2.10-8), whatever the values of the functions $\phi_0, \ldots, \phi_N$ on $\partial\Omega$ may be. However, two observations are worth making. First, we can obviate formal calculation of any term in the boundary integral if we can ensure that it vanishes. In the classical Galerkin case when each $w_j = \phi_j$, discussed in Section 2.11, we can accomplish this by ensuring that $\phi_j(\mathbf{x}) = 0$ when $\mathbf{x} \in \partial\Omega$, for $j = 1, \ldots, N$. These homogeneous boundary conditions on the basis functions $\phi_1, \ldots, \phi_N$ can therefore lead to significant computational convenience.

The second observation has more importance. In the Dirichlet case, the weighted-residual equations (2.10-8) alone admit no mechanism for guaranteeing that $\hat{u}(\mathbf{x}) = \gamma(\mathbf{x})$ on $\partial\Omega$, and an arbitrary choice of $\phi_0, \ldots, \phi_N$ will typically yield an approximate solution $\hat{u}$ that fails to satisfy the boundary conditions even approximately. Thus for Dirichlet boundary-value problems the trial function $\hat{u}(\mathbf{x}) = \phi_0(\mathbf{x}) + \sum_{i=1}^{N} u_i \phi_i(\mathbf{x})$ must be forced to satisfy the boundary conditions a priori. Assuming that the functions $\phi_1, \ldots, \phi_N$ vanish on $\partial\Omega$, as suggested previously, we can guarantee this by setting $\phi_0(\mathbf{x}) = \gamma(\mathbf{x})$ on $\partial\Omega$. At any rate, we must explicitly impose Dirichlet boundary conditions on the trial function to obtain an acceptable approximate solution. These boundary conditions are therefore called **essential** boundary conditions.

Now suppose we are solving a Neumann problem, for which we can write the boundary conditions as $\nabla u(\mathbf{x}) \cdot \mathbf{n}(\mathbf{x}) = \gamma(\mathbf{x})$ on $\partial\Omega$. In this case it is not necessary to impose boundary values on $\hat{u}$ a priori. Indeed, not knowing the values of $\hat{u}(\mathbf{x})$ on $\partial\Omega$, we need to leave them as unknown degrees of freedom. We impose the Neumann conditions simply by requiring that the boundary integral in Equation (2.10-8) have the same value as it would have were we to substitute the true solution u for the trial function $\hat{u}$. Specifically, we equate

$$\oint_{\partial\Omega} w_j(\mathbf{x}) p(\mathbf{x}) \nabla \left[\phi_0(\mathbf{x}) + \sum_{i=1}^{N} u_i \phi_i(\mathbf{x}) \right] \cdot \mathbf{n}(\mathbf{x}) d\mathbf{x}$$

$$= \oint_{\partial\Omega} w_j(\mathbf{x}) p(\mathbf{x}) \nabla u(\mathbf{x}) \cdot \mathbf{n}(\mathbf{x}) d\mathbf{x} = \oint_{\partial\Omega} w_j(\mathbf{x}) p(\mathbf{x}) \gamma(\mathbf{x}) d\mathbf{x}$$

and substitute the last integral, which we can compute since the integrand is known, for the boundary integral in Equation (2.10-8). We see, then, that the integral formulation (2.10-8) admits a natural mechanism for enforcing Neumann conditions without necessitating a priori restrictions on the functions $\phi_0, \ldots, \phi_N$. Thus for this problem we call Neumann boundary conditions **natural** boundary conditions.

Finally, a similar mechanism exists for imposing a Robin condition

$$\alpha(\mathbf{x})u(\mathbf{x}) + \beta(\mathbf{x})\nabla u(\mathbf{x}) \cdot \mathbf{n}(\mathbf{x}) = \gamma(\mathbf{x}), \qquad \mathbf{x} \in \partial\Omega,$$

via the boundary integral in Equation (2.10-8). Wherever $\beta(\mathbf{x}) \neq 0$ on $\partial\Omega$, we can rearrange this boundary condition to get

$$w_j(\mathbf{x})p(\mathbf{x})\nabla u(\mathbf{x}) \cdot \mathbf{n}(\mathbf{x}) = \frac{1}{\beta(\mathbf{x})}w_j(\mathbf{x})p(\mathbf{x})[\gamma(\mathbf{x}) - \alpha(\mathbf{x})u(\mathbf{x})].$$

By substituting $\hat{u}(\mathbf{x}) = \phi_0(\mathbf{x}) + \sum_{i=1}^{N} u_i\phi_i(\mathbf{x})$ for $u(\mathbf{x})$ in this last equation, we can rewrite the boundary integral in Equation (2.10-8) as follows

$$\oint_{\partial\Omega} \frac{1}{\beta(\mathbf{x})}w_j(\mathbf{x})p(\mathbf{x})\gamma(\mathbf{x})d\mathbf{x} - \oint_{\partial\Omega} \frac{\alpha(\mathbf{x})}{\beta(\mathbf{x})}w_j(\mathbf{x})p(\mathbf{x})\phi_0(\mathbf{x})d\mathbf{x}$$

$$-\sum_{i=1}^{N} \oint_{\partial\Omega} \frac{\alpha(\mathbf{x})}{\beta(\mathbf{x})}w_j(\mathbf{x})p(\mathbf{x})\phi_i(\mathbf{x})d\mathbf{x}.$$

Thus, regardless of our choice of $\phi_0, \ldots, \phi_N$, we can impose Robin boundary conditions through proper evaluation of the boundary integral arising from Green's theorem. Hence, as for Neumann conditions, Robin conditions impose no a priori restrictions on the trial function $\hat{u}$, and again they are natural boundary conditions.

So far, we have predicated our reasoning about boundary conditions on the assumption that the weighting functions w_j are differentiable, at least in the piecewise sense. This assumption holds for standard Galerkin methods, as we shall see in Section 2.11, but it fails in the case of finite-element collocation, which we shall discuss in Section 2.14. If the weighting functions are not differentiable, then Green's theorem (or integration by parts, in the one-dimensional case) is invalid, and no boundary integral arises. In such methods the only mechanism for forcing $\hat{u}$ to satisfy given boundary conditions is to impose them a priori, be they Dirichlet, Neumann, or Robin conditions.

2.11. The Galerkin Finite-Element Method.

The **Galerkin method** constitutes the mathematical foundation upon which most finite-element methods are built. This method is obtained easily from the method of weighted residuals by selecting the basis functions

$\phi_j(\mathbf{x})$, $j = 1, 2, \ldots, N$, as the weighting functions $w_j(\mathbf{x})$, $j = 1, 2, \ldots, N$. Thus we have

$$(2.11\text{-}1) \qquad \sum_{i=1}^{N} u_i \langle \mathcal{L}\phi_i(\mathbf{x}), \phi_j(\mathbf{x}) \rangle = -\langle \mathcal{L}\phi_0(\mathbf{x}), \phi_j(\mathbf{x}) \rangle + \langle f(\mathbf{x}), \phi_j(\mathbf{x}) \rangle,$$
$$j = 1, \ldots, N,$$

as the working equation for the Galerkin method. To complete the formulation requires only the selection of the functions $\phi_j(\mathbf{x})$, a task that should be guided by the information presented earlier in Sections 2.4 and 2.5 on polynomial approximation theory.

Example: a second-order differential equation.

• Consider, for the sake of comparison, the problem treated in Section 2.7 using a finite-difference approach. We have

$$\frac{d^2 u}{dx^2} - p(x)\frac{du}{dx} - q(x)u = f(x), \quad q(x) \geq 0, \ x \in (a, b),$$

with boundary conditions

$$u(a) = \alpha,$$
$$u(b) = \beta.$$

Our first step is to select a set of basis functions $\phi_j(x)$, $j = 0, 1, \ldots, N$, from the various piecewise polynomial bases presented in Sections 2.4 and 2.5. These basis functions will define an interpolation scheme for the approximate solution over a grid $(a =) x_0 < \cdots < x_{N+1} (= b)$. To minimize the algebraic complexity of the problem, we use piecewise Lagrange polynomials $\ell_i(x)$ of first degree as the basis functions $\phi_i(x)$. We can write these interpolating polynomials as

$$(2.11\text{-}2a) \qquad \hat{\ell}_0(\xi) = \frac{\xi - \xi_1}{\xi_0 - \xi_1} = \tfrac{1}{2}(1 - \xi),$$

$$(2.11\text{-}2b) \qquad \hat{\ell}_1(\xi) = \frac{\xi - \xi_0}{\xi_1 - \xi_0} = \tfrac{1}{2}(1 + \xi),$$

in local element coordinates $-1 \leq \xi \leq 1$. See Figure 2-11 for an explanation of the relationship between the functions ϕ_j and $\hat{\ell}_j$.

Using this basis, we must determine the N nodal values of the approximate solution $\hat{u}$ at the interior nodes $x_1, \ldots, x_N$. Notice that the Dirichlet boundary conditions, being essential, determine the value of $\hat{u}(x)$ at the end nodes x_0 and x_{N+1}. In fact, we can impose these conditions by taking

(a)

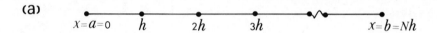

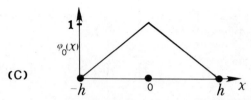

(b)

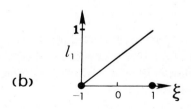

(C)

Figure 2-11. (a) Discretization for example problem of Section 2.11. (b) Lagrange polynomial of degree one on interval $(-1 \leq \xi \leq 1)$. (c) Basis function defined using Lagrange polynomials of degree one defined over interval $(-h \leq x \leq h)$.

$\phi_0(x) = x\ell_0(x) + \beta\ell_{N+1}(x)$, thus linearly interpolating the boundary data between the end nodes x_0 and x_{N+1} and their adjacent interior nodes x_1 and x_N.

Our second step is to employ the Galerkin method to obtain the integral equations

$$(2.11\text{-}3) \quad \left\langle \frac{d^2\hat{u}}{dx^2} - p(x)\frac{d\hat{u}}{dx} - q(x)\hat{u}, \phi_j \right\rangle = \langle f(x), \phi_j \rangle, \qquad j = 1, 2, \ldots, N.$$

Because $\hat{u}$ is a sum of piecewise linear Lagrange polynomials, the second derivatives appearing in Equations (2.11-3) vanish except at the interelement boundaries x_i, where they become infinite in some sense. To negotiate this apparent difficulty with a minimum of mathematical machinery, it is helpful to employ integration by parts to obtain

$$(2.11\text{-}4) \quad -\left\langle \frac{d\hat{u}}{dx}, \frac{d\phi_j}{dx} \right\rangle - \left\langle p(x)\frac{d\hat{u}}{dx} + q(x)\hat{u}, \phi_j \right\rangle + \frac{d\hat{u}}{dx}\phi_j \Big|_a^b$$

$$= \langle f(x), \phi_j \rangle, \qquad j = 1, 2, \ldots, N.$$

Substitution of the trial function $\hat{u}(x) = \phi_0(x) + \sum_{i=1}^{N} u_i \phi_j(x)$ into the integral equations (2.11-4) yields

$$
(2.11\text{-}5) \quad \sum_{i=1}^{N} u_i \left[\left\langle \frac{d\phi_i}{dx}, \frac{d\phi_j}{dx} \right\rangle + \left\langle p(x) \frac{d\phi_i}{dx} + q(x)\phi_i, \phi_j \right\rangle \right]
$$

$$
= -\alpha \left\langle \frac{d\ell_0}{dx}, \frac{d\phi_j}{dx} \right\rangle - \beta \left\langle \frac{d\ell_{N+1}}{dx}, \frac{d\phi_j}{dx} \right\rangle
$$

$$
- \alpha \left\langle p(x) \frac{d\ell_0}{dx} + q(x)\ell_0, \phi_j \right\rangle
$$

$$
- \beta \left\langle p(x) \frac{d\ell_{N+1}}{dx} + q(x)\ell_{N+1}, \phi_j \right\rangle
$$

$$
+ \frac{d\hat{u}}{dx} \phi_j \Big|_a^b - \langle f(x), \phi_j \rangle, \qquad j = 1, 2, \ldots, N.
$$

Observe that all quantities on the right side of Equation (2.11-5) can be computed from known boundary and forcing data. This expression therefore provides N equations in the N unknown values u_i at interior nodes.

The third step is to calculate the integrals appearing in Equation (2.11-5), taking advantage of the local (ξ) coordinate system. Observing that $d\xi/dx = 2/h$, we find

$$
\int_a^b \frac{d\phi_i}{dx} \frac{d\phi_j}{dx} \, dx
$$

$$
:= \begin{cases}
\displaystyle\int_{-1}^{1} \frac{d\ell_0}{d\xi} \frac{d\xi}{dx} \frac{d\ell_1}{d\xi} \frac{d\xi}{dx} \frac{dx}{d\xi} \, d\xi = -\frac{1}{h}, & i = j - 1 \\[2mm]
\displaystyle\int_{-1}^{1} \frac{d\ell_1}{d\xi} \frac{d\xi}{dx} \frac{d\ell_0}{d\xi} \frac{d\xi}{dx} \frac{dx}{d\xi} \, d\xi = -\frac{1}{h}, & i = j + 1 \\[2mm]
\displaystyle\int_{-1}^{1} \frac{d\ell_1}{d\xi} \frac{d\xi}{dx} \frac{d\ell_1}{d\xi} \frac{d\xi}{dx} \frac{dx}{d\xi} \, d\xi = \frac{2}{h}, & i = j \\[2mm]
0, & |i - j| > 1.
\end{cases}
$$

This procedure is now used to evaluate all of the integrals appearing in Table 2-4.

Assuming p, q, and f are constant, the integrals appearing in Table 2-4 can be used in Equation (2.11-5) to yield, for a typical internal node j,

$$
(2.11\text{-}6) \quad u_{j-1}\left(-\frac{1}{h} - \frac{p}{2} + \frac{qh}{6}\right) + u_j\left[\frac{1}{h} + \frac{1}{h} + p\left(\frac{1}{2} - \frac{1}{2}\right) + \frac{2qh}{3}\right]
$$

$$
+ u_{j+1}\left(-\frac{1}{h} + \frac{p}{2} + \frac{qh}{6}\right) = -f_j\left(\frac{h}{2} + \frac{h}{2}\right).
$$

TABLE 2-4. Piecewise linear finite-element integrals.*

INTEGRAL	$i = j - 1$	$i = j$	$i = j + 1$
$\displaystyle\int_{-h}^{0} \frac{d\phi_i}{dx}\frac{d\phi_j}{dx}\,dx$	$-\dfrac{1}{h}$	$\dfrac{1}{h}$	0
$\displaystyle\int_{0}^{h} \frac{d\phi_i}{dx}\frac{d\phi_j}{dx}\,dx$	0	$\dfrac{1}{h}$	$-\dfrac{1}{h}$
$\displaystyle\int_{-h}^{0} \frac{d\phi_i}{dx}\phi_j\,dx$	$-\dfrac{1}{2}$	$\dfrac{1}{2}$	0
$\displaystyle\int_{0}^{h} \frac{d\phi_i}{dx}\phi_j\,dx$	0	$-\dfrac{1}{2}$	$\dfrac{1}{2}$
$\displaystyle\int_{-h}^{0} \phi_i\,\phi_j\,dx$	$\dfrac{h}{6}$	$\dfrac{h}{3}$	0
$\displaystyle\int_{0}^{h} \phi_i\,\phi_j\,dx$	0	$\dfrac{h}{3}$	$\dfrac{h}{6}$
$\displaystyle\int_{-h}^{0} \phi_j\,dx$	$-$	$\dfrac{h}{2}$	$-$
$\displaystyle\int_{0}^{h} \phi_j\,dx$	$-$	$\dfrac{h}{2}$	$-$

*Observe that, when $|i - j| > 1$, there is no region in which the piecewise linear basis functions ϕ_i and ϕ_j are both nonzero, and hence in these cases the Galerkin integrals vanish. Also note that node j corresponds to $x = 0$ for the integrals in this table

Equation (2.11-6) can be divided by h and rearranged to give

(2.11-7) $$\frac{1}{h^2}(u_{j-1} - 2u_j + u_{j+1}) - \frac{p}{2h}(u_{j+1} - u_{j-1})$$
$$- q(\tfrac{1}{6}u_{j-1} + \tfrac{2}{3}u_j + \tfrac{1}{6}u_{j+1}) = f_j.$$

Recognizing that $u_j = \hat{u}(x_j)$, we see that the finite-element equation (2.11-7) is identical to the finite-difference approximation (2.7-3) to the same equation, with the exception of the third term. There, the finite-difference representation $u(x_i)$ is replaced by $\frac{1}{6}(u_{j-1} + 4u_j + u_{j+1})$, which turns out to be a Simpson's rule approximation to the integrated average of $\hat{u}$ over the interval $[x_{j-1}, x_{j+1}]$.

The fourth step is to assemble equations of the form (2.11-7) into matrix form. We get

(2.11-8)
$$
\begin{bmatrix}
a_1 & b_1 & & & & \\
c_2 & a_2 & b_2 & & & \\
& \ddots & \ddots & \ddots & & \\
& & c_{N-1} & a_{N-1} & b_{N-1} \\
& & & c_N & a_N
\end{bmatrix}
\begin{bmatrix}
u_1 \\
u_2 \\
\vdots \\
\\
u_N
\end{bmatrix}
=
\begin{bmatrix}
-c_1\alpha + f_1 \\
f_2 \\
\vdots \\
\\
-b_N\beta + f_N
\end{bmatrix},
$$

where

$$a_j = \frac{2}{h} + \frac{2qh}{3},$$

$$b_j = -\frac{1}{h} + \frac{p}{2} + \frac{qh}{6},$$

$$c_j = -\frac{1}{h} - \frac{p}{2} + \frac{qh}{6},$$

$$f_j = -fh.$$

Notice that this formulation automatically accommodates the Dirichlet boundary conditions. While Equation (2.11-8) is similar in form to its finite-difference counterpart, Equation (2.7-6), it changes slightly when Neumann boundary conditions are imposed. When this situation arises, we evaluate the boundary contribution $\left[(d\hat{u}/dx)\phi_j\right]_a^b$ in Equation (2.11-5) using the known boundary values of $d\hat{u}/dx$ and the fact that $\ell_{N+1}\big|_a = 1$, $\ell_0\big|_b = 1$ and all other basis functions vanish at the boundary. One should observe that, if Neumann boundary conditions only were imposed, the values of $\hat{u}(a)$ and $\hat{u}(b)$ would remain unknown. Thus the Galerkin formulation would yield a set of $N + 2$ equations for nodal values $u_0, \ldots, u_{N+1}$. We note in passing that the matrix equation would also be singular, because a Dirichlet condition at at least one point on the boundary is required for uniqueness.

Our fifth step is to solve the resulting matrix equation for the coefficients u_i and to use the interpolation scheme $\hat{u}(x) = \phi_0(x) + \sum_{i=1}^N u_i\,\phi_i(x)$ defining the trial function to obtain an approximation to $u(x)$ in the interval $[a, b]$.

2.12. The Galerkin Method in Two Space Dimensions.

In the preceding section we introduced all of the methodological concepts of the Galerkin finite-element method using a one-dimensional example. The

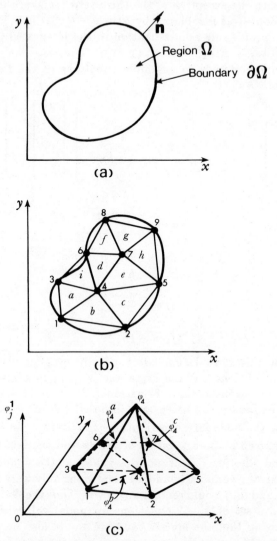

Figure 2-12. (a) Definition sketch for region Ω with boundary $\partial\Omega$ and outward directed normal **n**. (b) Region Ω discretized into nine elements denoted as a through h and employing nine nodes. (c) Representation of basis function ϕ_4 of (b) such that its segmentation over the six elements is clearly indicated.

extension of this presentation to two-dimensional elements is not difficult. More challenging, however, are the algorithmic aspects of implementing the procedure. We begin with the formulation of the approximating equations using the Galerkin method. As an example equation, let us assume an operator equation of the form $\mathcal{L}(u) = f$, where

$$(2.12\text{-}1) \qquad \mathcal{L} = \left[\frac{\partial^2}{\partial x^2} + \frac{\partial^2}{\partial y^2} + a(x,y)\frac{\partial}{\partial x} + b(x,y)\frac{\partial}{\partial y} + c(x,y)\right]$$

on a region Ω bounded by $\partial\Omega$ (see Figure 2-12).

For simplicity let us assume that $u = 0$ on the boundary $\partial\Omega$.

Let $\hat{u}(x,y)$ approximate $u(x,y)$ on $\Omega \cup \partial\Omega$ and have the form

$$(2.12\text{-}2) \qquad \hat{u}(x,y) = \sum_{i=1}^{N} u_i \, \phi_i(x,y).$$

Equation (2.12-2) is similar to that encountered earlier, only now the basis function $\phi_i(x,y)$ depends on both x and y. Observe that the terms accommodating the essential boundary conditions in Equation (2.12-2) vanish in our special case by the assumption that $u = 0$ on $\partial\Omega$.

Galerkin equations for triangular elements.

Consider now the approximation of $\mathcal{L}(u) = f$ using the Galerkin method. Integrating the weighted residual yields

$$(2.12\text{-}3) \qquad \int_{\Omega}\left\{\left(\frac{\partial^2\hat{u}}{\partial x^2} + \frac{\partial^2\hat{u}}{\partial y^2}\right)\phi_j(x,y)\right.$$

$$\left. + \left[a(x,y)\frac{\partial\hat{u}}{\partial x} + b(x,y)\frac{\partial\hat{u}}{\partial y} + c(x,y)\,\hat{u}\right]\phi_j(x,y)\right\}dx\,dy$$

$$= \int_{\Omega} f(x,y)\,\phi_j\,dx\,dy, \qquad j = 1,2,\ldots,N.$$

Application of Green's theorem to the first term in Equation (2.12-3) gives

$$(2.12\text{-}4) \qquad \int_{\Omega}\left\{-\frac{\partial\hat{u}}{\partial x}\frac{\partial\phi_j}{\partial x} - \frac{\partial\hat{u}}{\partial y}\frac{\partial\phi_j}{\partial y}\right.$$

$$\left. + \left[a(x,y)\frac{\partial\hat{u}}{\partial x} + b(x,y)\frac{\partial\hat{u}}{\partial y} + c(x,y)\,\hat{u}\right]\phi_j\right\}dx\,dy$$

$$= -\oint_{\partial\Omega}\frac{\partial\hat{u}}{\partial n}\phi_j\,dx\,dy + \int_{\Omega} f(x,y)\,\phi_j\,dx\,dy,$$

$$j = 1,2,\ldots,N,$$

where $\partial/\partial n$ is the outward directed normal derivative along $\partial\Omega$ (see Figure 2-12).

One is now faced with a choice of basis functions from among the alternatives presented in Section 2.5. In the present example we select the popular linear triangular element. Figure 2-12b shows the region Ω discretized into nine elements Ω_e, such that

(2.12-5)
$$\Omega = \bigcup_{e=1}^{9} \Omega_e.$$

Figure 2-12c illustrates the subdivision of basis function ϕ_4 into six components, one residing above each triangular element having node 4 as a vertex, so that one can write

$$\phi_4(x, y) = \sum_{e=1}^{6} \phi_4^e(x, y).$$

In writing this decomposition, we agree that each function $\phi_4^e = \phi_4$ on the triangle numbered e, while $\phi_4^e \equiv 0$ elsewhere. We know from Section 2.5 that $\phi_j^e(x, y)$ is normally expressed in terms of the area basis functions L_k, $k = 1, 2, 3$ (see Figure 2-8). To facilitate the use of the area basis function we decompose the Galerkin equations (2.12-4) elementwise, getting

(2.12-6)
$$\sum_{e=1}^{9} \left(\sum_{i=1}^{9} u_i \int_{\Omega_e} \left\{ \frac{\partial \phi_i^e}{\partial x} \frac{\partial \phi_j^e}{\partial x} + \frac{\partial \phi_i^e}{\partial y} \frac{\partial \phi_j^e}{\partial y} \right. \right.$$
$$\left. + \left[a^e(x, y) \frac{\partial \phi_i^e}{\partial x} + b^e(x, y) \frac{\partial \phi_i^e}{\partial y} + c^e(x, y) \phi_i^e \right] \phi_j^e \right\} dx \, dy$$
$$\left. - \oint_{\partial \Omega_e} \frac{\partial \hat{u}^e}{\partial n} \phi_j^e \, dx \, dy + \int_{\Omega_e} f^e(x, y) \phi_j^e \, dx \, dy \right) = 0,$$
$$j = 1, 2, \ldots, N.$$

The subscript e on a, b, c, $\partial \hat{u}/\partial n$, and f designates the restriction of these functions to the element Ω_e or its boundary $\partial \Omega_e$ as appropriate.

Calculation of integrals.

Let us now evaluate the integrals appearing in Equation (2.12-6) using area basis functions. Consider the first term I_1 of the integrand appearing in the first integral in Equation (2.12-6). Assuming element e has a local coordinate system such that global index i corresponds to local index 1 and global index j corresponds to local index 2 (see Figure 2-13),

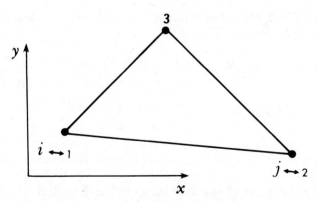

Figure 2-13. Typical triangular element with local numbering system. In this instance the global indices i and j correspond to 1 and 2, respectively.

$$(2.12\text{-}7) \quad I_1 \equiv \int_{\Omega_e} \left(\frac{\partial \phi_i^e}{\partial x} \frac{\partial \phi_j^e}{\partial x} + \frac{\partial \phi_i^e}{\partial y} \frac{\partial \phi_j^e}{\partial y} \right) dx\, dy$$

$$= \int_{\Omega_e} \left[\left(\frac{\partial L_1^e}{\partial L_1^e} \frac{\partial L_1^e}{\partial x} + \frac{\partial L_1^e}{\partial L_2^e} \frac{\partial L_2^e}{\partial x} \right) \left(\frac{\partial L_2^e}{\partial L_1^e} \frac{\partial L_1^e}{\partial x} + \frac{\partial L_2^e}{\partial L_2^e} \frac{\partial L_2^e}{\partial x} \right) \right.$$

$$\left. + \left(\frac{\partial L_1^e}{\partial L_1^e} \frac{\partial L_1^e}{\partial y} + \frac{\partial L_1^e}{\partial L_2^e} \frac{\partial L_2^e}{\partial y} \right) \left(\frac{\partial L_2^e}{\partial L_1^e} \frac{\partial L_1^e}{\partial y} + \frac{\partial L_2^e}{\partial L_2^e} \frac{\partial L_2^e}{\partial y} \right) \right] dx\, dy.$$

Notice that L_3^e does not appear in Equation (2.12-7). This is because, as first noted in Section 2.5, L_3^e depends on L_1^e and L_2^e according to the relationship $L_1^e + L_2^3 + L_3^e = 1$, so

$$\frac{\partial L_3}{\partial x} = -\frac{\partial L_1}{\partial x} - \frac{\partial L_2}{\partial x}.$$

The integral in Equation (2.12-7) reduces to

$$(2.12\text{-}8) \qquad I_1 = \int_{\Omega_e} \left(\frac{\partial L_1^e}{\partial x} \frac{\partial L_2^e}{\partial x} + \frac{\partial L_1^e}{\partial y} \frac{\partial L_2^e}{\partial y} \right) dx\, dy.$$

We can now express I_1 using the definition of L_i offered in Equation (2.5-8), that is,

$$(2.12\text{-}9) \qquad L_I(x, y) = \frac{[x(y_J - y_K) + y(x_K - x_J) + (x_J y_K - x_K y_J)]}{2A_e},$$

where (I, J, K) is an even permutation of the local coordinates $(1, 2, 3)$. If for example $I = 2$, then $J = 3$ and $K = 1$, and we can combine Equations (2.12-8) and (2.12-9) to yield

$$(2.12\text{-}10) \quad I_1 = \left(\frac{y_2 - y_3}{2A_e} \cdot \frac{y_3 - y_1}{2A_e} + \frac{x_3 - x_2}{2A_e} \cdot \frac{x_1 - x_3}{2A_e} \right) \int_{\Omega_e} dx\, dy$$

$$= \frac{1}{4A_e} [(y_2 - y_3)(y_3 - y_1) + (x_3 - x_2)(x_1 - x_3)].$$

One can readily evaluate the expression (2.12-10) when the vertices of the triangle are known. The area A_e of the triangle Ω_e is easily obtained from Equation (2.5-12).

The second term of the first integrand in Equation (2.12-6) will be evaluated in two parts. Consider first

$$(2.12\text{-}11) \quad I_2 \equiv \int_{\Omega_e} \left(a^e \frac{\partial \phi_i^e}{\partial x} + b^e \frac{\partial \phi_i^e}{\partial y} \right) \phi_j^e\, dx\, dy$$

$$= a^e \int_{\Omega_e} \left(\frac{\partial L_1^e}{\partial L_1^e} \frac{\partial L_1^e}{\partial x} + \frac{\partial L_1^e}{\partial L_2^e} \frac{\partial L_2^e}{\partial x} \right) L_2^e\, dx\, dy$$

$$+ b^e \int_{\Omega_e} \left(\frac{\partial L_1^e}{\partial L_1^e} \frac{\partial L_1^e}{\partial y} + \frac{\partial L_1^e}{\partial L_2^e} \frac{\partial L_2^e}{\partial y} \right) L_2^e\, dx\, dy$$

$$= a^e \int_{\Omega_e} \frac{\partial L_1^e}{\partial x} L_2^e\, dx\, dy + b^e \int_{\Omega_e} \frac{\partial L_1^e}{\partial y} L_2^e\, dx\, dy$$

$$= \left(a^e \frac{y_2 - y_3}{2A^e} + b^e \frac{x_3 - x_2}{2A^e} \right) \int_{\Omega_e} L_2^e\, dx\, dy.$$

We can evaluate this last integral using the identity (2.5-19), getting

$$\int_{\Omega_e} L_2^e\, dx\, dy = 2A_e \frac{1}{3!} = \frac{A_e}{3}.$$

Thus the integral I_2 becomes

$$I_2 = \left(a^e \frac{y_2 - y_3}{6} + b^e \frac{x_3 - x_2}{6} \right) A_e.$$

Now consider the integral

$$(2.12\text{-}12) \quad I_3 \equiv \int_{\Omega_e} c^e(x, y)\, \phi_i^e\, \phi_j^e\, dx\, dy = c^e \int_{\Omega_e} L_1^e L_2^e\, dx\, dy.$$

In light of Equation (2.5-19), we can evaluate this integral as follows:

$$I_3 = c^3 \frac{2A_e}{4!} = c^e \frac{A_e}{12}.$$

The next term in Equation (2.12-6) that we need to consider is the boundary integral,

$$I_B = \oint_{\partial\Omega_e} \frac{\partial \hat{u}^e}{\partial n} \phi_j^e \, dx \, dy.$$

In our particular case we have specified values of u on $\partial\Omega$, so only the basis functions ϕ_j^e that are nonzero along boundary segments of $\partial\Omega_e$ *interior* to Ω appear as weighting functions in the integral I_B. However, each interior boundary segment contributing to $\partial\Omega_e$ is a boundary segment to an adjacent element Ω_f, and the contribution of this line segment to integral over $\partial\Omega_e$ is the negative of its contribution to the integral over $\partial\Omega_f$, provided one consistently assigns the correct counterclockwise orientations to all element boundaries. Thus we need not explicitly compute the integral I_B, since we know in advance that it will be canceled by contributions to the boundary integrals for adjacent elements.

Finally, we need to evaluate

$$(2.12\text{-}13) \qquad I_4 = \int_{\Omega_e} f^e(x,y) \, \phi_j^e \, dx \, dy = f^e \int_{\Omega_e} L_2 \, dx \, dy,$$

which, once again in light of Equation (2.5-19), gives

$$I_4 = f^e \frac{2 A_e}{3!} = f^e \frac{A_e}{3}.$$

Computing the sum of integrals $I_1 + I_2 + I_3$ gives, for the term involving node i in the equation associated with weighting function j, the following coefficient of the unknown nodal value u_i:

$$(2.12\text{-}14) \quad E(j,i) = \frac{1}{4 A_e} \left[(y_2 - y_3)(y_3 - y_1) + (x_3 - x_2)(x_1 - x_3) \right]$$

$$- A_3 \left[a^e \frac{y_2 - y_3}{6} + b^3 \frac{x_3 - x_2}{6} \right] - c^e \frac{A_e}{12}.$$

Since the integral I_4 does not multiply an unknown, it will contribute to the right side of the matrix equation for the sought values u_i.

Assembly of the global coefficient matrix.

It may seem appropriate at this point to complete the integration for term i of equation j by calculating the contribution of each element containing the node i. This course of action would be equivalent to summing over index e in Equation (2.12-6). A perhaps less obvious alternative, and the one used in most finite-element computer codes, keeps the index e of the triangular element constant and permutes the indices i and j. This procedure gives rise to nine coefficients per triangular element. The resulting 3×3 array of values is known as the **element coefficient matrix**.

In the case of the example illustrated in Figure 2-12a there would arise nine element coefficient matrices. Because the global coordinates of the three element vertices are needed to compute the integrals (2.12-10) through (2.12-13), we must keep track of the relationship between the local and global coordinates of each element. A typical **element incidence list** for the example problem of Figure 2-12b appears in Table 2-5. The entries of a typical element coefficient matrix, such as that obtained for element a, are

$$(2.12\text{-}15) \qquad \mathbf{E}^a = \begin{bmatrix} E(1,1) & E(1,2) & E(1,3) \\ E(2,1) & E(2,2) & E(2,3) \\ E(3,1) & E(3,2) & E(3,3) \end{bmatrix},$$

where

$$(2.12\text{-}16) \qquad E(j,i) = \int_{\Omega_e} \left[\frac{\partial \phi_i^e}{\partial x} \frac{\partial \phi_j^e}{\partial x} + \frac{\partial \phi_i^e}{\partial y} \frac{\partial \phi_j^e}{\partial y} \right.$$
$$\left. - \left(a^e \frac{\partial \phi_i^e}{\partial x} + b^e \frac{\partial \phi_i^e}{\partial y} + c^e \phi_i^e \right) \phi_j^e \right] dx\, dy.$$

Once we have computed the element coefficient matrices for all nine elements, they are overlayed according to nodal incidences to give the **global**

Table 2-5. Element incidence list
for the example problem of Figure 2-12b.

ELEMENT	LOCAL COORDINATES		
	1	2	3
	GLOBAL COORDINATES		
a	1	4	3
b	1	2	4
c	4	2	5
d	6	4	7
e	4	5	7
f	6	7	8
g	8	7	9
h	7	5	9
i	3	4	6

TABLE 2-6. Assembly of the global matrix **G** for the triangular finite-element example.

$$
\mathbf{G}=\begin{bmatrix}
(E_{1,1}^a+E_{1,1}^b) & (E_{1,2}^b) & (E_{1,3}^a) & (E_{1,2}^a+E_{1,3}^b) & (0) & (0) & (0) & (0) & (0)\\[4pt]
(E_{2,1}^b) & (E_{2,2}^b+E_{2,2}^c) & (0) & (E_{2,3}^b+E_{2,1}^c) & (E_{2,3}^c) & (0) & (0) & (0) & (0)\\[4pt]
(E_{3,1}^a) & (0) & (E_{3,3}^a) & (E_{3,2}^a+E_{1,2}^i) & (0) & (E_{1,3}^i) & (0) & (0) & (0)\\[4pt]
(E_{2,1}^a+E_{3,1}^b) & (E_{3,2}^b+E_{1,2}^c) & (E_{2,3}^a+E_{2,1}^i) & \begin{pmatrix}E_{2,2}^a+E_{3,3}^b\\+E_{1,1}^c+E_{2,2}^d\\+E_{1,1}^e+E_{2,2}^i\end{pmatrix} & (E_{1,3}^c+E_{1,2}^e) & (E_{2,1}^d+E_{2,3}^e) & (E_{2,3}^d+E_{1,3}^e) & (0) & (0)\\[4pt]
(0) & (E_{3,2}^c) & (0) & (E_{3,1}^c+E_{2,1}^e) & \begin{pmatrix}E_{3,3}^c+E_{2,2}^e\\+E_{2,2}^h\end{pmatrix} & (0) & (E_{2,3}^e+E_{2,1}^h) & (0) & (E_{2,3}^h)\\[4pt]
(0) & (0) & (E_{3,1}^i) & (E_{1,2}^d+E_{3,2}^i) & (0) & \begin{pmatrix}E_{1,1}^d+E_{1,1}^f\\+E_{3,3}^i\end{pmatrix} & (E_{1,3}^d+E_{1,2}^f) & (E_{1,3}^f) & (0)\\[4pt]
(0) & (0) & (0) & (E_{3,2}^d+E_{3,1}^e) & (E_{3,2}^e+E_{1,2}^h) & (E_{3,1}^d+E_{2,1}^f) & \begin{pmatrix}E_{3,3}^d+E_{3,3}^e\\+E_{2,2}^f+E_{2,2}^g\\+E_{1,1}^h\end{pmatrix} & (E_{2,3}^f+E_{2,1}^g) & (E_{2,3}^g+E_{1,3}^h)\\[4pt]
(0) & (0) & (0) & (0) & (0) & (E_{3,1}^f) & (E_{3,2}^f+E_{1,2}^g) & (E_{3,3}^f+E_{1,1}^g) & (E_{1,3}^g)\\[4pt]
(0) & (0) & (0) & (0) & (E_{3,2}^h) & (0) & (E_{3,2}^g+E_{3,1}^h) & (E_{3,1}^g) & (E_{3,3}^g+E_{3,3}^h)
\end{bmatrix}
$$

coefficient matrix that will multiply the vector of unknown nodal values u_i. In our example, the global coefficient matrix is a 9×9 array. If we designate the elements of the element coefficient matrix $\mathbf{E}^a$ as $E_{k,\ell}^a$, then the element matrix information would appear in the global matrix **G** shown in Table 2-6. In programming the finite-element algorithm it is not necessary to store more than one element coefficient matrix at a time, inasmuch as the element-level information can be transferred directly to the global coefficient matrix as soon as each element-level integration has been completed.

Other basis functions and numerical quadrature.

While the assembly procedure just outlined for triangular elements is also employed with rectangular and isoparametric elements, elementwise integration over rectangular elements can often be achieved analytically. It is even possible to perform analytical integration over quadrilateral elements, provided the elements have straight sides and linear Lagrange basis functions are employed (Babu and Pinder,1984).

When elements with variable coefficients or curved sides are encountered the integrands can be quite complicated, and a numerical integration approach is required. The two integration procedures in common use are **Gauss-Legendre quadrature** and **Lobatto quadrature**. Both are approximate integration techniques which, for an integrand $f(x, y)$, $(x, y) \in [-1, 1]$, can be expressed as

$$(2.12\text{-}17) \qquad \int_{-1}^{1} \int_{-1}^{1} f(x, y)\, dx\, dy \simeq \sum_{i=1}^{n} \sum_{j=1}^{m} A_i\, A_j\, f(x_i, y_j),$$

where n and m are the number of sampling points in the x and y directions, respectively. The restriction $-1 \le x, y \le 1$ poses no real limitation if we transform the Galerkin integrals to local element coordinates before integrating. The coefficients A_k are weights and the points (x_i, y_j) are sampling points. Typical values for A_k and (x_i, y_j) are presented in Table 2-7; more extensive tabulations are available in standard references (see, for example, Stroud and Secrest, 1966). Note that in using Lobatto integration some of the sampling points correspond to nodal locations; this can result in computational efficiencies for many types of basis functions. However, a polynomial of degree $2n - 1$ can be integrated exactly using n integration points using Gauss-Legendre formulas, whereas Lobatto formulas require one more integration point to achieve the same accuracy. Numerical integration formulae for triangles are less straightforward, and the interested reader is referred to Lapidus and Pinder (1982).

Table 2-7. **Weighting coefficients and sampling points for numerical integration.**

GAUSS-LEGENDRE INTEGRATION		
n	x_i	A_i
2	± 0.577350	1.000000
3	± 0.774597 0.000000	0.555556 0.888889
4	± 0.861136 ± 0.339981	0.347855 0.652145
5	± 0.906180 ± 0.538469 0.000000	0.236927 0.478629 0.568889

LOBATTO INTEGRATION		
n	x_i	A_i
3	± 1.000000 0.000000	0.333333 1.333333
4	± 1.000000 ± 0.447214	0.166667 0.833333
5	± 1.000000 ± 0.654654 0.000000	0.100000 0.544444 0.711111

Boundary conditions.

Up to this point we have concentrated on the interior nodes of the two-dimensional domain at the expense of the boundary information. We have discussed the reduction of the matrix equation to accommodate Dirichlet boundary conditions in one dimension, and the extension to several dimensions is straightforward. Indeed, we simply impose the Dirichlet (or essential) boundary conditions on the trial function $\hat{u}$ in advance of forming the Galerkin integral equations, and as a result the integrals involving these boundary data become known information contributing to the right side of the final matrix equation.

However, we still have not considered Neumann boundaries in two dimensions. When Neumann boundary conditions occur, we use them to compute the first term on the right side of Equation (2.12-4), that is, the boundary integral

$$I_B = - \oint_{\partial \Omega} \frac{\partial \hat{u}}{\partial n} \, \phi_j \, dx \, dy.$$

Thus, given

$$\frac{\partial u}{\partial n} = B(x, y) \quad \text{on} \quad \partial \Omega,$$

we can write I_B as

$$I_B = - \sum_{e=1}^{E} \oint_{\partial \Omega_e} \frac{\partial \hat{u}}{\partial n} \, \phi_j^e \, dx \, dy = - \sum_{e=1}^{E} \oint_{\partial \Omega_e} B(x, y) \, \phi_j^e(x, y) \, dx \, dy,$$

since any contribution from a boundary segment interior to Ω will cancel a contribution from an adjacent element sharing that boundary segment. Because the weighting functions ϕ_j^e are generally simple polynomials along the boundaries, the integration along $\partial \Omega$ proper can often be performed analytically, particularly when $B(x, y)$ is assumed piecewise constant or piecewise linear. The resulting integral I_B is thus computable for each of the Galerkin equations (2.12-4), so it contributes to the right side of the matrix equation for the unknowns u_i. Notice that we can accommodate Neumann boundary conditions in this fashion without explicitly incorporating them into the definition of the trial function $\hat{u}$. This observation accords with our discussion in Section 2.10, where we identified Neumann conditions as natural boundary conditions for this type of problem.

2.13. Error Bounds on the Galerkin Finite-Element Method.

The objective in this section is to estimate the error of the finite-element approximation to an equation of the form $\mathcal{L}(u) = f$. We shall use the concepts introduced in Section 2.4, wherein we established the interpolation errors associated with the various polynomial approximations. This information will be combined with fundamental concepts of normed function spaces to arrive at a global discretization error similar to that obtained for the finite-difference approximation in Section 2.7. For conceptual ease we examine error estimates for one-dimensional operators, although similar reasoning is applicable to operators in several independent variables.

Let us consider the differential equation $\mathcal{L}(u) = f$, where $\mathcal{L}$ is a self-adjoint differential operator defined by

$$(2.13\text{-}1) \qquad \mathcal{L}(u) = -\frac{d}{dx}\left[p(x)\frac{du}{dx}\right] + q(x)u,$$

on the open interval (a, b). We assume that the coefficient $p(x) > 0$ and that $q(x)$ is bounded below by a positive constant, in the sense $q(x) \geq K > 0$. Associated with this operator is an inner product, which we write as

$$(2.13\text{-}2) \qquad \langle g, h \rangle_{\mathcal{L}} \equiv \int_a^b \left[p(x)\frac{dg}{dx}\frac{dh}{dx} + q(x)gh\right] dx$$

and denote as the **energy inner product**. This inner product gives rise to an associated norm

$$(2.13\text{-}3) \qquad \|g\|_{\mathcal{L}} = \sqrt{\langle g, g \rangle_{\mathcal{L}}},$$

which we call the **energy norm**.

Following the classical Galerkin approach, we first consider a trial function $\hat{u}(x)$, approximating the unknown solution $u(x)$, defined by

$$(2.13\text{-}4) \qquad \hat{u}(x) = \phi_0(x) + \sum_{i=1}^N u_i\,\phi_i(x).$$

Here the function $\phi_0(x)$ satisfies the same essential boundary conditions as $u(x)$, and the remaining terms in $\hat{u}$ satisfy homogeneous boundary conditions. Our goal then is to determine the function $\hat{u}(x)$ that is in some sense the best approximation to $u(x)$ in the function space containing all possible linear combinations of the form (2.13-4), where the coefficients u_i range over the set of real numbers. We shall denote this space by $\phi_0 + \text{span}\{\phi_1, \phi_2, \ldots, \phi_N\}$.

Substitution of $\hat{u}(x)$ for $u(x)$ in the operator equation (2.13-1) and application of Galerkin's approximation yields

$$(2.13\text{-}5) \qquad \int_a^b [\mathcal{L}(\hat{u})\phi_i - f\,\phi_i]\,dx = 0, \qquad i = 1, 2, \ldots, N.$$

Because $f = \mathcal{L}(u)$, (2.13-5) can be written in the form

$$(2.13\text{-}6) \qquad \int_a^b \mathcal{L}(\hat{u} - u)\,\phi_i\,dx = 0, \qquad i = 1, 2, \ldots, N.$$

We now wish to demonstrate that $\|\hat{u} - u\|_{\mathcal{L}} \leq \|v - u\|_{\mathcal{L}}$, where v is any other function belonging to the space $\phi_0 + \text{span}\{\phi_1, \phi_2, \ldots, \phi_N\}$. This is the sense in which $\hat{u}$ will be the best approximation to u possible using the basis $\{\phi_1, \phi_2, \ldots, \phi_N\}$ in the energy norm.

From the definition of the energy norm we write

$$(2.13\text{-}7) \qquad \|v - u\|_{\mathcal{L}}^2 = \langle v - u, v - u \rangle_{\mathcal{L}}.$$

Adding and subtracting $\hat{u}$ to each term on the right in Equation (2.13-7), we obtain the equivalent but more helpful expression

$$(2.13\text{-}8a) \qquad \|v - u\|_{\mathcal{L}}^2 = \langle v - \hat{u} + \hat{u} - u, v - \hat{u} + \hat{u} - u \rangle_{\mathcal{L}}.$$

Expansion of the right side of Equation (2.13-8a) yields

$$(2.13\text{-}8b) \qquad \|v - u\|_{\mathcal{L}}^2 = \langle v - \hat{u}, v - \hat{u} \rangle_{\mathcal{L}} + 2\langle \hat{u} - u, v - \hat{u} \rangle_{\mathcal{L}}$$

$$+ \langle \hat{u} - u, \hat{u} - u \rangle_{\mathcal{L}}.$$

Because the functions v and $\hat{u}$ satisfy the same boundary conditions, $v - \hat{u}$ satisfies homogeneous boundary conditions and thus belongs to the space spanned by $\{\phi_1, \phi_2, \ldots, \phi_N\}$. Hence there are coefficients $A_1, \ldots, A_n$ such that

$$(2.13\text{-}9) \qquad v - \hat{u} = \sum_{i=1}^{N} A_i \, \phi_i,$$

so the second term on the right side of Equation (2.13-8b) becomes

$$(2.13\text{-}10) \qquad 2\langle \hat{u} - u, v - \hat{u} \rangle_{\mathcal{L}} = 2 \sum_{i=1}^{N} A_i \, \langle \hat{u} - u, \phi_i \rangle_{\mathcal{L}}.$$

Revisiting Equation (2.13-6), we find, after applying integration by parts,

$$(2.13\text{-}11) \qquad \int_a^b \mathcal{L}(\hat{u} - u) \, \phi_i \, dx = \langle \hat{u} - u, \phi_i \rangle_{\mathcal{L}} = 0, \qquad i = 1, 2, \ldots, N.$$

Therefore the right side of Equation (2.13-10) vanishes, and Equation (2.13-8b) can now be written as follows:

$$(2.13\text{-}12) \qquad \|v - u\|_{\mathcal{L}}^2 = \langle v - \hat{u}, v - \hat{u} \rangle_{\mathcal{L}} + \langle \hat{u} - u, \hat{u} - u \rangle_{\mathcal{L}}$$

$$= \|v - \hat{u}\|_{\mathcal{L}}^2 + \|\hat{u} - u\|_{\mathcal{L}}^2.$$

Because $\|v - \hat{u}\|_{\mathcal{L}}^2 \geq 0$, we conclude that

$$(2\text{-}13.13) \qquad \|v - u\|_{\mathcal{L}}^2 \geq \|\hat{u} - u\|_{\mathcal{L}}^2.$$

Thus we have demonstrated that, in the energy norm, $\hat{u}$ is the best possible approximation to u using the basis $\{\phi_1, \phi_2, \ldots, \phi_N\}$.

It is possible, under our restrictions that the functions p, q be positive with q bounded away from zero, to relate this result to convergence in the Euclidean norm $\|\cdot\|_2$. Expansion of $\|\hat{u} - u\|_{\mathcal{L}}^2$ using integration by parts gives

$$(2.13\text{-}14) \qquad \|\hat{u} - u\|_{\mathcal{L}}^2 = \int_\Omega \left[p(x) \frac{d(\hat{u} - u)}{dx} \frac{d(\hat{u} - u)}{dx} + q(x)(\hat{u} - u)^2 \right] dx.$$

But $p > 0$, and since $q \geq K > 0$ for some constant K,

$$(2.13\text{-}15) \qquad \|\hat{u} - u\|_{\mathcal{L}}^2 = \int_a^b \left[p(x) \frac{d(\hat{u} - u)}{dx} \frac{d(\hat{u} - u)}{dx} + q(x)(\hat{u} - u)^2 \right] dx$$

$$\geq \int_a^b q(x)(\hat{u} - u)^2 \, dx \geq K \int_a^b (\hat{u} - u)^2 \, dx$$

$$= K\|\hat{u} - u\|_2^2 \geq 0.$$

Thus convergence in the energy norm implies convergence in the Euclidean norm for this class of operators.

Returning to the estimate (2.13-13), we can establish the order of the discretization error using the results of Sections 2.4 and 2.5. Consider, for example, the use of the one-dimensional Lagrange polynomials of degree n as the basis functions $\{\phi_1, \phi_2, \ldots, \phi_N\}$. Let v be the function in the space $\phi_0 + \mathrm{span}\{\phi_1, \phi_2, \ldots, \phi_N\}$ whose nodal values correspond to the values of u at the nodes, that is, v is the linear interpolate of the exact solution u. By what we have just shown, the energy norm of the Galerkin error is less than the energy norm of the interpolation error:

$$\|u - \hat{u}\|_{\mathcal{L}} \leq \|u - v\|_{\mathcal{L}} = \|E_n(x)\|_{\mathcal{L}}$$

$$= \left\| \frac{w_n(x)}{(n+1)!} \frac{d^{n+1}u}{dx^{n+1}} \right\|_{\mathcal{L}}$$

$$= Ch^{n+1} \left\| \frac{d^{n+1}u}{dx^{n+1}} \right\|_{\mathcal{L}},$$

where $w_n(x)$ is the function defined in Equation (2.4-5b) and C is a constant. Thus $\|\hat{u} - u\|_{\mathcal{L}} = \mathcal{O}(h^{n+1})$. For the case of Lagrange polynomials of degree one, that is, the chapeau functions, $n = 1$ and $\|\hat{u} - u\|_{\mathcal{L}} = \mathcal{O}(h^2)$.

Note that our hypotheses on the coefficients $p(x)$ and $q(x)$ impose rather stringent restrictions on our operator $\mathcal{L}$. In fact, these conditions force $\mathcal{L}$ to be **self-adjoint** in the sense that $\langle \mathcal{L}v_1, v_2 \rangle = \langle v_1, \mathcal{L}v_2 \rangle$ whenever $v_1, v_2 \in \mathrm{span}\{\phi_1, \ldots, \phi_N\}$. Moreover $\mathcal{L}$ is **positive definite**, meaning that if $v \in \mathrm{span}\{\phi_1, \ldots, \phi_N\}$, then $\langle v, \mathcal{L}v \rangle \geq 0$ with equality holding only when $v = 0$ identically. Under selected constraints, it is possible to show that this order of approximation can be established for a wider class of operators. The interested reader is referred to Strang and Fix (1973). It is also worth repeating that the extension of the argument given here to several dimensions (PDEs) is straightforward.

2.14. The Method of Collocation.

The method of **collocation** is very attractive because it is theoretically simple and exhibits a small discretization error. In spite of these advantages it is not widely used in practice, possibly because of the additional smoothness requirements that are imposed on the basis functions. The method of collocation can be viewed as a method of weighted residuals in which the weighting functions $w_i(\mathbf{x})$ are taken to be Dirac delta distributions, that is, $w_i(\mathbf{x}) = \delta(\mathbf{x} - \mathbf{x}_i)$ for a set of **collocation points** $\mathbf{x}_i$. The Dirac delta distribution has the property that, for any integrable function f defined on an open domain Ω and for any point $\mathbf{x}_i \in \Omega$, $\int_\Omega f(\mathbf{x})\delta(\mathbf{x} - \mathbf{x}_i)d\mathbf{x} = f(\mathbf{x}_i)$.

Let us assume that we have a problem on a domain Ω with homogeneous boundary conditions. We can approximate the unknown solution $u(\mathbf{x})$ by the trial function

$$(2.14\text{-}1) \qquad \hat{u}(\mathbf{x}) = \sum_{i=1}^{N} \mathbf{u}_i^\top \boldsymbol{\phi}_i(\mathbf{x}).$$

We take the basis functions $\boldsymbol{\phi}_i(\mathbf{x})$ to be piecewise Hermite polynomials, so that $\hat{u} \in C^1(\Omega)$. While collocation is readily applicable to multidimensional problems, for the sake of clarity of presentation we consider a one-dimensional case. Under this assumption the vectors $\mathbf{u}_i$ and $\boldsymbol{\phi}_i$ have the forms

$$(2.14\text{-}2) \qquad \mathbf{u}_i = \left(u_i, u_i'\right)^\top, \qquad \boldsymbol{\phi}_i = \left(h_i^0(x), h_i^1(x)\right)^\top,$$

where $h_i^0(x)$ and $h_i^1(x)$ are the Hermite polynomials defined by Equation (2.4-13) and u_i and u_i' denote the nodal values of $\hat{u}(x)$ and $\hat{u}'(x)$, respectively. Our objective is to determine these nodal values.

The approximating equations are obtained by introducing the trial function (2.14-1) into the weighted-residual equations. Assuming $\mathcal{L}(u) - f = 0$ with

$$(2.14\text{-}3) \qquad \mathcal{L}(u) = -\frac{d}{dx}\left[p(x)\frac{du}{dx}\right] + q(x)u$$

on a domain $\Omega = (a, b)$, we obtain

$$(2.14\text{-}4) \quad \int_a^b [\mathcal{L}(\hat{u}) - f]\, w_j(x)\, dx = \int_a^b [\mathcal{L}(\hat{u}) - f]\, \delta(x - x_j)\, dx$$

$$= [\mathcal{L}(\hat{u}) - f]\Big|_{x_j} = 0, \quad j = 1, 2, \ldots, N,$$

where N is the total number of collocation points.

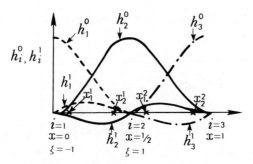

Figure 2-14. Discretization and Hermite basis functions used to solve $\mathcal{L}(u) - f = 0$ using collocation.

For example, let $(a, b) = (0, 1)$, $p(x) \equiv 1$, $q(x) \equiv 1 + x^2$ and $f = -1$ in Equation (2.14-3), and impose the boundary conditions $u(1) = u'(0) = 0$. Consider the three-node discretization using Hermite cubic basis functions, as shown in Figure 2-14. The resulting equation set has the form

$$(2.14\text{-}5) \quad [\mathcal{L}\hat{u} - f]_{x_j^e} = \left[\mathcal{L}\left(\sum_{i=1}^{3} \mathbf{u}_i^{\mathsf{T}} \boldsymbol{\phi}_i\right) - f\right]_{x_j^e}$$

$$= \left[\sum_{i=1}^{3} \mathbf{u}_i^{\mathsf{T}} \mathcal{L}\boldsymbol{\phi}_i - f\right]_{x_j^e} = 0, \qquad e = 1, 2, \ j = 1, 2.$$

In matrix form we can write Equation (2.14-5) as

$$\begin{bmatrix} [\mathcal{L}h_1^0]_{x_1^1} & [\mathcal{L}h_1^1]_{x_1^1} & [\mathcal{L}h_2^0]_{x_1^1} & [\mathcal{L}h_2^1]_{x_1^1} & 0 & 0 \\ [\mathcal{L}h_1^0]_{x_2^1} & [\mathcal{L}h_1^1]_{x_2^1} & [\mathcal{L}h_2^0]_{x_2^1} & [\mathcal{L}h_2^1]_{x_2^1} & 0 & 0 \\ 0 & 0 & [\mathcal{L}h_2^0]_{x_1^2} & [\mathcal{L}h_2^1]_{x_1^2} & [\mathcal{L}h_3^0]_{x_1^2} & [\mathcal{L}h_3^1]_{x_1^2} \\ 0 & 0 & [\mathcal{L}h_2^0]_{x_2^2} & [\mathcal{L}h_2^1]_{x_2^2} & [\mathcal{L}h_3^0]_{x_2^2} & [\mathcal{L}h_3^1]_{x_2^2} \end{bmatrix} \begin{bmatrix} u_1 \\ u_1' \\ u_2 \\ u_2' \\ u_3 \\ u_3' \end{bmatrix}$$

$$= \begin{bmatrix} f(x_1^1) \\ f(x_2^1) \\ f(x_1^2) \\ f(x_2^2) \end{bmatrix} .$$

Because the boundary conditions imply $\hat{u}'(0) = u_1' = 0$ and $\hat{u}(1) = u_3 = 0$, the second and fifth columns of the coefficient matrix in Equation (2.14-6) can be eliminated. We are left with a square matrix equation of the form

$$
\begin{bmatrix}
[\mathcal{L}h_1^0]_{x_1^1} & [\mathcal{L}h_2^0]_{x_1^1} & [\mathcal{L}h_2^1]_{x_1^1} & 0 \\
[\mathcal{L}h_1^0]_{x_2^1} & [\mathcal{L}h_2^0]_{x_2^1} & [\mathcal{L}h_2^1]_{x_2^1} & 0 \\
0 & [\mathcal{L}h_2^0]_{x_1^2} & [\mathcal{L}h_2^1]_{x_1^2} & [\mathcal{L}h_3^1]_{x_1^2} \\
0 & [\mathcal{L}h_2^0]_{x_2^2} & [\mathcal{L}h_2^1]_{x_2^2} & [\mathcal{L}h_3^1]_{x_2^2}
\end{bmatrix}
\begin{bmatrix}
u_1 \\ u_2 \\ u_2' \\ u_3'
\end{bmatrix}
=
\begin{bmatrix}
f(x_1^1) \\ f(x_2^1) \\ f(x_1^2) \\ f(x_2^2)
\end{bmatrix}.
$$

To complete the analysis we must evaluate the matrix entries by selecting appropriate collocation points. The zeros of the Legendre polynomials defined over each element furnish a theoretically attractive choice (see Douglas and Dupont, 1973), although other points may be more appropriate for some operators. (The n-th Legendre polynomial $p_n(x)$ on any interval $[a, b]$ is defined to be the polynomial of degree n that is orthogonal to all lower-degree polynomials in the inner product $\langle f, g \rangle = \int_a^b fg\, dx$ and that is normalized so that $\langle p_n, p_n \rangle = 1$.) The zeros for the Legendre polynomials on any interval are precisely the sample or "Gauss" points used in Gauss-Legendre quadrature on that interval, and they are tabulated in local coordinates in Table 2-7. For this problem the Gauss points are located at the global coordinates $x_1^1 = 0.1057$, $x_2^1 = 0.3943$, $x_1^2 = 0.6057$, $x_2^2 = 0.8943$. The resulting matrix equation is

$$
\begin{bmatrix}
-12.96 & 13.97 & -1.482 & 0 \\
13.99 & -12.83 & 5.388 & 0 \\
0 & -12.65 & -5.374 & -1.488 \\
0 & 14.06 & 1.496 & 5.346
\end{bmatrix}
\begin{bmatrix}
u_1 \\ u_2 \\ u_2' \\ u_3'
\end{bmatrix}
=
\begin{bmatrix}
-1 \\ -1 \\ -1 \\ -1
\end{bmatrix}
$$

with the solution

(2.14-6)
$$
\left(u_1, u_1', u_2, u_2', u_3, u_3' \right)
$$

$$
= \left(0.9311, 0, 0.6905, -0.9583, 0, -1.735 \right).
$$

Substitution of the values given in (2.14-6) into the interpolating trial function (2.14-1) provides a solution for any location $x \in [0, 1]$.

The extension of the collocation approach to multiple dimensions is not difficult conceptually if we use tensor-product bases. The number of unknowns per node using tensor-product Hermite cubic basis functions increases to four in two dimensions and eight in three dimensions. To minimize computational effort, alternating direction algorithms have been developed and found to be effective over a broad range of applications.

Details regarding collocation procedures in higher dimensions can be found in Lapidus and Pinder (1982), Celia (1983), and Allen and Murphy (1986).

2.15. Error Bounds on the Collocation Method.

To develop error bounds on the collocation method, we consider a one-dimensional problem of the form $\mathcal{L}(u) = f$ on an open interval (a, b) with homogeneous, Dirichlet boundary conditions. Let $\hat{u}(x)$ be a trial function approximating the true solution $u(x)$ for $x \in [a, b]$, and let

(2.15-1) $\qquad \Delta_x : (a =) \ x_0 < x_1 < x_2 < \cdots < x_n \ (= b)$

be a grid partitioning the closed interval $[a, b]$ into n elements. The approximate solution $\hat{u}(x)$ will satisfy the operator equation at the collocation points, so that

(2.15-2) $\qquad [\mathcal{L}(\hat{u}) - f]_{x_i^e} = 0, \qquad i = 1, 2, \ldots, m, \ e = 1, 2, \ldots, n,$

where m is the number of collocation points per element and depends on the degree of the basis functions selected for $\hat{u}(x)$. The solution $\hat{u}(x)$ obtained from the collocation equations (2.15-2) can be substituted into the operator to yield

(2.15-3) $\qquad \mathcal{L}(\hat{u}(x)) = \hat{f}(x),$

where $\hat{f}(x)$ generally differs from the true forcing function $f(x)$ in $[a, b]$. From Equation (2.15-2) we know, however, that $\hat{f}(x_i^e) = f(x_i^e)$, that is, the residual $R(x) \equiv \hat{f}(x) - f(x)$ vanishes at the collocation points.

Consider now the behavior of $R(x)$ over one element, say $[x_p, x_{p+1}]$, and assume there are two collocation points located on each element. This assumption is appropriate for the case when $\mathcal{L}$ is a second-order operator and the trial function $\hat{u}$ is a piecewise cubic Hermite polynomial. If the collocation points are zeros of the quadratic Legendre polynomial, the product $(x - x_1^e)(x - x_2^e)$ constitutes a polynomial of degree two that differs from the quadratic Legendre polynomial by at most a multiplicative constant on $[x_p, x_{p+1}]$. The residual $R(x)$ also has roots at $x = x_1^e$ and $x = x_2^e$ but may not be a polynomial of degree two on $[x_p, x_{p+1}]$. Thus there is an unknown function $r(x)$ defined through the relationship

(2.15-4) $\qquad R(x) = r(x)(x - x_1^e)(x - x_2^e).$

From Equation (2.15-3) we can write

(2.15-5) $\qquad \mathcal{L}\big(\hat{u}(x) - u(x)\big) = \hat{f}(x) - f(x)$

or, defining the **error** $e(x) \equiv \hat{u}(x) - u(x)$,

$$(2.15\text{-}6) \qquad\qquad\qquad \mathcal{L}(e(x)) = R(x).$$

Now we assume that the properties of $\mathcal{L}$ are such that the homogeneous boundary-value problem for $e(x)$ has a Green's function $G(x, \alpha)$. This means that $G(x, \alpha)$ satisfies the equation $\mathcal{L}(G(x, \alpha)) = \delta(x - \alpha)$, with the same homogeneous Dirichlet boundary conditions as we imposed on the original problem. Here we treat $\mathcal{L}$ as a differential operator with respect to the variable α. Thus

$$\int_a^b \mathcal{L}(e(\alpha)) G(x, \alpha)\, d\alpha = \int_a^b R(\alpha) G(x, \alpha)\, d\alpha.$$

The reader should check that integrating by parts twice yields

$$\int_a^b e(\alpha) \mathcal{L}(G(x, \alpha))\, d\alpha = \int_a^b e(\alpha) \delta(x - \alpha)\, d\alpha = \int_a^b R(\alpha) G(x, \alpha)\, d\alpha.$$

Therefore, using the definition of $\delta(x - \alpha)$, we have

$$(2.15\text{-}7) \qquad\qquad\qquad e(x) = \int_a^b G(x, \alpha)\, R(\alpha)\, d\alpha.$$

(We discuss Green's functions in more detail in Section 3.8.)

Recognizing that the partition of $[a, b]$ allows us to write the representation (2.15-7) as a sum of n integrals, we have

$$(2.15\text{-}8) \qquad e(x) = \sum_{e=1}^n \int_{x_{e-1}}^{x_e} G(x, \alpha)\, R(\alpha)\, d\alpha$$

$$= \sum_{e=1}^n \int_{x_{e-1}}^{x_e} G(x, \alpha)\, r(\alpha)\, (\alpha - x_1^e)\, (\alpha - x_2^e)\, d\alpha.$$

Let $K(x, \alpha) \equiv G(x, \alpha) r(\alpha)$. Consider a point x_k^e distinct from x_1^e and x_2^e in the interval $[x_{e-1}, x_e]$. Fixing x for the moment and writing a Taylor series for K with respect to α about x_k^e, we have

$$(2.15\text{-}9) \qquad K(x, \alpha) = K\bigg|_{x_k^e} + \frac{\partial K}{\partial \alpha}\bigg|_{x_k^e} (\alpha - x_k^e)$$

$$+ \frac{1}{2!} \frac{\partial^2 K}{\partial \alpha^2}\bigg|_{x_k^e} (\alpha - x_k^e)^2 + \mathcal{O}(h^3),$$

where h is the mesh of the partition Δ_x. Substitution of the series (2.15-9) into the integral representation (2.15-8) yields

$$(2.15\text{-}10) \quad e(x) = \sum_{e=1}^{n} \int_{x_{e-1}}^{x_e} \left[K \bigg|_{x_k^e} + \frac{\partial K}{\partial \alpha} \bigg|_{x_k^e} (\alpha - x_k^e) \right.$$
$$\left. + \frac{1}{2!} \frac{\partial^2 K}{\partial \alpha^2} \bigg|_{x_k^e} (\alpha - x_k^e)^2 + O\left(h^3\right) \right] (\alpha - x_1^e)(\alpha - x_2^e) \, d\alpha.$$

But, because the multiple $(\alpha - x_1^e)(\alpha - x_2^e)$ of the quadratic Legendre polynomial is orthogonal to any polynomial of degree less than two in x on the interval $[x_{e-1}, x_e]$, the first two terms in the integral vanish. Thus the error $e(x)$ can be written

$$(2.15\text{-}11) \qquad e(x) = \sum_{e=1}^{n} \int_{x_{e-1}}^{x_e} \left[\frac{1}{2!} \frac{\partial^2 K}{\partial \alpha^2} \bigg|_{x_k^e} (\alpha - x_k^e)^2 \right.$$
$$\left. + O\left(h^3\right) \right] (\alpha - x_1^e)(\alpha - x_2^e) \, d\alpha.$$

Because $(\alpha - x_k^e) < h$, Equation (2.15-11) implies

$$(2.15\text{-}12) \qquad e(x) = \sum_{e=1}^{n} \int_{x_{e-1}}^{x_e} O\left(h^2\right) \cdot O\left(h\right) \cdot O\left(h\right) \, d\alpha.$$

Since the integration over $[x_{e-1}, x_e]$ yields an additional factor of $O\left(h\right)$, we have

$$(2.15\text{-}13) \qquad e(x) = \sum_{e=1}^{n} O\left(h^2\right) \cdot O\left(h\right) \cdot O\left(h\right) \cdot O\left(h\right).$$

But the sum contains $(b - a)/h = n$ terms, so we have

$$e(x) = \frac{b - a}{h} O\left(h^5\right) = O\left(h^4\right).$$

Thus the collocation method has $O\left(h^4\right)$ error when cubic Hermite polynomials are used as a basis and the zeros of the Legendre polynomials are designated as collocation points.

2.16. Boundary-Element Methods.

Boundary-element methods are currently used extensively in the area of solid and, to a more limited degree, fluid mechanics. The method is based upon the concept of reducing the dimensionality of a problem by generating approximating equations that have unknown parameters associated with only the boundary of a region. In one dimension the technique is of marginal advantage. It is particularly well suited to problems of elliptic type, and consequently we postpone discussion of this method until Chapter Three.

2.17. Problems for Chapter Two.

1. State whether the following PDEs are linear or nonlinear, and classify them as elliptic, parabolic, or hyperbolic. (The classification may depend on values of the independent variables.)

(a) $\quad e^{xy}\dfrac{\partial^2 u}{\partial x^2} + \cos^2 y\dfrac{\partial^2 u}{\partial y^2} + \dfrac{\partial u}{\partial x} - \dfrac{\partial u}{\partial y} = 0.$

(b) $\quad \ln(1+x^2)\dfrac{\partial^2 u}{\partial t^2} - 2\dfrac{\partial^2 u}{\partial x^2} = e^u.$

(c) $\quad (\cosh x)\dfrac{\partial^2 u}{\partial x^2} + u\dfrac{\partial u}{\partial x} + \dfrac{\partial u}{\partial y} = f(x,y).$

(d) $\quad \dfrac{\partial^2 u}{\partial t^2} + u^2\dfrac{\partial^2 u}{\partial x^2} = 0.$

2. Interpolation on uniform grids can sometimes produce interpolates that exhibit large excursions from the interpolated function between nodes, an occurrence called the **Runge phenomenon**. Plot the function $f(x) = (1+25x^2)^{-1}$ on the interval $[-1,1]$. Plot the Lagrange interpolating polynomial $P_n(x)$ determined by $f(x)$ on a grid $-1 = x_0 < x_1 < \cdots < x_n = 1$ of uniformly spaced nodes, say with $n = 8$ or 10. Plot the Lagrange interpolating polynomial $P_n^*(x)$ of the same degree with the nodes $x_0, \cdots, x_n$ located at the **Tchebycheff abscissae**,

$$x_i = \cos\left(\frac{2i+1}{n+1}\frac{\pi}{2}\right).$$

Finally, plot the piecewise linear Lagrange interpolate for f on the uniform grid.

3. Prove the error estimate (2.4-5a) for Lagrange interpolation, namely, if $f \in C^{n+1}[a,b]$ and $P_n(x)$ is the Lagrange polynomial of degree n interpolating f on the grid $a = x_0 < x_1 < \ldots < x_n = b$, then for any $x \in [a,b]$ we can find some number $\varsigma \in (a,b)$ such that

$$E_n \equiv f(x) - P_n(x) = \frac{w_n(x)}{(n+1)!}\frac{d^{n+1}f}{dx^{n+1}}(\varsigma).$$

Here, $w_n(x) = (x - x_0)(x - x_1) \cdots (x - x_n)$. (Hint: For $\bar{x} \in [a, b]$ with $\bar{x} \neq x_0, \ldots, x_n$, define

$$g(x) = f(x) - P_n(x) - \frac{w_n(x)}{w_n(\bar{x})}[P_n(\bar{x}) - f(\bar{x})].$$

Observe that g has $n + 2$ zeros in the interval $[a, b]$. Apply Rolle's theorem to successive derivatives of g to deduce that there is some $\varsigma \in (a, b)$ such that

$$\frac{d^{n+1}g}{dx^{n+1}}(\varsigma) = 0.$$

Rolle's theorem asserts that if a function φ is continuous on an interval $[\alpha, \beta]$ and differentiable on (α, β), with $\varphi(\alpha) = \varphi(\beta) = 0$, then $\varphi'(\xi) = 0$ for some $\xi \in (\alpha, \beta)$.)

4. Many numerical schemes for PDEs in one dimension lead to tridiagonal matrix equations of the form

$$\begin{bmatrix} b_1 & c_1 & & & & \\ a_2 & b_2 & c_2 & & & \\ & a_3 & b_3 & c_3 & & \\ & & & \ddots & & \\ & & & a_{N-1} & b_{N-1} & c_{N-1} \\ & & & & a_N & b_N \end{bmatrix} \begin{bmatrix} u_1 \\ u_2 \\ u_3 \\ \vdots \\ u_{N-1} \\ u_N \end{bmatrix} = \begin{bmatrix} r_1 \\ r_2 \\ r_3 \\ \vdots \\ r_{N-1} \\ r_N \end{bmatrix}.$$

The **Thomas algorithm** for solving such systems is as follows:

$\beta_1 = b_1$
$\gamma_1 = r_1/\beta_1$
For $i = 2, \ldots, N$:
$\quad \beta_i = b_i - (a_i c_{i-1}/\beta_{i-1})$
$\quad \gamma_i = (r_i - a_i \gamma_{i-1})/\beta_i$
$u_N = \gamma_N$
For $j = 1, \ldots, N - 1$:
$\quad u_{N-j} = \gamma_{N-j} - c_{N-j} u_{N-j+1}/\beta_{N-j}$
End.

Write a computer program using the Thomas algorithm to solve the equation set arising from a fully implicit, spatially centered difference approximation to the following initial-boundary-value problem:

$$\frac{\partial u}{\partial t} = \frac{\partial^2 u}{\partial x^2}, \quad t > 0, \quad 0 < x < 1,$$
$$u(x, 0) = (1 - x)^4, \quad 0 < x < 1,$$
$$u(0, t) = 1, \quad u(1, t) = 0, \quad t > 0.$$

5. Consider the initial-boundary-value problem

$$\frac{\partial u}{\partial t} - k\frac{\partial^2 u}{\partial x^2} = 0, \quad k > 0,$$

$$u(x,0) = \cos x, \quad 0 < x < 2\pi,$$

$$u(0,t) = u(2\pi,t), \quad t > 0.$$

If you were to use a centered-in-space, fully implicit finite-difference scheme to solve this problem numerically, what would be the structure of the matrix? (Hint: Think of the spatial nodes $0 = x_0, \ldots, x_N = 2\pi$ as lying on a circle with $x_0 = x_N$.)

6. Consider the first-order PDE

$$\frac{\partial u}{\partial t} + v\frac{\partial u}{\partial x} = 0, \quad v > 0,$$

and the two finite-difference approximations

$$u_j^{n+1} = u_j^n - C(u_j^n - u_{j-1}^n) \quad \text{(explicit)},$$
$$u_j^{n+1} = u_j^n - C(u_j^{n+1} - u_{j-1}^{n+1}) \quad \text{(implicit)}.$$

Here, $C = vk/h$, k being the time step and h being the spatial grid mesh, and u_j^n denotes the approximate value of $u(jh, nk)$. Use von Neumann stability analysis to show that the explicit scheme is stable when $C \leq 1$, while the implicit scheme is unconditionally stable.

7. For vectors $\mathbf{x} = (x_1, \ldots, x_n)$, the quantity $\|\mathbf{x}\|_\infty = \max_{1 \leq i \leq n} |x_i|$ defines a norm. Show that the associated matrix norm

$$\|\mathbf{A}\|_\infty = \sup_{\mathbf{x} \neq 0} \frac{\|\mathbf{A}\mathbf{x}\|_\infty}{\|\mathbf{x}\|_\infty}$$

can be computed as the "maximum row sum,"

$$\|\mathbf{A}\|_\infty = \max_{1 \leq i \leq n} \sum_{i=1}^n |A_{ij}|,$$

where A_{ij} denotes the (i,j)-th entry of $\mathbf{A}$. (Hint: First show

$$\|\mathbf{A}\mathbf{x}\|_\infty \leq \max_{1 \leq i \leq n} \sum_{j=1}^n |A_{ij}| \|\mathbf{x}\|_\infty.$$

Then find a vector $\mathbf{x}$ so that $\|\mathbf{A}\mathbf{x}\|_\infty$ attains this upper bound.)

8. Use the matrix norm $\| \cdot \|_\infty$, discussed in Problem 7, to perform a matrix stability analysis of the scheme

$$u_i^{n+1} = u_i^n - \frac{vk}{h}(u_i^n - u_{i-1}^n)$$

for solving $\partial u/\partial t + v\partial u/\partial x = 0$ with $v > 0$.

9. Consider the boundary-value problem

$$\nabla \cdot [K(x,y)\nabla u] = 0, \quad 0 < x < L, \quad 0 < y < M,$$

$$\frac{\partial u}{\partial x}(0,y) = \frac{\partial u}{\partial x}(L,y) = 1,$$

$$u(x,0) = u(x,M) = 2,$$

to be solved numerically on the two-dimensional grid $\{0, L/3, 2L/3, L\} \times \{0, M/3, 2M/3, M\}$.

(a) Using the tensor-product basis formed by the one-dimensional Lagrange piecewise linear basis functions $\ell_i(x)$, $\ell_j(y)$, write the trial function $\hat{u}(x,y)$ that you would use in Galerkin's method. Be sure to incorporate boundary conditions as appropriate.

(b) Define the residual $R(x,y)$ in terms of $\hat{u}$.

(c) Write a typical Galerkin integral equation in terms of the residual R. Indicate *explicitly* which (and how many) functions you would use as weighting functions.

(d) Using Green's theorem, briefly discuss how you would handle the Neumann boundary conditions for this problem.

10. One method for computationally checking the order of accuracy of a numerical scheme producing approximations $\hat{u}$ to an exact solution u is to compare the error $E = \|\hat{u} - u\|$ for various grid meshes h. If $E = \mathcal{O}(h^p)$, then we reason heuristically as follows: $E \simeq Ch^p$, where C is some constant, if we neglect higher-order terms in h. Thus $\log E \simeq \log C + p\log h$, so a plot of $\log E$ versus $\log h$ should roughly yield a straight line having slope p.

The PDE $\nabla \cdot (K\nabla u) = 0$, with $K(x,y) = e^{xy}$ and $u(x,y) = x^2 - y^2$ on the boundary of the unit square $(0,1) \times (0,1)$, has exact solution $u(x,y) = x^2 - y^2$. Develop a numerical scheme for this problem and computationally estimate its order of accuracy.

11. Consider the initial-boundary-value problem

$$\frac{\partial u}{\partial t} + \frac{\partial u}{\partial x} = -u,$$

$$u(x, 0) = u_I, \quad x > 0,$$

$$u(0, t) = u_L, \quad t \geq 0.$$

We shall use a piecewise linear trial function $\hat{u} = \sum_{i=1}^{N} u_i(t)\ell_i(x)$ on a uniform grid $0 = x_0 < x_1 < \cdots < x_N = L$ covering a subset of the spatial domain $x > 0$. Notice that the nodal coefficients $u_i(t)$ are functions of time t.

(a) Write the collocation equation associated with a typical collocation point $\bar{x}_i = (x_{i-1} + x_i)/2$, assuming the grid has mesh h. (Your answer should be an ordinary differential equation.)

(b) Convert the ordinary differential equation from part (a) to an algebraic equation using an implicit finite-difference approximation.

(c) Display the matrix equation giving the nodal values u_i^n at time level n in terms of the values u_i^{n-1} at the previous time step.

12. Consider Laplace's equation $\nabla^2 u = 0$ on the unit square $\Omega = (0, 1) \times (0, 1)$ together with Dirichlet boundary conditions $u(x, y) = f(x, y)$ for $(x, y) \in \partial\Omega$. If we partition Ω into four square elements by bisecting its edges, then the resulting grid will have nine nodes. The appropriate tensor-product Hermite cubic trial function has the form

$$\hat{u}(x, y) = \sum_{i=1}^{9} [u_\ell \phi_{00\ell}(x, y) + u_\ell^{(x)} \phi_{10\ell}(x, y)$$

$$+ u_\ell^{(y)} \phi_{01\ell}(x, y) + u_\ell^{(xy)} \phi_{11\ell}(x, y)],$$

where $\phi_{ij\ell}(x, y) = h_\ell^i(x) h_\ell^j(y)$, and u_ℓ, $u_\ell^{(x)}$, $u_\ell^{(y)}$, $u_\ell^{(xy)}$ represent the values of $\hat{u}$, $\partial\hat{u}/\partial x$, $\partial\hat{u}/\partial y$, $\partial^2\hat{u}/\partial x\partial y$, respectively, at node ℓ. This trial function has 36 coefficients, eight of which we know directly given the Dirichlet boundary conditions. Differentiating the boundary data $f(x, y)$ tangentially along $\partial\Omega$ gives values of $\partial\hat{u}/\partial x$ at six nodes along horizontal boundary segments and values of $\partial\hat{u}/\partial y$ at six nodes along vertical boundary segments. Thus differentiating the Dirichlet data determines 12 additional nodal coefficients. Collocating the residual at four collocation points in each of the four square elements then gives enough equations to determine the remaining unknown nodal coefficients of $\hat{u}$. Show the *structure* of the 16 × 16 matrix multiplying these coefficients. (You need only indicate which entries are nonzero.)

2.18. References.

Allen, M.B. and Murphy, C.L., "A finite-element collocation method for variably saturated flow in two space dimensions," *Water Resour. Res.,* *22:11* (1986), 1537-1542.

Babu, D.K. and Pinder, G.F., "A finite-element-finite-difference alternating-direction algorithm for three dimensional groundwater transport," in J.P. Laible et al., Eds., *Proceedings of the Fifth International Conference on Finite Elements in Water Resources,* June 18-22, 1984, Burlington, Vermont, Berlin: Springer-Verlag, 1984, pp. 165-174.

Barnhill, R.E. and Mansfield, L., "Error bounds for smooth interpolation in triangles," *J. Approx. Theory, 11* (1974), 306-318.

Botha, J.F. and Pinder, G.F., *Fundamental Concepts in the Numerical Solution of Differential Equations,* John Wiley, New York, 1983.

Celia M.A., "Collocation on Deformed Finite Elements and Alternating Direction Collocation Methods," Ph.D. Dissertation, Department of Civil Engineering, Princeton University, Princeton, New Jersey, 1983.

Charney, J.G., Fjörtoft, R., and von Neumann, J., "Numerical integration of the barotropic vorticity equation," *TELLUS, 2* (1950), 237-254.

Douglas, J. and Dupont, T., "A finite-element collocation method for quasilinear parabolic equations," *Math. Comp. 27:121* (1973), 17-28.

Garabedian, P.R., *Partial Differential Equations,* New York: Wiley, 1964.

Lapidus, L. and Pinder, G.F., *Numerical Solution of Partial Differential Equations in Science and Engineering,* New York: Wiley, 1982.

Mitchell, A.R. and Wait, R., *The Finite Element Method in Partial Differential Equations,* Chichester: Wiley, 1977.

Oden, J.T. and Reddy, J.N., *An Introduction to the Mathematical Theory of Finite Elements,* New York: Wiley, 1976.

Prenter, P.M., *Splines and Variational Methods,* New York: Wiley, 1975.

Strang, G. and Fix, G.F., *An Analysis of the Finite Element Method,* Englewood Cliffs, New Jersey: Prentice-Hall, 1973.

Stroud, A.H. and Secrest, D., *Gaussian Quadrature Formulas,* Englewood Cliffs, New Jersey: Prentice-Hall, 1966.

Zauderer, E., *Partial Differential Equations of Applied Mathematics,* New York: Wiley, 1983.

CHAPTER THREE
STEADY-STATE SYSTEMS

3.1. Introduction.

In Section 2.2 we classified PDEs into three types: elliptic, parabolic, and hyperbolic. The present chapter describes some powerful techniques for solving elliptic differential equations, while Chapters Four and Five are concerned with parabolic and hyperbolic differential equations, respectively. In typical applications, these latter two types of equations describe time evolutionary phenomena, while elliptic equations are commonly associated with the steady states of physical systems. We begin this chapter by explaining how elliptic equations occur in physical problems of concern in engineering. This is useful not only to motivate the interest of the reader, but also because relating the equations to specific physical phenomena allows the use of physical intuition as a heuristic guide for the understanding of the results and for suggesting conjectures about further properties of the equations.

In the absence of jumps, the general balance law (1.3-5) can be written as follows:

$$(3.1.-1) \qquad \frac{\partial(\rho\Psi)}{\partial t} + \nabla \cdot (\rho\Psi\mathbf{v}) - \nabla \cdot \boldsymbol{\tau} - \rho g = 0.$$

When the system under investigation is in a steady state, all partial derivatives with respect to time vanish, and we have

$$(3.1-2) \qquad \nabla \cdot (\rho\Psi\mathbf{v}) - \nabla \cdot \boldsymbol{\tau} - \rho g = 0.$$

For example, given the definitions of Chapter One, we obtain the energy balance (1.3-11) in its steady-state form by setting $\partial(\rho E + \mathbf{v} \cdot \mathbf{v}/2)/\partial t = 0$:

$$(3.1-3) \qquad \nabla \cdot \left[(\rho E + \tfrac{1}{2}\mathbf{v} \cdot \mathbf{v})\mathbf{v}\right] - \nabla \cdot (\mathbf{q} + \mathbf{t} \cdot \mathbf{v}) - \rho(h + \mathbf{b} \cdot \mathbf{v}) = 0.$$

As another example, consider the mass balance equation for multiphase mixtures, as discussed in Section 1.5. For a particular phase α with vol-

126

ume fraction ϕ^α, density ρ^α, and velocity $\mathbf{v}^\alpha$, the steady-state condition $\partial(\phi^\alpha \rho^\alpha)/\partial t = 0$ reduces the mass balance to the following:

$$(3.1\text{-}4) \qquad \nabla \cdot (\phi^\alpha \rho^\alpha \mathbf{v}^\alpha) = 0.$$

Let us see how equations such as these lead to elliptic PDEs.

3.2. Laplace's Equation in Physics and Engineering.

Laplace's equation describes certain steady-state problems, and thus it arises in the analysis of equilibrium states of many physical systems. As specific examples, we mention the following:

- the temperature of a homogeneous and isotropic body in thermal equilibrium,

- the hydraulic head associated with the steady state of the flow of a homogeneous fluid through a homogeneous and isotropic porous medium,

- the electrostatic and magnetostatic potentials in empty space,

- the potential of the irrotational flow of an incompressible fluid in free space,

- the gravitational potential in empty space, and

- the vertical deflection of a nearly horizontal membrane (or drumhead) in the absence of forces in its interior.

While most of these physical systems involve some rather significant simplifying assumptions, their behaviors reflect many important features of systems governed by more complicated physics. To see what sorts of simplifying assumptions typically lead to Laplace's equation, let us review the first two of these applications.

Steady-state heat flow.

The equation governing thermal equilibrium of a homogeneous body is a special case of the steady-state energy balance equation (3.1-3). For a body at rest without internal heat sources, in thermal equilibrium, the velocity $\mathbf{v}$ and the heat sources h vanish. Let us assume, moreover, that the density ρ is a constant. Equation (3.1-3) then reduces to

$$(3.2\text{-}1) \qquad -\nabla \cdot \mathbf{q} = 0.$$

Now recall from Section 1.4 that Fourier's law of heat flux states that heat flux is proportional to the temperature gradient, the factor of proportionality being the coefficient of thermal conductivity. In the interest of simplicity, we assumed this coefficient to be a scalar quantity in Chapter One. In general, however, the dependence of heat fluxes on directional properties of the material may require us to use a tensor form. In this case Fourier's law becomes

$$(3.2\text{-}2) \qquad\qquad \mathbf{q} = \mathbf{k}_H \nabla T,$$

where $\mathbf{k}_H$ is the tensor form of the coefficient of thermal conductivity and T is the temperature. Equations (3.2-1) and (3.2-2) together yield the elliptic PDE

$$(3.2\text{-}3) \qquad\qquad \nabla \cdot (\mathbf{k}_H \nabla T) = 0.$$

Equation (3.2-3) can be simplified further when the solid body is homogeneous, so that $\mathbf{k}_H$ does not vary with position $\mathbf{x}$, and is isotropic, so that the thermal conductivity is independent of direction. In this case $\mathbf{k}_H$ is a constant multiple of the unit tensor, and Equation (3.2-3) reduces to Laplace's equation:

$$(3.2\text{-}4) \qquad\qquad \nabla^2 T = 0.$$

Here, the symbol ∇^2 stands for the Laplace operator, which in three-dimensional Cartesian coordinates takes the form

$$(3.2\text{-}5\text{a}) \qquad\qquad \nabla^2 = \frac{\partial^2}{\partial x^2} + \frac{\partial^2}{\partial y^2} + \frac{\partial^2}{\partial z^2}.$$

In two dimensions this reduces to

$$(3.2\text{-}5\text{b}) \qquad\qquad \nabla^2 = \frac{\partial^2}{\partial x^2} + \frac{\partial^2}{\partial y^2}.$$

Steady flow in uniform porous media.

An identical equation arises in steady fluid flows through porous media. As discussed in Section 1.5, a porous medium is a solid, such as sandstone, possessing interconnected voids in which a fluid can flow. Figure 3-1 illustrates such a mixture, at least conceptually. The interconnected void space is termed the **effective pore space**. In what follows we assume that the porous medium is macroscopically homogeneous and that it is **saturated** with a homogeneous fluid. The latter assumption means that the effective pore space is completely filled with the fluid. Let ρ^F be the density of the fluid. Then the **bulk density** (that is, the mass of fluid per unit volume of the space occupied by the fluid-solid mixture) is $\phi \rho^F$, where $\phi = \phi^F$ is

Figure 3-1. Conceptual picture of a multiphase mixture of rock and fluid forming a fluid-saturated porous medium.

the porosity of the solid. The local mass balance equation for steady states therefore becomes

$$(3.2\text{-}6) \qquad \qquad \nabla \cdot (\phi \rho^F \mathbf{v}^F) = 0.$$

According to Darcy's law (Equation 1.5-6), $\mathbf{v}^F$ is related to gradients in the fluid pressure and the depth via the permeability of the solid matrix. As with the coefficient of thermal conductivity, the most general form of the permeability allows for directional dependencies in the solid by assuming a tensor version of Darcy's law:

$$(3.2\text{-}7) \qquad \qquad \mathbf{v}^F = -\frac{\mathbf{k}}{\mu^F \phi^F} (\nabla p^F - \rho^F g \nabla Z).$$

The notation and terminology adopted for this law varies among the fields of petroleum engineering, groundwater hydrology, and soil physics. Among groundwater hydrologists it is common to reduce Darcy's law to a simpler form taking advantage of the nearly constant chemical composition and viscosity of underground water. Provided the fluid density varies at most as a function of fluid pressure, so that $\rho^F = \rho^F(p^F)$, we can define the **hydraulic head** h by the equation

$$gh = -gZ + \int_{p_{\text{ref}}}^{p^F} \frac{d\pi}{\rho(\pi)},$$

where p_{ref} is any reference pressure. (Observe that we use the letter h in this context, even though we reserved this symbol for internal heat sources in Chapter One. The use of h for hydraulic head is so widespread in the groundwater literature that we prefer to stick to it.) In terms of this new variable, it is easy to show that Darcy's law in the form (3.2-7) simplifies to

$$(3.2\text{-}8) \qquad \qquad \mathbf{v}^F = -\mathbf{K}_F \nabla h,$$

where the tensor coefficient

(3.2-9) $$\mathbf{K}_F = \frac{\rho^F g \mathbf{k}}{\mu^F}$$

is called the **hydraulic conductivity**.

Substituting Darcy's law in the hydrologists' form (3.2-8) into the steady-state mass balance (3.2-6), we arrive at the steady groundwater flow equation,

$$\nabla \cdot (\mathbf{K}_F \nabla h) = 0.$$

From this equation it is clear that, when the porous medium is macroscopically homogeneous and isotropic, the flow of groundwater in a steady-state aquifer reduces to Laplace's equation,

(3.2-10) $$\nabla^2 h = 0.$$

3.3. Well Posed Boundary-Value Problems.

Functions that satisfy Laplace's equation are said to be **harmonic**. In any given region of space there are many harmonic functions, and consequently demanding that a physical quantity satisfy Laplace's equation in a region Ω is not sufficient to specify it uniquely. Usually, however, one considers physical and engineering problems to have unique solutions. As discussed in Section 2.3, we specify the unique physical solution to a PDE from among the possibly infinite number of general solutions by imposing additional auxiliary conditions, such as boundary and initial conditions. To ensure that the resulting mathematical solution is indeed physically realistic, we must check that the original PDE and the auxiliary conditions combine to give a **well posed** problem.

Generally, a well posed problem is one that possesses one and only one solution, which in turn depends continuously on the auxiliary conditions. This latter requirement is essential for the results to be physically meaningful. The reason for this is that the physical parameters defining boundary and initial conditions cannot be measured exactly. One expects, though, that a mathematical model of the physics will nevertheless be robust against small errors in measurement. More precisely, we demand that one can reduce the magnitude of the error in the mathematical solution below any prescribed tolerance by making the errors in the auxiliary conditions small enough.

For elliptic equations, most well posed problems are boundary-value problems. This means that the additional constraints on the unknown function are specified at every point of the boundary $\partial \Omega$ of a region Ω, depicted in Figure 3-2. As mentioned in Section 2.3, such problems are classified into Dirichlet, Neumann, and Robin boundary-value problems.

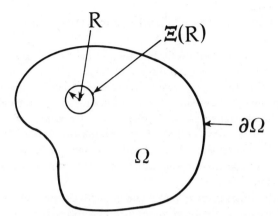

Figure 3-2. Geometry of the region Ω used in deriving integral representation theorems.

The data specified at the boundary $\partial\Omega$ for these problems give the value of the function, the normal derivative or a linear combination of the function and its normal derivative, respectively. A problem prescribing one type of data on one part of the boundary and a different type on another part of the boundary is usually called a **mixed boundary-value problem**.

Some of the physical processes governed by Laplace's equations, such as those discussed in Section 3.2, can be used to supply physical interpretations of the different types of boundary conditions. For example, let T be the temperature of a homogeneous and isotropic body in thermal equilibrium. In this case T satisfies Laplace equation (3.2-4). The Dirichlet problem corresponds to the physical problem of finding the steady-state temperature of a body whose external temperature is known. To see how Neumann problems arise, recall that, according to Fourier's law (3.2-2), when the body is isotropic, the heat flux is as follows:

$$(3.3\text{-}1) \qquad\qquad \mathbf{q} = k_H \nabla T.$$

Thus, the flow of heat per unit area going out of the body is proportional to the outward normal derivative of T:

$$(3.3\text{-}2) \qquad\qquad \mathbf{q}\cdot\mathbf{n} = k_H \nabla T \cdot \mathbf{n} = k_H \frac{\partial T}{\partial n}.$$

Hence, the Neumann problem for Laplace's equation corresponds to the physical problem of finding the equilibrium temperature of a body when the flow of heat through its boundary is prescribed. The interpretation of Robin boundary-value problems for these physics is more complicated, and we leave it as an exercise for the reader.

3.4. General Properties of the Laplace Operator.

The operator ∇^2 possesses many mathematical properties that have useful implications for the numerical solution of Laplace's equation and of steady-state problems in general. We shall therefore spend some time reviewing some mathematical background before focusing on numerical methods. After listing a set of rather general properties, we shall examine variational and maximum principles for elliptic problems involving ∇^2. We close our review of the mathematics of the Laplace operator in Section 3.8 by discussing several integral representation theorems that will prove useful in our later numerical work.

Some of the most important of the general properties of ∇^2 on a region Ω in two- or three-dimensional space are the following:

(i) ∇^2 is linear;

(ii) ∇^2 is formally symmetric;

(iii) ∇^2 is negative definite for functions satisfying suitable boundary conditions, and

(iv) ∇^2 has constant coefficients.

Property (i) means that, when u and v are given functions in $C^2(\Omega)$ while a and b are two scalar constants, one has

$$(3.4\text{-}1) \qquad\qquad \nabla^2(au + bv) = a\nabla^2 u + b\nabla^2 v.$$

The linearity of ∇^2 implies that we can construct solutions to Laplace's equation using the principle of superposition, which we discuss in the next subsection.

Property (ii) means that, given two functions $u, v \in C^2(\Omega)$,

$$(3.4\text{-}2) \qquad \int_\Omega v\nabla^2 u \, d\mathbf{x} = \int_\Omega u\nabla^2 v \, d\mathbf{x}$$

$$+ \text{ an integral involving boundary terms.}$$

To prove this assertion, and to see what we mean by "an integral involving boundary terms," observe that $v\nabla^2 u - u\nabla^2 v = \nabla\cdot(v\nabla u - u\nabla v)$. Integrating this identity over Ω and using the divergence theorem, we get **Green's second identity**,

$$(3.4\text{-}3) \qquad \int_\Omega v\nabla^2 u \, d\mathbf{x} = \int_\Omega u\nabla^2 v \, d\mathbf{x} + \oint_{\partial\Omega} (v\frac{\partial u}{\partial n} - u\frac{\partial v}{\partial n})d\mathbf{x}.$$

This equation shows explicitly the integral involving boundary terms referred to in Equation (3.4-2).

Property (iii) means that, whenever $u \equiv 0$ on the boundary $\partial\Omega$, $\int_\Omega u\nabla^2 u\,dx \le 0$, equality holding only in the case that $u \equiv 0$ throughout Ω. This property can be shown as follows. From the identity $u\nabla^2 u = \nabla \cdot (u\nabla u) - \nabla u \cdot \nabla u$ we can deduce the equation

$$(3.4\text{-}4) \qquad \int_\Omega u\nabla^2 u\,d\mathbf{x} = \oint_{\partial\Omega} u\frac{\partial u}{\partial n}d\mathbf{x} - \int_\Omega \nabla u \cdot \nabla u\,d\mathbf{x},$$

where we have used the divergence theorem to convert the volume integral of $\nabla \cdot u(\nabla u)$ to a boundary integral. Since the first integral on the right side vanishes by the boundary conditions on u, while the integrand of the second integral is a squared quantity,

$$(3.4\text{-}5) \qquad \int_\Omega u\nabla^2 u\,d\mathbf{x} = -\int_\Omega \nabla u \cdot \nabla u\,d\mathbf{x} \le 0.$$

Equality holds in Equation (3.4-5) only when $u \equiv 0$ in Ω. To see why, observe that the integrand $\nabla u \cdot \nabla u$ is never negative. Hence, if equality holds, one has $\nabla u \equiv \mathbf{0}$ in Ω. Thus u is a constant, whose only possible value is zero because it vanishes on the boundary $\partial\Omega$.

Property (iv) of the Laplace operator is obvious; we shall discuss some of its implications in Section 3.7.

The principle of superposition.

When an operator $\mathcal{L}$ is linear, any linear combination of solutions of the homogeneous equation $\mathcal{L}u = 0$ is itself a solution of the same equation. This implies that any linear combination of harmonic functions is harmonic. Indeed, if $\nabla^2 u = 0$ and $\nabla^2 v = 0$, then for any two numbers a and b we have $\nabla^2(au + bv) = a\nabla^2 u + b\nabla^2 v = 0$. To derive a continuous extension of this argument, let $w(\xi,\mathbf{x})$ be a family of harmonic functions that depends on the parameter ξ. By this we mean that $w(\xi,\mathbf{x})$ is harmonic with respect to the variable $\mathbf{x}$ for every fixed value of ξ. Also, let $K(\xi)$ be an integrable function of ξ, where ξ ranges over some measurable set Γ of real numbers. Then the function defined by the "continuous superposition"

$$(3.4\text{-}6) \qquad u(\mathbf{x}) \equiv \int_\Gamma K(\xi)w(\xi,\mathbf{x})d\xi$$

is itself a harmonic function. To see this, notice that

$$(3.4\text{-}7) \qquad \nabla^2 u = \nabla^2 \int_\Gamma K(\xi)w(\xi,x)d\xi = \int_\Gamma K(\xi)\nabla^2 w(\xi,\mathbf{x})d\xi = 0.$$

Rigorously speaking, for Equation (3.4-7) to be valid, it is necessary for the integrand to satisfy certain regularity assumptions required for interchanging the integration and differentiation operations. We shall not explore this issue here. Integral representations analogous to Equation (3.4-6) play an important role in the theory of elliptic PDEs and will be developed further later in this chapter.

The nonhomogeneous equation associated with Laplace's equation, namely,

$$(3.4\text{-}8) \qquad\qquad \nabla^2 u = f,$$

is called **Poisson's equation.** The linearity of the Laplace operator implies that the most general solution of Poisson's equation can be written as $v + w_0$, where v is harmonic, while w_0 is a particular solution of Poisson equation. That is,

$$(3.4\text{-}9) \qquad\qquad \nabla^2 w_0 = f.$$

Because of this property it is possible, to a large extent, to reduce the study of Poisson's equation to that of Laplace's equation.

3.5. Variational Principles.

Many elliptic problems possess equivalent variational formulations stating that the solution to the PDE together with the given boundary conditions is a stationary (or "equilibrium") value of some functional. Such **variational principles** are useful both as theoretical tools in the study of existence and uniqueness of solutions and as the foundations of numerical techniques such as the finite-element method. Throughout this section and the next, we assume Ω is a bounded region in two- or three-dimensional space.

The formulation of variational principles for Laplace's and Poisson's equations depends in an essential way on the formal symmetry and negative definiteness of the Laplace operator. Following Herrera (1980, 1984), we begin the formulation of variational principles for equations involving ∇^2 by rearranging the terms in Green's second identity, Equation (3.4-3). Generally, let $T_1(v)$ be some linear combination of the function v and its normal derivative on $\partial\Omega$, and let $T_2(v)$ be another such combination. Thus $T_1(u) = \gamma_1$ and $T_2(u) = \gamma_2$ might each be a suitable boundary condition on a function u satisfying a PDE. If, for any functions $u, v \in C^2(\Omega)$, these boundary operators are related by the equation

$$(3.5\text{-}1) \qquad T_1(u)\,T_2(v) - T_1(v)\,T_2(u) = v\frac{\partial u}{\partial n} - u\frac{\partial v}{\partial n},$$

then we say that the boundary operators T_1 and T_2 are **complementary.**

In this case the bilinear functional $\langle \mathcal{P}u, v \rangle$ defined on functions $u, v \in C^2(\Omega)$ by

$$\langle \mathcal{P}u, v \rangle = \int_\Omega v\nabla^2 u \, d\mathbf{x} - \oint_{\partial\Omega} T_1(u)\,T_2(v)\,d\mathbf{x} \qquad (3.5\text{-}2)$$

$$= \int_\Omega u\nabla^2 v \, d\mathbf{x} - \oint_{\partial\Omega} T_1(v)\,T_2(u)\,d\mathbf{x}$$

is **symmetric**; that is, $\langle \mathcal{P}u, v \rangle = \langle \mathcal{P}v, u \rangle$. This fact follows directly from Green's second identity and the relationship (3.5-1).

The bilinear functional just defined is suitable for formulating Poisson's equation (and Laplace's equation) variationally. Consider the equation $\nabla^2 u = f$, subject to the boundary conditions

$$(3.5\text{-}3) \qquad\qquad T_1(u) = g \qquad \text{on} \quad \partial\Omega,$$

where g is a prescribed function. Observe that, if u is a solution of this problem, then

$$(3.5\text{-}4) \qquad \langle \mathcal{P}u, v \rangle = \int_\Omega fv \, d\mathbf{x} - \oint_{\partial\Omega} g\,T_2(v)\,d\mathbf{x}$$

for every function $v \in C^2(\Omega)$. From these equations, it follows that

$$(3.5\text{-}5) \qquad \int_\Omega (\nabla^2 u - f)v \, d\mathbf{x} - \oint_{\partial\Omega} [T_1(u) - g]\,T_2(v)\,d\mathbf{x} = 0.$$

The fundamental theorem of the calculus of variations (see Courant and Hilbert, 1953, Section IV.3) implies that Equation (3.5-5) is satisfied for every **test function** $v \in C^2(\Omega)$ if and only if u satisfies Poisson's equation subject to the boundary conditions (3.5-3). Thus Equation (3.5-5) constitutes an equivalent integral formulation of our Poisson boundary-value problem.

To see how this integral formulation leads to a variational principle, we make a specific choice for the test function. In particular, let us define the **variation** δu of a function $u(\mathbf{x}) \in C^2(\Omega)$ to be the difference between u and an arbitrary function $u + \delta u$ in the space. If we take v to be the variation δu of u, we can write Equation (3.5-5) in the form

$$(3.5\text{-}6) \qquad \langle \mathcal{P}u, \delta u \rangle - \int_\Omega f\,\delta u \, d\mathbf{x} + \oint_{\partial\Omega} g\,T_2(\delta u)\,d\mathbf{x} = 0.$$

An important implication of the symmetry of the bilinear functional $\langle \mathcal{P}u, v \rangle$ is that the left side of Equation (3.5-6) is the variation of a single functional. Indeed, define

$$(3.5\text{-}7) \qquad J(u) \equiv \tfrac{1}{2}\langle \mathcal{P}u, u \rangle - \int_\Omega fu \, d\mathbf{x} + \oint_{\partial\Omega} g\,T_2(u)\,d\mathbf{x}.$$

Then, if we denote the variation of u by $\delta u(\mathbf{x}) = \varepsilon w(\mathbf{x})$, where ε stands for an arbitrary parameter and $w(\mathbf{x})$ is an arbitrary function, we can define the variation in the functional J by

$$\delta J(u) = \lim_{\varepsilon \to 0} [J(u + \varepsilon w) - J(u)].$$

The reader should check that the variation $\delta J(u)$ of $J(u)$ is precisely the left side of (3.5-6). Therefore we have the following:

Variational Principle. *A smooth function u is a solution of Poisson's equation, subject to the boundary conditions (3.5-3), if and only if the total variation $\delta J(u)$ of $J(u)$ vanishes, that is, if and only if u is a stationary value of the functional J.*

Specific variational principles for ∇^2 with the three types of boundary conditions considered here can be derived by defining the boundary operators T_1 and T_2 of Equation (3.5-1) in appropriate ways. If, for example,

$$(3.5\text{-}8a) \qquad\qquad T_1(u) \equiv u, \quad T_2(v) \equiv -\frac{\partial v}{\partial n},$$

then Equation (3.5-3) becomes a Dirichlet boundary condition. Similarly,

$$(3.5\text{-}8b) \qquad\qquad T_1(u) \equiv \frac{\partial u}{\partial n}, \quad T_2(v) \equiv v$$

yields a Neumann boundary condition, while the choice

$$(3.5\text{-}9) \qquad T_1(u) \equiv \frac{\partial u}{\partial n} + \alpha u, \quad T_2(v) = (\alpha^2 + 1)^{-1}\left(v - \alpha\frac{\partial v}{\partial n}\right)$$

is suitable for Robin boundary conditions. Notice that Neumann boundary conditions are actually a special case of Equation (3.5-9) corresponding to the choice $\alpha \equiv 0$.

Thus, we see that solving the PDE $\nabla^2 u = f$ with given boundary conditions reduces to the task of finding a stationary value of the functional $J(u)$ introduced in Equation (3.5-7), subject to appropriate choice of the boundary operators T_1 and T_2. The general form of the functional J is given by

$$(3.5\text{-}10a) \qquad 2J(u) = \int_\Omega u(\nabla^2 u - 2f)d\mathbf{x} - \oint_{\partial\Omega} [T_1(u) - 2g]T_2(u)d\mathbf{x}.$$

For example, for Dirichlet problems Equations (3.5-8a) and (3.5-10a) imply that finding the solution to $\nabla^2 u = f$ is equivalent to finding a stationary value of

$$(3.5\text{-}10b) \qquad 2J(u) = \int_\Omega u(\nabla^2 u - 2f)d\mathbf{x} + \oint_{\partial\Omega} (u - 2g)\frac{\partial u}{\partial n}d\mathbf{x}.$$

It is possible to derive "reduced forms" of the previous variational principles. These are obtained by restricting the admissible functions u to those that already satisfy the prescribed boundary conditions (3.5-3). For such functions Equation (3.5-10a) becomes

$$(3.5\text{-}11) \qquad 2J(u) = \int_\Omega u(\nabla^2 u - 2f)dx + \oint_{\partial\Omega} T_1(u)T_2(u)dx.$$

If the variation of J is taken over functions $u + \delta u$ that satisfy the boundary conditions, then, by linearity of T_1, $T_1(\delta u) = T_1(u + \delta u - u) = T_1(u + \delta u) - T_1(u) = 0$. Thus,

$$2\delta J(u) = \int_\Omega [\delta u(\nabla^2 u - 2f) + u\nabla^2 \delta u]dx + \oint_{\partial\Omega} T_1(u)T_2(\delta u)dx.$$

Hence, the variation of J vanishes,

$$\delta J(u) = \int_\Omega (\nabla^2 u - f)\delta u\, dx = 0,$$

if and only if Poisson's equation holds.

More explicit expressions of the reduced functionals are as follows. For Dirichlet boundary conditions, solving the Poisson problem is equivalent to setting the variation of

$$(3.5\text{-}12a) \qquad 2J(u) = \int_\Omega u(\nabla^2 u - 2f)dx - \oint_{\partial\Omega} u\frac{\partial u}{\partial n}dx$$

equal to zero. For Neumann boundary conditions, the functional whose variation vanishes at the solution to the Poisson problem is the following:

$$(3.5\text{-}12b) \qquad 2J(u) = \int_\Omega u(\nabla^2 u - 2f)dx + \oint_{\partial\Omega} u\frac{\partial u}{\partial n}dx.$$

Finally, for Robin boundary conditions, the solution to Poisson's equation forces the variation of

$$2J(u) = \int_\Omega u(\nabla^2 u - 2f)dx$$
$$+ (1 + \alpha^2)^{-1} \oint_{\partial\Omega} (\frac{\partial u}{\partial n} + \alpha u)(u - \alpha\frac{\partial u}{\partial n})dx$$

to vanish. Notice that the boundary integrals in Equations (3.5-12) no longer explicitly contain the prescribed boundary data. For Dirichlet problems the following form frequently appears in the literature:

$$(3.5\text{-}13) \qquad 2J(u) = -\int_\Omega (\nabla u \cdot \nabla u + 2fu)dx.$$

One can derive this form by applying the divergence theorem in Equation (3.5-12a). Similar expressions can be obtained for the other types of boundary conditions.

3.6. Maximum Principles.

A **maximum principle** for a PDE states that the solution of the PDE, subject to boundary conditions, can be characterized as the function that maximizes a certain functional over a specified class of **admissible functions**. This definition differs from another also in common use. The latter notion of maximum principle, which we shall not discuss, concerns the magnitudes of functions satisfying a PDE. For many PDEs there is a deep connection between maximum (or minimum) principles in the sense of interest here and certain finite-element approximations. We shall briefly explore this connection later in this chapter. In the case of Poisson's equation, the functional to be maximized is J, defined in the previous section. However, as we shall see, the possibility of formulating maximum principles for particular boundary-value problems depends on whether the quadratic form $\langle Pv, v \rangle$ is negative definite. Once we know that the quadratic form is nonpositive, we can apply a rather classical argument to demonstrate that the solution u to the PDE maximizes J. In developing maximum principles we shall encounter several variants of this argument.

Let us first investigate when $\langle Pv, v \rangle$ can be expected to be nonpositive. Applying integration by parts to Equation (3.5-2), we find

$$(3.6\text{-}1) \qquad \langle Pv, v \rangle = -\int_{\Omega} \nabla v \cdot \nabla v \, d\mathbf{x} + \oint_{\partial \Omega} [v \frac{\partial v}{\partial n} - T_1(v) T_2(v)] d\mathbf{x}.$$

The bilinear form

$$(3.6\text{-}2) \qquad D(u, v) = \int_{\Omega} \nabla u \cdot \nabla v \, d\mathbf{x}$$

appearing in the right side of Equation (3.6-1) is known as the **Dirichlet inner product**. Its associated quadratic form $D(v, v)$ is nonnegative (since the integrand is a perfect square) and vanishes only when v is a constant. However, in the general case the boundary integral appearing in the right side of (3.6-1) does not have a definite sign. In fact, the only case when $\langle Pv, v \rangle$ is nonpositive for arbitrary functions $v \in C^2(\Omega)$ is that for which

$$(3.6\text{-}3) \qquad v \frac{\partial v}{\partial n} = T_1(v) T_2(v).$$

This case corresponds to the choice of T_1 and T_2 specified in Equations (3.5-8b), that is, to Neumann boundary conditions.

This observation suggests that we can state a maximum principle for Poisson's equation with Neumann boundary conditions without making any further restrictions. The maximum principle is actually a version of the variational principle stated in the previous section. In this case the class of admissible functions contains all functions that are sufficiently smooth for the integral defining J to make sense. The argument runs as follows: Assume that the function u is a solution of $\nabla^2 u = f$ with Neumann boundary conditions. We can write any other admissible function as $u + v$, with no restriction on v except that v must be sufficiently smooth. We wish to show that $J(u+v) \leq J(u)$. Applying the definition of J in Equation (3.5-10a), the identity given in Equation (3.5-2), and the stipulation (3.5-1) on the linear boundary operators T_1 and T_2, we find, after some manipulation,

$$(3.6\text{-}4) \qquad J(u+v) = J(u) - \frac{1}{2} \int_{\Omega} (2\nabla^2 u - 2f) v \, d\mathbf{x}$$
$$- \frac{1}{2} \oint_{\partial\Omega} [2T_1(u) - 2g] T_2(v) d\mathbf{x} + \frac{1}{2} \langle \mathcal{P}v, v \rangle.$$

Since u solves Poisson's equation with the given boundary conditions, the second and third terms on the right vanish, leaving us with

$$(3.6\text{-}5) \qquad J(u+v) - J(u) = \frac{1}{2} \langle \mathcal{P}v, v \rangle.$$

We have already shown for this problem that $\langle \mathcal{P}v, v \rangle < 0$ unless v is a constant, and consequently $J(u) > J(u+v)$ unless v is a constant. This line of reasoning shows, by the way, that any two solutions of the same Neumann problem differ at most by an additive constant.

To deduce maximum principles for more general problems, we must restrict the class of admissible functions to those that already satisfy the boundary conditions. As a result we must impose extra conditions on the perturbations v to guarantee that $u+v$ will be admissible, and we must also make sure that the quadratic form $\langle \mathcal{P}v, v \rangle$ is nonpositive. For this purpose, if the original boundary-value problem requires $T_1(u) = g$, we shall restrict v to satisfy the homogeneous boundary conditions $T_1(v) = 0$ on $\partial\Omega$. This ensures that the perturbed function $u + v$ will remain admissible. Assume further that $\alpha \geq 0$ in the general expression (3.5-9). Under these conditions the quadratic form $\langle \mathcal{P}v, v \rangle$ is nonpositive for Dirichlet and Robin boundary conditions (as well as for the Neumann conditions already considered). For example, if v satisfies a homogeneous Dirichlet problem we have, from Equations (3.6-3) and (3.5-8a),

$$(3.6\text{-}6) \qquad v \frac{\partial v}{\partial n} - T_1(v) T_2(v) = 2v \frac{\partial v}{\partial n} \equiv 0,$$

when $v \equiv 0$ on $\partial\Omega$. For Neumann and Robin problems, nonpositivity of $\langle \mathcal{P}v, v \rangle$ holds because, from Equation (3.5-9),

$$\frac{\partial v}{\partial n} = -\alpha v + T_1(v) = -\alpha v,$$

when $T_1(v) \equiv 0$ on $\partial\Omega$. In view of Equation (3.6-1), this shows that

$$(3.6\text{-}7) \qquad \langle \mathcal{P}v, v \rangle = -\int_\Omega \nabla v \cdot \nabla v \, dx - \oint_{\partial\Omega} \alpha v^2 \, dx,$$

which is nonpositive when $\alpha \geq 0$.

Now let us see in detail how maximum principles for Dirichlet and Robin problems follow when we constrain the functions v to satisfy homogeneous boundary conditions. We shall see in the process that, in this restricted case, the maximum principle for Neumann problems amounts to a special case of that for Robin problems. Suppose u satisfies Poisson's equation subject to the boundary conditions $T_1(u) = g$ on $\partial\Omega$. Any other admissible function can be written as $u + v$, where v satisfies $T_1(v) = 0$. For the Dirichlet problem, Equation (3.6-1) together with Equation (3.6-5) imply that $J(u + v) - J(u) = -\frac{1}{2}\mathcal{D}(v,v)$, showing that $J(u) > J(u + v)$ unless v is a constant. Actually, this constant must be zero because $v \equiv 0$ on $\partial\Omega$. This argument, by the way, establishes uniqueness for Dirichlet problems.

For the Robin problem, Equations (3.6-5) and (3.6-7) imply that

$$J(u + v) - J(u) = -\frac{1}{2}\mathcal{D}(v,v) - \oint_{\partial\Omega} \alpha v^2 \, dx.$$

Hence $J(u) > J(u+v)$ unless $\alpha = 0$. The case $\alpha = 0$ is precisely the special case corresponding to the Neumann problem, which we have already treated more generally. In that case we found that $u+v$ also maximizes J provided v is a constant, a fact that remains valid in this more restricted setting. More important is the case $\alpha \neq 0$. Here our argument yields the maximum principle for Robin problems, as desired. It also establishes uniqueness for Robin problems.

The converses of these extremal principles state that, if J attains a maximum at u, then u is a solution of the boundary-value problem. They can be established in a straightforward manner using the fact that, when J attains a maximum at u, the variation $\delta J(u)$ vanishes. Then, using the variational principles of the previous sections proves the desired results. To be more precise, for any fixed function v and any real number ε, define a function

$$F(\varepsilon) = J(u) - J(u + \varepsilon v).$$

By appealing to Equation (3.5-7), we can rewrite this function as follows:

$$(3.6\text{-}8) \qquad F(\varepsilon) = \varepsilon \left[\langle \mathcal{P}u, v \rangle - \int_\Omega fv \, dx + \oint_{\partial\Omega} gT_2(v)\,dx \right] + \frac{\varepsilon^2}{2}\langle \mathcal{P}v, v \rangle.$$

If J attains a maximum at u, then $F(\varepsilon)$ attains a maximum at $\varepsilon = 0$. Therefore the derivative of F with respect to ε must vanish when $\varepsilon = 0$:

$$(3.6\text{-}9) \qquad F'(0) = \langle \mathcal{P}u, v \rangle - \int_{\Omega} fv\, d\mathbf{x} + \oint_{\partial\Omega} g\mathcal{T}_2(v)\, d\mathbf{x} = 0.$$

This equation is the same as Equation (3.5-4). Note that the perturbation function v either is arbitrary (for the unrestricted maximum principle) or satisfies zero boundary conditions (for the restricted maximum principle). We can thus apply the fundamental theorem of the calculus of variations to conclude that $\nabla^2 u = f$ in Ω and $\mathcal{T}_1(u) = g$ on $\partial\Omega$.

3.7. Invariance Under Translation and Fundamental Solutions.

Let us now leave the topic of variational and maximum principles and discuss certain properties of *linear* elliptic operators. The next two sections culminate in the development of integral representations that will form an important point of departure for the boundary-element methods discussed in Section 3.17.

Invariance under translation.

Whenever a differential operator $\mathcal{L}$ has constant coefficients, the set of solutions of the homogeneous equation $\mathcal{L}u = 0$ is **invariant under translation**. To see what this means, let $u(\mathbf{x})$ be a solution to $\nabla^2 u = 0$, and let $\mathbf{a}$ be any fixed vector. Then the **translate** $v(\mathbf{x}) = u(\mathbf{x} + \mathbf{a})$ is also a solution. In particular, the translate of a harmonic function is harmonic: If $\nabla^2 u = 0$ and v is defined as above, then $\nabla^2 v = 0$. It is especially profitable to combine this property with the principle of superposition explained in Section 3.4. Starting from simple solutions one can construct very rich families by translation and then use the principle of superposition to represent any other solution. The integral representations presented in the next section furnish an important class of examples. These representations will require the operator to have fundamental solutions, which we define below. In general, the construction of fundamental solutions is difficult for operators with variable coefficients, but the process is greatly simplified when the solutions are invariant under translation. We illustrate this next.

Fundamental solutions.

Let $\boldsymbol{\xi}$ be a fixed point of two- or three-dimensional Euclidean space. A function $K(\boldsymbol{\xi}, \mathbf{x})$ is said to be a **fundamental solution** for the Laplace operator, with singularity at $\boldsymbol{\xi}$, when four conditions hold. First, $K(\boldsymbol{\xi}, \mathbf{x})$

must be harmonic as a function of $\mathbf{x}$, except when $\mathbf{x} = \boldsymbol{\xi}$. Second, if Ξ is any arbitrary sphere of radius R centered at $\mathbf{x} = \boldsymbol{\xi}$, then the surface integral

$$(3.7\text{-}1a) \qquad \oint_{\Xi} |\, K(\boldsymbol{\xi},\mathbf{x})\,|\, dx \to 0 \qquad \text{as} \qquad R \to 0.$$

Third, the function $K(\boldsymbol{\xi},\mathbf{x})$ has a definite sign (either positive or negative) on the surface Ξ when the radius R is sufficiently small. Fourth, K must satisfy

$$(3.7\text{-}1b) \qquad \oint_{\Xi} \frac{\partial K}{\partial n}(\boldsymbol{\xi},\mathbf{x})dx \to M \neq 0 \quad \text{as} \quad R \to 0.$$

It is worth remarking that fundamental solutions for ∇^2 can also be defined, using the Dirac distribution centered at $\mathbf{x} = \boldsymbol{\xi}$, as solutions to $\nabla^2 K(\mathbf{x},\boldsymbol{\xi}) = \delta(\mathbf{x} - \boldsymbol{\xi})$, where we consider ∇^2 to differentiate with respect to $\mathbf{x}$.

For example, in three dimensions take $K(\mathbf{0},\mathbf{x}) = 1/r$, where $r = \|\mathbf{x}\|_2$, the Euclidean length of $\mathbf{x}$. A direct computation shows that $K(\mathbf{0},\mathbf{x})$ is harmonic except at the origin $\mathbf{x} = \mathbf{0}$ of the coordinate system. On the sphere Ξ of radius R centered at $\boldsymbol{\xi} = \mathbf{0}$, we have $K(\mathbf{0},\mathbf{x}) = 1/R$ and

$$\frac{\partial K}{\partial n}(\mathbf{0},\mathbf{x}) = -\frac{1}{R^2}.$$

Therefore

$$\oint_{\Xi} |\, K(\mathbf{0},\mathbf{x})\,|\, dx = 4\pi R \to 0 \quad \text{as} \quad R \to 0$$

and

$$(3.7\text{-}2) \qquad \oint_{\Xi} \frac{\partial K(\mathbf{0},\mathbf{x})}{\partial n} dx = -4\pi \to -4\pi \quad \text{as} \quad R \to 0.$$

Thus, $K(\mathbf{0},\mathbf{x})$ is a fundamental solution for the Laplace operator with singularity at the origin $\mathbf{x} = \mathbf{0}$ of the coordinate system. Applying the invariance under translations, we see that $K(\boldsymbol{\xi},\mathbf{x}) = 1/\|\mathbf{x} - \boldsymbol{\xi}\|_2$ is a fundamental solution with singularity at an arbitrary point $\boldsymbol{\xi}$ for the Laplace operator in three dimensions.

Similarly, the reader can verify that $K(\boldsymbol{\xi},\mathbf{x}) = \ln\|\mathbf{x} - \boldsymbol{\xi}\|_2$ is a fundamental solution for the Laplace operator in two dimensions with singularity at $\boldsymbol{\xi}$. In this case Equation (3.7-1b) holds, with $M = 2\pi$.

3.8. Integral Representation Theorems.

Integral representations similar to that presented in Equation (3.4-6) can be derived from Green's second identity (3.4-3). The usefulness of these representations will become apparent when we discuss boundary-element methods in Section 3.17. Let $K(\boldsymbol{\xi},\mathbf{x})$ be a fundamental solution of Laplace's equation with singularity at $\boldsymbol{\xi}$, where $\boldsymbol{\xi}$ is an interior point of a problem domain Ω, as shown in Figure 3-2. Let $\Xi(R)$ be a sphere or circle as in Section 3.7, and suppose the radius R is sufficiently small that Ξ lies in the interior of the region Ω. Denote by Ω_0 the open region obtained from Ω by deleting the sphere Ξ and its interior. If we apply Green's third identity (3.4-3) on Ω_0, taking $v(\mathbf{x}) = K(\boldsymbol{\xi}, \mathbf{x})$ and recalling that K is harmonic with respect to the variable $\mathbf{x}$ in Ω_0, we find

$$(3.8\text{-}1) \qquad \oint_{\partial\Omega_0} u(\mathbf{x})\frac{\partial K}{\partial n}(\boldsymbol{\xi}, \mathbf{x})\,dx = \oint_{\partial\Omega_0} K(\boldsymbol{\xi}, \mathbf{x})\frac{\partial u}{\partial n}(\mathbf{x})\,dx$$
$$- \int_{\Omega_0} K(\boldsymbol{\xi}, \mathbf{x})\nabla^2 u(\mathbf{x})\,dx.$$

Clearly $\partial\Omega_0 = \Xi \cup \partial\Omega$, where Ξ and $\partial\Omega$ are disjoint, as Figure 3-2 implies. Also, observe that the unit normal vector on Ξ points outward from Ω_0, that is, toward the center of the sphere Ξ. This sense is opposite to that taken in Equation (3.7-1b). In view of these facts and the definition of fundamental solutions, we see that, as $R \to 0$,

$$(3.8\text{-}2a) \qquad \oint_{\partial\Omega_0} u(\mathbf{x})\frac{\partial K}{\partial n}(\boldsymbol{\xi}, \mathbf{x})\,dx \to \oint_{\partial\Omega} u(\mathbf{x})\frac{\partial K}{\partial n}(\boldsymbol{\xi}, \mathbf{x})\,dx - Mu(\boldsymbol{\xi}),$$

and

$$(3.8\text{-}2b) \qquad \oint_{\partial\Omega_0} K(\boldsymbol{\xi}, \mathbf{x})\frac{\partial u}{\partial n}(\mathbf{x})\,dx \to \oint_{\partial\Omega} K(\boldsymbol{\xi}, \mathbf{x})\frac{\partial u}{\partial n}(\mathbf{x})\,dx.$$

[Rigorously speaking, to establish the limits (3.8-2) we must assume that u is continuous and ∇u bounded.] Taking the limit as $R \to 0$ in Equation (3.8-1) now yields an integral representation for the solution u:

$$(3.8\text{-}3) \qquad u(\boldsymbol{\xi}) = \frac{1}{M}\oint_{\partial\Omega}\left[u(\mathbf{x})\frac{\partial K}{\partial n}(\boldsymbol{\xi}, \mathbf{x}) - K(\boldsymbol{\xi}, \mathbf{x})\frac{\partial u}{\partial n}(\mathbf{x})\right]dx$$
$$+ \frac{1}{M}\int_{\Omega} K(\boldsymbol{\xi}, \mathbf{x})\nabla^2 u(\mathbf{x})\,dx,$$

for points $\boldsymbol{\xi}$ belonging to the interior of Ω.

When $\boldsymbol{\xi}$ lies on the boundary $\partial\Omega$ of Ω, the previous argument can also be applied. However, in that case, in the limit when $R \to 0$, only half of the surface Ξ contributes to integral over $\partial\Omega_0$, and as a result we must replace M by $M/2$ in Equation (3.8-2a). The same happens in Equation (3.8-3). Thus when $\boldsymbol{\xi} \in \partial\Omega$ one has the following integral representation:

$$(3.8\text{-}4) \qquad u(\boldsymbol{\xi}) = \frac{2}{M}\oint_{\partial\Omega}\left[u(\mathbf{x})\frac{\partial K}{\partial n}(\boldsymbol{\xi}, \mathbf{x}) - K(\boldsymbol{\xi}, \mathbf{x})\frac{\partial u}{\partial n}(\mathbf{x})\right]dx$$
$$+ \frac{2}{M}\int_{\Omega} K(\boldsymbol{\xi}, \mathbf{x})\nabla^2 u(\mathbf{x})\,dx.$$

In particular, using the fundamental solutions $K(\boldsymbol{\xi}, \mathbf{x})$ derived in Section 3.7, we have the following integral representation formulas, which hold for points $\boldsymbol{\xi}$ in the interior of the region Ω. In two dimensions,

$$(3.8\text{-}5a) \qquad u(\boldsymbol{\xi}) = \frac{1}{2\pi} \oint_{\partial\Omega} \left[\frac{u}{r} \frac{\partial r}{\partial n} - (\ln r) \frac{\partial u}{\partial n} \right] d\mathbf{x} + \frac{1}{2\pi} \int_{\Omega} (\ln r) \nabla^2 u \, d\mathbf{x},$$

and, in three dimensions,

$$(3.8\text{-}5b) \qquad u(\boldsymbol{\xi}) = \frac{1}{4\pi} \oint_{\partial\Omega} \left[\frac{1}{r} \frac{\partial u}{\partial n} - u \frac{\partial}{\partial n} \left(\frac{1}{r} \right) \right] d\mathbf{x}$$
$$- \frac{1}{4\pi} \int_{\Omega} \frac{1}{r} \nabla^2 u \, d\mathbf{x}.$$

Here, as before, $r = \| \mathbf{x} - \boldsymbol{\xi} \|_2$. When $\boldsymbol{\xi}$ lies on the boundary, we must multiply the right sides of Equations (3.8-5) by two.

Although it may not be apparent yet, the integral representation formulas (3.8-3) can be used to solve boundary-value problems. When the boundary conditions are given by Equation (3.5-3), it is convenient to transform Equation (3.8-3) into the following:

$$(3.8\text{-}6) \quad u(\boldsymbol{\xi}) = \frac{1}{M} \oint_{\partial\Omega} [T_1(K) T_2(u) - T_1(u) T_2(K)] d\mathbf{x} + \frac{1}{M} \int_{\Omega} K \nabla^2 u \, d\mathbf{x}.$$

In this equation, the differential operators in T_1 and T_2 act on the $\mathbf{x}$-dependencies in K. Hence, if u is solution of the boundary-value problem, then

$$(3.8\text{-}7) \qquad u(\boldsymbol{\xi}) = \frac{1}{M} \oint_{\partial\Omega} [T_1(K) T_2(u) - g \, T_2(K)] d\mathbf{x} + \frac{1}{M} \int_{\Omega} K f \, d\mathbf{x}.$$

This representation does not generally give the solution to the problem explicitly, because the boundary value $T_2(u)$ is not known. However, the fundamental solution is not unique, and we can select it cleverly to force $T_1(K) = 0$ on $\partial\Omega$. In other words, we can impose on K the homogeneous version of the boundary conditions on the unknown solution u. With this choice of boundary conditions, the fundamental solution $K(\boldsymbol{\xi}, \mathbf{x})$ becomes a **Green's function** for the boundary value problems. The significance of this choice lies in the observation that, if $T_1(K) = 0$ in the boundary integral in Equation (3.8-7), then this equation gives $u(\boldsymbol{\xi})$ entirely in terms of known information. This reasoning furnishes a highly useful exact solution technique, provided we can find $K(\boldsymbol{\xi}, \mathbf{x})$. We shall revisit the concept of fundamental solutions in Section 3.17.

3.9. Finite-Difference Approximations.

Having established some mathematical background for Laplace's equation, we now turn to its numerical approximation. To formulate finite-difference analogs for Laplace's equation, we can apply any of the procedures described in Sections 2.6 through 2.9 to approximate the PDE. To illustrate the procedure, consider the two-dimensional version

$$(3.9-1) \qquad \nabla^2 u \equiv \frac{\partial^2 u}{\partial x^2} + \frac{\partial^2 u}{\partial y^2} = 0 \quad \text{on} \quad \Omega,$$

subject to the Dirichlet boundary condition

$$(3.9\text{-}2) \qquad u(\mathbf{x}) = g(\mathbf{x}) \quad \text{on} \quad \partial\Omega.$$

For simplicity, let us take Ω to be the unit square, drawn in Figure 3-3. Let us subdivide Ω by means of a uniform grid made of n^2 squares of side length $h = 1/n$, whose vertices, or grid points, lie at $(x_i, y_j) = (ih, jh)$, with

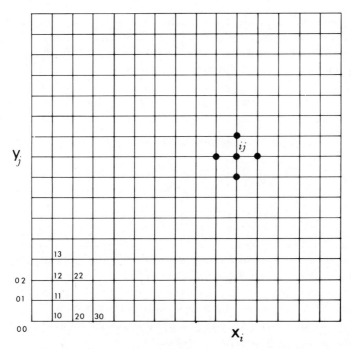

Figure 3-3. Discretization of the domain Ω for Laplace's equation in two dimensions.

$i, j = 0, 1, ..., n$. We shall use the notation $u_{i,j}$ to signify the approximations to nodal values $u(x_i, y_j)$.

To approximate Equation (3.9-1), we can use the central difference approximation given in Equation (2.6-16), applying it in each coordinate direction. Thus,

$$\frac{\partial^2 u}{\partial x^2}(x_i, y_j) = (u_{i-1,j} + u_{i+1,j} - 2u_{i,j})/h^2 + \mathcal{O}(h^2).$$

Similarly, in the y-direction,

$$\frac{\partial^2 u}{\partial y^2}(x_i, y_j) = (u_{i,j-1} + u_{i,j+1} - 2u_{i,j})/h^2 + \mathcal{O}(h^2).$$

Adding these equations, one gets

$$(3.9\text{-}3) \qquad \nabla^2(x_i, x_j) = (u_{i+1,j} + u_{i,j+1} + u_{i-1,j} \\ + u_{i,j-1} - 4u_{i,j})/h^2 + \mathcal{O}(h^2).$$

Observe that Equation (3.9-3) gives a five-point approximation of the Laplace operator that has an $\mathcal{O}(h^2)$ truncation error. This approximation can be applied at any interior node, and doing so replaces Equation (3.9-1) by the approximate finite-difference equation

$$(3.9\text{-}4) \qquad 4u_{i,j} - (u_{i+1,j} + u_{i,j+1} + u_{i-1,j} + u_{i,j-1}) = 0.$$

The boundary conditions (3.9-2) imply that

$$(3.9\text{-}5) \qquad u_{i,j} = g_{i,j} \quad \text{for} \quad (x_i, y_j) \in \partial\Omega.$$

This latter set of equations reduces the number of unknowns to the number of interior nodes, namely $(n-1)^2$. Equation (3.9-4) then supplies $(n-1)^2$ equations because it applies to the interior nodes only.

In most cases, central difference approximations give the highest order of accuracy that can be achieved without going to larger or more complicated stencils. Thus, for the Laplace operator, it is impossible to attain more than $\mathcal{O}(h^2)$ accuracy by a five-point approximation. However, using nine-point approximations, one can obtain smaller truncation errors. We shall review two such approximations.

One very simple approach is to use higher-order approximations for $\partial^2 u/\partial x^2$ and $\partial^2 u/\partial y^2$. Thus, using the $\mathcal{O}(h^4)$ central difference approximation given in Table 2-3, we have

(3.9-6a) $\dfrac{\partial^2 u}{\partial x^2}(x_i, y_j) = \dfrac{-1}{12h^2}(u_{i-2,j} - 16u_{i-1,j} + 30u_{i,j}$

$\qquad\qquad - 16u_{i+1,j} + u_{1+2,j}) + \mathcal{O}(h^4)$

and

(3.9-6b) $\dfrac{\partial^2 u}{\partial y^2}(x_i, y_j) = \dfrac{-1}{12h^2}(u_{i,j-2} - 16u_{i,j-1} + 30u_{i,j}$

$\qquad\qquad - 16u_{i,j+1} + u_{i,j+2}) + \mathcal{O}(h^4).$

These approximations, taken together, yield the nine-point approximation

(3.9-7) $u_{i,j} = \dfrac{-1}{60}(u_{i-2,j} + u_{i,j-2} - 16u_{i-1,j} - 16u_{i+1,j}$

$\qquad\qquad - 16u_{i,j-1} - 16u_{i,j+1} + u_{i+2,j} + u_{i,j+2}) + \mathcal{O}(h^4)$

for Laplace's equation. Figure 3-4a illustrates the stencil for this analog. However, such an approximation is computationally inconvenient, specially near the boundary of the region, since it calls for values of u at grid points $2h$ distant from the central point (x_i, y_j).

For efficient computation, it is more convenient to use a stencil of nine points made of the central node together with the eight nearest neighboring nodes, as shown in Figure 3-4b. To derive an approximation for the Laplace operator using values at these nodes, consider a Taylor expansion about (x_i, y_j), having the form

(3.9-8) $u(x_i + h, y_j + h) = u_{i,j} + \left(\dfrac{\partial u}{\partial x}h + \dfrac{\partial u}{\partial y}h\right)_{i,j}$

$\qquad\qquad + \dfrac{1}{2!}\left(\dfrac{\partial^2 u}{\partial x^2}h^2 + 2\dfrac{\partial^2 u}{\partial x\partial y}h^2 + \dfrac{\partial^2 u}{\partial y^2}h^2\right)_{i,j}$

$\qquad\qquad + \dfrac{1}{3!}\left(\dfrac{\partial^3 u}{\partial x^3}h^3 + 3\dfrac{\partial^3 u}{\partial x^2\partial y^2}h^3\right.$

$\qquad\qquad \left. + 3\dfrac{\partial^3 u}{\partial x\partial y^2}h^3 + \dfrac{\partial^3 u}{\partial y^3}h^3\right)_{i,j} + \mathcal{O}(h^4).$

We can use this expansion to establish a system of eight equations relating the values of the function u at the points of the stencil to the partial derivatives of u at (x_i, y_j). Additional equations can be obtained by taking partial derivatives of the differential equation satisfied by u. For Laplace's equation this procedure yields, for example,

(3.9-9a) $\dfrac{\partial^3 u}{\partial x^2 \partial y} = -\dfrac{\partial^3 u}{\partial y^3}, \qquad \dfrac{\partial^3 u}{\partial x^3} = -\dfrac{\partial^3 u}{\partial x\partial y^2},$

and

$$(3.9\text{-}9b) \qquad \frac{\partial^4 u}{\partial x^4} = -\frac{\partial^4 u}{\partial x^2 \partial y^2} = \frac{\partial^4 u}{\partial y^4}, \qquad \frac{\partial^4 u}{\partial x^3 \partial y} = \frac{\partial^4 u}{\partial x \partial y^3}.$$

For the stencil of points illustrated in Figure 3-4b, one can verify that

$$\nabla^2 u = \frac{u_1 + u_2 + u_3 + u_4 - 4u_0}{h^2} - \frac{\partial^2 u}{\partial x^2 \partial y^2} \frac{2h^2}{4!} + O(h^6)$$

and

$$\nabla^2 u = \frac{u_5 + u_6 + u_7 + u_8 - 4u_0}{2h^2} + \frac{\partial^2 u}{\partial x^2 \partial y^2} \frac{4h^2}{4!} + O(h^6).$$

By combining these two equations, one can eliminate the $O(h^2)$ terms. The corresponding $O(h^6)$ approximation for Laplace's equation is then

$$(3.9\text{-}10) \qquad u_0 = \frac{1}{20} \left[4(u_1 + u_2 + u_3 + u_4) + u_5 + u_6 + u_7 + u_8 \right].$$

Thus far we have considered a uniform grid only. Generally, we may wish to develop finite-difference approximations for a grid that has variable spacing, either throughout Ω or in some specific subregion of Ω. A rather simple possibility is that the mesh h of the grid in the x-direction is different

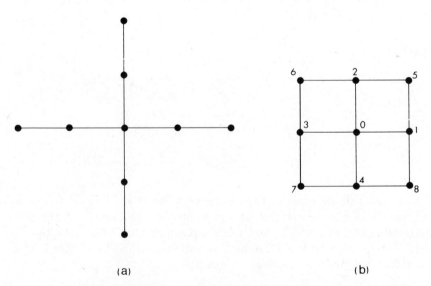

(a)   (b)

Figure 3-4. Finite-difference stencils for two different nine-point approximations to Laplace's equation.

from the mesh k in the y-direction. In this case, the argument used to obtain Equation (3.9-4) yields

$$(3.9\text{-}11) \quad 2\left(1+\frac{h^2}{k^2}\right)u_{i,j} - (u_{i+1,j}+u_{i-1,j}) - \frac{h^2}{k^2}(u_{i,j+1}+u_{i,j-1}) = 0.$$

The local truncation error remains the same, namely $\mathcal{O}(h^2)$, when k is kept fixed.

It is possible to obtain a five-point approximation for more general types of variable spacing, as illustrated in Figure 3-5, using once again the argument that yielded the approximation (3.9-3). In this case the approximation is

$$(3.9\text{-}12) \quad \frac{\partial^2 u}{\partial x^2} + \frac{\partial^2 u}{\partial y^2} \simeq \alpha_1 u_1 + \alpha_2 u_2 + \alpha_3 u_3 + \alpha_4 u_4 - \alpha_0 u_0 = 0,$$

where

$$\alpha_0 = \frac{2}{h^2}\left(\frac{1}{s_1 s_3} + \frac{1}{s_2 s_4}\right), \quad \alpha_1 = \frac{2}{s_1 h^2(s_1 + s_3)}, \quad \alpha_3 = \frac{2}{s_3 h^2(s_1 + s_3)},$$

$$\alpha_2 = \frac{2}{s_2 h^2(s_2 + s_4)}, \quad \alpha_4 = \frac{2}{s_4 h^2(s_2 + s_4)}.$$

However, this approximation is in general accurate only to $\mathcal{O}(h)$. The special case of equal spacing corresponds to $s_1 = s_2 = s_3 = s_4$, for which

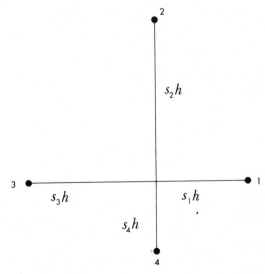

Figure 3-5. Nodal geometry for a finite-difference grid with nonuniform spacing.

the truncation error reduces to $\mathcal{O}(h^2)$. We conclude that, when one is dealing with Laplace's operator, the use of equal spacing often has definite advantages.

3.10. Boundary Conditions via Finite Differences.

Regarding the approximation of boundary conditions, there are two aspects that must be considered. One is the methodology for transforming the specified boundary conditions into finite-difference analogs. The other aspect concerns the treatment of boundary data when the boundary $\partial\Omega$ is curved or, more generally, whenever the boundary does not coincide with the grid lines. We deal with these issues separately, assuming for the time being that the domain Ω is a square.

We have considered three kinds of boundary conditions: Dirichlet, Neumann, and Robin. When Dirichlet boundary conditions apply, their discretization is straightforward. Indeed, these boundary conditions simply imply that the nodal values $u_{i,j}$ are known at boundary nodes. Thus, for example, if the region Ω and the grid are as in Figure 3-3, the boundary conditions (3.9-5) reduce the number of unknowns in Equation (3.9-4) from $n^2 - 4$ (excluding the four corner nodes) to $(n-1)^2$.

Now consider a Neumann boundary condition along the line segment $x = 0$, say

$$(3.10\text{-}1) \qquad\qquad \frac{\partial u}{\partial x} = g(0, y).$$

Returning to Figure 3-3 we see that, in this case, the values $u_{0,1}, u_{0,2}, \ldots$ are unknown. One way in which the boundary condition (3.10-1) can be handled is via finite differences, using a forward difference scheme. As an illustration, let us write the approximation at the node (x_0, y_2):

$$(3.10\text{-}2) \qquad\qquad u_{1,2} - u_{0,2} = g_{0,2}h = g(x_0, y_2)h.$$

This relationship allows us to eliminate $u_{0,2}$ from the five-point approximation at the node (x_1, y_2), yielding

$$(3.10\text{-}3) \qquad\qquad u_{1,2} = \frac{u_{0,2} + u_{1,1} + u_{2,2} + u_{1,3}}{4}$$

$$= \frac{u_{1,2} - g_{0,2}h + u_{1,1} + u_{2,2} + u_{1,3}}{4}.$$

Then a complete system of equations results when we apply Equation (3.10-3) only at internal nodes. However, in this case the boundary approximation (3.10-2) has truncation error $\mathcal{O}(h)$, in contrast to the $\mathcal{O}(h^2)$ error achieved in the interior.

A more refined procedure, having truncation error $\mathcal{O}(h^2)$, requires us to extend the grid beyond the actual domain of definition of the problem. Applying central differences at the node (x_0, y_2), one has

(3.10-4) $\dfrac{u_{1,2} - u_{-1,2}}{2h} = g_{0,2},$ or $u_{-1,2} = u_{1,2} - 2hg_{0,2},$

where we consider the "fictitious" node (x_{-1}, y_2) to lie directly across the y-axis from the "real" node (x_1, y_2). We now apply the five-point approximation at the node (x_0, y_2), and we eliminate $u_{-1,2}$ from the resulting difference formula by means of the approximation (3.10-4). Thus we obtain an equation involving unknown values at actual nodes only and having truncation error consistent with that of the difference equations in the interior. The reader should verify that Robin boundary conditions can be treated in essentially the same manner.

Finally, consider the task of approximating boundary conditions in regions whose boundaries do not coincide with grid lines. Using as before a rectangular grid, we now supplement the set of nodes with the set of intersections of the rectangular grid with the true boundary $\partial\Omega$ of the region. In general this leads to the introduction of irregular stencils, as depicted in Figure 3-6. To produce finite-difference approximations, we simply use standard techniques for irregular grid spacing wherever boundary nodes enter the stencil.

Another approach is to use some form of interpolation to provide values at the regular grid points. To illustrate this technique we show in Figure 3-7 a square grid next to a curved boundary. Then, if

$$u_0^1 = \frac{ku_2' + s_2 k u_4}{k + s_2 k},$$

$$u_0^2 = \frac{ku_1' + s_1 k u_2}{k + s_1 k},$$

$$u_0 = \frac{s_1 k u_0' + s_2 k u_0^2}{k(s_1 + s_2)},$$

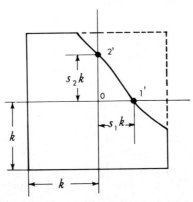

Figure 3-6. Irregular difference stencil near a nonrectangular boundary.

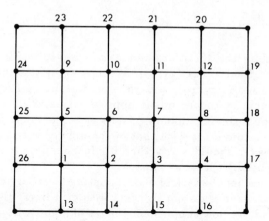

Figure 3-7. Square grid used to illustrate the interpolation of boundary data.

we obtain the following difference formula:

$$u = \frac{ks_2k}{(k+s_1k)(s_1k+s_2k)}u_1' + \frac{ks_1k}{(k+s_2k)(s_1k+s_2k)}u_2'$$
$$+ \frac{s_1ks_2k}{(k+s_1k)(s_1k+s_2k)}u_3 + \frac{s_1ks_2k}{(k+s_2k)(s_1k+s_2k)}u_4.$$

In this form of interpolation, which can be shown to be $\mathcal{O}(k^2)$, we retain the diagonal dominance in the matrix formulation, a feature we shall discuss in the next section. For further details and higher-order difference approximations, the reader should consult Tee (1963), Bramble and Hubbard (1962, 1965), and van Linde (1974).

3.11. Matrix Form of the Finite-Difference Equations.

In the last two sections we discussed finite-difference procedures for transforming boundary-value problems into systems of linear algebraic equations. The general matrix form of the resulting system of equations is

$$(3.11\text{-}1) \qquad\qquad\qquad \mathbf{Au} = \mathbf{b},$$

where $\mathbf{A}$ is a square matrix, $\mathbf{u}$ is a vector of unknown nodal values of the approximate solution, and $\mathbf{b}$ is a vector containing known boundary data. We shall exhibit the structure of this matrix equation shortly. The choice of methods for solving the system (3.11-1) depends on specific properties of the matrix $\mathbf{A}$. In this section we examine the various forms of $\mathbf{A}$ that arise for Laplace's equation.

For simplicity, let us apply the five-point approximation (3.9-4) to a Dirichlet problem in the region of Figure 3-7, with $S = 4$ interior nodes in the x-direction and $T = 3$ interior nodes in the y-direction. Using the notation implied by the figure, we see that the interior values $u_1, u_2, \ldots, u_{12}$ are unknown, while the boundary values $u_{13}, u_{14}, \ldots, u_{26}$ are prescribed. Arranging the difference equations to keep known boundary data on the right side shows (Lapidus and Pinder, 1982) that

$$\mathbf{u} = \begin{bmatrix} u_1 \\ u_2 \\ \vdots \\ u_{12} \end{bmatrix}, \quad \mathbf{b} = \begin{bmatrix} u_{13} + u_{26} \\ u_{14} \\ \vdots \\ u_{19} + u_{20} \end{bmatrix},$$

and $\mathbf{A}$ is a 12×12 (in general, $ST \times ST$) matrix with the structure

$$(3.11\text{-}2) \qquad \mathbf{A} = \begin{bmatrix} \mathbf{T} & \mathbf{B} & & & \\ \mathbf{B} & \mathbf{T} & \mathbf{B} & & \\ & \ddots & \ddots & \ddots & \\ & & \mathbf{B} & \mathbf{T} & \mathbf{B} \\ & & & \mathbf{B} & \mathbf{T} \end{bmatrix}.$$

Here $\mathbf{B}$ and $\mathbf{T}$ are square matrices of size 4×4 (in general, $S \times S$), with tridiagonal structures, namely,

$$(3.11\text{-}3) \qquad \mathbf{T} = \begin{bmatrix} 4 & -1 & & & \\ -1 & 4 & -1 & & \\ & \ddots & \ddots & \ddots & \\ & & -1 & 4 & -1 \\ & & & -1 & 4 \end{bmatrix} = \mathrm{tri}(-1, 4, -1)$$

and

$$(3.11\text{-}4) \qquad \mathbf{B} = -\mathbf{I} = \mathrm{tri}(0, -1, 0).$$

Equation (3.11-2) shows that $\mathbf{A}$ is actually **block-tridiagonal**, having the form

$$(3.11\text{-}5) \qquad \mathbf{A} = \mathrm{tri}(-\mathbf{I}, \mathbf{T}, -\mathbf{I}).$$

If we had not used equal spacing in the x- and y-directions, forcing $h = k$, but rather had chosen $h/k = \beta \neq 1$, the only change would have been that

$$(3.11\text{-}6) \qquad \mathbf{T} = \mathrm{tri}(-\beta, 2(1 + \beta), -\beta).$$

Suppose we had used Neumann boundary conditions with a central difference approximation. Then some of the structure shown above would have been different. In fact one would, on renumbering, obtain

$$
\mathbf{A} = \begin{bmatrix}
\mathbf{T} & 2\mathbf{B} & & & \\
\mathbf{B} & \mathbf{T} & \mathbf{B} & & \\
 & \ddots & \ddots & \ddots & \\
 & & \mathbf{B} & \mathbf{T} & \mathbf{B} \\
 & & & 2\mathbf{B} & \mathbf{T}
\end{bmatrix},
$$

where **B** is as before but

$$
\mathbf{T} = \begin{bmatrix}
4 & -2 & & & \\
-1 & 4 & -1 & & \\
 & \ddots & \ddots & \ddots & \\
 & & -1 & 4 & -1 \\
 & & & -2 & 4
\end{bmatrix}.
$$

Thus it is apparent that the boundary conditions influence the structure of the matrix **A**.

Although we shall shortly show some alternative forms of these matrix equations, it is of interest first to note certain properties of the matrix **A**:

(i) **A** can be of large order since, to obtain accuracy in the finite-difference approximation, S and T may be on the order of 100, 1000, or larger.

(ii) Even though **A** is of large order, it is sparse; that is, only a small fraction of the entries are nonzero.

(iii) Depending on the boundary conditions, **A** is frequently symmetric in the case of Laplace's equation.

(iv) **A** is an **irreducible** matrix. This implies that no sequence of operations interchanging a pair of rows and the identically numbered pair of columns in succession can reduce the matrix **A** to the partitioned form

$$
\mathbf{A} = \begin{bmatrix}
\mathbf{A}_1 & \mathbf{0} \\
\mathbf{A}_2 & \mathbf{A}_3
\end{bmatrix}.
$$

This property has physical significance: If the matrix **A** were reducible, some of the unknown nodal values $u_{s,t}$ would be independent of the boundary conditions.

(v) **A** is **weakly diagonally dominant**; that is, the magnitude of the main diagonal entry is greater than or equal to the sum of the magnitudes of the off-diagonal entries in any row, strict inequality holding in at least one row. (In cases where strict inequality holds in all rows, the matrix is called **strongly diagonally dominant**.)

Although both the boundary conditions and the ordering of the nodes influence the structure of the matrix **A**, it is possible to show that, for the five-point difference analog with Dirichlet boundary conditions, **A** is symmetric and positive definite for any ordering of the nodes. We now turn to the issue of how to solve the matrix equation (3.11-1).

3.12. Direct Methods of Solution.

By **direct methods** for matrix equations $\mathbf{Au} = \mathbf{b}$ we mean methods that, neglecting roundoff, give the solution $\mathbf{u}$ in a finite number of steps. While we cannot give an exhaustive treatment of direct matrix methods here, we shall give a brief overview of some of the most popular options. The interested reader should consult, for example, Golub and Van Loan (1983) for a more thorough discussion.

Gauss elimination.

Perhaps the most naive approach to direct solution is ordinary Gauss elimination. This algorithm, however, is not very efficient, since a general system of N equations, with N large, requires storage of all N^2 entries of **A** and about $N^3/3$ arithmetic operations. Nevertheless, one can reasonably regard Gauss elimination as a starting point in the theory of many more sophisticated direct methods.

Band elimination.

When one uses natural "lexicographic" ordering of the nodes, considerable savings in both storage and computation can be achieved using band elimination. This algorithm consists in triangularizing **A** by forward elimination, operating only on the matrix entries inside a central band containing all the nonzero elements. Thus row j (beginning with $j = 1$) is multiplied by $m_{i,j} = a_{i,j}/a_{j,j}$ and the result subtracted from row i for $i = j + 1, j + 2, \ldots$. The result is an upper triangular matrix **U** that has zeros below the diagonal element in every column. If no row exchanges are required (for example, to avoid division by zero in computing $m_{i,j}$), then the entries in the upper triangular matrix **U** will be zero outside the central band as well.

LU decomposition.

Let **U** be the upper triangular matrix obtained from **A** by band elimination. The reader can verify that this is $\mathbf{U} = \mathbf{L}^{-1}\mathbf{A}$, where **L** is lower traingular with 1's along the main diagonal. The entries of $\mathbf{L}^{-1}$ can be derived from the coefficients $m_{i,j}$ defined earlier and can be stored in place of the strictly lower-diagonal entries of **A** as these are eliminated. This gives the **LU factorization** of **A**. Being closely related to Gauss elimination, it requries about $N^3/3$ operations for a full matrix, but only about Nb^2 operations for a band matrix with half-bandwidth b.

Cholesky decomposition.

When **A** is symmetric and positive definite, as in the central difference approximation to the Dirichlet problem for Laplace's equation, one has the LU decomposition $\mathbf{A} = \mathbf{LU} = \mathbf{R}^\top\mathbf{R}$, where $\mathbf{R} = \sqrt{\mathbf{D}}\mathbf{L}^\top$ is upper triangular and $\mathbf{D} = \mathrm{diag}(\mathbf{U})$ is diagonal with positive diagonal entries. The main advantage of this decomposition is that, in cases where it is applicable, the cost of computation is greatly reduced.

There are other direct methods that present advantages when the matrices treated possess some additional special features or in situations where, by special circumstances, some of the methods just mentioned are less efficient. For example, in rectangular domains, band elimination is reasonably efficient. However, in some other regions (such as in a diamond-shaped one), when band elimination is applied, many zero multiples of rows are subtracted from later rows during elimination. In such cases, **profile elimination** can considerably improve the efficiency of the procedure (George and Liu, 1981). Also, reordering the nodal unknowns may produce considerable reduction in computation costs in certain geometries.

3.13. Iterative Methods.

In contrast to direct methods for $\mathbf{Au} = \mathbf{b}$, iterative methods start with an initial guess for the solution vector **u** and produce better approximations via a sequence of iterations. While these methods do not formally yield **u** in a finite number of steps, one can terminate such a procedure after a finite number of iterations when it has produced a sufficiently good approximate answer. Most iterative schemes possess the attractive feature that they require arithmetic operations only on the nonzero entries of the matrix **A**. The first class of iterative methods that we shall discuss will be **one-step stationary methods**. A one-step method for $\mathbf{Au} = \mathbf{b}$ computes from an initial vector $\mathbf{u}^{(0)}$ a sequence of successive approximations $\mathbf{u}^{(1)}, \mathbf{u}^{(2)}, ...,$ each approximation depending only on the approximation computed at the previous stage in the iteration. The term **stationary** applies when the process used to produce the approximating sequence does not change from one step to the next. Stationary, linear, one-step iterative methods have the general form

$$(3.13\text{-}1) \qquad\qquad \mathbf{u}^{(m+1)} = \mathbf{Gu}^{(m)} + \mathbf{k}.$$

(It is worth mentioning that this matrix form, while providing a convenient vehicle for theory, is *not* an efficient formulation for actual coding.) Here

$u^{(m)}$ is an estimate of the solution u at the most recently known iterative level, while $u^{(m+1)}$ is the "improved" estimate of u to be computed at the next unknown iterative level. In the notation of Equation (3.13-1), the improvement at each step is the following:

$$(3.13\text{-}2) \qquad u^{(m+1)} - u^{(m)} = (G - I)u^{(m)} + k.$$

We shall consider two procedures for generating a new iterative method from a given one. The first is by making use of improved entries of the vector u as soon they become available within an iteration. The second procedure is **overrelaxation**, that is, multiplying the increment appearing on the left side of Equation (3.13-2) by a scalar $\omega \neq 1$ at each iterative step. The first four methods to be presented are derived from a very simple scheme, called **Jacobi's method**, via successive application of these procedures, as schematized below:

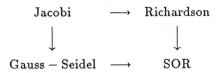

In this diagram, the horizontal arrows indicate the use of overrelaxation, while the vertical arrows indicate the use of information generated within iterations.

Jacobi's method.

Suppose for the moment that the matrix A has unit diagonal entries, that is, $\text{diag}(A) = I$. If we decompose $A = I - B$, then the system of equations (3.11-1) can be written as follows:

$$(3.13\text{-}3) \qquad (I - B)u = b.$$

Here, as usual, I is the identity matrix. For the time being, assume $k = b$. Equation (3.13-3) can also be written more suggestively as follows:

$$(3.13\text{-}4) \qquad u = Bu + k.$$

Given an initial guess $u^{(0)}$, we can obtain improved estimates by replacing u on the right side of (3.13-4) by its most recently computed approximation $u^{(m)}$. This is the basic idea behind Jacobi's method. The iterative process is as follows: Pick $u^{(0)}$, and compute successive improved approximations by the formula

$$(3.13\text{-}5) \qquad u^{(m+1)} = Bu^{(m)} + k.$$

The application of Jacobi's method to equations of the form

$$(3.13\text{-}6) \qquad\qquad \mathbf{Au} = \mathbf{b}$$

with more general matrices $\mathbf{A}$ is straightforward so long as $\mathbf{A}$ has nonzero diagonal entries. In this case, multiplying Equation (3.13-6) by $\mathbf{D}^{-1} = [\text{diag}(\mathbf{A})]^{-1}$ gives Equation (3.13-4), with

$$(3.13\text{-}7) \qquad\qquad \mathbf{B} = \mathbf{I} - \mathbf{D}^{-1}\mathbf{A}, \qquad \mathbf{k} = \mathbf{D}^{-1}\mathbf{b}.$$

Gauss-Seidel iteration.

This scheme is the modification to Jacobi's method that one obtains by making use of improved values $u_i^{(m+1)}$ as soon as they are available. Thus Equation (3.13-5), written componentwise, becomes

$$(3.13\text{-}8) \qquad u_i^{(m+1)} = \sum_{j<i} b_{ij} u_j^{(m+1)} + \sum_{j>i} b_{ij} u_j^{(m)} + k_i.$$

Observe that the Gauss-Seidel method requires the same number of arithmetic operations per iteration as Jacobi's method. However, the storage required is half that of Jacobi's method, because we can overwrite $u_i^{(m)}$ with $u_i^{(m+1)}$ as soon as we have constructed the latter. The Gauss-Seidel method can be written in matrix form as

$$(3.13\text{-}9) \qquad\qquad \mathbf{u}^{(m+1)} = \mathbf{Eu}^{(m+1)} + \mathbf{Fu}^{(m)} + \mathbf{k},$$

where $\mathbf{E}$ and $\mathbf{F}$ are strictly lower and upper triangular, respectively. One can then give the form (3.13-1) to this equation, taking $\mathbf{G} = (\mathbf{I} - \mathbf{E})^{-1}\mathbf{F}$ and replacing $\mathbf{k}$ in Equation (3.13-1) by $(\mathbf{I} - \mathbf{E})^{-1}\mathbf{k}$. This exhibits the Gauss-Seidel method as a stationary one-step iterative method.

Richardson's method.

Let us write Jacobi's method as follows:

$$(3.13\text{-}10) \qquad\qquad \mathbf{u}^{(m+1)} - \mathbf{u}^{(m)} = \mathbf{b} - \mathbf{Au}^{(m)}.$$

Richardson's method arises from this scheme if we incorporate the strategy of overrelaxation. Thus we try to improve convergence by multiplying the increment defined in Equation (3.13-10) by a scalar ω. This gives

$$\mathbf{u}^{(m+1)} = \mathbf{u}^{(m)} + \omega(\mathbf{b} - \mathbf{Au}^{(m)}),$$

which is Richardson's method. It can also be defined by

(3.13-11) $$u^{(m+1)} = (\mathbf{I} - \omega\mathbf{A})u^{(m)} + \omega\mathbf{b}.$$

Successive overrelaxation (SOR).

From the Gauss-Seidel method, one can use overrelaxation to derive the following scheme:

(3.13-12) $$u_i^{(m+1)} = (1 - \omega)u_i^{(m)}$$
$$+ \omega\left[\sum_{j=1}^{i-1} b_{ij}u_j^{(m+1)} + \sum_{j=i+1}^{N} b_{ij}u_j^{(m)} + k_i\right].$$

According to the Ostrowski-Reich theorem (see Birkhoff and Lynch, 1984, Section 4.8), when $\mathbf{A}$ is a symmetric matrix with positive diagonal entries, the SOR method converges if and only if $\mathbf{A}$ is positive-definite and $0 < \omega < 2$. Observe that, when $\mathbf{B} = \mathbf{I} - \mathbf{A}$, SOR can also be derived from Richardson's method by making use of new iterative values as soon as they become available.

Matrix splitting.

Matrix splitting is actually a general form of iterative scheme that includes the four methods just mentioned but admits infinitely many other schemes. Whenever $\mathbf{A} = \mathbf{M} - \mathbf{N}$, and $\mathbf{M}$ is "easily invertible", one can define

(3.13-13) $$\mathbf{M}\mathbf{u}^{(m+1)} = \mathbf{N}\mathbf{u}^{(m)} + \mathbf{b}.$$

This scheme has the form given in Equation (3.13-1), with $\mathbf{G} = \mathbf{M}^{-1}\mathbf{N}$. Using this general format, one can generate many one-step stationary iterative schemes, perhaps taking advantage of the particular matrix structure of interest.

Block-iterative methods.

Block-iterative methods can, in part, be viewed as extensions of the point-iterative approaches presented above. As a point of departure, consider the linear system derived from a second-order accurate five-point difference approximation to Poisson's equation,

(3.13-14) $$u_{i,j} - \tfrac{1}{4}(u_{i+1,j} + u_{i-1,j}) - \tfrac{1}{4}(u_{i,j+1} + u_{i,j-1}) = \gamma f_{i,j},$$

where $\gamma = -h^2/4$ and $1 \leq i \leq I$, $1 \leq j \leq J$. Here I and J denote the number of internal nodes in the x- and y-coordinate directions, respectively. Written in the form (3.13-14), this difference approximation results in a set of IJ equations in IJ unknowns, assuming we have properly accommodated the boundary conditions. The idea behind the block-iterative approach is to change the formulation to yield J sets of equations, each containing I unknowns. Because the scheme is iterative, these J sets of equations must be solved repetitively. The objective is to develop an algorithm that requires fewer arithmetic operations to solve the iterative problem than to solve the original direct problem

Although the Jacobi method described earlier in this section is a point-iterative algorithm, it can readily be extended to obtain a block-iterative counterpart, the **block Jacobi method**. To achieve this we rewrite Equation (3.13-14) as follows:

$$(3.13\text{-}15) \quad u_{i,j}^{(m+1)} - \tfrac{1}{4}\big(u_{i+1,j}^{(m+1)} + u_{i-1,j}^{(m+1)}\big) = \tfrac{1}{4}\big(u_{i,j+1}^{(m)} + u_{i,j-1}^{(m)}\big) + \gamma f_{i,j}.$$

Notice that the iteration index m plays the same role as it did for the point-iterative schemes. In this case, however, we treat the entire j-th line of nodal values as a set of simultaneous unknowns, in contrast to the point iterative schemes, which require the solution of only one nodal equation at a time.

We implement the algorithm described by Equation (3.13-15) as follows. First, write the set of equations for $j = 1$ at the $m + 1$ iteration level for $i = 1, 2, \ldots, I$. The right side of Equation (3.13-15) is assumed known because the solution $u_{ij}^{(m)}$ is provided, either through an initial guess or from the solution of the m-th iteration. After solving the set of I simultaneous equations for $u_{i,j}^{(m+1)}$, we increment the index j and then solve for the next line of I unknowns. This procedure continues for $j = 1, 2, \ldots, J$. When the J-th equation set has been solved, the values $u_{i,j}^{(m)}$ have been updated to $u_{i,j}^{(m+1)}$. This completes one iteration. We now reset $j = 1$ and repeat the process. Note that, in this procedure, the values of $u_{i,j-1}^{(m)}$ and $u_{i,j+1}^{(m)}$ are obtained from the last iteration, and no dynamic updating takes place as values at the $(m + 1)$-st level become available.

An obvious improvement is to use the results of these most recent iterative calculations as soon as we have computed them. This is the strategy that we used earlier for point-iterative methods, and the result was the Gauss-Seidel method. Extending this concept to the block-iterative class of procedures we obtain the **block-Gauss-Seidel** method,

$$(3.13\text{-}16) \quad u_{i,j}^{(m+1)} - \tfrac{1}{4}\big(u_{i+1,j}^{(m+1)} + u_{i-1,j}^{(m+1)}\big) = \tfrac{1}{4}\big(u_{i,j+1}^{(m)} + u_{i,j-1}^{(m+1)}\big) + \gamma f_{i,j}.$$

This algorithm is similar to the block-Jacobi method, except now we update $u_{i,j}^{(m)}$ to $u_{i,j}^{(m+1)}$ as soon as the latter values are known. Thus, although the right side of Equation (3.13-16) contains a nodal value at the $(m+1)$-st level, it is a known value because the $(m+1)$-st level calculations have already been completed for the $(j-1)$-st row of nodes.

While the block-Jacobi and block-Gauss-Seidel methods are readily formulated and applied, they are not nearly as popular as **line-successive overrelation (LSOR)**. By analogy with the point-SOR method, we write the LSOR method as follows:

$$(3.13\text{-}17) \quad u_{i,j}^{(m+1)} - \frac{1}{4}(u_{i+1,j}^{(m+1)} + u_{i-1,j}^{(m+1)}) = \frac{\omega}{4}(u_{i,j+1}^{(m)} + u_{i,j-1}^{(m+1)})$$

$$+ (1-\omega)[u_{ij}^{(m)} - \frac{1}{4}(u_{i+1,j}^{(m)} + u_{i-1,j}^{(m)})] + \omega\gamma f_{ij}$$

The application of this method is analogous to the earlier block-Gauss-Seidel algorithm in that we employ the most recent calculations in the equation formulation. Among the common methods employed in obtaining good values for ω in practical applications are those based upon a feedback strategy. One approach is to assume $\omega = 1$ initially and compute an iterative sequence of solutions $\mathbf{u}^{(m)}$, each time computing $\mathbf{S}^{(m)} = \mathbf{u}^{(m+1)} - \mathbf{u}^{(m)}$. The ratio $\|\mathbf{S}^{(m+1)}\|_2/\|\mathbf{S}^{(m)}\|_2$ will, for large iteration numbers, approach the largest eigenvalue of the iteration matrix $\mathbf{G}$. This eigenvalue, in turn, dictates the optimal choice of ω. For additional information regarding the selection of ω we refer the reader to Wachspress (1966). Also, an extension of the LSOR procedure that incorporates an additive correction to accelerate convergence is presented in detail in Peaceman (1977).

One should bear in mind the fact that block-iterative methods can be further extended to accommodate several lines per solution step. In other words, the concepts that we have just applied one line at a time can be extended to multiple lines. In fact, two-line methods are quite common in practical applications.

Alternating-direction iteration (ADI).

This class of methods is similar to the block-iterative methods just outlined. The approach, first introduced by Peaceman and Rachford (1955), can be developed from the point-Jacobi algorithm, which for a constant-coefficient equation is as follows:

$$(3.13\text{-}18) \qquad u_{i,j}^{(m+1)} = \tfrac{1}{4}(u_{i+1,j}^{(m)} + u_{i-1,j}^{(m)} + u_{i,j+1}^{(m)} + u_{i,j-1}^{(m)}) + \gamma f_{i,j}.$$

We now add and subtract $u_{i,j}^{(m)}$ from this equation to yield

$$(3.13\text{-}19) \qquad u_{i,j}^{(m+1)} = u_{ij}^{(m)} + \tfrac{1}{4}(u_{i+1,j}^{(m)} + u_{i-1,j}^{(m)}$$

$$+ u_{i,j+1}^{(m)} + u_{i,j-1}^{(m)} + u_{i,j}^{(m)}) + \gamma f_{i,j},$$

or, equivalently,

3.13-20 $u_{i,j}^{(m+1)} - u_{i,j}^{(m)} = \frac{1}{4}[(u_{i+1,j}^{(m)} - 2u_{i,j}^{(m)} + u_{i-1,j}^{(m)})$
$+ u_{i,j+1}^{(m)} - 2u_{i,j}^{(m)}) + u_{i-1,j}^{(m)})] + \gamma f_{i,j}.$

The right side of Equation (3.13-20) contains approximations to second derivatives. Let us consider each iteration as a two-step procedure wherein the first step advances to the $(m + \frac{1}{2})$ level and the second step to the $(m + 1)$ level. Each of these two steps allows the derivative approximations in one direction to lag by a half step. Thus, for example, we assign the iteration level for each term in the first step as follows:

(3.13-21a) $\beta_m(u_{i,j}^{(m+1/2)} - u_{i,j}^{(m)})$
$= \frac{1}{4}[(u_{i+1,j}^{(m+1/2)} - 2u_{i,j}^{(m+1/2)} + u_{i-1,j}^{(m+1/2)})$
$+ (u_{i,j+1}^{(m)} - 2u_{i,j}^{(m)} + u_{i,j-1}^{(m)})] + \gamma f_{i,j}.$

The assignment for each term in the second step is then

(3.13-21b) $\beta_m(u_{ij}^{(m+1)} - 2u_{i,j}^{(m+1/2)})$
$= \frac{1}{4}[(u_{i+1,j}^{(m+1/2)} - 2u_{i,j}^{(m+1/2)} - u_{i-1,j}^{(m+1/2)})$
$+ (u_{i,j+1}^{(m+1)} - 2u_{i,j}^{(m+1)} + u_{i,j-1}^{(m+1)})] + \gamma f_{i,j},$

where β_m is an iteration parameter to be discussed below.

To devise a computational procedure, one writes Equation (3.13-21a) for all nodes in a row. This gives rise to a tridiagonal set of equations at the $(m + \frac{1}{2})$ level. We then solve the equations along all rows of the grid, one row at a time. Once all nodes have been elevated to the $(m + \frac{1}{2})$ level, one follows a similar procedure for the columns of nodes using Equation (3.13-2b). Having calculated the new values $u_{i,j}^{(m+1)}$ columnwise, we have completed one two-step iteration.

The procedure described by Equations (3.13-21) is sensitive to the choice of the iteration parameter β_m. A sequence of values for β_m has been found preferable to a single parameter value. There exist several strategies for selecting β_m; we refer the interested reader to Peaceman (1977) for further details.

Numerous variants on the ADI procedure have been developed. Noteworthy are the algorithms of Douglas and Rachford (1956) and Mitchell and Fairweather (1964). It is also possible to formulate ADI algorithms for other numerical procedures, such as the finite-element Galerkin and collocation methods. The relationship between the various ADI techniques has been recently summarized by Celia and Pinder (1985).

3.14. Convergence and Related Topics.

In the previous section we reviewed five iterative methods for solving systems of equations having the form $\mathbf{Au} = \mathbf{b}$ typically arising from finite-difference approximations to elliptic PDEs. Here we analyze several questions regarding the effectiveness of iterative methods. Among these questions are consistency, convergence, and, in the case of SOR, criteria for an optimal choice of the parameter ω. We shall assume throughout this section that the matrix $\mathbf{A}$ is nonsingular.

Consistency.

In Equation (3.11-1) we wrote the general form of a stationary, linear, one-step method as $\mathbf{u}^{(m+1)} = \mathbf{Gu}^{(m)} + \mathbf{k}$. Such a scheme is **consistent** with $\mathbf{Au} = \mathbf{b}$ if, whenever the sequence $\{\mathbf{u}^{(m)}\}_{m=0}^{\infty}$ of iterative vectors converges, it converges to the true solution $\mathbf{u}$ of the original matrix equation. We claim that this is true if and only if

$$(3.14\text{-}1) \qquad\qquad \mathbf{A}(\mathbf{I} - \mathbf{G})^{-1}\mathbf{k} = \mathbf{b}.$$

Notice that we are assuming $\mathbf{I} - \mathbf{G}$ to be nonsingular. To see that Equation (3.14-1) is necessary for convergence, suppose that $\mathbf{u}^{(m)} \rightarrow \mathbf{u}$ as $m \rightarrow \infty$. Then $\mathbf{u} = \mathbf{Gu} + \mathbf{k}$, so $\mathbf{k} = (\mathbf{I} - \mathbf{G})\mathbf{u} = (\mathbf{I} - \mathbf{G})\mathbf{A}^{-1}\mathbf{b}$. Given that $\mathbf{A}$ and $\mathbf{I} - \mathbf{G}$ are invertible, we can solve for $\mathbf{b}$ to conclude that $\mathbf{A}(\mathbf{I} - \mathbf{G})^{-1}\mathbf{k} = \mathbf{b}$. To prove sufficiency, assume that Equation (3.14-1) holds and that there is a vector $\mathbf{u}^*$ such that $\mathbf{u}^{(m)} \rightarrow \mathbf{u}^*$ as $m \rightarrow \infty$. We must demonstrate that $\mathbf{u}^* = \mathbf{u}$, that is, that $\mathbf{u}^* = \mathbf{A}^{-1}\mathbf{b}$. The fact that $\mathbf{u}^{(m)} \rightarrow \mathbf{u}^*$ implies $\mathbf{u}^* = \mathbf{Gu}^* + \mathbf{k}$, or $(\mathbf{I} - \mathbf{G})\mathbf{u}^* = \mathbf{k}$. Now, premultiplying both sides by $\mathbf{A}(\mathbf{I} - \mathbf{G})^{-1}$ and applying Equation (3.14-1), we see that

$$\mathbf{A}(\mathbf{I} - \mathbf{G})^{-1}(\mathbf{I} - \mathbf{G})\mathbf{u}^* = \mathbf{Au}^* = \mathbf{b},$$

so that $\mathbf{u}^* = \mathbf{A}^{-1}\mathbf{b}$, as desired. Therefore, provided $\mathbf{I} - \mathbf{G}$ is nonsingular, Equation (3.14-1) is both necessary and sufficient for the iterative scheme to be consistent.

Convergence.

A consistent scheme is **convergent** if the sequence $\{\mathbf{u}^{(m)}\}_{m=0}^{\infty}$ converges for every initial vector $\mathbf{u}^{(0)}$. Given any vector norm $\| \cdot \|$ and its associated matrix norm, a sufficient condition for the scheme defined by

$\mathbf{u}^{(m+1)} = \mathbf{Gu}^{(m)} + \mathbf{k}$ to be convergent is $\|\mathbf{G}\| < 1$. To see why, observe that, by repeated application of the iteration equation,

$$(3.14\text{-}2) \qquad\qquad \mathbf{u}^{(m+1)} = \mathbf{G}^{m+1}\mathbf{u}^{(0)} + \sum_{j=1}^{m} \mathbf{G}^j\mathbf{k}.$$

We wish to show that, when $\|\mathbf{G}\| < 1$, the sequence $\{\mathbf{u}^{(m)}\}_{m=0}^{\infty}$ is a **Cauchy sequence**, that is, that $\|\mathbf{u}^{(m+1)} - \mathbf{u}^{(m)}\| \to 0$ as $M \to \infty$. This will, in turn, establish that the sequence of iterative vectors converges since, by the completeness of Euclidean spaces, every Cauchy sequence of such vectors converges. Observe that each term in the sum on the right side of Equation (3.14-2) obeys the bound

$$\| \mathbf{G}^j\mathbf{k} \| \leq \| \mathbf{G} \|^j \| \mathbf{k} \|$$

by the Cauchy-Schwarz inequality. Also, as $m \to \infty$, $\mathbf{G}^{m+1}\mathbf{u}^{(0)} \to 0$. Thus $\|\mathbf{u}^{(m+1)} - \mathbf{u}^{(m)}\| \to \|\mathbf{G}\|^{m+1}\|\mathbf{k}\| \to 0$ as $m \to \infty$; hence $\{\mathbf{u}^{(m)}\}_{m=0}^{\infty}$ is a Cauchy sequence, and convergence is proved.

While $\|\mathbf{G}\| < 1$ suffices for convergence, it is necessary only when $\mathbf{G}$ is symmetric. To get a criterion that is both necessary and sufficient, we must appeal to the spectral radius $\rho(\mathbf{G}) = \max_i\{|\lambda_i|\}$, where the numbers λ_i range over the eigenvalues of $\mathbf{G}$. In terms of this quantity, a necessary and sufficient condition for a consistent scheme to be convergent is $\rho(\mathbf{G}) < 1$. While the formal proof of this fact requires somewhat more matrix theory than we wish to delve into here, we can sketch a proof of sufficiency as follows. Define the error $\boldsymbol{\epsilon}^{(m)} = \mathbf{u}^{(m)} - \mathbf{u}$, and expand the (arbitrary) initial vector $\boldsymbol{\epsilon}^{(0)}$ in terms of the eigenvectors $\mathbf{e}_1, \ldots, \mathbf{e}_N$ of $\mathbf{G}$ as follows:

$$\boldsymbol{\epsilon}^{(0)} = \sum_{i=1}^{N} a_i \mathbf{e}_i.$$

(This expansion is what makes the argument a "sketch" rather than a complete proof.) Since $\mathbf{u}^{(m+1)} = \mathbf{Gu}^{(m)} + \mathbf{k}$ and $\mathbf{u} = \mathbf{Gu} + \mathbf{k}$, we see that $\boldsymbol{\epsilon}^{(m+1)} = \mathbf{G}^{m+1}\boldsymbol{\epsilon}^{(0)}$. Using the eigenvector expansion we get, in any norm,

$$(3.14\text{-}3) \qquad \|\boldsymbol{\epsilon}^{(m+1)}\| = \left\| \sum_{i=1}^{N} a_i \lambda_i^{m+1} \mathbf{e}_i \right\| \leq \sum_{i=1}^{N} |a_i||\lambda_i|^{m+1}\|\mathbf{e}_i\|.$$

When $\rho(\mathbf{G}) < 1$, $|\lambda_i|^{m+1} \to 0$ as $m \to \infty$ for each eigenvalue λ_i, so each term on the right side of the inequality (3.14-3) tends to zero as $m \to \infty$. Therefore $\|\boldsymbol{\epsilon}^{(m+1)}\| \to 0$ as $m \to \infty$, and hence the scheme converges. To show necessity, assume $\rho(\mathbf{G}) \geq 1$, so that $\mathbf{G}$ has an eigenvalue λ_I such that $|\lambda_I| \geq 1$. If $\mathbf{e}_I$ is the associated eigenvector, we can pick the initial vector

$\mathbf{u}^{(0)} = \mathbf{e}_I$, which produces a sequence $\{\mathbf{u}^{(m)}\}_{m=0}^{\infty}$ for which $\|\boldsymbol{\epsilon}^{(m)}\| \not\to 0$ as $m \to \infty$. Therefore, if $\rho(\mathbf{G}) \geq 1$, the scheme fails to converge for arbitrary $\mathbf{u}^{(0)}$.

As mentioned, the arguments just given are sketchy in that they rely on an expansion of an arbitrary vector $\boldsymbol{\epsilon}^{(0)}$ in eigenvectors of $\mathbf{G}$. In making this expansion, we have assumed implicitly that the eigenvectors are linearly independent, an assumption that is not always valid and is in fact unnecessary. We refer the reader to Golub and Van Loan (1983, Chapter 10) for a more rigorous treatment.

The inequality (3.14-3) motivates a way of measuring the speed with which an iterative scheme converges. To reduce the magnitude of any component $a_i \mathbf{e}_i$ in the eigenvector expansion of the error by one order of magnitude, we need to take n iterations, where n is the smallest number such that $|\lambda_i| \leq 10^{-1}$. As the iteration count grows large, only the error component associated with the eigenvalue of largest magnitude makes a signifigant contribution to the expansion of $\boldsymbol{\epsilon}^{(m)}$, and for this reason we want $[\rho(\mathbf{G})]^n \leq 10^{-1}$, implying $n \log_{10} \rho(\mathbf{G}) \leq -1$. Since a convergent scheme has $\rho(\mathbf{G}) < 1$, we conclude that $n \geq [-\log_{10} \rho(\mathbf{G})]^{-1}$. In other words, asymptotically it takes at least $-\log_{10} \rho(\mathbf{G})$ iterations to reduce the norm of the error by one order of magnitude. Therefore we often call $-\log_{10} \rho(\mathbf{G})$ the **asymptotic rate of convergence** of the iterative scheme.

As an example of the application of these ideas to an actual iterative scheme, let us show that SOR is not convergent unless $0 < \omega < 2$. To do this, we must first find the form of the iteration matrix $\mathbf{G}_{\mathrm{SOR}}$. Let us decompose the original matrix as $\mathbf{A} = \mathbf{D}(\mathbf{L} + \mathbf{I} + \mathbf{U})$, where $\mathbf{D} = \mathrm{diag}(\mathbf{A})$, $\mathbf{L}$ is strictly lower triangular, $\mathbf{I}$ is the identity matrix, and $\mathbf{U}$ is strictly upper triangular. Given this decomposition, it is easy to verify that

$$\mathbf{G}_{\mathrm{SOR}} = (\mathbf{I} - \omega\mathbf{L})^{-1}[(1-\omega)\mathbf{I} - \omega\mathbf{U}].$$

Thus,

(3.14-4) $$\det(\mathbf{G}_{\mathrm{SOR}}) = \det(\mathbf{I} + \omega\mathbf{L})^{-1}\det[(1-\omega)\mathbf{I} - \omega\mathbf{U}].$$

The determinants on the right are determinants of triangular matrices and hence are just the products of their eigenvalues. Consequently, Equation (3.14-4) reduces to $\det(\mathbf{G}_{\mathrm{SOR}}) = (1-\omega)^N$, where N is the order of $\mathbf{G}_{\mathrm{SOR}}$. But the determinant of any matrix is also the product of its eigenvalues. We have, then, $\det(\mathbf{G}_{\mathrm{SOR}}) = \lambda_1 \lambda_2 \cdots \lambda_N$, from which it follows that $\max_i\{|\lambda_i|\} \geq |1-\omega|$, that is, $\rho(\mathbf{G}_{\mathrm{SOR}}) \geq |\omega - 1|$. Therefore $\rho(\mathbf{G}_{\mathrm{SOR}}) < 1$ only if $0 < \omega < 2$.

Optimal choice of ω.

Finally, we can settle the question regarding the optimal choice of ω in SOR, at least for the case when $\mathbf{A}$ is symmetric. In this case, all eigenvalues

λ_n of **A** are real, and three possibilities occur. Either (a) all eigenvalues are positive, (b) all eigenvalues are negative, or (c) both positive and negative eigenvalues exist. (None of the eigenvalues is zero; otherwise **A** would be singular.) In case (c), SOR diverges for every choice of ω because at least one of the eigenvalues $1 - \omega\lambda_n$ of $\mathbf{G}_{SOR} = \mathbf{I} - \omega\mathbf{A}$ is not less than unity in magnitude. For the other two possibilities, (a) and (b), the greatest rate of convergence occurs when

$$(3.14\text{-}5) \qquad\qquad \omega = \omega_{\text{opt}} = 2/(\alpha + \beta),$$

where α and β are the magnitudes of the eigenvalues of **A** whose absolute values are smallest and largest, respectively. For this choice,

$$(3.14\text{-}6) \qquad\qquad \rho(\mathbf{G}_{SOR}) = \frac{\beta - \alpha}{\beta + \alpha}.$$

Observe that, for symmetric matrices **A**, the ratio β/α coincides with the condition number cond(**A**). Therefore Equation (3.14-6) states that, with the optimal choice of ω,

$$\rho(\mathbf{G}_{SOR}) = \frac{\text{cond}(\mathbf{A}) - 1}{\text{cond}(\mathbf{A}) + 1}.$$

To prove that Equation (3.14-5) gives the optimum value of ω, assume first that all eigenvalues of **A** are positive. Observe that the eigenvalues of $\mathbf{G}_{SOR} = \mathbf{I} - \omega\mathbf{A}$ lie between $\lambda_\alpha = 1 - \omega\alpha$ and $\lambda_\beta = 1 - \omega\beta$. Thus $\rho(\mathbf{G}_{SOR}) = \max\{|\lambda_\alpha, \lambda_\beta|\}$, which has its smallest value when $1 - \omega\alpha = \omega\beta - 1$. Equation (3.14-5) follows. The proof for the case when all eigenvalues of **A** are negative is similar.

3.15. Other Iterative Schemes.

While stationary schemes like those just reviewed are the most common iterative schemes used in applications, there are several other types of schemes that have attracted attention. Two of the most important of these in current vogue are multigrid methods and methods of the conjugate-gradient type. This section briefly discusses these two classes of schemes.

Multigrid methods.

In recent years interest has grown in a set of nonstationary iterative schemes based on the "relaxation" schemes discussed in Sections 3.13 and 3.14. Since these new schemes exploit the idea of performing, say, Jacobi or Gauss-Seidel iteration on a sequence of grids of varying fineness to optimize convergence rates, they are called **multigrid** iterative techniques.

Although fairly new, their popularity has led to a wide variety of applications and thus to a large body of literature. We shall merely review the overall concepts here, referring the reader to an article by Brandt (1977) for a more complete introduction.

Consider the finite-difference solution of Laplace's equation, $\nabla^2 u = 0$, on a two-dimensional domain Ω with appropriate boundary conditions. Given a rectangular grid Δ_{h_M} of uniform spacing h_M in both coordinate directions, a Gauss-Seidel iterative scheme applied to the standard central-difference approximation yields

$$(3.15\text{-}1) \qquad \frac{1}{h_M}(u_{i+1,j}^{(k)} - 2u_{i,j}^{(k+1)} + u_{i-1,j}^{(k+1)})$$
$$+ \frac{1}{h_M}(u_{i,j+1}^{(k)} - 2u_{i,j}^{(k+1)} + u_{i,j-1}^{(k+1)}) = 0$$

as the algebraic equation for the new iterate $u_{i,j}^{(k+1)}$. If $u_{i,j}$ denotes the exact solution to the difference equation centered at node (x_i, y_j), then we can measure the progress of the scheme by comparing the errors $\boldsymbol{\epsilon}^{(k+1)} = \mathbf{u} - \mathbf{u}^{(k+1)}$ and $\boldsymbol{\epsilon}^{(k)} = \mathbf{u} - \mathbf{u}^{(k)}$ at successive iterations. Here, the boldface letters $\mathbf{u}$ denote vectors containing the nodal values $u_{i,j}$. The **convergence factor**

$$\mu_{k+1} = \frac{\|\boldsymbol{\epsilon}^{(k+1)}\|_\infty}{\|\boldsymbol{\epsilon}^{(k)}\|_\infty}$$

furnishes a convenient index of the rate at which the scheme is converging at stage $k+1$. For problems like Equation (3.15-1), one can show that $\mu_k \to 1 - \mathcal{O}(h_M^2)$ as $k \to \infty$, implying that, asymptotically, $\mathcal{O}(h_M^{-2})$ iterations are needed to reduce the error $\boldsymbol{\epsilon}$ by an order of magnitude. This rate can be quite slow for problems requiring fine grids.

However, for most schemes of the relaxation type the error reduction occurs much more rapidly than the asymptotic rate for the first few iterations. Figure 3-8 shows plots of $-\log_{10}(\|\boldsymbol{\epsilon}^{(k+1)}\|_\infty / \|\boldsymbol{\epsilon}^{(k)}\|_\infty)$ versus k for Gauss-Seidel and Jacobi schemes solving the central-difference approximation to $\nabla^2 u = 0$ on $\Omega = (0, 10) \times (0, 5)$ with $h_M = 1$ and Dirichlet boundary conditions. Observe that, for both schemes, the error reduction occurs very quickly for the first five or so iterations, then settles down to its asymptotic rate. (In this example, roughly, $\mu_k \to 0.77$ for the Gauss-Seidel scheme, while $\mu_k \to 0.94$ for the Jacobi scheme.)

The explanation for this behavior is as follows. The error $\boldsymbol{\epsilon}^{(k)}$ at a given iteration on Δ_{h_M} is a superposition of components having various wavelengths, the smallest being $2h_M$. Relaxation schemes are highly efficient at reducing error components having wavelengths that are small compared with the grid spacing h_M, but they reduce the longer-wavelength or "smooth" components at the slower asymptotic rate. Thus, after five iterations in Figure 3-8, each of the schemes has eliminated most of the rapidly

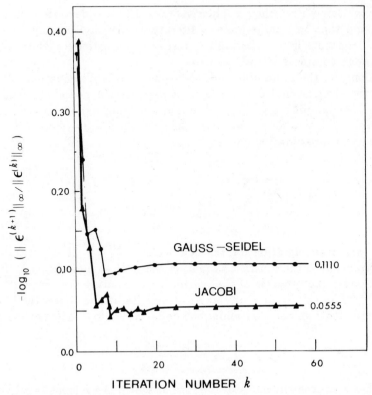

Figure 3-8. Decay in convergence rate of the Jacobi and Gauss-Seidel iterative schemes for Laplace's equation.

fluctuating error components and has settled down to the slow process of reducing the smooth error components.

The idea behind multigrid methods is to solve a problem like Equation (3.15-1) on a family $\{\Delta_{h_1}, \ldots, \Delta_{h_M}\}$ of grids with increasingly fine spacings, so that $h_1 > \cdots > h_M$. Thus we envision a family

$$\left\{ \{\mathbf{u}_1^{(k)}\}, \ldots, \{\mathbf{u}_M^{(k)}\} \right\}$$

of solution sequences, each pertaining to a different grid on Ω. To see how such a scheme can lead to improved convergence, suppose we have iterated on the fine grid Δ_{h_M} to the stage where the convergence factor has begun to approach its asymptotic value. Rather than continuing with the standard scheme, we map the current approximate solution $\mathbf{u}_M^{(k)}$ onto a coarser grid Δ_{h_m}, $m < M$, thus computing a vector $\mathbf{u}_m^{(k)}$ of coarse-grid values. On this new grid some of the error components that were smooth with respect to the fine grid Δ_{h_M} will now be rapidly varying. Therefore, relaxation on the coarse grid will give rapid error reduction for k' iterations, after which the remaining error will be smooth with respect to Δ_{h_m}. At this point we can map the solution $\mathbf{u}_m^{(k+k')}$ either to still coarser grids or else back to finer grids, depending on whether reduction of rapidly varying error or the acheivement of fine-grid accuracy will yield the fastest overall reduction in error. Thus, by optimally switching among coarser and finer grids in the iteration, one can attain high rates of convergence using standard stationary techniques on each grid.

Of course, many technical issues remain. For example, how many grids Δ_{h_m} are best for a given problem? What are the best techniques for mapping discrete solution vectors from fine grids to coarse grids and vice versa? What computable criteria should we use to determine when to map to a new grid? These questions are subjects of active research. We refer the interested reader to the growing technical literature for further details on these issues and on the applicability of multigrid methods to different discretization techniques and nonelliptic PDEs.

Conjugate-gradient methods.

The method of conjugate gradients and its extensions follow a different line of reasoning. At the basis of the standard conjugate-gradient method lies the equivalence between the following two problems: (i) given a symmetric, positive-definite matrix $\mathbf{A}$ and a vector $\mathbf{b}$, find a vector $\mathbf{u}$ such that $\mathbf{Au} = \mathbf{b}$; (ii) given the same $\mathbf{A}$ and $\mathbf{b}$, find the vector $\mathbf{u}$ that minimizes the functional $J(\mathbf{u}) = \frac{1}{2}\mathbf{u} \cdot \mathbf{Au} - \mathbf{u} \cdot \mathbf{b}$. While conjugate-gradient algorithms have been known since the work of Hestenes and Stiefel (1952), Reid (1971) revived interest within the computational community by casting the method as an iterative technique.

The idea behind the standard conjugate-gradient method is as follows. Given an iterative estimate $\mathbf{u}^{(k)}$ leaving a nonzero **residual** $\mathbf{r}^{(k)} = \mathbf{b} - \mathbf{Au}^{(k)}$, we seek a search-direction vector $\mathbf{p}^{(k+1)}$ and a scalar $\alpha^{(k+1)}$ satisfying two conditions. First, $\mathbf{p}^{(k+1)}$ must be nonzero and orthogonal to all search directions used at previous stages, that is, $\mathbf{p}^{(k+1)} \cdot \mathbf{p}^j = 0$ for $j \leq k$. Second, $\alpha^{(k+1)}$ must minimize $J(\mathbf{u}^{(k)} + \alpha^{(k+1)}\mathbf{p}^{(k+1)})$, so that the choice $\mathbf{u}^{(k+1)} = \mathbf{u}^{(k)} + \alpha^{(k+1)}\mathbf{p}^{(k+1)}$ finds the smallest value of J along the line passing through $\mathbf{u}^{(k)}$ with direction $\mathbf{p}^{(k+1)}$. One can show (Golub and Van Loan, 1983, Section 10.2) that, if we call

$$\beta^{(k+1)} = \frac{\mathbf{p}^{(k)} \cdot \mathbf{Ar}^{(k)}}{\mathbf{p}^{(k)} \cdot \mathbf{Ap}^{(k)}},$$

then the choices

$$
\mathbf{p}^{(k+1)} = \begin{cases} \mathbf{r}^{(0)} = \mathbf{b} - \mathbf{A}\mathbf{u}^{(0)}, & \text{if } k = 0, \\ \mathbf{r}^{(k)} - \beta^{(k+1)}\mathbf{p}^{(k)}, & \text{if } k > 0, \end{cases}
$$

$$
\alpha^{(k+1)} = \frac{\mathbf{r}^{(k)} \cdot \mathbf{r}^{(k)}}{\mathbf{p}^{(k+1)} \cdot \mathbf{A}\mathbf{p}^{(k+1)}}
$$

satisfy both constraints. Since each new search direction is orthogonal to the previous ones, we should eventually minimize J over all directions and thus find the solution to $\mathbf{A}\mathbf{u} = \mathbf{b}$. This reasoning leads to the following algorithm: given $\mathbf{A}$ and $\mathbf{b}$ as above, an initial guess $\mathbf{u}^{(0)}$, and a tolerance $\varepsilon > 0$,

$\mathbf{r}^{(0)} = \mathbf{b} - \mathbf{A}\mathbf{u}^{(0)}$

$\mathbf{p}^{(1)} = \mathbf{r}^{(0)}$

$k = 0$

Do while $\|\mathbf{r}^{(k)}\|_\infty \geq \varepsilon$:

$\quad \beta^{(k+1)} = \mathbf{p}^{(k)} \cdot \mathbf{A}\mathbf{r}^{(k)}/(\mathbf{p}^{(k)} \cdot \mathbf{A}\mathbf{p}^{(k)})$

$\quad \mathbf{p}^{(k+1)} = \mathbf{r}^{(k)} - \beta^{(k+1)}\mathbf{p}^{(k)}$

$\quad \alpha^{(k+1)} = \mathbf{r}^{(k)} \cdot \mathbf{r}^{(k)}/(\mathbf{p}^{(k+1)} \cdot \mathbf{A}\mathbf{p}^{(k+1)})$

$\quad \mathbf{u}^{(k+1)} = \mathbf{u}^{(k)} + \alpha^{(k+1)}\mathbf{p}^{(k+1)}$

$\quad \mathbf{r}^{(k+1)} = \mathbf{b} - \mathbf{A}\mathbf{u}^{(k+1)} = \mathbf{r}^{(k)} - \alpha^{(k+1)}\mathbf{A}\mathbf{p}^{(k+1)}$

$\quad k \leftarrow k + 1.$

Two computational aspects of this algorithm are worth mentioning. First, it appears that the algorithm should produce the exact solution in a finite number of steps. Indeed, if we seek an n-dimensional vector $\mathbf{u}$ then, theoretically, after n steps we will have exhausted the possibilities for nonzero search-direction vectors $\mathbf{p}^{(k)}$ orthogonal to all previous search directions. However, in practice machine roundoff usually destroys exact orthogonality of the search directions at each step, so the conjugate-gradient method is best viewed as a true iterative procedure. In this light the rate of convergence becomes a crucial consideration. One can show (again, see Golub and Van Loan, 1983, Section 10.2) that, if $\mathbf{A}$ has condition number cond$(\mathbf{A})$ and $\|\cdot\|_A$ denotes the energy norm associated with $\mathbf{A}$ and defined by $\|\mathbf{u}\|_A^2 = \mathbf{u} \cdot \mathbf{A}\mathbf{u}$, then

$$
(3.15-2) \qquad \|\mathbf{u} - \mathbf{u}^{(k)}\|_A \leq \|\mathbf{u} - \mathbf{u}^{(0)}\|_A \left[\frac{1 - \sqrt{\text{cond}(\mathbf{A})}}{1 + \sqrt{\text{cond}(\mathbf{A})}}\right]^{2k}.
$$

The estimate (3.15-2) leads to the second computational consideration. When cond($\mathbf{A}$) is large, the conjugate-gradient method will usually
be unacceptably slow. This fact poses difficulties in applications to PDEs,
since $\mathcal{O}(h^p)$ discretization procedures commonly yield matrices for which
cond($\mathbf{A}$) $= \mathcal{O}(h^{-p})$. The usual way to avoid the resulting slow convergence
in fine-grid problems is to **precondition** the original problem $\mathbf{A}\mathbf{u} = \mathbf{b}$ by
a **preconditioner** $\hat{\mathbf{A}}$ to yield

$$\hat{\mathbf{A}}^{-1}\mathbf{A}\mathbf{u} = \hat{\mathbf{A}}^{-1}\mathbf{b}.$$

Heuristically, we can view $\hat{\mathbf{A}}^{-1}$ as an "approximate inverse" to $\mathbf{A}$. In more
concrete terms, any matrix $\hat{\mathbf{A}}$ having the properties that $\hat{\mathbf{A}}^{-1}\mathbf{A}$ has a small
condition number and $\hat{\mathbf{A}}^{-1}\mathbf{b}$ is easy to compute for any vector $\mathbf{b}$ is a candi-
date preconditioner for $\mathbf{A}$. A considerable literature has grown around the
invention of preconditioners for various finite-difference and finite-element
discretizations; see, for example, Concus et al. (1976) and Meijerink and
Van der Vorst (1977).

Finally, while the development outlined here applies to symmetric,
positive-definite matrices only, analogous methods exist for nonsymmet-
ric matrices $\mathbf{A}$ whose symmetric parts $(\mathbf{A} + \mathbf{A}^\top)/2$ are positive-definite.
These methods have extended the applicability of conjugate-gradient-like
schemes to parabolic, hyperbolic, and coupled PDEs as well as to the symmetric elliptic case. For examples, see Axelsson (1980), Elman (1981), and
Obeysekare et al. (1986).

3.16. Finite-Element Methods.

Connection to variational principles.

The finite-element method uses an integral equation formulation as its
theoretical point of departure. In Chapter Two we employed a Galerkin approach to develop these integral equations. In this section we shall explore
the connection between the Galerkin approach and the variational formulation of certain elliptic PDEs, as discussed in Section 3.5. For simplicity,
let us consider Poisson's equation,

(3.16-1) $$\nabla^2 u(\mathbf{x}) = f(\mathbf{x}) \quad \text{in} \quad \Omega,$$

subject to Neumann boundary conditions on $\partial\Omega$. As explained in Section
2.10, the appropriate trial function for this problem has the form $\hat{u}(\mathbf{x}) = \sum_{i=1}^{N} u_i \phi_i(\mathbf{x})$. The Galerkin formalism then leads to the equations

(3.16-2) $$\sum_{i=1}^{N} u_i \int_\Omega [\nabla\phi_i(\mathbf{x}) \cdot \nabla\phi_j(\mathbf{x}) + f(\mathbf{x})\phi_j(\mathbf{x})] \, d\mathbf{x}$$

$$= \oint_{\partial\Omega} \frac{\partial\hat{u}}{\partial n} \phi_j(\mathbf{x}) \, d\mathbf{x}, \qquad j = 1, 2, \dots, N,$$

where the parameters u_i are undetermined coefficients. Using inner-product notation (see Section 2.10), we can write this equation somewhat more concisely as follows:

$$(3.16\text{-}3) \qquad \sum_{i=1}^{N} u_i \langle \nabla \phi_i(\mathbf{x}), \nabla \phi_j(\mathbf{x}) \rangle = -\langle f(\mathbf{x}), \phi_j(\mathbf{x}) \rangle$$

$$+ \oint_{\partial \Omega} \frac{\partial \hat{u}}{\partial n} \phi_j(\mathbf{x}) d\mathbf{x}, \qquad j = 1, 2, \ldots, N.$$

By following what is known as the **Rayleigh-Ritz procedure**, we can also use the variational principles described in Section 3.5 to obtain the Galerkin equations (3.16-3). An analogous argument will apply to a large class of differential operators sharing with ∇^2 the property of being formally symmetric. Recall that for the Laplace operator this property means that, for two functions u and w satisfying homogeneous boundary conditions,

$$(3.16\text{-}4) \qquad \langle \nabla^2 u, w \rangle = \langle u, \nabla^2 w \rangle.$$

We begin with the reduced functionals presented in Equations (3.5-12). Using the expression appropriate for Neumann conditions, we see that our Poisson problem is equivalent to finding a stationary value of

$$(3.5\text{-}12\text{b}) \qquad 2J(u) = \int_{\Omega} u(\nabla^2 u - 2f) d\mathbf{x} + \oint_{\partial \Omega} u \frac{\partial u}{\partial n} d\mathbf{x}.$$

Application of Green's theorem to the second-order term in the integral over Ω yields

$$(3.16\text{-}5) \qquad 2J(u) = -\int_{\Omega} (\nabla u \cdot \nabla u + 2uf) d\mathbf{x} + 2 \oint_{\partial \Omega} u \frac{\partial u}{\partial n} d\mathbf{x}.$$

The variational principle of Section 3.5 states that the true solution $u(\mathbf{x})$ to our Poisson problem is the function that maximizes J over all admissible functions.

The fundamental idea of the Rayleigh-Ritz procedure is to maximize J over a finite subspace of the entire collection of admissible functions, namely, the subspace spanned by a set of linearly independent basis functions $\phi_i(\mathbf{x}), i = 1, 2, \ldots, N$. An arbitrary function in this space can be written as follows:

$$(3.16\text{-}6) \qquad \hat{u}(\mathbf{x}) = \sum_{i=1}^{N} u_i \phi_i(\mathbf{x}).$$

By forcing $\hat{u}$ to maximize J over this subspace, we shall arrive at an approximation to the true solution $u(\mathbf{x})$.

The equations determining the unknown coefficients u_i of $\hat{u}$ are obtained by first substituting the expansion (3.16-6) into Equation (3.16-5) and subsequently maximizing the result with respect to the coefficients u_i. The substitution step yields

$$
(3.16\text{-}7) \qquad 2J(\hat{u}) = -\int_\Omega \left[\sum_{i=1}^N u_i \nabla\phi_i(\mathbf{x}) \cdot \sum_{j=1}^N u_j \nabla\phi_j(\mathbf{x}) \right.
$$
$$
\left. + 2f(\mathbf{x}) \sum_{j=1}^N u_j \phi_j(\mathbf{x}) \right] d\mathbf{x} + 2\oint_{\partial\Omega} \hat{u}\frac{\partial\hat{u}}{\partial n}d\mathbf{x},
$$

where the boundary integral is deliberately left unexpanded. We maximize the functional J in Equation (3.16-7) with respect to the coefficients u_j by setting $\partial J(\hat{u})/\partial u_j = 0$. The reader should verify that this procedure yields

$$
(3.16\text{-}8) \qquad \sum_{i=1}^N u_i \int_\Omega [\nabla\phi_i(\mathbf{x}) \cdot \nabla\phi_j(\mathbf{x}) + f(\mathbf{x})\phi_j(\mathbf{x})]d\mathbf{x}
$$
$$
= \oint_{\partial\Omega} \phi_j(\mathbf{x})\frac{\partial\hat{u}}{\partial n}(\mathbf{x})d\mathbf{x}, \quad j = 1, 2, \ldots, N,
$$

which is identical to the Galerkin system (3.16-2).

Petrov-Galerkin methods.

Consider a nonsymmetric variant on Equation (3.16-1), such as

$$
(3.16\text{-}9) \qquad -\nabla^2 u(\mathbf{x}) + \mathbf{p}(\mathbf{x}) \cdot \nabla u(\mathbf{x}) = f(\mathbf{x}) \quad \text{in} \quad \Omega,
$$

where $\mathbf{p}(\mathbf{x})$ is some continuous vector field. For boundary-value problems involving this equation the maximum principle needed for the Rayleigh-Ritz approach no longer directly applies. Hence, we must resort to Galerkin's method, or a variant on Galerkin's method, to generate the required integral equations. Straightforward application of Galerkin's method to Equation (3.16-9) yields

$$
(3.16\text{-}10) \qquad \sum_{i=1}^N u_i [\langle \nabla\phi_i(\mathbf{x}), \nabla\phi_j(\mathbf{x}) \rangle + \langle \mathbf{p}(\mathbf{x}) \cdot \nabla\phi_i(\mathbf{x}), \phi_j(\mathbf{x}) \rangle]
$$
$$
= \langle f(\mathbf{x}), \phi_j(\mathbf{x}) \rangle + \oint_{\partial\Omega} \frac{\partial\hat{u}}{\partial n}\phi_j(\mathbf{x})d\mathbf{x}, \qquad j = 1, 2, \ldots, N.
$$

(Here we have already applied Green's theorem to the second-order term.)

Equations (3.16-10) have the same form as Equations (2.12-4). It is therefore possible to apply the finite-element methodology of Chapter Two to solve this system. However, it is important to recognize possible numerical difficulties peculiar to the first-order term appearing in Equation (3.16-10). In particular, it is sometimes advantageous to use **upstream weighting** to ameliorate pathologic behavior that can arise when the coefficient **a(x)** is large and a coarse grid spacing is used. These difficulties are by no means endemic to the finite-element approach; indeed, similar considerations arise in finite differences. The upstream finite-difference algorithm of choice is attributed to Allen and Southwell (1955). In the case of finite-element methods we consider, instead of the traditional Galerkin method, the more general **Petrov-Galerkin** formulation.

This formulation is most readily understood in one space dimension. We begin with the equation

$$(3.16\text{-}11) \qquad p\frac{du}{dx} - \frac{d^2u}{dx^2} = 0, \qquad 0 < x < 1,$$

subject to Neumann boundary conditions. Application of the Galerkin finite-element approach yields the integral equations

$$(3.16\text{-}12) \qquad \sum_{i=1}^{N} u_i \int_0^1 \left(p\frac{d\phi_i}{dx}\phi_j + \frac{d\phi_i}{dx}\frac{d\phi_j}{dx} \right) dx - \left[\frac{d\hat{u}}{dx}\phi_j \right]_0^1 = 0,$$

where $\hat{u}(x) = \sum_{i=1}^{N} u_i\phi_i(x)$ is the trial function. When the basis functions ϕ_i are chosen as the piecewise linear Lagrange polynomials, Equation (3.16-12) yields the difference formula

$$\frac{ph}{2}(u_{k+1} - u_{k-1}) - (u_{k-1} - 2u_k + u_{k+1}) = 0,$$

$$k = 2, \ldots, N-1,$$

for all interior nodes. This expression is the classic central difference approximation. While it appears to be a perfectly reasonable discretization, the numerical solutions produced by this scheme exhibit spurious oscillatory behavior whenever $p/2h \geq 1$ (see, for example, Jensen and Finlayson, 1980).

Having recognized the source of oscillatory behavior, analysts have attempted to mimic upstream differencing by using asymmetric weighting functions in the method of weighted residuals. Thus Equation (3.16-12) is modified to read

$$(3.16\text{-}13) \qquad \sum_{i=1}^{N} u_i \int_0^1 \left(p\frac{d\phi_i}{dx}\psi_j + \frac{d\phi_i}{dx}\frac{d\psi_j}{dx} \right) dx - \left[\frac{d\hat{u}}{dx}\psi_j \right]_0^1 = 0,$$

where $\psi_k(x) = \phi_k(x) + \alpha\gamma_k(x)$, that is, the weighting function ψ_k differs from the original piecewise linear basis function ϕ_k by a higher-degree perturbation $\alpha\gamma_k(x)$. For example, in local (ξ) coordinates we might adopt a quadratic perturbation given by

$$(3.16\text{-}14) \qquad \gamma_k^{-1}(\xi) = \tfrac{3}{4}(\xi^2 - 1),$$

$$(3.16\text{-}15) \qquad \gamma_k^{1}(\xi) = -\tfrac{3}{4}(\xi^2 - 1),$$

where the superscript ± 1 indicates the node for which the function is defined (see Figure 3-9).

The significance of using ψ_k in lieu of ϕ_k as a weighting function can be seen in Figure 3-9. The upstream weighting effect obviously increases as the parameter α increases. However, because the functions shown in parts (a) and (b) of the figure are antisymmetric for any value of α, the second-order term in Equation (3.16-11) is not influenced by the second-degree perturbation γ_k. When we use this Petrov-Galerkin scheme, the difference formula that is obtained for the model equation (3.16-11) is (Fletcher, 1984)

$$(3.16\text{-}16) \quad \frac{ph}{2}(u_{k+1} - u_{k-1}) - \alpha\frac{ph}{2}(u_{k-1} - 2u_k + u_{k+1})$$
$$- (u_{k-1} - 2u_k + u_{k+1}) = 0, \qquad k = 2, \ldots, N - 1.$$

Thus the introduction of the asymmetry in the weighting function leads to an additional term whose form is that of a difference analog to a second-order "diffusion" term and whose magnitude is controlled by the parameter α.

The solution to Equation (3.16-16) is (Christie et al. 1976)

$$(3.16\text{-}17) \qquad u_k = A_0 + B_0 \left[\frac{1 + \beta(\alpha + 1)}{1 + \beta(\alpha - 1)}\right]^k,$$

where A_0 and B_0 are selected to satisfy the boundary conditions imposed on the problem and $\beta = ph/2$. Examination of Equation (3.16-17) reveals that, when $\alpha = 0$, u_k will oscillate as a function of the nodal index k whenever $\beta \geq 1$. This oscillation occurs independently of the behavior of the true solution $u(x)$. However, if $\alpha \geq 1$, u_k will not oscillate, nor will it oscillate if $\beta(1 - \alpha) > 1$ in the case $-\infty < \alpha < 1$. The reader should be aware that we pay a price for this nonoscillatory behavior. By introducing asymmetric weighting functions, we have reduced the method from second-order to first-order accurate in the grid mesh h (Christie et al., 1976). We shall revisit the issue of asymmetric weighting functions in Chapter Four.

The extension of the asymmetric weighting function concept to higher

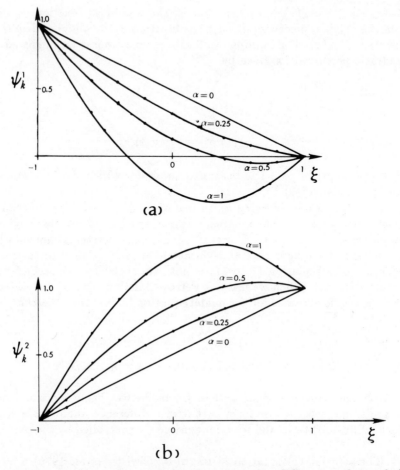

Figure 3-9. Asymmetric weighting functions for different values of the parameter α [see Equations (3.16-14) and (3.16-15)]. Note that $\psi_k^{\xi_o} = \phi_k^{\xi_o} + \alpha \gamma_k^{\xi_o}$.

dimension is straightforward when a tensor-product formulation is used. In the case of triangular elements, however, the extension is not so obvious. Huyakorn (1976) proposed one method of introducing asymmetry into linear triangular elements as follows: Let the weighting function be given by

$$(3.16\text{-}18) \qquad \psi_i(\mathbf{x}) = L_i(\mathbf{x}) + F_i(\mathbf{x}), \qquad i = 1, 2, 3,$$

where

$$F_i = 3(\alpha_j L_k L_i - \alpha_k L_j L_i).$$

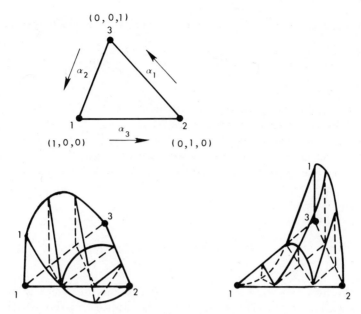

Figure 3-10. Asymmetric weighting functions for linear triangular elements. Here, $\alpha_1 = \alpha_2 = \alpha_3 = 1$ (from Huyakorn, 1976.)

Here, α_i is the upstream weighting parameter, defined in the sense of Figure 3-10, $L_i(\mathbf{x})$ is the usual triangular basis function defined in Section 2.12, and (i, j, k) is an even permutation of the indices $(1, 2, 3)$.

Representation of variable coefficients.

While physical problems involving homogeneous material properties are often encountered, in general one must be prepared to accommodate variable properties. There are two approaches to this problem. Consider the governing equation

$$(3.16\text{-}19) \qquad \nabla \cdot [k(\mathbf{x})\nabla u(\mathbf{x})] = f(\mathbf{x}) \quad \text{in} \quad \Omega.$$

Application of Galerkin's method and Green's theorem on a grid consisting of E elements, assuming a trial function $\hat{u}$ having N nodal degrees of freedom, yields

$$(3.16\text{-}20) \quad \sum_{e=1}^{E} \left[\sum_{i=1}^{N} u_i \int_{\Omega_e} k(\mathbf{x})\nabla\phi_i(\mathbf{x}) \cdot \nabla\phi_j(\mathbf{x})dx + \int_{\Omega_e} f(\mathbf{x})\phi_j(\mathbf{x})dx \right.$$

$$\left. - \int_{(\partial\Omega)_e} k(\mathbf{x})\frac{\partial\hat{u}}{\partial n}(\mathbf{x})\phi_j(\mathbf{x})dx \right] = 0, \qquad j = 1, 2, \ldots, N.$$

In writing Equation (3.16-20), we have employed the decompositions

$$\Omega = \bigcup_{e=1}^{E} \Omega_e, \qquad \partial\Omega = \bigcup_{e=1}^{E} (\partial\Omega)_e,$$

where Ω_e is the region interior to element e and $(\partial\Omega)_e$ is the segment of $\partial\Omega$ intersecting element e (see Figure 2-12).

To evaluate the integrals appearing in the Galerkin equations (3.16-20), we must choose a suitable numerical representation of the coefficient $k(\mathbf{x})$. Often the form of $k(\mathbf{x})$ is such as to make direct integration awkward if we use its exact values. To facilitate the integration process we can approximate $k(\mathbf{x})$ as

$$(3.16\text{-}21) \qquad k(\mathbf{x}) \simeq \hat{k}(\mathbf{x}) = \sum_{\ell=1}^{N} k_\ell \phi_\ell(\mathbf{x}),$$

where the numbers k_ℓ, $\ell = 1, 2, \ldots, N$, denote nodal values of $k(\mathbf{x})$. This approximation is sometimes called a **functional representation** of k. Substitution of the approximation (3.16-21) into Equations (3.16-20) yields

$$(3.16\text{-}22) \quad \sum_{e=1}^{E}\left[\sum_{i=1}^{N} u_i \sum_{\ell=1}^{N} k_\ell \int_{\Omega_e} \phi_\ell(\mathbf{x})\nabla\phi_i(\mathbf{x})\cdot\nabla\phi_j(\mathbf{x})d\mathbf{x}\right.$$
$$\left. + \int_{\Omega_e} f(\mathbf{x})\phi_j(\mathbf{x})d\mathbf{x} - \int_{(\partial\Omega)_e} k(\mathbf{x})\frac{\partial u}{\partial n}(\mathbf{x})\phi_j(\mathbf{x})d\mathbf{x}\right] = 0,$$
$$j = 1, 2, \ldots, N.$$

We do not discretize $k(\mathbf{x})$ in the last term in this equation because one typically prescribes Neumann boundary conditions by specifying the normal flux $-k(\mathbf{x})\partial u/\partial n$ rather than the normal derivative $\partial u/\partial n$. The integrals in Equation (3.16-22) can now be readily evaluated, the only complication over the case when k is uniform being the necessity to integrate higher-degree polynomials.

A very simple case of this approach to the treatment of variable coefficients is to assume that they are elementwise constant. In this case we can rewrite (3.16-20) as follows:

$$(3.16\text{-}23) \quad \sum_{e=1}^{E}\left[\sum_{i=1}^{N} u_i k_e \int_{\Omega_e} \nabla\phi_i(\mathbf{x})\nabla\phi_j(\mathbf{x})d\mathbf{x} + \int_{\Omega_e} f(\mathbf{x})\phi_j(\mathbf{x})d\mathbf{x}\right.$$
$$\left. - \int_{(\partial\Omega)_e} k(\mathbf{x})\frac{\partial u}{\partial n}\phi_j(\mathbf{x})d\mathbf{x}\right] = 0, \qquad j = 1, 2, \ldots, N,$$

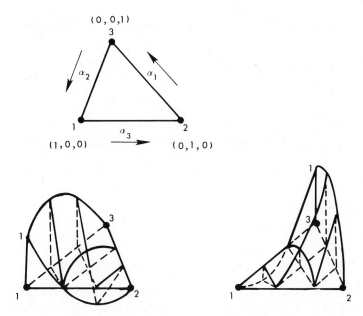

Figure 3-10. Asymmetric weighting functions for linear triangular elements. Here, $\alpha_1 = \alpha_2 = \alpha_3 = 1$ (from Huyakorn, 1976.)

Here, α_i is the upstream weighting parameter, defined in the sense of Figure
3-10, $L_i(\mathbf{x})$ is the usual triangular basis function defined in Section 2.12,
and (i, j, k) is an even permutation of the indices $(1, 2, 3)$.

Representation of variable coefficients.

While physical problems involving homogeneous material properties
are often encountered, in general one must be prepared to accommodate
variable properties. There are two approaches to this problem. Consider
the governing equation

(3.16-19) $$\nabla \cdot [k(\mathbf{x})\nabla u(\mathbf{x})] = f(\mathbf{x}) \quad \text{in} \quad \Omega.$$

Application of Galerkin's method and Green's theorem on a grid consisting of E elements, assuming a trial function $\hat{u}$ having N nodal degrees of
freedom, yields

(3.16-20) $$\sum_{e=1}^{E}\left[\sum_{i=1}^{N} u_i \int_{\Omega_e} k(\mathbf{x})\nabla\phi_i(\mathbf{x}) \cdot \nabla\phi_j(\mathbf{x})d\mathbf{x} + \int_{\Omega_e} f(\mathbf{x})\phi_j(\mathbf{x})d\mathbf{x}\right.$$

$$\left. - \int_{(\partial\Omega)_e} k(\mathbf{x})\frac{\partial\hat{u}}{\partial n}(\mathbf{x})\phi_j(\mathbf{x})d\mathbf{x}\right] = 0, \qquad j = 1, 2, \ldots, N.$$

In writing Equation (3.16-20), we have employed the decompositions

$$\Omega = \bigcup_{e=1}^{E} \Omega_e, \qquad \partial\Omega = \bigcup_{e=1}^{E} (\partial\Omega)_e,$$

where Ω_e is the region interior to element e and $(\partial\Omega)_e$ is the segment of $\partial\Omega$ intersecting element e (see Figure 2-12).

To evaluate the integrals appearing in the Galerkin equations (3.16-20), we must choose a suitable numerical representation of the coefficient $k(\mathbf{x})$. Often the form of $k(\mathbf{x})$ is such as to make direct integration awkward if we use its exact values. To facilitate the integration process we can approximate $k(\mathbf{x})$ as

$$(3.16\text{-}21) \qquad k(\mathbf{x}) \simeq \hat{k}(\mathbf{x}) = \sum_{\ell=1}^{N} k_\ell \phi_\ell(\mathbf{x}),$$

where the numbers k_ℓ, $\ell = 1, 2, \ldots, N$, denote nodal values of $k(\mathbf{x})$. This approximation is sometimes called a **functional representation** of k. Substitution of the approximation (3.16-21) into Equations (3.16-20) yields

$$(3.16\text{-}22) \quad \sum_{e=1}^{E} \left[\sum_{i=1}^{N} u_i \sum_{\ell=1}^{N} k_\ell \int_{\Omega_e} \phi_\ell(\mathbf{x}) \nabla\phi_i(\mathbf{x}) \cdot \nabla\phi_j(\mathbf{x}) dx \right.$$

$$\left. + \int_{\Omega_e} f(\mathbf{x})\phi_j(\mathbf{x}) dx - \int_{(\partial\Omega)_e} k(\mathbf{x})\frac{\partial u}{\partial n}(\mathbf{x})\phi_j(\mathbf{x}) dx \right] = 0,$$

$$j = 1, 2, \ldots, N.$$

We do not discretize $k(\mathbf{x})$ in the last term in this equation because one typically prescribes Neumann boundary conditions by specifying the normal flux $-k(\mathbf{x})\partial u/\partial n$ rather than the normal derivative $\partial u/\partial n$. The integrals in Equation (3.16-22) can now be readily evaluated, the only complication over the case when k is uniform being the necessity to integrate higher-degree polynomials.

A very simple case of this approach to the treatment of variable coefficients is to assume that they are elementwise constant. In this case we can rewrite (3.16-20) as follows:

$$(3.16\text{-}23) \quad \sum_{e=1}^{E} \left[\sum_{i=1}^{N} u_i k_e \int_{\Omega_e} \nabla\phi_i(\mathbf{x}) \nabla\phi_j(\mathbf{x}) dx + \int_{\Omega_e} f(\mathbf{x})\phi_j(\mathbf{x}) dx \right.$$

$$\left. - \int_{(\partial\Omega)_e} k(\mathbf{x})\frac{\partial u}{\partial n}\phi_j(\mathbf{x}) dx \right] = 0, \qquad j = 1, 2, \ldots, N,$$

where k_e is a constant value associated with element e. We have assumed, once again, that the boundary flux $k\partial u/\partial n$ will be specified wherever the boundary integral must be evaluated. In practice the assumption of elementwise constant coefficients results in an algorithm that requires essentially the same computational effort as an algorithm developed for a homogeneous medium, where k is a global constant independent of $\mathbf{x}$.

3.17. Boundary-Element Methods.

The idea behind the finite-element methods discussed so far is to discretize the interior of the computational domain, forcing the trial function to satisfy certain of the boundary conditions a priori. The idea behind the **boundary-element method** is in a sense just the opposite: We discretize the boundary, forcing the trial function to satisfy the interior problem through the use of fundamental solutions, which we discussed in Section 3.7.

Let us consider Laplace's equation $\nabla^2 u = 0$ on a domain Ω, subject to the mixed boundary conditions

(3.17-1a) $$u = \overline{u} \quad \text{on} \quad \Gamma_1 \subset \partial\Omega$$

(3.17-1b) $$\frac{\partial u}{\partial n} = \mathcal{T}(u) = \overline{q} \quad \text{on} \quad \Gamma_2 \subset \partial\Omega.$$

Here, $\Gamma_1 \cup \Gamma_2 = \partial\Omega$, with Γ_1 and Γ_2 being disjoint boundary segments. $\mathcal{T}$ denotes the operator of normal differentiation, $\partial/\partial n$, on the boundary segment $\Gamma_2 \subset \partial\Omega$. Thus, given a fundamental solution for ∇^2 on the interior of Ω, the integral representation (3.8-4) for points $\boldsymbol{\xi} \in \partial\Omega$ collapses to the following:

(3.17-2) $$u(\boldsymbol{\xi}) + \oint_{\partial\Omega} u(\mathbf{x})\mathcal{T}(G(\boldsymbol{\xi},\mathbf{x}))d\mathbf{x} = \oint_{\partial\Omega} \mathcal{T}(u(\mathbf{x}))G(\boldsymbol{\xi},\mathbf{x})d\mathbf{x},$$

where, in the notation of Section 3.8, $G(\boldsymbol{\xi},\mathbf{x}) = 2K(\boldsymbol{\xi},\mathbf{x})/M$. Observe that, according to the boundary conditions (3.17-1), u is not known on Γ_2 while $\mathcal{T}(u)$ is not known on Γ_1.

We interpret Equation (3.17-2) as a **boundary integral equation** for the function u on Γ_2 and for its normal derivative $\mathcal{T}(u)$ on Γ_1. To discretize this integral equation, we replace the boundary $\partial\Omega$ by straight line segments or **boundary elements,** as drawn in Figure 3-11. There are two commonly used procedures for interpolating the approximate solution along this discretized boundary. In one, the approximate solution is constant on each segment, and in the other it is a linear polynomial on each segment.

To see how this idea leads to a set of algebraic equations, let the total number of boundary elements be n, with n_1 elements belonging to Γ_1 and

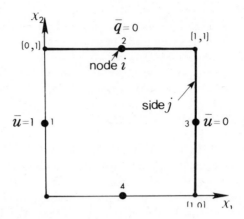

Figure 3-11. Illustration of the sample problem for the boundary-element method (after Lapidus and Pinder, 1982).

n_2 to Γ_2. Consider the interpolation scheme in which the values of u and $T(u)$ are assumed to be constant on each element. Applying Equation (3.17-2) at the midpoint of each boundary element, one gets

$$(3.17\text{-}3) \qquad u_i + \sum_{j=1}^{n} u_j \int_{\Gamma_j} q_i^* \, d\mathbf{x} = \sum_{j=1}^{n} q_j \int_{\Gamma_j} G_i \, d\mathbf{x}.$$

Here u_i and q_i signify the values of u and $T(u)$, respectively, at "node" i, that is, at the midpoint of the boundary element numbered i. The symbols q_i^* and G_i denote the functions $T(G)$ and G, respectively, when the singularity of G is located at node i. Now define the $n \times n$ matrix $\mathbf{H}$ as that having entries

$$(3.17\text{-}4a) \qquad H_{ij} = \int_{\Gamma_j} q_i^* \, d\mathbf{x},$$

when $i \neq j$, and

$$(3.17\text{-}4b) \qquad H_{ii} = 1 + \int_{\Gamma_i} q_i^* \, d\mathbf{x}.$$

Similarly, define the matrix $\mathbf{J}$ by

$$(3.17\text{-}5) \qquad J_{ij} = \int_{\Gamma_j} G_i \, d\mathbf{x}.$$

Then Equation (3.17-3) yields the following equation at each boundary element i:

(3.17-6)
$$\sum_{j=1}^{n} H_{ij}u_j = \sum_{j=1}^{n} J_{ij}q_j.$$

This constitutes a system of n equations in $n_1 + n_2 = n$ unknowns, since the index i ranges over the integers $1, ..., n$. The unknowns are q_j for $j = 1, ..., n_1$, and u_j for $j = n_1 + 1, ..., n_1 + n_2 = n$. Rearranging the equations in such a way that all the unknowns remain on the left side, one can write, in matrix notation

(3.17-7)
$$\mathbf{Au = f},$$

where the components of the vector $\mathbf{u}$ include the n_1 unknown boundary values q_i of the normal derivative $T(u)$ and the n_2 unknown boundary values u_i of the solution. Thus, like the finite-element method, the boundary-element method converts differential equations to discrete algebraic analogs.

Once we have solved for the approximate nodal values of u and $T(u)$ over the whole boundary, we can calculate an approximation to u at any interior point by means of the discretized version of the integral representation (3.8-3):

(3.17-8)
$$2u_i = \sum_{j=1}^{n} q_j J_{ij} - \sum_{j=1}^{n} u_j H_{ij}.$$

Let us now consider a simple example to illustrate the method. Consider Laplace's equation $\nabla^2 u = 0$ on the unit square $\Omega = (0, 1) \times (0, 1)$ in the (x_1, x_2)-plane, subject to the boundary conditions

(3.17-9a) $u(0, x_2) \equiv \bar{u}(0, x_2) = 1$

(3.17-9b) $u(1, x_2) \equiv \bar{u}(1, x_2) = 0$

(3.17-9c) $\dfrac{\partial u}{\partial n}(x_1, 0) \equiv \bar{q}(x_1, 0) = 0$

(3.17-9d) $\dfrac{\partial u}{\partial n}(x_1, 1) \equiv \bar{q}(x_1, 1) = 0$

Figure 3-11 illustrates the example problem. We assume that u and $\partial u/\partial n$ are constant along each boundary element. The algebraic system of equations for this case is given by Equation (3.17-6) for the special case of four nodes and four sides, that is,

(3.17-10)
$$\sum_{j=1}^{4} H_{ij}u_j + \sum_{j=1}^{4} J_{ij}q_j = 0, \qquad i = 1, 2, 3, 4.$$

The matrix entries in this equation have the following definitions:

$$H_{ij} = \int_{\Gamma_j} \frac{\partial G_i}{\partial n}\, d\mathbf{x}, \quad i \neq j;$$

$$H_{ii} = -\pi + \int_{\Gamma_i} \frac{\partial G_i}{\partial n}\, d\mathbf{x};$$

$$J_{ij} = \int_{\Gamma_j} G_i\, d\mathbf{x}.$$

Here, Γ_j denotes the j-th edge of the discretized boundary $\partial\Omega$, and u_j and q_j are, as defined earlier, values of the function u and its normal derivative $\partial u/\partial n$ at node j. Recall that G_i and $\partial G_i/\partial n$ are the functions G and $\partial G/\partial n$ when the singularity of G is located at node i. In Section 3.7 we observed that, when Ω is two dimensional, the fundamental solution for the Laplace operator is as follows:

$$G(\boldsymbol{\xi}, \mathbf{x}) = \frac{1}{2\pi} \ln \|\mathbf{x} - \boldsymbol{\xi}\|_2 = \ln r.$$

Thus, from Equation (3.17-10), we have

(3.17-11a)
$$H_{ij} = \int_{\Gamma_j} \left[\frac{\partial}{\partial n} (\ln r) \right]_i d\mathbf{x} = \int_{\Gamma_j} \left(\frac{1}{r} \frac{\partial r}{\partial n} \right)_i d\mathbf{x},$$

and

(3.17-11b)
$$J_{ij} = \int_{\Gamma_j} \ln r_i\, d\mathbf{x}.$$

The integrations indicated in Equations (3.17-11) can be performed in closed form using the local coordinate systems shown in Figure 3-12 (see Huyakorn and Pinder, 1983). The results are as follows:

$$H_{ij} = \pm\beta_{ij}, \quad i \neq j,$$
$$H_{ii} = -\pi,$$
$$J_{ij} = [-\ell_2 \ln(r_{i2}) + \ell_1 \ln(r_{i1}) + L_j - p_j\beta_{ij}], \quad i \neq j,$$
$$J_{ii} = L_i[\ln(2/L_i) + 1].$$

The parameter β_{ij} denotes the angle, measured in radians, subtended by the j-th boundary element at the midpoint of the i-th boundary element. The choice of signs indicated above depends on the orientation of the (ℓ, n)-coordinate axes: The positive sign applies when (ℓ, n) forms a right-handed coordinate system, as drawn. We now use these results to construct the matrix equation corresponding to the algebraic system (3.17-10). Evaluating the coefficients H_{ij} and J_{im}, we have

$$\mathbf{H} = \begin{bmatrix} -\pi & 1.11 & 0.927 & 1.11 \\ 1.11 & -\pi & 1.11 & 0.927 \\ 0.927 & 1.11 & -\pi & 1.11 \\ 1.11 & 0.927 & 1.11 & -\pi \end{bmatrix},$$

$$\mathbf{J} = \begin{bmatrix} 1.69 & 0.337 & -0.0389 & 0.337 \\ 0.337 & 1.69 & 0.337 & -0.0389 \\ -0.0389 & 0.337 & 1.69 & 0.337 \\ 0.337 & -0.0389 & 0.337 & 1.69 \end{bmatrix}.$$

Employing the known information, we see that Equation (3.17-10) yields

$$\begin{bmatrix} 1.69 & 1.11 & -0.0389 & 1.11 \\ 0.337 & -\pi & 0.337 & 0.927 \\ -0.0389 & 1.11 & 1.69 & 1.11 \\ 0.337 & 0.927 & 0.337 & -\pi \end{bmatrix} \begin{bmatrix} q_1 \\ u_2 \\ q_3 \\ u_4 \end{bmatrix} = \begin{bmatrix} \pi \\ -1.11 \\ -0.927 \\ -1.11 \end{bmatrix}$$

The solution to this system of equations is

$$(q_1, u_2, q_3, u_4) = (1.17, 0.500, -1.17, 0.500),$$

which compares with the exact solution values

$$(q_1, u_2, q_3, u_4) = (1.00, 0.500, -1.00, 0.500).$$

As with the finite-element method, we could have achieved better accuracy by using a finer discretization of the boundary, therefore solving for more nodal unknowns.

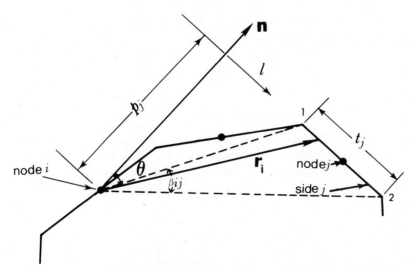

Figure 3-12. Local coordinate system for element j (after Huyakorn and Pinder, 1983).

3.18. Problems for Chapter Three.

1. The momentum balance equation of linear elasticity was derived in Section 1.4. Using it, show that in linear elastostatics, in the absence of body forces, the displacement field **U** satisfies the equation

$$\mu\nabla^2\mathbf{U} + (\mu + \lambda)\nabla\nabla \cdot \mathbf{U} = 0.$$

It is well known that any sufficiently smooth vector field **U** can be represented as the sum of two vector fields **V** and **W** such that $\nabla \cdot \mathbf{V} = 0$ and $\nabla \times \mathbf{W} = 0$.

a. Use this fact to prove that the functions $\Phi_i = \mu V_i + (\lambda + 2\mu)W_i$, $i = 1, 2, 3$, are harmonic.

b. The condition $\nabla \times \mathbf{W} = 0$ implies that **W** can be derived from a potential φ. That is, there exists a function φ such that $\mathbf{W} = \nabla\varphi$. Prove that

$$\nabla^2\varphi = \frac{1}{\lambda + 2\mu}\nabla \cdot \boldsymbol{\Phi},$$

and that any solution of the displacement field equation derived in Problem 1 can be written as follows:

$$\mathbf{U} = \frac{1}{\mu}\boldsymbol{\Phi} - \frac{\lambda + \mu}{2\mu(\lambda + 2\mu)}[\nabla\psi + \nabla(\mathbf{x} \cdot \boldsymbol{\Phi})],$$

where ψ and the components Φ_1, Φ_2, and Φ_3 are harmonic functions. This equation is the **Neuber-Papkovich representation**.

2. Consider the Dirichlet problem for Laplace's equation in the unit square $(0, 1) \times (0, 1)$, with boundary conditions

$$u(0, y) = g_1(y), \quad u(1, y) = g_2(y), \quad 0 \le y \le 1,$$

and

$$u(x, 0) = g_3(x), \quad u(x, 1) = g_4(x), \quad 0 \le x \le 1.$$

Use linearity to decompose this problem into two problems, one for which $g_1 \equiv g_2 \equiv 0$ and another one for which $g_3 \equiv g_4 \equiv 0$. Such a decomposition is useful when applying the method of separation of variables, which can be used only when the boundary conditions are homogeneous in two of the square sides.

3. Consider Laplace's equation in three dimensions:

$$\nabla^2 u \equiv \frac{\partial^2 u}{\partial x^2} + \frac{\partial^2 u}{\partial y^2} + \frac{\partial^2 u}{\partial z^2} = 0.$$

a. Establish the three-dimensional analog of the five-point approximation (3.9-3).

b. Similarly, establish the three-dimensional analogs of the nine-point approximations (3.9-7) and (3.9-10).

State the number of nodes involved in each of these approximations.

4. Solve a Dirichlet problem for Poisson's equation

$$\nabla^2 u + 1 = 0$$

in the unit square $(0,1) \times (0,1)$, using a five-point difference approximation. Reduce this problem to Laplace's equation, and write a program to solve this latter equation. To invert the resulting matrix use:

a. Gauss elimination,

b. Band elimination, and

c. LU decomposition.

Verify the Nb^2 estimate given for the number of operations required by LU decomposition.

5. Write a computer program to express any positive definite symmetric matrix $\mathbf{A}$ in the form $\mathbf{A} = \mathbf{L}\mathbf{L}^T$, where $\mathbf{L}$ is lower triangular. To this end use the following algorithm:

$$\ell_{11} = \sqrt{a_{11}}$$
For $i = 1, \ldots, N$:
$$\ell_{ij} = \left(a_{ij} - \sum_{k=1}^{j-1} \ell_{ik}\ell_{jk}\right)/\gamma_{ij}, \; j = 1, 2, \ldots, j - i$$
$$\ell_{ii} = \left(a_{ii} - \sum_{k=1}^{i-1} \ell_{ik}^2\right)^{1/2}$$
$$\ell_{i-1,j} = 0, \; j = i, i + 1, \ldots, N.$$

Here, the sum is taken to be zero if the upper limit is less than the lower limit.

6. The aim of this problem is to solve Problem 4 using Cholesky decomposition. To invert a lower triangular matrix $\mathbf{L}$ (with nonzero diagonal entries) apply the following algorithm:

$$\ell_{ii}^I = 1/\ell_{ii}, \; i = 1, 2, \ldots, N$$
$$\ell_{ij}^I = \left(-\sum_{k=j+1} \ell_{ik}^I \ell_{kj}\right)/\ell_{jj}, \; j = 1, \ldots, i - 1, \; i = 2, 3, \ldots, N$$
$$\ell_{ij}^I = 0 \text{ when } j > 1.$$

Here ℓ_{ij}^I denotes the (i, j)-th element of the inverse matrix $\mathbf{L}^{-1}$.

7. Solve Problem 4 using SOR. Note that while the optimal value of ω is given by Equation (3.14-5), generally it is difficult to find this optimum value. The difficulty arises because application of Equation (3.14-5) requires knowledge of the eigenvalues of the matrix **A**. Optimal relaxation factors for sets of equations that result from the application of finite-difference methods to elliptic problems have been studied extensively by, for example, Forsythe and Wasow (1960), Varga (1962), and Fox (1962). For Problem 4, Forsythe and Wasow (1960) give

$$\omega_{\text{opt}} = \frac{2}{\sqrt{1 + (1 - q^2)}},$$

where q^2 is determined as follows: Define

$$\Delta_{j,k}^{(m+1)} u = u_{j,k}^{(m+1)} - u_{j,k}^{(m)}.$$

Given a vector $\mathbf{\Delta}^{(m+1)}$ containing such differences, define the norm

$$\|\mathbf{\Delta}^{(m+1)}\|_1 = \sum_j \sum_k |\Delta_{j,k}^{(m+1)}|.$$

Then

$$\frac{\|\mathbf{\Delta}^{(m+1)}\|_1}{\|\mathbf{\Delta}^{(m)}\|_1} \to q^2 \quad \text{when} \quad m \to \infty.$$

8. Formulate the homogeneous Dirichlet problem for Poisson's equation (3.16-1) in the unit square $(0, 1) \times (0, 1)$ using the finite-element method. More specifically, develop the system equations (3.16-8) for the case when $\Omega = (0, 1) \times (0, 1)$, the finite elements are squares, and the basis functions are tensor products of Lagrange linear interpolation basis functions. Compare your results with those obtained for Problem 4.

9. An alternative formuation of boundary methods is that of Trefftz (Herrera, 1984), which is based on Green's second identity (3.4-3). For harmonic functions u and v, the Trefftz method reduces to the following:

$$\oint_{\partial\Omega} \left(v \frac{\partial u}{\partial n} - u \frac{\partial v}{\partial n} \right) d\mathbf{x} = 0.$$

Trefftz method consists in replacing the function v in this equation by a system $\{\psi_1, \psi_2, \ldots, \psi_N\}$ of weighting functions that are themselves harmonic. Solve the sample problem for the boundary-element method in Section 3.17 using the Trefftz approach. Choose $\psi_1 \equiv 1$, $\psi_2 \equiv x$, $\psi_3 \equiv y$, and $\psi_4 \equiv x^2 - y^2$. Evaluate the Trefftz integral using Gauss quadrature on each side of the square, with only one Gauss point (the midpoint) on each side. (We warn the reader against drawing general conclusions from the fact that, for this simple example, the Trefftz procedure yields the exact solution.)

10. The upwinding procedure used in Petrov-Galerkin method given by Equation (3.16-13) depends on the value chosen for the paramัter α. However, there are no definite criteria for choosing α. A systematic procedure for upwinding has been proposed by Herrera et al. (1985), Herrera (1987), Celia and Herrera (1987), and Celia et al. (1987). The procedure is related to the Trefftz method sketched in Problem 9. Let

$$\mathcal{L}u \equiv \frac{d^2u}{dx^2} - p\frac{du}{dx}, \quad \text{and} \quad \mathcal{L}^*v \equiv \frac{d^2v}{dx^2} + \frac{d}{dx}(pv).$$

Then, integrating by parts in any interval (x_{k-1}, x_k), we have

$$\int_{x_{k-1}}^{x_k} (v\mathcal{L}u - u\mathcal{L}^*v)dx = \left[v\frac{du}{dx} - u\left(\frac{dv}{dx} + pv\right)\right]_{x_{k-1}}^{x_k}.$$

Assume u satisfies $\mathcal{L}u = 0$ (i.e., Equation 3.16-11), while $\{\psi^1, \psi^2\}$ are two linearly independent solutions of $\mathcal{L}^*\psi = 0$. Let $\Delta : 0 = x_0 < x_1 < \cdots < x_{N-1} < x_N = 1$ be a partition of $(0,1)$, and apply the equation just written at each of the N subintervals of the partition Δ. Thus, obtain the following finite-difference scheme:

$$\psi_k^\alpha \frac{du_k}{dx} - \psi_{k-1}^\alpha \frac{du_{k-1}}{dx} - u_k q_k^\alpha + u_{k-1}q_{k-1}^\alpha = 0, \quad k = 1, \ldots, N, \quad \alpha = 1, 2,$$

where $q^\alpha = d\psi^\alpha/dx + p\psi^\alpha$. The subscripts indicate the node at which each function is evaluated. Observe that this is a system of $2N$ equations for $2N$ unknowns (since u_0 and u_N are the prescribed boundary values).

a. For constant p, obtain the exact solutions of the differential equation

$$\frac{d^2\psi^\alpha}{dx^2} + p\frac{d\psi^\alpha}{dx} = 0.$$

b. For nonconstant p, develop two approximate solutions of $\mathcal{L}^*\psi = 0$, separately at each subinterval $(x_{\alpha-1}, x_\alpha)$.

We suggest the use of polynomial approximations and collocation at the Gauss points, as described in Section 2.14. Obtain a particular form for the finite-difference equations just derived, and develop a computer code to solve them. Observe that the number of collocation points n used in each subinterval is related to the degree d of the polynomial used to represent the weighting functions ψ by the equation $d = n + 1$.

3.19. References.

Allen, D. and Southwell, R., "Relaxation methods applied to determine the motion in two dimensions of a viscous fluid past a fixed cylinder," *Quart. J. Mech. Appl. Math., 8* (1955), 129-145.

Axelsson, O., "Conjugate gradient type methods for unsymmetric and inconsistent systems of linear equations," *Lin. Alg. Its Applic., 29* (1980), 1-16.

Birkhoff, G. and Lynch, R.E., *Numerical Solution of Elliptic Problems,* Philadelphia: SIAM, 1984.

Bramble, J. and Hubbard, B.E., "On the formulation of finite difference analogues of the Dirichlet problem for Poisson's equation," *Numerische Mathematik, 4* (1962), 383-390.

Bramble, J. and Hubbard, B.E., "A finite difference analog of the Neumann problem for Poisson's equation," *SIAM J. Numer. Anal., 2* (1965). 105-118.

Brandt, A., "Multi-level adaptive solutions to boundary-value problems," *Math. Comp., 31:138* (1977), 333-390.

Celia, M. and Herrera, I., "Solution of general ordinary differential equations using the algebraic theory approach," *Numer. Methods Partial Differential Equations, 3* (1987), 117-129.

Celia, M., Herrera, I., and Kindred, J.S., "A new numerical approach for the advective-diffusive transport equation, *Comunicaciones Técnicas No. 7,* Instituto de Geofísica, National University of Mexico, Mexico City, 1987.

Celia, M.A. and Pinder, G.F., "An analysis of alternating-direction methods for parabolic equations," *Numer. Methods Partial Differential Equations, 1* (1985), 57-70.

Christie, I., Griffiths, D.F., Mitchell, A.R., and Zienkiewicz, O.C., "Finite-element methods for second-order differential equations with significant first derivatives," *Int. J. Num. Meth. Eng., 10* (1976), 1389-1396.

Concus, P., Golub, G.H., and O'Leary, D.P., "A generalized conjugate gradient method for the numerical solution of elliptic partial differential equations," in J.R. Bunch and D.J. Rose, Eds., *Sparse Matrix Computations,* New York: Academic Press, 1976.

Courant, R. and Hilbert, D., *Methods of Mathematical Physics,* vol.I, New York: Wiley, 1953.

Douglas, J., Jr. and H.H. Rachford, "On the numerical solution of heat conduction problems in two or three space variables," *Trans. Am. Math. Soc., 82* (1956), 421-439.

Elman, H.C., "Preconditioned conjugate-gradient methods for nonsymmetric systems of linear equations," Research Report No. 203, Yale University Department of Computer Science, New Haven, Connecticut, 1981.

Fletcher, C. A., *Computational Galerkin Methods*, New York: Springer-Verlag, 1984.

Forsythe, G.E. and Wasow W.R., *Finite Difference Methods for Partial Differential Equations*, New York: Wiley, 1960.

Fox, L., *Numerial Solution of Ordinary and Partial Differential Equations*, Oxford: Pergamon Press, 1962.

George, A. and Liu, J., *Computer Solution of Large, Sparse, Positive Definite Problems*, Englewood Cliffs, New Jersey: Prentice-Hall, 1981.

Golub, G.H. and Van Loan, C.F., *Matrix Computations*, Baltimore: Johns Hopkins University Press, 1983.

Herrera, I., "Variational principles for problems with linear constraints," *IMA J., 25* (1980), 67-96.

Herrera, I., *Boundary Methods: An Algebraic Theory*, New York: Pitman, 1984.

Herrera, I., "The algebraic theory approach for ordinary differential equations: Highly accurate finite differences," *Numer. Methods Partial Differential Equations, 3* (1987), 117-129.

Herrera, I., Chargoy, L., and Alduncin, G., "A unified formulation of numerical methods. III. Finite differences and ordinary differential equations," *Numer. Methods Partial Differential Equations, 1* (1985), 241-258.

Hestenes, M.R. and Stiefel, E., "Methods of conjugate gradients for solving linear systems," *J. Res. Natl. Bur. Stand., 49* (1952), 409-436.

Huyakorn, P. S., "An upwind finite element scheme for improved solution of the convection-diffusion equation," Report Number 76-WR-2, Department of Civil Engineering, Princeton University, Princeton, New Jersey, 1976.

Huyakorn, P.S. and Pinder, G.F., *Computational Methods in Subsurface Flow*, New York: Academic Press, 1983.

Jensen, O.K. and Finlayson, B.A., "Oscillation limits for weighted residual methods applied to convection diffusion equations," *Int. J. Num. Meth. Engrg., 15* (1980), 1681-1689.

Lapidus, L. and Pinder, G.F., *Numerical Solution of Partial Differential Equations in Science and Engineering*, New York: Wiley, 1982.

Meijerink, J.A. and Van der Vorst, H.A., "An iterative solution method for linear equation systems of which the coefficient matrix is a symmetric *M*-matrix," *Math. Comp., 31* (1977), 148-162.

Mitchell, A.R. and Fairweather, G., "Improved forms of the alternating direction methods of Douglas, Peaceman and Rachford for solving parabolic and elliptic equations," *Numer. Math., 6* (1964), 285-292.

Obeysekare, U.R.B., Allen, M.B., Ewing, R.E., and George, J.H., "Application of conjugate-gradient-like methods to a hyperbolic problem in porous-media flow," *Int. J. Numer. Methods in Fluids, 7* (1987), 551-566.

Peaceman, D.W., *Fundamentals of Numerical Reservoir Simulation*, Amsterdam: Elsevier, 1977.

Peaceman, D.W. and Rachford, H. H., "The numerical solution of parabolic and elliptic differential equations," *SIAM J., 3* (1955), 28-41.

Reid, J.K., "On the method of conjugate gradients for the solution of large sparse systems of linear equations," in J.K. Reid, Ed., *Large Sparse Sets of Linear Equations*, New York: Academic Press (1971), pp. 231-254.

Strang, G. and Fix, G., *An Analysis of the Finite Element Method*, Englewood Cliffs, New Jersey: Prentice-Hall, 1973.

Tee, G.J., "Finite-difference approximation to the biharmonic operator," *Comput. J., 6* (1963), 85-100.

van Linde, H.J., "High-order finite difference methods for Poisson's equation," *Math. Comp., 28* (1974), 426-442.

Varga, R.S., *Matrix Iterative Analysis*, Englewood Cliffs, New Jersey: Prentice-Hall, 1962.

Wachspress, E. L., *Iterative Solution of Elliptic Systems*, Englewood Cliffs, New Jersey: Prentice-Hall, 1966.

CHAPTER FOUR
DISSIPATIVE SYSTEMS

4.1. Introduction.

The qualitative behavior of time-dependent macroscopic systems depends rather strongly on whether the system is dissipative or nondissipative. In this chapter we shall consider dissipative systems, leaving for Chapter 5 the discussion of nondissipative ones. Our development will be motivated by several physical examples, including heat flow, diffusion of matter obeying Fick's law, and stochastic processes in certain mechanical systems.

Heat flow.

Consider the energy balance equation (1.3-14). When the rate of heating by compression and viscous dissipation ($\mathbf{t} : \nabla \mathbf{v}$) can be neglected, the equation reduces to

$$(4.1\text{-}1) \qquad \frac{DE}{Dt} - \frac{1}{\rho} \nabla \cdot \mathbf{q} = h \ .$$

In Section 1.4 we outlined an argument leading to a PDE governing the transient temperature distribution in a body undergoing dissipative heat transfer. That argument treated the general case when the effects of compressive heating could contribute to temperature changes in the body. We now review an analogous argument valid for the case when compressive heating is absent and the motion is isochoric.

One step in our earlier argument was the adoption of Fourier's law of heat conduction as a constitutive relationship for the heat flux. In the present derivation, we initially assume a more general version of this constitutive relationship, allowing for anisotropic heat flux through a tensor form of the thermal conductivity. We have

$$(4.1\text{-}2) \qquad \mathbf{q} = \mathbf{k}_H \cdot \nabla T.$$

The energy balance (4.1-1) thus reduces to

(4.1-3)
$$\frac{DE}{Dt} - \frac{1}{\rho} \nabla \cdot (\mathbf{k}_H \cdot \nabla T) = h.$$

As we discussed in Section 1.4, we can use the mass balance in the form

(4.1-4)
$$\frac{D\rho^{-1}}{Dt} = \rho^{-1} \nabla \cdot \mathbf{v},$$

together with the chain rule, to convert Equation (4.1-3) to a form in which the temperature T appears as the principal unknown. If we consider the internal energy E to be a function of T and the specific volume ρ^{-1}, the chain rule implies

(4.1-5)
$$\frac{DE}{Dt} = \frac{\partial E}{\partial T} \frac{DT}{Dt} + \frac{\partial E}{\partial \rho^{-1}} \frac{D\rho^{-1}}{Dt}.$$

The function $\partial E/\partial T = c_v$ is the **heat capacity** of the material measured under conditions of constant volume. This coefficient differs slightly from the heat capacity c_p at constant pressure introduced in Section 1.4 in that c_p accounts for the thermal effects of compression while c_v neglects them. [In practice, for many common substances the two functions have nearly the same values; for example, for pure water at atmospheric pressure and 15° C, $(c_p - c_v)/c_p \simeq 0.003$.] In view of the mass balance (4.1-4), and since we have restricted our attention to isochoric motions for which $\nabla \cdot \mathbf{v} = 0$, $D\rho^{-1}/Dt = 0$. Therefore, the second term on the right in Equation (4.1-5) vanishes. Making these substitutions in Equation (4.1-3) and expanding the material derivative DT/Dt, we find

(4.1-6)
$$\rho c_v \left(\frac{\partial T}{\partial t} + \mathbf{v} \cdot \nabla T \right) - \nabla \cdot (\mathbf{k}_H \cdot \nabla T) = 0.$$

In the absence of advection, $\mathbf{v} = \mathbf{0}$, and we have

$$\rho c_v \frac{\partial T}{\partial t} - \nabla \cdot (\mathbf{k}_H \cdot \nabla T) = 0.$$

In simple cases the heat flux is isotropic, so that $\mathbf{k}_H = k_H \mathbf{1}$, and we can write

$$\rho c_v \frac{\partial T}{\partial t} - \nabla \cdot (k_H \nabla T) = 0.$$

Since the motion is isochoric, it is reasonable to assume as well that the density remains uniform in space. If we further assume that c_v is spatially uniform and, again, that the heat flux is isotropic, then Equation (4.1-6) reduces to

(4.1-7)
$$\frac{\partial T}{\partial t} + \mathbf{v} \cdot \nabla T - \nabla \cdot (\kappa_H \nabla T) = 0.$$

Here, $\kappa_H = k_H/(\rho c_v)$ is the **thermal diffusivity,** similar to that defined in Section 1.4 except that we have used c_v instead of c_p. When advection is absent and κ_H is uniform, Equation (4.1-7) collapses to the classic heat equation,

(4.1-8)
$$\frac{\partial T}{\partial t} - \kappa_H \nabla^2 T = 0.$$

Advection-diffusion equation.

Consider a substance S dissolved in a moving fluid F. As discussed in Section 1.5, the mass balance for each species S and F arises from the general mixture balance laws. From Equation (1.5-2) we have, for the solute,

$$\frac{\partial \rho^S}{\partial t} + \nabla \cdot (\rho^S \mathbf{v}^S) = r^S.$$

Here, ρ^S is the concentration of the solute, measured as mass per unit volume; $\mathbf{v}^S$ is the solute's velocity, and r^S represents a reaction rate. If we rewrite this equation in terms of the barycentric velocity $\mathbf{v}$ of the mixture, we get

(4.1-9)
$$\frac{\partial \rho^S}{\partial t} + \nabla \cdot (\rho^S \mathbf{v}) + \nabla \cdot \mathbf{j}^S = r^S,$$

where $\mathbf{j}^S$ is the diffusive flux of the solute with respect to the mixture velocity. In Section 1.5 we introduced Fick's law as a constitutive relationship for $\mathbf{j}^S$. The tensor form of this law is

(4.1-10)
$$\mathbf{j}^S = -\mathbf{K}^S \cdot \nabla \rho^S,$$

where $\mathbf{K}^S$ is the diffusion coefficient. Thus we have

(4.1-11)
$$\frac{\partial \rho^S}{\partial t} + \nabla \cdot (\rho^S \mathbf{v}) - \nabla \cdot (\mathbf{K}^S \cdot \nabla \rho^S) = r^S.$$

Now let us consider the special case when the solute undergoes linear decay. This happens, for example, when the solute is an unstable radionuclide. In this case $r^S = -r\rho^S$, where r stands for a positive decay constant.

If, moreover, the diffusive flux is isotropic, then $\mathbf{K}^S = K^S \mathbf{1}$, and Equation (4.1-11) reduces to

$$(4.1\text{-}12) \qquad \frac{\partial \rho^S}{\partial t} + \nabla \cdot (\rho^S \mathbf{v}) - \nabla \cdot (K^S \nabla \rho^S) = -r\rho^S .$$

This is the advection-diffusion equation in the presence of linear decay.

Stochastic processes of mechanical systems.

Stochastic processes of mechanical systems, under very general conditions, lead to equations analogous to those governing dissipative continua. In fact, in a sense we have just treated such a stochastic process. Stochastic Brownian motion at the corpuscular scale is the mechanism responsible for macroscopic dissipation in the form of diffusion. Another example, and one that we shall briefly explore here, is that of a mechanical system, such as a solid structure, subjected to random excitations. Applications to stochastic systems admittedly differ from the deterministic applications we have treated heretofore. Nevertheless, because of their importance in structural dynamics and other fields, we shall illustrate the procedure used to derive the basic equations.

To analyze stochastic processes in any mechanical system, we first establish a **phase space** for the system. The coordinate directions in this phase space correspond to the variables that completely specify the system. For example, for a single particle in three-space governed by Newton's laws of motion, the state is completely specified once we know the three components of the vector giving the particle's position and the three components of its velocity. Thus, the phase space for a single particle has dimension six. We describe such a system stochastically by postulating a **probability density** $\vec{\rho}(\vec{x}, t)$, where $\vec{x}$ is the position of the system in phase space. We interpret $\vec{\rho}$ as follows: The probability of finding the system's state inside any region $\mathcal{U}$ of phase space at time t is $\int_{\mathcal{U}} \vec{\rho}(\vec{x}, t) d\vec{x}$.

The probability density $\vec{\rho}(\vec{x}, t)$ obeys a conservation law very similar in form to the balance laws we have discussed so far, namely,

$$(4.1\text{-}13) \qquad \frac{\partial \vec{\rho}}{\partial t} + \vec{\nabla} \cdot (\vec{\rho}\vec{v}) - \vec{\nabla} \cdot \mathbf{T} = 0 .$$

Here, $\vec{v}$ denotes the time rate of change of the system's location in phase space, that is, its velocity in phase space, and $\mathbf{T}$ accounts for the spreading in phase space attributable to stochastic behavior. The notation $\vec{\nabla}$ indicates the operator of differentiation with respect to phase-space coordinates. Generally, for mechanical systems the velocity in phase space is a function of position in phase space, $\vec{v} = \vec{v}(\vec{x})$, even if the mechanical system is nonconservative. On the other hand, for a large class of stochastic

processes, the random dissipation **T** is a linear function of the gradient of the probability density $\vec{\rho}$, that is,

(4.1-14)
$$\mathbf{T} = \mathbf{A} \cdot \vec{\nabla}\vec{\rho} \ .$$

The specific form of the matrix **A** depends on the nature of the stochastic process.

For example, consider a one-dimensional harmonic oscillator, drawn schematically in Figure 4-1, whose equation of motion is the second-order ordinary differential equation

(4.1-15)
$$m\frac{d^2 x}{dt^2} + \eta\frac{dx}{dt} + k^2 x = 0 \ .$$

Here, x represents the position of the particle in physical space, η stands for the viscous damping, and k^2 is the spring constant. For such a mechanical system, the coordinates of phase space can be taken to be the particle's position and normalized velocity (see Figure 4-2),

(4.1-16)
$$\vec{x}_1 = x, \quad \vec{x}_2 = \frac{1}{\omega_0}\frac{dx}{dt} \ ,$$

the constant ω_0 being the particle's angular frequency, $\omega_0^2 = k^2/m$. We can rewrite the equation of motion (4.1-15) in terms of these phase-space coordinates as a coupled set of first-order equations. These equations give the velocity $\vec{v} = (\vec{v}_1, \vec{v}_2)$ of the system in phase space as follows:

(4.1-17a)
$$\vec{v}_1 = \frac{d\vec{x}_1}{dt} = \omega_0\vec{x}_2 \ ,$$

(4.1-17b)
$$\vec{v}_2 = \frac{d\vec{x}_2}{dt} = -\omega_0(\vec{x}_1 + \xi\vec{x}_2),$$

where $\xi = \eta/k^2$. When the oscillator undergoes random accelerations, the system experiences random changes in $\vec{x}_2$, and diffusion in phase space occurs along the $\vec{x}_2$-axis. In terms of the formalism of Equation (4.1-14), we can model this diffusion by assuming the existence of a positive parameter α such that

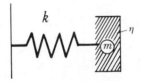

Figure 4-1. Schematic diagram of a harmonic oscillator with mass m, viscosity coefficient η, and spring constant k.

(4.1-18) $\mathbf{T} = \begin{bmatrix} T_1 \\ T_2 \end{bmatrix} = \begin{bmatrix} 0 & 0 \\ 0 & \alpha \end{bmatrix} \begin{bmatrix} \partial\vec{\rho}/\partial\vec{x}_1 \\ \partial\vec{\rho}/\partial\vec{x}_2 \end{bmatrix} = \begin{bmatrix} 0 \\ \alpha\partial\vec{\rho}/\partial\vec{x}_2 \end{bmatrix}.$

The equation (4.1-13) for the probability density is therefore

(4.1-19) $\dfrac{\partial\vec{\rho}}{\partial t} + \dfrac{\partial}{\partial\vec{x}_1}(\vec{\rho}\,\vec{v}_1) + \dfrac{\partial}{\partial\vec{x}_2}(\vec{\rho}\,\vec{v}_2) - \dfrac{\partial T_1}{\partial\vec{x}_1} - \dfrac{\partial T_2}{\partial\vec{x}_2}$

$\qquad\qquad = \dfrac{\partial\vec{\rho}}{\partial t} + \omega_0\vec{x}_2\dfrac{\partial\vec{\rho}}{\partial\vec{x}_1} - \omega_0(\vec{x}_1 + \xi\vec{x}_2)\dfrac{\partial\vec{\rho}}{\partial\vec{x}_2} - \omega_0\xi\vec{\rho} - \alpha\dfrac{\partial^2\vec{\rho}}{\partial\vec{x}_2^2}$

$\qquad\qquad = 0.$

When the viscosity vanishes, we are left with

(4.1-20) $\dfrac{\partial\vec{\rho}}{\partial t} + \omega_0\vec{x}_2\dfrac{\partial\vec{\rho}}{\partial\vec{x}_1} - \omega_0\vec{x}_1\dfrac{\partial\vec{\rho}}{\partial\vec{x}_2} - \alpha\dfrac{\partial^2\vec{\rho}}{\partial\vec{x}_2^2} = 0\ .$

Thus the probability density for this simple system obeys a parabolic PDE, the second-order term of which has a mathematical form similar to those of the heat flux and diffusion terms of our earlier examples.

Well posed problems.

There are many physical applications in which the mathematical model to be developed must be able to predict deterministic phenomena. That is, at any time the model must predict the physical state of the system by furnishing a unique mathematical solution. Furthermore, the model must be robust in view of the fact that we can never measure physical quantities exactly. If the prediction of the mathematical model is to be useful, it is essential that states predicted by the model using erroneous values of the

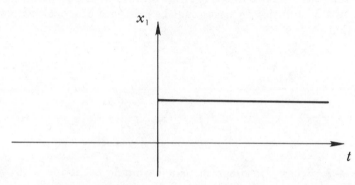

Figure 4-2. Solution $x_1(t)$ for the motion of a free particle ($k = 0$) in a viscous medium.

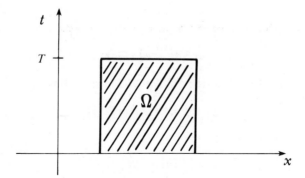

Figure 4-3. Example of a problem domain Ω in space-time for a parabolic
PDE.

parameters approach the actual state of the system when the parameter
errors tend to zero. As discussed in Section 3.3, mathematical models satisfying these criteria are said to be well posed problems. Such problems
possess unique solutions that depend continuously on the auxiliary conditions. As with elliptic PDE's, a given parabolic PDE typically possesses
many solutions in any region of space-time such as that drawn in Figure 4-3.
Therefore, to formulate well posed problems, it will be necessary to impose
additional boundary and initial conditions. In general, for parabolic PDE's,
well posed problems can be initial-value problems, initial-boundary-value
problems, or pure initial-value problems. Before beginning our discussion
of numerical schemes, we shall examine some of these possibilities.

4.2. The Heat Equation.

For simplicity, let us consider the one-dimensional heat equation, assuming
for mathematical convenience that $\kappa_H = 1$:

$$(4.2\text{-}1) \qquad \frac{\partial u}{\partial t} - \frac{\partial^2 u}{\partial x^2} = 0.$$

Initial-boundary-value problems for this simple PDE can be treated using
the technique of **separation of variables**. For example, consider Equation
(4.2-1) on the spatial interval $(0,1)$ subject to the initial conditions

$$(4.2\text{-}2) \qquad u(x,0) = u_I(x), \quad 0 < x < 1,$$

and the Dirichlet boundary conditions

$$(4.2\text{-}3) \qquad u(0,t) = u(1,t) = 0, \quad t \geq 0.$$

The method has the name "separation of variables" because we seek solutions having the form of products of functions of space and time:

$$(4.2\text{-}4) \qquad\qquad u(x,t) = \varphi(x)\psi(t).$$

Substituting this form into the PDE (4.2-1) yields the equation

$$(4.2\text{-}5) \qquad\qquad \varphi(x)\psi'(t) = \varphi''(x)\psi(t),$$

which we can rearrange to get

$$(4.2\text{-}6) \qquad\qquad \frac{\varphi''(x)}{\varphi(x)} = \frac{\psi'(t)}{\psi(t)}.$$

Here is the observation that is the crux of the method: For a function of x to equal one of t for all values of x and t, it is necessary that both functions be constant. Hence both sides of Equation (4.2-6) equal a number λ, and we can write

$$(4.2\text{-}7) \qquad\qquad \varphi'' - \lambda\varphi = 0; \quad \psi' - \lambda\psi = 0.$$

Thus we have reduced the solution of the original PDE to the task of solving two ordinary differential equations, the main difficulty now being that the constant λ remains unspecified.

We now examine the original boundary data to see which values of λ yield nontrivial (that is, nonzero) product solutions (4.2-4). For $u(x,t)$ to satisfy the Dirichlet boundary conditions (4.2-3), the function $\varphi(x)$ must satisfy these conditions. Thus $\varphi(0) = \varphi(1) = 0$. The general solution for φ in Equation (4.2-7) is a linear combination of $\sin\sqrt{\lambda}x$ and $\cos\sqrt{\lambda}x$. The boundary conditions (4.2-3) require the coefficient of the cosine term to vanish. However, $\sin\sqrt{\lambda}x$ satisfies the boundary conditions whenever λ assumes any of the **eigenvalues** $\lambda_n = n^2\pi^2$ for $n = 1, 2, \ldots$. We exclude the possibility $n = 0$, since this choice yields the trivial solution $\varphi(x) \equiv 0$. Thus the **eigenfunctions** for the spatial part of our initial-boundary-value problem are the following:

$$(4.2\text{-}8a) \qquad\qquad \varphi_n(x) = \sin(n\pi x), \quad n = 1, 2, \ldots.$$

Substituting the eigenvalues λ_n into the time-dependent differential equation in the pair (4.2-7) yields one solution $\psi_n(t)$ for each positive integer n, namely,

$$(4.2\text{-}8b) \qquad\qquad \psi_n(t) = e^{-n^2\pi^2 t}.$$

Because the heat equation is linear, any linear combination of solutions is also a solution. Thus, the general form of $u(x,t)$ for the given boundary data is the **Fourier sine series**

$$(4.2\text{-}9) \qquad u(x,t) = \sum_{n=1}^{\infty} A_n e^{-n^2\pi^2 t} \sin(n\pi x),$$

where the coefficients A_n remain to be determined. Observe that the boundary conditions (4.2-3) are satisfied independently of the choice of the constants A_n. However, we have not yet imposed the initial conditions (4.2-2). These conditions imply that

$$(4.2\text{-}10) \qquad u_I(x) = \sum_{n=1}^{\infty} A_n \sin(n\pi x).$$

We can use this relationship and our knowledge of $u_I(x)$ to determine the coefficients A_n.

To accomplish this task, we need to show that the spatial eigenfunctions $\varphi(x) = \sin(n\pi x)$ form an **orthogonal set**, that is,

$$(4.2\text{-}11) \qquad \int_0^1 \varphi_m(x)\varphi_n(x)\,dx = 0, \quad n \neq m.$$

We establish this fact as follows. If $\mu_n = n^2\pi^2$ and $\mu_m = m^2\pi^2$ for $m \neq n$, then clearly

$$(4.2\text{-}12a) \qquad \varphi_n'' + \mu_n\varphi_n = 0, \quad 0 < x < 1,$$

and

$$(4.2\text{-}12b) \qquad \varphi_m'' + \mu_m\varphi_m = 0, \quad 0 < x < 1.$$

Multiplying Equation (4.2-12a) by φ_m and Equation (4.2-12b) by φ_n and subtracting the resulting equations, we see that

$$(4.2\text{-}13) \qquad (\mu_n - \mu_m)\int_0^1 \varphi_n\varphi_m\,dx = \int_0^1 (\varphi_m''\varphi_n - \varphi_n''\varphi_m)\,dx.$$

However, if we integrate the right side of this equation by parts and use the fact that $\varphi_n(0) = \varphi_m(0) = 0$ and $\varphi_n(1) = \varphi_m(1) = 0$, we find

$$(4.2\text{-}14) \qquad \int_0^1 (\varphi_m''\varphi_n - \varphi_n\varphi_m)\,dx = \int_0^1 (\varphi_m'\varphi_n - \varphi_n'\varphi_m)'\,dx$$

$$= (\varphi_m'\varphi_n - \varphi_n'\varphi_m)\big|_0^1 = 0.$$

This proves

$$(4.2\text{-}15) \qquad (\mu_n - \mu_m)\int_0^1 \varphi_n\varphi_m\,dx = 0,$$

which implies the orthogonality relationship (4.2-11) since $\mu_n \neq \mu_m$.

Now we can compute the coefficients A_n from the initial condition (4.2-10). If we multiply that equation by a typical eigenfunction $\sin(n\pi x)$ and integrate, then the orthogonality relationship (4.2-11) ensures that all terms in the series vanish except for the n-th one. Thus,

$$(4.2\text{-}16) \qquad A_n = \frac{\int_0^1 u_I(x)\sin(n\pi x)\,dx}{\int_0^1 \sin^2(n\pi x)\,dx} = 2\int_0^1 u_I(x)\sin(n\pi x)\,dx,$$

since $\int_0^1 \sin^2(n\pi x)\,dx = \frac{1}{2}$.

For more general boundary conditions, the spatial functions $\sin(n\pi x)$ are not sufficient to represent the solution. However, it is possible to show (Zauderer, 1983, Chapter 4) that, under quite general conditions, Fourier series involving both sines and cosines do suffice for the representation of solutions to the spatial boundary value problem. Such series can be written in compact form if we use the relationship $e^{i\theta} = \cos\theta + \hat{\imath}\sin\theta$. Thus, in general, we can expect the solution to an initial-boundary-value problem for the one-dimensional heat equation on $(0, 1)$ to have the form

$$u(x,t) = \sum_{n=-\infty}^{\infty} C_n e^{-n^2\pi^2 t} e^{\hat{\imath}n\pi x},$$

where C_n is a complex coefficient determined from the initial data. Similar series representations arise when the lower-order terms are present, as, for example, in the advection-diffusion transport equation. In Section 4.3 we shall make use of such a representation in our analysis of numerical schemes.

If the initial values are prescribed on the entire x-axis, then no boundary conditions are required. In this case, we can obtain the solution using a Green's function approach similar to that outlined in Chapter Three. For the one-dimensional heat equation, the appropriate Green's function for the spatial domain $(-\infty, \infty)$ is (Zauderer, 1983, Chapter Five)

$$(4.2\text{-}17) \qquad G(x,t) = \frac{e^{-x^2/4t}}{2\sqrt{\pi t}}.$$

This function has the following properties. First, its limit as $t \to 0$ vanishes away from the spatial origin:

$$(4.2\text{-}18a) \qquad \lim_{t\to 0} G(x,t) = 0, \quad x \neq 0.$$

Second, its integral over the entire x-axis equals unity for all positive times:

$$(4.2\text{-}18b) \qquad \int_{-\infty}^{\infty} G(x,t)\,dx = 1, \quad t > 0.$$

Equations (4.2-18) together may interpreted as follows:

$$\text{(4.2-19)} \qquad\qquad \lim_{t \to 0} G(x,t) = \delta(x),$$

where $\delta(x)$ denotes the Dirac distribution centered at $x = 0$.

The heat equation (4.2-1) has constant coefficients. Thus, as explained in Section 3.7, its solutions are invariant under translation. Using this fact together with Equation (4.2-19), we see that the solution of the initial-value problem on the entire x-axis is given by the integral representation

$$\text{(4.2-20)} \qquad\qquad u(x,t) = \int_{-\infty}^{\infty} \frac{u_I(\xi)e^{-(x-\xi)^2/4t}}{2\sqrt{\pi t}} d\xi,$$

where $u_I(x)$ is the initial function defined for $-\infty < x < \infty$.

4.3. Finite-Difference Methods.

We now turn our attention to numerical methods. Because the advection-diffusion transport equation reviewed in Section 4.1 reduces to the heat equation under conditions of no advection, we focus our attention in this section on the approximation of the transport equation. When advection is small, the approximation of the spatial derivatives in the transport equation requires fairly straightforward extension of the approaches used in treating the steady-state systems described in the Chapter Three. The added complexity, of course, is the treatment of the time derivative. While the approximation of this term is easily accomplished using standard procedures, alternative approaches are often attractive, owing to their computational efficiency. We shall therefore expend some effort in deriving such alternative methods. In addition, as we shall see, when advection becomes the dominant transport mechanism, standard discretizations often produce unexpected and undesirable results. Thus we also devote time to novel methods for the simulation of advection-dominated transport.

We begin our presentation with a brief discussion of the advection-diffusion transport equation developed in Section 4.1, that is,

$$\text{(4.3-1)} \qquad\qquad \frac{\partial u}{\partial t} + \nabla \cdot (\mathbf{v}u) + ru - \nabla \cdot (\mathbf{K} \cdot \nabla u) = 0 \, ,$$

with u representing concentration of some species. We shall work in Cartesian coordinate systems, so that in two space dimensions, for example, we can write $\mathbf{K}$ in the matrix form

$$\mathbf{K} = \begin{bmatrix} K_{xx} & K_{xy} \\ K_{yx} & K_{yy} \end{bmatrix}.$$

As we noted earlier, numerical problems can arise when the advective term $\nabla \cdot (\mathbf{v}u)$ appearing in Equation (4.3-1) has a dominant influence. In one sense, these difficulties arise from the fact that, as $\mathbf{K}$ becomes small, Equation (4.3-1) formally approaches a first-order hyperbolic PDE. As we shall discuss further in Chapter Five, the numerical solution of such equations demands some rather special considerations. In particular, numerical methods applicable to second-order equations can produce serious qualitative errors when applied to first-order equations. These same errors arise in the numerical solution of Equation (4.3-1) when the advection term dominates the diffusion term.

Even in the explicit absence of the advective term in Equation (4.3-1) a large first-order derivative can arise. In particular, a "pseudo-advection" term proportional to ∇u may lurk in the diffusion term of Equation (4.3-1). To see this, notice that when we expand $\nabla \cdot (\mathbf{K} \cdot \nabla u)$ in two dimensions we get

$$(4.3\text{-}2) \quad \nabla \cdot (\mathbf{K} \cdot \nabla u) = \left(K_{xx} \frac{\partial^2 u}{\partial x^2} + K_{xy} \frac{\partial^2 u}{\partial x \partial y} + K_{yx} \frac{\partial^2 u}{\partial y \partial x} + K_{yy} \frac{\partial^2 u}{\partial y^2} \right)$$
$$+ \left(\frac{\partial K_{xx}}{\partial x} \frac{\partial u}{\partial x} + \frac{\partial K_{xy}}{\partial x} \frac{\partial u}{\partial y} + \frac{\partial K_{yx}}{\partial y} \frac{\partial u}{\partial x} + \frac{\partial K_{yy}}{\partial y} \frac{\partial u}{\partial y} \right).$$

The second term on the right has the form $(\nabla \cdot \mathbf{K}) \cdot \nabla u$, which is first-order in u. This term can be troublesome when the coefficient $\mathbf{K}$ varys rapidly in space, since then the coefficient $\nabla \cdot \mathbf{K}$ may be large enough to cause the term proportional to ∇u to dominate the second-order terms.

Difference approximations.

We shall consider finite-difference analogs to Equation (4.3-1) in two steps. First we consider the time derivative. The most common technique used to approximate the time derivative is the two-point approximation written about a point $t = (n+\theta)k$, where $0 \leq \theta \leq 1$ and k is the time step. Thus we have the semidiscrete equation

$$(4.3\text{-}3) \qquad \frac{u^{n+1} - u^n}{k} = [-\nabla \cdot (\mathbf{v}u) - ru + \nabla \cdot (\mathbf{K} \cdot \nabla u)]^{n+\theta},$$

where superscripts denote time levels. In the second step we approximate the spatial derivatives. We initially consider a problem posed in two space dimensions. To simplify notation, let us introduce the following definitions. Given a grid having uniform x- and y-increments h_1 and h_2, respectively, let

$$(4.3\text{-}4\text{a}) \qquad\qquad x_i = ih_1, \quad i = 0, 1, \ldots, N_x ,$$

(4.3-4b) $$y_j = jh_2, \quad j = 0, 1, \ldots, N_y ,$$

where N_x and N_y denote the number of grid intervals in the x-and y-direction, respectively. Also, recall the central difference operators giving

$$h_1 \frac{\partial u}{\partial x}\bigg|_{i,j} \simeq \delta_x u_{i,j} = u_{i+1/2,j} - u_{i-1/2,j} ,$$

$$h_1^2 \frac{\partial^2 u}{\partial x^2}\bigg|_{(i,j)} \simeq \delta_x^2 u_{i,j} = \delta_x(\delta_x u_{i,j}) = u_{i+1,j} - 2u_{i,j} + u_{i-1,j},$$

$$h_1 h_2 \frac{\partial^2 u}{\partial x \partial y}\bigg|_{(i,j)} \simeq \delta_{xy}^2 u_{i,j} = \delta_x(\delta_y u_{i,j})$$

$$= (\delta_y u_{i+1/2,j}) - (\delta_y u_{i-1/2,j})$$
$$= \tfrac{1}{4}(u_{i+1,j+1} - u_{i+1,j-1} - u_{i-1,j+1}$$
$$+ u_{i-1,j-1}) .$$

Combination of the semidiscrete scheme (4.3-3) with the spatial approximations just defined yields

(4.3-5)
$$\frac{u_{i,j}^{n+1} - u_{i,j}^n}{k} = -\frac{\delta_x}{h_1}(v_x u_{i,j}^{n+\theta}) - \frac{\delta_y}{h_2}(v_y u)_{i,j}^{n+\theta} - r u_{i,j}^{n+\theta}$$
$$+ \frac{\delta_x}{h_1^2}(K_{xx} \delta_x u)_{i,j}^{n+\theta} + \frac{\delta_y}{h_2^2}(K_{yy} \delta_y u)_{i,j}^{n+\theta}$$
$$+ \frac{\delta_y}{h_2}\left(K_{yx}\frac{\delta_x u}{h_1}\right)_{i,j}^{n+\theta} + \frac{\delta_x}{h_1}\left(K_{xy}\frac{\delta_y u}{h_2}\right)_{i,j}^{n+\theta} ,$$

where $\mathbf{v} = (v_x, v_y)$. Depending on the value of θ selected, Equation (4.3-5) may be either implicit or explicit. When $\theta = 0$, the right side of Equation (4.3-5) is known, and one can solve the equations one at a time. The price one pays for this simplicity is a stability constraint. That is, one must abide by a constraint on the relationship between the increments h_1, h_2 and k to assure a solution that does not exhibit unbounded growth in machine error as $n \to \infty$. On the other hand, if $\theta \neq 0$, all nine nodal values in the difference stencil around node (i, j) appear as unknowns, and it is necessary to solve a set of simultaneous equations at each time level.

Stability of difference approximations.

Let us now consider the stability of the one-dimensional analog of Equation (4.3-1) given by

(4.3-6) $$\frac{\partial u}{\partial t} = a_0(x,t)\frac{\partial^2 u}{\partial x^2} + a_1(x,t)\frac{\partial u}{\partial x} + a_2(x,t)u + d(x,t).$$

Following the development of Forsythe and Wasow (1960), we write the difference analog for Equation (4.3-6) in compact form as

(4.3-7a) $$u(x_0, (n+1)k) = \sum_{r=-1}^{1} C_r(x,t)u(x_0 + rh, nk) + kd(x,t),$$

where h and k are the time and space increments, respectively. If we assume a forward difference representation for the time derivative and a centered representation of the space derivative and define $\bar{\rho} = k/h^2$, then the coefficients in Equation (4.3-7a) are given by

(4.3 − 7b) $$C_{-1}(x,t) = \bar{\rho}a_0(x,t) - \frac{\bar{\rho}h}{2}a_1(x,t),$$

(4.3 − 7c) $$C_0(x,t) = 1 - 2\bar{\rho}a_0(x,t) + \bar{\rho}h^2 a_2(x,t),$$

(4.3 − 7d) $$C_1(x,t) = \bar{\rho}a_0(x,t) + \bar{\rho}\frac{h}{2}a_1(x,t).$$

Forsythe and Wasow (1960) define those approximations for which all the coefficients C_r are nonnegative whenever h is sufficiently small as approximations of **positive type**. They further show that all difference approximations of positive type are stable. Thus stability is assured in our example if

(4.3-8a) $$\bar{\rho}a_0(x,t) - \bar{\rho}\frac{h}{2}a_1(x,t) > 0,$$

(4.3-8b) $$1 - 2\bar{\rho}a_0(x,t) + \bar{\rho}h^2 a_2(x,t) > 0,$$

(4.3-8c) $$\bar{\rho}a_0(x,t) + \bar{\rho}\frac{h}{2}a_1(x,t) > 0.$$

Let us now focus on the advection-diffusion transport equation. If we use a centered-in-space difference scheme to discretize this equation, then the coefficients a_0, a_1, and a_2 are given by $a_0 = K_{xx} \equiv K$, $a_1 = -v_x \equiv -v$, and $a_2 = 0$. The stability constraints become

(4.3-9a) $$\bar{\rho}K + \bar{\rho}\frac{h}{2}v > 0,$$

(4.3-9b) $$1 - 2\bar{\rho}K > 0,$$

(4.3-9c) $$\bar{\rho}K - \bar{\rho}\frac{h}{2}v > 0,$$

and therefore we require $\bar{\rho} < 1/2K$ and $h < 2K/v$. The first constraint, $\bar{\rho} < 1/2K$, is that normally encountered in standard stability analyses of the advection-free equation, such as those based upon the von Neumann approach discussed in Section 2.9. The second constraint, $h < 2K/v$, is the generally recognized requirement for a solution free of spurious oscillations. Such oscillations are one form of the undesirable behavior associated with advection-dominated flows as mentioned above. We discuss this problem in the next subsection.

Numerous other analyses have been made for the stabilty of the transport equation. A summary of the constraints associated with many of the more generally encountered difference formulas for the transport equation can be found in Paolucci and Chenoweth (1982).

Solution pathology at sharp fronts: Fourier analysis.

The numerical analysis literature is replete with examples of the pathologic behavior of finite-difference schemes applied to transport equations. The pathology arises with versions of the transport equation that exhibit solutions with sharp concentration fronts. Consider the one-dimensional example of Figure 4-4. Here the dimensionless velocity is $\overline{V} \equiv vk/h = 0.369$, and the dimensionless diffusion coefficient is $\overline{K} \equiv Kk/h^2 = 0.0069$. The ratio of these two dimensionless groups, which is the **grid Peclet number** denoted by Pe, is $\overline{V}/\overline{K} = vh/K = 53.48$.

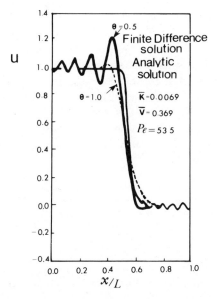

Figure 4-4. Propagation of a concentration front by the finite-difference method (after Gray and Pinder, 1976).

While there are many possibilities for the numerical approximation of the transport equation, consider for a moment the solution generated by the centered-in-space scheme defined above. Figure 4-4 shows that, for Pe=53.48, the numerical solution is oscillatory and does not exhibit the monotonic behavior characteristic of the exact solution. As we shall see, other difference approximations eliminate these spurious numerical oscillations but produce artificially smeared fronts. To understand these types of pathologic behavior we can examine the numerical and exact solutions from the point of view of Fourier series. Drawing on the work of Gray and Pinder (1976), we consider the exact and numerical propagation of the Fourier components arising from a given initial condition. A comparison of the results serves to explain, at least in part, how numerical procedures lead to undesirable oscillations or numerical smearing in the neighborhood of sharp fronts. Before examining particular numerical schemes in this way, let us establish a general framework for the analysis.

In the following we examine the one-dimensional, constant-coefficient form of the advection-diffusion transport equation, that is,

$$(4.3\text{-}10) \qquad \frac{\partial u}{\partial t} + v\frac{\partial u}{\partial x} - K\frac{\partial^2 u}{\partial x^2} = 0.$$

As reviewed in Section 4.2, the general solution of Equation (4.3-10) will have a Fourier-series representation,

$$(4.3\text{-}11) \qquad u(x,t) = \sum_{m=-\infty}^{\infty} C_m \exp(\hat{\imath}\beta_m t + \hat{\imath}\sigma_m x),$$

where $-1 \leq x - vt \leq 1$, β_m is the temporal frequency of the m-th Fourier component, $\sigma_m = 2\pi/L_m$ is its spatial frequency or wave number, and L_m is the wavelength of the m-th component. C_m is a complex Fourier coefficient dependent on initial conditions. Figure 4-5 displays two conventional means of depicting the propagation of a wave.

Because Equation (4.3-10) is linear, it is possible to characterize the exact solution behavior by examining the propagation of one typical term in the series (4.3-11), say

$$(4.3\text{-}12) \qquad u_m(x,t) = C_m \exp(\hat{\imath}\beta_m t + \hat{\imath}\sigma_m x).$$

Substitution of this Fourier component into the PDE (4.3-10) yields, after some algebra,

$$(4.3\text{-}13) \qquad \beta_m = \sigma_m(\hat{\imath}K\sigma_m - v).$$

Therefore, a typical Fourier term in the exact solution to the equation has the following form:

$$(4.3\text{-}14) \qquad u_m(x,t) = C_m \exp[\hat{\imath}\sigma_m(x - vt)]\exp(-K\sigma_m^2 t).$$

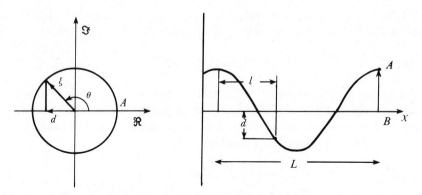

	Phase	Amplitude	
Wave Plane	l/L	d	$\tan \theta = \dfrac{\Im(\xi)}{\Re(\xi)}$
Complex Plane	$\theta/2\pi$	$\Re(\xi)$	

Figure 4-5. Diagrammatic representation of wave propagation in the wave
plane and the complex plane.

The factor $\exp[\hat{\imath}\sigma_m(x - vt)]$ in this term describes the translation of the
component via advection with velocity v; the factor $\exp(-K\sigma_m^2 t)$ describes
the change in the wave's amplitude as a result of diffusion.

Now consider the behavior of the Fourier components of $u(x, t)$ when
they are propagated by a numerical algorithm. Employing the strategy
used above for the exact solution, we write the value of the approximate
solution at a grid point $x = ih$ as a superposition of waves,

$$(4.3\text{-}15) \qquad u(ih) \simeq u_i = \sum_{m=-p}^{p} C_m \exp[\hat{\imath}\beta'_m t + \hat{\imath}\sigma_m(ih)].$$

Here β'_m is the temporal frequency of the m-th component of the numerical
wave, and p is the wave number associated with the smallest wavelength
that can be propagated by the numerical scheme, usually $2h$. Once again
taking advantage of the linearity of the PDE, we examine one term of the
series (4.3-14), say

$$(4.3\text{-}16) \qquad u_{i_m} = C_m \exp[\hat{\imath}\beta'_m t + \hat{\imath}\sigma_m(ih)].$$

When we substitute this component into the finite-difference scheme, we
find a complex relationship between β'_m and σ_m that is different from Equa-

tion (4.3-13). The particular relationship will depend on the numerical scheme used, but generically we can write this relationship as follows

$$(4.3\text{-}17) \qquad \beta'_m = \beta'_m(h, k, v, K, \sigma_m).$$

Assuming we can represent the initial conditions exactly in our numerical scheme, the Fourier coefficients C_m are the same in both the exact and numerical solutions. After an elapsed time period of length k, the m-th component of the exact solution has the form

$$(4.3-18) \qquad \begin{aligned} u_m^{t+k} &= C_m \exp[\hat{\imath}\beta_m(t+k) + \hat{\imath}\sigma_m x] \\ &= C_m \exp(\hat{\imath}\beta_m t + \hat{\imath}\sigma_m x)\exp(\hat{\imath}\beta_m k) \\ &= u_m^t \exp(\hat{\imath}\beta_m k) \equiv u_m^t \xi_m, \end{aligned}$$

while the m-th component of the numerical solution is as follows:

$$(4.3-19) \qquad \begin{aligned} u_{i_m}^{t+k} &= C_m \exp[\hat{\imath}\beta'_m(t+k) + \hat{\imath}\sigma_m(ih)] \\ &= C_m \exp(\hat{\imath}\beta'_m t + \hat{\imath}\sigma_m ih)\exp(\hat{\imath}\beta'_m k) \\ &= u_{i_m}^t \exp(\hat{\imath}\beta'_m k) \equiv u_{i_m}^t \xi'_m. \end{aligned}$$

We call the parameters ξ_m and ξ'_m the exact and numerical **amplification factors**, respectively.

One can establish a relationship between the magnitude of the amplification factor and the amplitude of the m-th Fourier component. Expanding the definition of ξ_m using Equation (4.3-13), we obtain

$$(4.3-20) \qquad \begin{aligned} \xi_m &= \exp(\hat{\imath}\beta_m k) = \exp\{\hat{\imath}[\sigma_m(\hat{\imath}K\sigma_m - v)]k\} \\ &= \exp[(-K\sigma_m^2 - \hat{\imath}\sigma_m v)k]. \end{aligned}$$

The magnitude of ξ_m is, therefore,

$$(4.3\text{-}21) \qquad |\xi_m| = |\exp(\hat{\imath}\beta_m k)| = \exp(-\sigma_m^2 K k) .$$

Recalling that the right side of Equation (4.3-21) is the amplitude modification of the m-th component over the time interval $[t, t+k]$, we can interpret $|\xi_m|$ as the ratio of the amplitude of the m-th Fourier component after an elapsed time k to its amplitude at the beginning of the time step. Since the number of time steps required to propagate the wave through one wavelength is $N_m = L_m/vk$, the dissipation in the exact solution that occurs over one wavelength can be written

$$(4.3\text{-}22) \qquad |\xi_m|^{N_m} = [\exp(-\sigma_m^2 K k)]^{N_m} .$$

The corresponding dissipation in the numerical solution over one wavelength is $|\xi'_m|^{N_m}$. Because truncation errors will be amplified by the factor $|\xi'_m|^{N_m}$, a necessary condition for the stability of the numerical scheme is $|\xi'_m| \leq 1$ for all wave numbers m.

We can now introduce a measure R_m of the error in the amplification of component m by defining

$$(4.3-23) \qquad R_m = \frac{|\xi'_m|^{N_m}}{|\xi_m|^{N_m}} = \frac{|\xi'_m|^{N_m}}{[\exp(-\sigma_m^2 K k)]^{N_m}}.$$

When $R_m < 1$, the amplitude of the m-th component in the numerical solution after propagation over one wavelength is less than that in the exact solution. When $R_m > 1$, the opposite behavior is implied.

There remains the question of the phase error associated with a numerical scheme. To establish the algebraic relationship for this error, consider once again the representation of the m-th Fourier component in the complex plane, as presented in Figure 4-5. The phase angle θ_m of this component in the Fourier expansion after one time step is given by the trigonometric relationship

$$\tan \theta_m = \frac{\Im(\xi_m)}{\Re(\xi_m)} \, ,$$

where $\Im(\xi_m)$ and $\Re(\xi_m)$ denote the imaginary and real parts of the complex variable ξ_m. (Observe that this subscripted "theta" differs from the variable θ used above to indicate time levels.) Thus the phase angle θ_m is as follows:

$$(4.3-24) \qquad \theta_m = \tan^{-1}\left[\frac{\Im(\xi_m)}{\Re(\xi_m)}\right].$$

After N_m time steps, the value of the phase angle is given by $N_m \theta_m$. For the exact solution, $N_m \theta_m = 2\pi$ radians, since the component will have traveled exactly one wavelength. On the other hand, the phase angle for the numerical solution after N_m time steps will be $N_m \theta'_m$, where

$$(4.3-25) \qquad \theta'_m = \tan^{-1}\left[\frac{\Im(\xi'_m)}{\Re(\xi'_m)}\right].$$

The **phase-lag error**

$$(4.3-26) \qquad \Theta_m = N_m \theta'_m - 2\pi$$

provides a useful indicator of the failure of the numerical scheme to propagate Fourier components of the solution with the correct phase. If $\Theta_m = 0$, then the scheme propagates the m-th Fourier component with no error in phase. An alternative indicator of the phase error is the **phase ratio**,

$$(4.3-27) \qquad \hat{\Theta}_m = \frac{N_m \theta'_m}{2\pi}.$$

In this case, correct propagation of the m-th Fourier component corresponds to the case $\hat{\Theta}_m = 1$.

While the above phase-error expressions measure the cummulative error over N_m time steps, sometimes it is convenient to consider the error per time step. In this case the exact phase is given by

$$(4.3\text{-}28) \qquad \qquad \theta_m = \frac{2\pi}{N_m},$$

and the numerical phase is given by Equation (4.3-25). The **phase error per time step** is then

$$(4.3\text{-}29) \qquad \qquad \bar{\Theta}_m = \theta'_m - \frac{2\pi}{N_m}.$$

Solution pathology at sharp fronts: example.

Now, at last, we are in a position to analyze the solution pathology at sharp fronts. The finite-difference equation we shall use as an example is the one-dimensional, constant-coefficient form of Equation (4.3-5), that is

$$(4.3\text{-}30) \qquad \frac{u_i^{n+1} - u_i^n}{k} + v\frac{\delta_x}{h}u_i^{n+\theta} - K\frac{\delta_x}{h^2}(\delta_x u_i^{n+\theta}) = 0 \; ,$$

where, as introduced above, θ used as a superscript denotes the position of the spatial discretization in the time interval $[nk, (n+1)k]$. One can show (Gray and Pinder, 1976) that the amplification factor for the difference approximation (4.3-30) is

$$(4.3\text{-}31) \qquad \xi'_m = \frac{1 - (1-\theta)\{\overline{V}\hat{\imath}\sin(\sigma_m h) + 2\overline{K}[1 - \cos(\sigma_m h)]\}}{1 + \theta\{\overline{V}\hat{\imath}\sin(\sigma_m h) + 2\overline{K}[1 - \cos(\sigma_m h)]\}} \; ,$$

where, as earlier, $\overline{V} \equiv vk/h$ and $\overline{K} \equiv Kk/h^2$. We shall use this expression to examine the amplitude and phase errors associated with the difference scheme.

Consider first the use of Equation (4.3-31) to establish algorithmic stability. A necessary condition for stability is that $|\xi'_m|$ never exceed unity. The values of $|\xi'_m|$ corresponding to selected values of the wave number m and time weighting parameter θ are given in Figure 4-6. Note that for the values $\theta = 0.5$ and $\theta = 1.0$ corresponding to implicit schemes, $|\xi_m|$ remains less than unity for all wavelengths examined. Thus, over this range of values for θ, the algorithm is stable. This is consistent with stability analyses performed using other methodologies and with practical experience.

Next we examine the phase-lag error, Θ_m, for a range of values of m. This information is provided for temporal weights $\theta = 0.5$ and $\theta = 1.0$ in Figure 4-7. With $\overline{V} = 0.369$ and $\overline{K} = 0.069$, the grid Peclet number for this example is 5.35. The plotted curves illustrate that components with wavelengths of $2h$ are stationary, unable to propagate on a grid of mesh h. The phase error decreases as the wavelength increases until, for wavelengths of approximately $20h$, the error is quite small. Components with wavelengths between $2h$ and $20h$ propagate with phase errors ranging monotonically from $-360°$ to less than $10°$. Thus the small-wavelength components, that is, those with the highest spatial frequencies, are propagated incorrectly. These are precisely the waves needed to resolve the sharp front, and when they travel with incorrect phase they show up as small-wavelength oscillations in the Fourier series of the numerical solution. Consequently, when the various components are summed to form a complete numerical solution, the result will exhibit spurious oscillations related to the phase errors.

The story does not end at this point, however. We must also consider the amplitude errors to ascertain whether the oscillatory effects of phase errors will be noticeable. Figure 4-8 presents the amplitude errors for the same set of parameters as considered in the preceding paragraphs. For $\theta = 0.5$, all waves exhibiting significant phase errors are underdamped. As in the case of phase errors, these amplitude errors are primarily associated with the smaller wavelengths. Indeed, the scheme underdamps waves as long as $100h$. It is not suprising, then, to observe oscillatory solution behavior as illustrated in Figure 4-4.

When $\theta = 1.0$, the amplitude error behaves quite differently. As can be seen in Figure 4-8, the shortest waves are once again underdamped. Wavelengths larger than about $4h$ are severely overdamped relative to their correct amplitude. This is both good news and bad news. The good news is that many of the waves that were getting out of phase in Figure 4-7

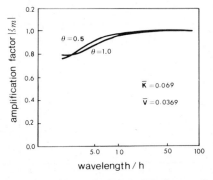

Figure 4-6. Magnitude of the amplification factor ξ_m in a finite-difference method for the advection-diffusion equation (after Gray and Pinder 1976).

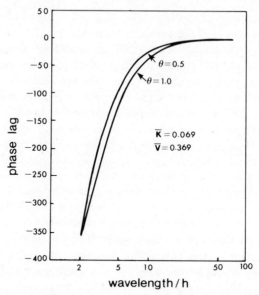

Figure 4-7. Phase lag Θ_m for the finite-difference equation (after Gray and Pinder, 1976).

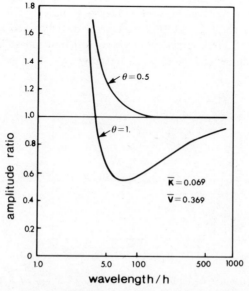

Figure 4-8. Amplitude ratio for the finite-difference equation (after Gray and Pinder, 1976).

are now numerically damped and thus do not cause spurious oscillations. The bad news is that these waves are needed to resolve the sharp front, and they are now virtually absent. In addition, many of the remaining longer waves are also artificially damped. The result is a smooth solution that exhibits unrealistic smearing near the sharp front, as shown in Figure 4-4. This effect is sometimes called **numerical dispersion, numerical dissipation, or numerical diffusion**. It is worth remarking that numerical dispersion is not a very good term because phase error is also called numerical dispersion by researchers in computational fluid dynamics.

In summary, numerical methods for the advection-diffusion transport equation typically exhibit phase errors and amplitude errors. Phase errors appear in the form of oscillatory numerical approximations to monotonic exact solutions. Amplitude errors tend to ameliorate the oscillation problem, but they do so at the expense of numerical smearing at fronts that should, in fact, be sharp. The two types of error described here are of greatest concern when advection dominates diffusion, since then the exact solution depends particularly strongly upon high-frequency spatial components to resolve the solution. When the numerical problem has a low Peclet number, fronts tend to be less steep, and the amplitude and phase errors should be of less concern.

Nonstandard methods.

Several algorithms have been formulated to circumvent inaccuracies introduced by the numerical dispersion and dissipation. Possibly the most common technique for overcoming the oscillatory behavior associated with numerical phase error is **upstream weighting**. There are several strategies that are used to achieve upstream weighting. The simplest, in two space dimensions, is to approximate advective terms, such as $\partial(v_x u)/\partial x$, using a first-order difference approximation involving the spatial nodes (i, j) and $(i-1, j)$ when the difference equation is centered at node (i, j) and the x-component v_x of the velocity is positive. This is a special case of a more general formulation that relies on a weighted average of the approximation using spatial nodes (i, j) and $(i-1, j)$ and the approximation using (i, j) and $(i+1, j)$. The latter formulation leads to the following generalized counterpart to Equation (4.3-5):

$$
(4.3\text{-}32) \quad \frac{u_{i,j}^{n+1} - u_{i,j}^n}{k} = -\sigma \frac{\delta_x}{h_1}(v_x u)_{i-1/2,j}^{n+\theta} - (1-\sigma)\frac{\delta_x}{h_1}(v_x u)_{i+1/2,j}^{n+\theta}
$$

$$
- \gamma \frac{\delta_y}{h_2}(v_y u)_{i,j-1/2}^{n+\theta} - (1-\gamma)\frac{\delta_y}{h_2}(v_y u)_{i,j+1/2}^{n+\theta}
$$

$$
+ \frac{\delta_x}{h_1^2}(K_{xx}\,\delta_x u)_{i,j}^{n+\theta} + \frac{\delta_y}{h_2^2}(K_{yy}\,\delta_y u)_{i,j}^{n+\theta}
$$

$$
+ \frac{\delta_y}{h_2}\left(K_{yx}\frac{\delta_x u}{h_1}\right)_{i,j}^{n+\theta} + \frac{\delta_x}{h_1}\left(K_{xy}\frac{\delta_y u}{h_2}\right)_{i,j}^{n+\theta} .
$$

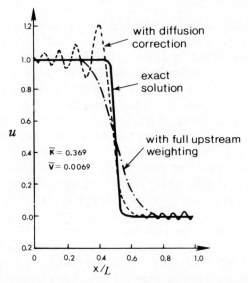

Figure 4-9. Numerical and analytic solution to the advective transport
equation illustrating the effects of diffusion correction and
upstream weighting (after Pinder and Gray, 1977).

When $\sigma = 1$ and $\gamma = 1$, the equation is fully upstream weighted. The
centered approximation (4.3-5) is obtained when $\sigma = \frac{1}{2}$ and $\gamma = \frac{1}{2}$. The
influence of upstream weighting is illustrated in Figure 4-9 for the extreme
case of $\sigma = 1$.

Recognizing that the truncation error in the Taylor expansion used to
generate the upstream-weighted finite-difference approximation to the advective term would generate a smeared solution at a sharp front, Chaudhari
(1971) introduced a correction term. He formulated a numerical diffusion
term in one space dimension as

$$K^*(x,t) \equiv \frac{vh}{2}\left(1 - \frac{vk}{h}\right) ,$$

which he subtracted from the physical diffusion to obtain a solution corrected for numerical diffusion. The resulting one-dimensional form of Equation (4.3-32) is

$$\frac{u_i^{n+1} - u_i^n}{k} = -\frac{\delta_x}{h}(vu)_i^{n+\theta} + \frac{\delta_x}{h^2}[(K - K^*)\delta_x u]_i^{n+\theta} .$$

The effect of this diffusion correction is illustrated in Figure 4-10. While
the standard finite-difference equation solution is not plotted in this figure,

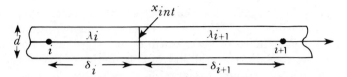

Figure 4-10. Finite-difference grid with variable properties λ_i and λ_{i+1} in series.

it is virtually identical to the diffusion-corrected scheme in this instance. More sophisticated corrections for numerical diffusion can be found in the early work of Boris and Book (1973) and, more recently in Morrow and Cram (1985).

Predictor-corrector methods.

In this subsection we investigate multistep procedures for solving parabolic problems. We focus on the advection-free equation. Consider, as a point of departure, the classic explicit approximation to the heat equation,

$$(4.3\text{-}33a) \qquad\qquad u_i^{n+1} - u_i^n = \bar{\rho}\delta_x^2 u_i^n \;,$$

where $\bar{\rho} \equiv k/h^2$, and the fully implicit or backwards difference approximation

$$(4.3\text{-}33b) \qquad\qquad u_i^{n+1} - u_i^n = \bar{\rho}\delta_x^2 u_i^{n+1}.$$

The two-step procedure we wish to consider uses Equation (4.3-33a) to advance to the time level $t = (n + \frac{1}{2})k$ and Equation (3.3-33b) to advance from $t = (n + \frac{1}{2})k$ to $t = (n + 1)k$. Thus the second formula will employ imformation generated by the first. The algorithm then reads

$$\text{step} \quad 1: \quad u_i^{n+1/2} - u_i^n = \frac{\bar{\rho}}{2}\delta_x^2 u_i^n,$$

$$\text{step} \quad 2: \quad u_i^{n+1} - u_i^{n+1/2} = \frac{\bar{\rho}}{2}\delta_x^2 u_i^{n+1}.$$

In step 1 we solve an explicit equation; in step 2 we solve an implicit equation.

One may ask, why not just solve step 2, since it constitutes an unconditionally stable algorithm and step 1 is only condtionally stable? To answer this question, let us add steps 1 and 2 together. We obtain

$$(4.3\text{-}34) \qquad\qquad u_i^{n+1} - u_i^n = \frac{\bar{\rho}}{2}\left(\delta_x^2 u_i^n + \delta_x^2 u_i^{n+1}\right) \;.$$

Equation (3.3-34) is none other than the Crank-Nicolson approximation, which has truncation error $\mathcal{O}(h^2 + k^2)$. Thus the two-step procedure yields an algorithm that is more accurate in time than step 1 or step 2 alone.

Let us now consider the stability of this method. Recall that, in the preceding section, we require the amplification factor ξ_m' associated with every Fourier component of the numerical scheme to have a magnitude less than unity to assure stability. The amplification factors for steps 1 and 2 above are found through Fourier analysis to be the following:

$$\text{step} \quad 1: \quad \xi_{mE}' = 1 - 4\bar{\rho}\sin^2\left(\frac{\sigma_m h}{2}\right),$$

$$\text{step} \quad 2: \quad \xi_{mI}' = \frac{1}{1 + 4\bar{\rho}\sin^2(\sigma_m h/2)}.$$

The amplification factor for the m-th mode in the two-step procedure is the product of these, that is, $\xi_m' = \xi_{mE}'\xi_{mI}'$. While step 1 is only conditionally stable ($|\xi_{mE}'| \leq 1$ can be assured only if $\bar{\rho} \leq \frac{1}{2}$), the product ξ_m' is unconditionally stable, provided the time step is held constant over the two-step process.

Stability of two-dimensional approximations.

A general stability analysis of the transport equation in two or more space dimensions is difficult. However, a limited analysis that leads to sufficient stability conditions has been done by Noye (1982). He begins with the special case of Equation (4.3-5),

$$(4.3\text{-}35) \qquad u_{i,j}^{n+1} - u_{i,j}^n = -\overline{V}_x \delta_x u_{i,j}^n - \overline{V}_y \delta_y u_{i,j}^n$$
$$+ \overline{K}_{xx}\,\delta_x^2 u_{i,j}^n + \overline{K}_{yy}\,\delta_y^2 u_{i,j}^n,$$

where, in a manner analogous to that employed in the one-dimensional analysis of the preceeding section, we define

$$\overline{V}_x \equiv \frac{v_y k}{h_1}, \quad \overline{K}_{xx} \equiv \frac{K_{xx} k}{h_1^2},$$

$$\overline{V}_y \equiv \frac{v_x k}{h_2}, \quad \overline{K}_{yy} \equiv \frac{K_{yy} k}{h_2^2}.$$

The amplification factor for the scheme (4.3-35) is given (Noye, 1982) as follows:

$$(4.3\text{-}36) \quad |\xi_m'|^2 = 1 - 4[(\overline{K}_{xx} + \overline{K}_{yy})$$
$$- (\overline{K}_{xx}\cos\sigma_{xm} + \overline{K}_{yy}\cos\sigma_{ym})]$$
$$+ 4[(\overline{K}_{xx} + \overline{K}_{yy}) - (\overline{K}_{xx}\cos\sigma_{xm} + \overline{K}_{yy}\cos\sigma_{ym})]^2$$
$$+ 4(\overline{V}_x\sin\sigma_{xm})^2 + 2\overline{V}_x\overline{V}_y\sin\sigma_{xm}\sin\sigma_{ym}$$
$$+ (\overline{V}_y\sin\sigma_{ym})^2.$$

Here, σ_{xm} and σ_{ym} denote the spatial frequencies of the m-th Fourier component of the numerical solution in the x- and y-directions, respectively.

Consider first the case when we have no advection, that is $\overline{V}_x = \overline{V}_y = 0$. Equation (4.3-36) becomes

$$(4.3\text{-}37) \qquad \xi_m = 1 - 4\overline{K}_{xx}\sin^2\left(\frac{\sigma_{xm}}{2}\right) - 4\overline{K}_{yy}\sin^2\left(\frac{\sigma_{ym}}{2}\right).$$

Because stability requires $|\xi_m| \leq 1$, Equation (4.3-37) leads to the requirement

$$(4.3\text{-}38) \qquad -1 \leq 1 - 4\overline{K}_{xx}\sin^2\left(\frac{\sigma_{xm}}{2}\right) - 4\overline{K}_{yy}\sin^2\left(\frac{\sigma_{ym}}{2}\right) \leq 1.$$

Taking the worst case for values of $\sin^2(\sigma_{xm}/2)$ and $\sin^2(\sigma_{ym}/2)$, we see that this pair of inequalities holds if

$$(4.3\text{-}39) \qquad\qquad\qquad \overline{K}_{xx} + \overline{K}_{yy} = \tfrac{1}{2}.$$

In the event that $\overline{K}_{xx} = \overline{K}_{yy} = \overline{K}$, the requirement reduces to $\overline{K} \leq \tfrac{1}{4}$. This criterion is twice as restrictive as the stability constraint encountered in the corresponding one-dimensional case.

Another special case considered by Noye (1982) assumes $\overline{V}_x \leq 2\overline{K}_{xx}$ and $\overline{V}_y \leq 2\overline{K}_{yy}$. Given this constraint, one obtains from Equation (4.3-36) the inequality

$$(4.3\text{-}40) \quad |\xi_m|^2 \leq 1 - 4[(\overline{K}_{xx} + \overline{K}_{yy})$$
$$- (\overline{K}_{xx}\cos\sigma_{xm} + \overline{K}_{yy}\cos\sigma_{ym})] + 4[(\overline{K}_{xx} + \overline{K}_{yy})^2$$
$$- 2(\overline{K}_{xx} + \overline{K}_{yy})(\overline{K}_{xx}\cos\sigma_{xm} + \overline{K}_{yy}\cos\sigma_{ym})$$
$$+ (\overline{K}_{xx})^2 + 2\overline{K}_{yx}\overline{K}_{yx}\cos(\sigma_{xm} - \sigma_{ym}) + (\overline{K}_{yy})^2].$$

Considering the range of values taken by the cosine function, one can deduce from Equation (4.3-40) that

$$(4.3\text{-}41) \quad (\overline{K}_{xx})^2 + 2\overline{K}_{xx}\overline{K}_{yy}\cos(\sigma_{xm} - \sigma_{ym}) + (\overline{K}_{yy})^2 \leq (\overline{K}_{xx} + \overline{K}_{yy})^2\ ,$$

which, when combined with Equation (4.3-40), yields

$$(4.3\text{-}42) \qquad |\xi_m|^2 \leq 1 - 4[\overline{K}_{xx}(1 - \cos\sigma_{xm})$$
$$+ \overline{K}_{yy}(1 - \cos\sigma_{ym})][1 - 2(\sigma_{xm} + \sigma_{ym})]\ .$$

Again considering that $|\cos\sigma_{xm}| \leq 1$ and $|\cos\sigma_{ym}| \leq 1$, we find that $|\xi|^2 \leq 1$ provided $\overline{K}_{xx} + \overline{K}_{yy} \leq \tfrac{1}{2}$. This is the same criterion as the one obtained above for the advection-free equation. If we further assume that $\overline{K}_{xx} = \overline{K}_{yy} = \overline{K}$ and $\overline{V}_x = \overline{V}_y = \overline{V}$, we have

$$(4.3\text{-}43) \qquad\qquad\qquad \overline{V} \leq 2\overline{K} \leq \tfrac{1}{2}.$$

One can obtain analogous results through an analysis of the upstream weighted equations. Furthermore, a similar development is possible with the implicit formulation. Such a development reveals that, as expected, the fully implicit formulation is unconditionally stable.

Convergence.

The convergence of a finite-difference algorithm is most easily established indirectly using Lax's equivalence theorem as outlined in Chapter Two. This theorem (see Lax and Richtmeyer, 1956) states the following:

Lax's Equivalence Theorem. *Given a well posed, linear initial-value problem and a consistent finite-difference approximation to it, stability is the necessary and sufficient condition for convergence.*

The practical importance of this theorem stems from the ease with which consistency and stability can be demonstrated and the relative difficulty that is associated with directly demonstrating convergence. Because the preceding difference formula can be shown to be consistent through straightforward examination of their truncation errors, it follows that, for the region over which they are stable, they are also convergent.

Alternating-direction implicit methods.

The alternating direction implicit (ADI) procedure was introduced by Peaceman and Rachford in their classic paper published in 1955. The fundamental idea, as introduced in Chapter Three, is to replace a two-dimensional problem with a series of one-dimensional problems in such a way as to generate a computationally efficient algorithm. In this subsection we shall explore the application of ADI methods to time-dependent problems.

We begin with a somewhat restricted version of the advection-diffusion equation written (see Celia and Pinder, 1985) in the form

$$(4.3\text{-}44) \qquad \frac{\partial u}{\partial t} - \mathcal{L}_x u - \mathcal{L}_y u = f(x, y),$$

where

$$\mathcal{L}_x u \equiv \frac{\partial}{\partial x}\left(K_{xx}\frac{\partial u}{\partial x}\right) - v_x\frac{\partial u}{\partial x},$$

and

$$\mathcal{L}_y u \equiv \frac{\partial}{\partial y}\left(K_{yy}\frac{\partial u}{\partial y}\right) - v_y\frac{\partial u}{\partial y}.$$

The first step is to discretize the time derivative, giving

$$(4.3\text{-}45) \qquad \frac{u^{n+1} - u^n}{k} = \mathcal{L}_x u^{n+\theta} + \mathcal{L}_y u^{n+\theta} + f^{n+\theta},$$

where the space derivatives are now being evaluated at time level $t = (n + \theta)k$. Now we rewrite Equation (4.3-45) as

(4.3-46) $[I - \theta k(\mathcal{L}_x + \mathcal{L}_y)](u^{n+1} - u^n) = k[f^{(n+\theta)} + (\mathcal{L}_x + \mathcal{L}_y)u^n]$,

where I denotes the identity operator. The left side of Equation (4.3-46) can be modified by completing the square to yield

(4.3-37) $[I - \theta k \mathcal{L}_x][I - \theta k \mathcal{L}_y](u^{n+1} - u^n)$
$= k[f^{n+\theta} + (\mathcal{L}_x + \mathcal{L}_y)u^n] + \theta^2 k^2 \mathcal{L}_x \mathcal{L}_y(u^{x+1} - u^n)$.

The difference between Equations (4.3-46) and (4.3-47) is the last term, which also appears implicitly on the left side of the equation. If we assume terms of magnitude $\mathcal{O}(k^2)$ to be negligible, then this equation can be simplified to yield

(4.3-48) $[I - \theta k \mathcal{L}_x][I - \theta k \mathcal{L}_y](u^{n+1} - u^n)$
$= k[f^{n+\theta} + (\mathcal{L}_x + \mathcal{L}_y)u^n] \equiv F(x, y, t)$.

We now define the two-step solution procedure as follows:

(4.3-49a) $[I - \theta k \mathcal{L}_x]z = F$,
(4.3-49b) $[I - \theta k \mathcal{L}_y](u^{n+1} - u^n) = z.$

The operators on the left side of Equation (4.3-49a) involve derivatives with respect to x only, and the left side of Equation (4.3-49b) involves derivatives with respect to y only. Various numerical schemes can now be employed to approximate the spatial operators $\mathcal{L}_x$ and $\mathcal{L}_y$. Notice that the unknown variable in Equation (4.3-49b) is the difference $u^{n+1} - u^n$. Viewed from the standpoint of coding, solving for time increments in this way helps to avoid unnecessary roundoff error.

We now employ a finite-difference representation of $\mathcal{L}_x u$ and $\mathcal{L}_y u$, namely,

(4.3-50a) $\mathcal{L}_x u^{n+\theta} \simeq \frac{1}{h_1^2}\delta_x(K_{xx}\delta_x)u_{i,j}^{n+\theta} - v_x \frac{\delta_x}{h_1}u_{i,j}^{n+\theta}$,

$i = 1, 2, \ldots I - 1, \quad j = 1, 2, \ldots J - 1;$

(4.3-50b) $\mathcal{L}_y u^{n+\theta} \simeq \frac{1}{h_2^2}\delta_y(K_{yy}\delta_y)u_{i,j}^{n+\theta} - v_y \frac{\delta_y}{h_2}u_{i,j}^{n+\theta}$,

$j = 1, 2, \ldots J - 1, \quad i = 1, 2, \ldots I - 1,$

where the x- and y-coordinates of the grid points are $x_i = ih_1$ and $y_j = jh_2$. When Equations (4.3-50) are assembled into a matrix equation one obtains for Equation (4.3-49a), for the i-th row of the grid, the tridiagonal system

$$
\begin{bmatrix}
A_{11} & A_{12} & & & & & \\
A_{21} & A_{22} & A_{23} & & & & \\
 & A_{32} & A_{33} & A_{34} & & & \\
 & & & \ddots & & & \\
 & & & A_{J-2,J-3} & A_{J-2,J-2} & A_{J-2,J-1} & \\
 & & & & A_{J-1,J-2} & A_{J-1,J-1}
\end{bmatrix}
\begin{bmatrix}
z_{i,1} \\
z_{i,2} \\
z_{i,3} \\
\vdots \\
z_{i,J-2} \\
z_{i,J-1}
\end{bmatrix}
$$

(4.3-51)
$$
=
\begin{bmatrix}
F_{i,1} \\
F_{i,2} \\
F_{i,3} \\
\vdots \\
F_{i,J-2} \\
F_{i,J-1}
\end{bmatrix}
+
\begin{bmatrix}
G_{i,1} \\
G_{i,2} \\
G_{i,3} \\
\vdots \\
G_{i,J-2} \\
G_{i,J-1}
\end{bmatrix} .
$$

We can abbreviate this equation as $\mathbf{Az} = \mathbf{F} + \mathbf{G}$, where the vector $\mathbf{G}$ contains the boundary information and the nonzero entries of the matrix $\mathbf{A}$ are as follows:

$$
A_{\ell,\ell-1} = 1 - \theta k \left(\frac{K_{xx}}{h_1^2} + \frac{v_x}{2h_1} \right),
$$

$$
A_{\ell,\ell} = 1 + \theta k \left(\frac{2K_{xx}}{h_1^2} \right),
$$

$$
A_{\ell,\ell+1} = 1 - \theta k \left(\frac{K_{xx}}{h_1^2} - \frac{v_x}{2h_1} \right).
$$

Equation (4.3-51) constitutes the **x-direction sweep**.

Equation (4.3-49b) can also be expressed in matrix form. For the j-th column of nodes, we get

(4.3-52)
$$
\begin{bmatrix}
B_{11} & B_{12} & & & & & \\
B_{21} & B_{22} & B_{23} & & & & \\
 & B_{32} & B_{33} & B_{34} & & & \\
 & & & \ddots & & & \\
 & & & B_{I-2,I-3} & B_{I-2,I-2} & B_{I-2,I-1} & \\
 & & & & B_{I-1,I-2} & B_{I-1,I-1}
\end{bmatrix}
$$

$$
\begin{bmatrix}
u_{1,j}^{n+1} - u_{1,j}^n \\
u_{2,j}^{n+1} - u_{2,j}^n \\
u_{3,j}^{n+1} - u_{3,j}^n \\
\vdots \\
u_{I-2,j}^{n+1} - u_{I-2,j}^n \\
u_{I-1,j}^{n+1} - u_{I-1,j}^n
\end{bmatrix}
=
\begin{bmatrix}
z_{1,j} \\
z_{2,j} \\
z_{3,j} \\
\vdots \\
z_{I-2,j} \\
z_{I-1,j}
\end{bmatrix} ,
$$

or $\mathbf{B}(\mathbf{u}^{n+1} - \mathbf{u}^n) = \mathbf{z}$. The nonzero entries of the matrix $\mathbf{B}$ are as follows:

$$B_{\ell,\ell-1} = 1 - \theta k \left(\frac{K_{yy}}{h_2^2} + \frac{v_y}{2h_2} \right),$$

$$B_{\ell,\ell} = 1 + \theta k \left(\frac{2K_{yy}}{h_2^2} \right),$$

$$B_{\ell,\ell+1} = 1 - \theta k \left(\frac{K_{yy}}{h_2^2} + \frac{v_y}{2h_2} \right).$$

Equation (4.3-52) is the **y-direction sweep**. Procedurally, one first solves $I-1$ systems of equations of the form (4.3-51) to obtain $(I-1)(J-1)$ values $z_{i,j}$. Then one solves Equation (4.3-52) for $(I-1)(J-1)$ values $(u_{i,j}^{n+1} - u_{i,j}^n)$ using the computed values of $z_{i,j}$. The final step is to compute the values of $u_{i,j}^{n+1}$, given the values $u_{i,j}^n$ from the preceding time step. Notice that each step involves the solution of a tridiagonal matrix, which we can do quite efficiently.

Stability of the ADI procedure.

While the ADI formulation is attractive because it is readily generalized to other numerical procedures, such as collocation, it is not easily analyzed for stability, consistency and, concomitantly, convergence. To perform these analyses it is better to write the split equations at the local level rather than the global or matrix level. Moreover, to simplify the algebra, we consider here the advection-free form of the transport equation, that is the diffusion equation. The ADI form for the x-direction sweep in this case is

$$(4.3\text{-}53a) \qquad \frac{u_{i,j}^{n+1/2} - u_{i,j}^n}{k/2} = \frac{\delta_x(K_{xx}\,\delta_x u_{i,j}^{n+1/2})}{h_1^2} + \frac{\delta_y(K_{yy}\,\delta_y u_{i,j}^n)}{h_2^2},$$

while for the y-direction sweep we have

$$(4.3\text{-}53b) \qquad \frac{u_{i,j}^{n+1} - u_{i,j}^{n+1/2}}{k/2} = \frac{\delta_x(K_{xx}\,\delta_x u_{i,j}^{n+1/2})}{h_1^2} + \frac{\delta_y(K_{yy}\,\delta_y u_{i,j}^{n+1})}{h_2^2}.$$

Note that we attribute no physical significance to the intermediate value $u_{i,j}^{n+1/2}$. Summing Equations (4.3-53) yields

$$(4.3\text{-}54) \qquad \frac{u_{i,j}^{n+1} - u_{i,j}^n}{k} = \frac{\delta_x(K_{xx}\,\delta_x u_{i,j}^{n+1/2})}{h_1^2} + \frac{\delta_y[K_{yy}\,\delta_y(u_{i,j}^{n+1} + u_{i,j}^n)]}{2h_2^2}$$

$$= \frac{\delta_x(K_{xx}\,\delta_x u_{i,j}^{n+1/2})}{h_1^2} + \frac{\delta_y(K_{yy}\,\delta_y u_{i,j}^{n+1/2})}{h_2^2}.$$

This is the well known Crank-Nicolson form, which is second-order accurate in both time and space.

To establish stability we first calculate the amplification factors for Equations (4.3-53). Let us simplify notation by writing ξ_x, ξ_y, ξ, σ_x, and σ_y in place of ξ'_{xm}, ξ'_{ym}, ξ'_m, σ_{xm}, and σ_{ym}, respectively. For Equation (4.3-53a) we have (Noye, 1982)

$$(4.3\text{-}55a) \qquad \xi_x = \frac{1 - 2K_{yy}\sin^2(\sigma_y h_2/2)}{1 + 2K_{xx}\sin^2(\sigma_x h_1/2)} \, ,$$

and for Equation (4.3-53b) we have

$$(4.3\text{-}55b) \qquad \xi_y = \frac{1 - 2K_{xx}\sin^2(\sigma_x h_1/2)}{1 + 2K_{yy}\sin^2(\sigma_y h_2/2)} \, .$$

For some values of σ_y and σ_x, $|\xi_x|$ and $|\xi_y|$ may exceed unity for certain values of K_{xx} and K_{yy}, and hence each of Equations (4.3-53) is only conditionally stable. However, the amplification factor ξ for the two-step procedure is the product of ξ_x and ξ_y, that is,

$$(4.3\text{-}56) \qquad \xi = \frac{1 - 2K_{yy}\sin^2(\sigma_y h_2/2)}{1 + 2K_{xx}\sin^2(\sigma_x h_1/2)} \cdot \frac{1 - 2K_{xx}\sin^2(\sigma_x h_1/2)}{1 + 2K_{yy}\sin^2(\sigma_y h_2/2)} .$$

It is evident that $|\xi|$ cannot exceed unity for any values of K_{xx} and K_{yy}, and consequently the two-step procedure is uncondtionally stable. Because the algorithm is consistent and stable, it is also convergent according to Lax's equivalence theorem.

Locally one-dimensional (LOD) methods .

An alternating-direction procedure can also be formulated using a splitting procedure attributed to D'Yakonov (1963) and described at length in Yanenko (1971). Considering once again the diffusion equation, the splitting method yields

$$(4.3\text{-}57a) \qquad \frac{1}{2}\frac{\partial u}{\partial t} = \frac{\partial}{\partial x}\left(K_{xx}\frac{\partial u}{\partial x}\right)$$

and

$$(4.3\text{-}57b) \qquad \frac{1}{2}\frac{\partial u}{\partial t} = \frac{\partial}{\partial y}\left(K_{yy}\frac{\partial u}{\partial y}\right).$$

The name **locally one-dimensional** comes from the fact that each equation in the pair (4.3-57) is one-dimensional in space.

While there are many ways to exploit this concept numerically, we shall consider a rather straightforward Crank-Nicolson formulation. We obtain from Equations (4.3-57) the difference forms

$$(4.3\text{-}58\text{a}) \qquad \frac{1}{2}\frac{u_{i,j}^{n+1/2} - u_{i,j}^{n}}{k/2} = \frac{1}{2}\frac{\delta_x(K_{xx}\,\delta_x u_{i,j}^{n+1/2})}{h_1^2} + \frac{1}{2}\frac{\delta_x(K_{xx}\,\delta_x u_{i,j}^{n})}{h_1^2}$$

and

$$(4.3\text{-}58\text{b}) \qquad \frac{1}{2}\frac{u_{i,j}^{n+1} - u_{i,j}^{n}}{k/2} = \frac{1}{2}\frac{\delta_y(K_{yy}\,\delta_y u_{i,j}^{n+1})}{h_2^2} + \frac{1}{2}\frac{\delta_y(K_{yy}\,\delta_y u_{i,j}^{n+1/2})}{h_2^2}.$$

Noye (1982) gives the amplification factors for the difference equations (4.3-58a) and (4.3-58b) as follows:

$$(4.3\text{-}59\text{a}) \qquad \xi_x = \frac{1 - 2K_{xx}\sin^2(\sigma_x h_1/2)}{1 + 2K_{xx}\sin^2(\sigma_x h_1/2)}$$

and

$$(4.3\text{-}59\text{b}) \qquad \xi_y = \frac{1 - 2K_{yy}\sin^2(\sigma_y h_2/2)}{1 + 2K_{yy}\sin^2(\sigma_y h_2/2)},$$

respectively. The amplification factor for the combined two-step procedure is just the product $\xi = \xi_x \xi_y$, that is,

$$(4.3\text{-}60) \qquad \xi = \frac{1 - 2K_{xx}\sin^2(\sigma_x h_1/2)}{1 + 2K_{xx}\sin^2(\sigma_x h_1/2)} \cdot \frac{1 - 2K_{yy}\sin^2(\sigma_y h_2/2)}{1 + 2K_{yy}\sin^2(\sigma_y h_2/2)}.$$

Because $|\xi_x| \le 1$ and $|\xi_y| \le 1$ for all values of K_{xx}, σ_x, K_{yy}, and σ_y, we conclude that $|\xi| \le 1$ also. Thus the LOD approximation of Equation (4.3-58) is unconditionally stable. This algorithm can be extended in a straightforward way to three space dimensions by writing

$$\frac{1}{3}\frac{\partial u}{\partial t} = \frac{\partial}{\partial x}\left(K_{xx}\frac{\partial u}{\partial x}\right),$$

$$\frac{1}{3}\frac{\partial u}{\partial t} = \frac{\partial}{\partial y}\left(K_{yy}\frac{\partial u}{\partial y}\right),$$

$$\frac{1}{3}\frac{\partial u}{\partial t} = \frac{\partial}{\partial z}\left(K_{zz}\frac{\partial u}{\partial z}\right).$$

Enhanced ADI algorithms.

It is possible to enhance the second-order accuracy of ADI approximations through the careful selection of difference formula. Mitchell and Fairweather (1964) present such an algorithm. Consider the model equation

$$\frac{\partial u}{\partial t} - \frac{\partial^2 u}{\partial x^2} - \frac{\partial^2 u}{\partial y^2} = 0.$$

Mitchell and Fairweather show that an unconditionally stable fourth-order accurate approximation is given by

$$\left(-\delta_y^2 + \frac{2}{\bar{\rho} - 1/6}\right) u^{n+1/2} = -\frac{\bar{\rho}(\bar{\rho} + 1/6)}{(\bar{\rho} + 5/6)^2(\bar{\rho} - 1/6)}\left(\delta_x^2 + \frac{2}{\bar{\rho} + 1/6}\right) u^n,$$

for the y-direction sweep and, for the x-direction sweep,

$$\left(-\delta_x^2 + \frac{2}{\bar{\rho} - 1/6}\right) u^{n+1} = -\frac{(\bar{\rho} + 5/6)^2(\bar{\rho} + 1/6)}{\bar{\rho}(\bar{\rho} - 1/6)}\left(\delta_y^2 + \frac{2}{\bar{\rho} + 1/6}\right) u^{n+1/2},$$

where $\bar{\rho} = k/h^2$. Other higher-order ADI algorithms are also possible; we refer the reader to Mitchell and Fairweather (1964) for details.

ADI algorithms with mixed partial derivatives.

While the ADI and LOD methods presented heretofore are very effective for parabolic equations devoid of mixed derivatives such as $\partial^2 u/\partial x \partial y$, the presence of such derivatives adds complexity to the split formulation. Let us consider the two-dimensional diffusion equation including the mixed derivative:

$$(4.3\text{-}61) \qquad \frac{\partial u}{\partial t} - K_{xx}\frac{\partial^2 u}{\partial x^2} - 2K_{xy}\frac{\partial^2 u}{\partial x \partial y} - K_{yy}\frac{\partial^2 u}{\partial y^2} = 0.$$

Here we have used the full tensor form for the diffusion coefficient **K**, assuming $K_{xy} = K_{yx}$. Mitchell and Griffiths (1980) report two ADI algorithms to solve this equation. One, which they attribute to Samarskii (1964), is written

$$(4.3\text{-}62\text{a}) \qquad (1 - \overline{K}_{xx}\,\delta_x^2)u_{i,j}^{n+1/2} = u_{i,j}^n,$$

for the x-direction sweep and

$$(4.3\text{-}62\text{b}) \quad (1 - \overline{K}_{yy}\,\delta_y^2)u_{i,j}^{n+1} = [1 + \overline{K}_{xy}(u_{i,j+1}^{n+1/2} - u_{i-1,j+1}^{n+1,2} - 2u_{i,j}^{n+1/2}$$
$$+ u_{i-1,j}^{n+1/2} + u_{i+1,j}^{n+1/2} - u_{i+1,j-1}^{n+1/2} + u_{i,j-1}^{n+1/2})],$$

for the y-direction sweep. Here, $\overline{K}_{xx} \equiv K_{xx}k/h_1^2$, $\overline{K}_{xy} \equiv K_{xy}k/h_1h_2$, and $\overline{K}_{yy} \equiv K_{yy}k/h_2^2$.

Offering another approach, McKee and Mitchell (1970) propose the scheme

(4.3-63a)
$$\left(1 - \frac{\overline{K}_{xx}}{2}\delta_x^2\right)u_{i,j}^{n+1*}$$

$$= \left(1 + \frac{\overline{K}_{xx}}{2}\delta_x^2 + \frac{\overline{K}_{yy}}{2}\delta_y^2 + \frac{\overline{K}_{xy}}{2}\delta_x\delta_y\right)u_{i,j}^n,$$

(4.3-63b)
$$\left(1 - \frac{\overline{K}_{yy}}{2}\delta_y^2\right)u_{i,j}^{n+1} = u_{i,j}^{n+1*} - \frac{\overline{K}_{yy}}{2}\delta_y^2 u_{i,j}^n,$$

where intermediate Dirichlet values are given by

$$u_{i,j}^{n+1*} = \left(\frac{1 - \overline{K}_{yy}}{2}\delta_y^2\right)u_{i,j}^{n+1} + \frac{K_{yy}}{2}\delta_y^2 u_{i,j}^n.$$

As is apparent from each of these approximations, the operator is not formally split, but rather is carried explicitly on the right side of the equation.

Treatment of boundary conditions.

Let us examine how to accommodate boundary conditions in the finite-difference method. For simplicity we shall treat PDEs on a two-dimensional spatial domain Ω. Boundary conditions appropriate to spatially second-order differential equations can be written in the general form

(4.3-64a)
$$\beta(x, y, t)\frac{\partial u}{\partial n} - \alpha(x, y, t)u = f(x, y, t), \quad (x, y) \in \partial\Omega,$$

where α and β are known coefficients and $f(x, y, t)$ is a specified, possibly time-dependent, function. When α and β are both nonzero, Equation (4.3-64a) specifies a Robin boundary condition. When $\alpha \equiv 0$, Equation (4.3-64a) reduces to

(4.3-64b)
$$\beta\frac{\partial u}{\partial n} = f(x, y, t),$$

which is a Neumann boundary condition, and when $\beta \equiv 0$ we get

(4.3-64c)
$$\alpha(x, y, t)u = f(x, y, t),$$

which is a Dirichlet boundary condition.

The treatment of Dirichlet conditions is straightforward. We simply replace any nodal approximation u_{ijk} on the boundary by its exact value $u(ih_1, jh_2, nk)$ in the appropriate difference equation. This value being known, we transfer the term containing it to the right side of the difference equation. This tactic reduces the number of equations in the matrix formulation by one for each Dirichlet node.

The Neumann and Robin conditions contain derivatives, and thus their treatment is slightly more complicated. We can obtain a second-order approximation for the Robin condition (4.3-64a) along the boundary where $i = 0$, for example, by writing

$$(4.3\text{-}65) \qquad \beta_{0,j}^{n+\theta} \frac{\delta_x}{h_1} u_{0,j}^{n+\theta} - \alpha_{0,j}^{n+\theta} u_{0,j}^{n+\theta} = f_{0,j}^{n+\theta},$$

or, upon expansion,

$$(4.3\text{-}66) \qquad \beta_{0,j}^{n+\theta} \left(\frac{u_{1,j}^{n+\theta} - u_{-1,j}^{n+\theta}}{2h_1} \right) - \alpha_{0,j}^{n+\theta} u_{0,j}^{n+\theta} = f_{0,j}^{n+\theta}.$$

Note that the approximation (4.3-66) calls for a nodal value lying outside of the finite-difference grid, namely $u_{-1,0}^{n+\theta}$. To eliminate this "fictitious" unknown it is necessary to combine Equation (4.3-66) with the approximation of the operator written for the location $(0h_1, jh_2)$, as described in Section 3.10. Thus the number of equations in the matrix formulation is not reduced as a result of imposing the boundary condition. This technique applies equally well to Neumann conditions if we set $\alpha = 0$.

An alternative, less accurate, approximation is obtained by writing a first-order approximation to the Neumann or Robin condition about the boundary node, that is,

$$(4.3\text{-}67) \qquad \beta_{0,j}^{n+\theta} \left(\frac{u_{1,j}^{n+\theta} - u_{0,j}^{n+\theta}}{h_1} \right) - \alpha_{0,j} u_{0,j}^{n+\theta} = f_{0,j}^{n+\theta}.$$

In this case there is no fictitious node in the expression. Thus, by substituting the expression (4.3-67) into the finite-difference approximation written about node $(1, j)$, it is possible to decrease the number of equations by one. It is also relatively convenient to use the approximation (4.3-67) from the point of view of computer programming.

While we have focused on the special case when the x-derivative coincides with the normal derivative on $\partial\Omega$, the same strategy can be employed along any regular boundary. For curved boundaries a more complex strategy is required. We refer the reader to Lapidus and Pinder (1982) or Noye (1982) for details.

Treatment of variable coefficients.

While we have not considered heretofore the treatment of variable coefficients, they do not normally cause additional complications in linear equations. Consider, for example, the simple case

$$(4.3\text{-}68) \qquad \frac{\partial u}{\partial t} = K(x)\frac{\partial^2 u}{\partial x^2} \ .$$

Using a fully explicit approximation yields the following difference equation centered at node $x_i = ih_1$:

$$(4.3\text{-}69) \qquad \frac{u_i^{n+1} - u_i^n}{k} = \frac{K(ih_1)}{h_1^2}\delta_x^2 u_i^n \equiv \frac{K_i}{h_1^2}\delta_x^2 u_i^n \ .$$

It is sometimes possible to formulate specific rules for assigning values to variable coefficients based on physical considerations. Consider, for example, coefficients associated with a mass or energy flux. Let $q(x)$ represent the flux along the x-direction of Figure 4-10, let u be an unknown potential (such as fluid pressure), and suppose $\lambda(x)$ is the constitutive coefficient governing the flux according to a gradient law such as $q = -\lambda\partial u/\partial x$. (Darcy's law for flow in a porous medium, introduced in Section 1.5, is an example of such a gradient law when gravitational effects are absent.) One can write generalized difference approximations for the flux across an interface separating nodes i and $i + 1$ (see Aziz and Settari, 1979) as

$$(4.3\text{-}70) \qquad -q = \lambda_i d\frac{u_{\text{int}} - u_i}{\delta_i} = \lambda_{i+1}d\frac{u_{i+1} - u_{\text{int}}}{\delta_{i+1}},$$

where u_{int} is the potential at the interface separating the two nodes; d is a geometric factor giving the cross-sectional area of that interface, and δ_i signifies the distance from node i to the interface. Our objective is to select a value of λ that will accurately represent the flow between nodes i and $i + 1$, that is, a number $\lambda_{i+1/2}$ such that

$$(4.3\text{-}71) \qquad -q = \lambda_{i+1/2}d\frac{u_{i+1} - u_i}{\delta_{i+1/2}} \ ,$$

where $\delta_{i+1/2} = (\delta_i + \delta_{i+1})/2$. Elimination of u_{int} from Equation (4.3-70) and subsequent comparison of the result with Equation (4.3-71) reveals that the value of $\lambda_{i+1/2}/\delta_{i+1/2}$ must be

$$(4.3\text{-}72) \qquad \frac{\lambda_{i+1/2}}{\delta_{i+1/2}} = \frac{\lambda_i\lambda_{i+1}/(\delta_i\delta_{i+1})}{(\lambda_i\delta_i) + (\lambda_{i+1}\delta_{i+1})} = \frac{\lambda_i\lambda_{i+1}}{\lambda_i\delta_{i+1} + \lambda_{i+1}\delta_i}.$$

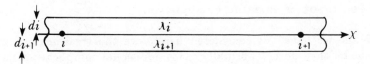

Figure 4-11. Finite-difference grid with variable properties λ_i and λ_{i+1} in parallel.

Thus the appropriate coefficient is the **harmonic mean** of the coefficients associated with each of the nodes.

As a second example, consider the case illustrated in Figure 4-11. The total flux through the system is given in this case by

$$(4.3\text{-}73) \qquad q = q_i + q_{i+1} = -\frac{\lambda_i d_i (u_{i+1} - u_i)}{\delta_{i+1/2}} - \frac{\lambda_{i+1} d_{i+1} (u_{i+1} - u_i)}{\delta_{i+1/2}} .$$

Comparison of this expression with Equation (4.3-71) reveals that the average coefficient should be

$$(4.3\text{-}74) \qquad \lambda_{i+1/2} = \frac{\lambda_i d_i + \lambda_{i+1} d_{i+1}}{d_i + d_{i+1}} ,$$

which is a weighted **arithmethic mean** of the coefficients λ_i and λ_{i+1}.

4.4. Finite-Element Methods.

In this section we use the general principles presented in Section 3.16 to solve diffusion and advection-diffusion problems. Because, in general, we shall be working with nonself-adjoint operators, we use the Galerkin method of approximation rather than a formal variational approach. As in the case of the finite-difference method described in the preceding section, we begin with problems of one space dimension and then extend the resulting concepts to higher dimensions.

Galerkin-in-space approximation.

We begin with the one-dimensional transport equation, which we write in the form

$$(4.3\text{-}6) \qquad \frac{\partial u}{\partial t} = a_o(x, t)\frac{\partial^2 u}{\partial x^2} + a_1(x, t)\frac{\partial u}{\partial x} + a_2(x, t)u + d(x, t),$$

$$x \in (0, 1).$$

In approximating Equation (4.3-6) we can choose between two distinct avenues of approach. The first discretizes the time derivative using finite differences, whereas the second applies Galerkin's method to the equation in its entirety. We begin with the forumlation based on finite differences in time. Discretization of Equation (4.3-6) about the time level $t = (n + \theta)k$ yields the approximate equation

$$(4.4\text{-}1) \quad \mathcal{L}(u^{n+\theta}(x)) \equiv \frac{u^{n+1}(x) - u^n(x)}{k} - a_0^{n+\theta}(x)\frac{\partial^2 u^{n+\theta}}{\partial x^2}(x)$$
$$- a_1^{n+\theta}(x)\frac{\partial u^{n+\theta}}{\partial x}(x) - a_2^{n+\theta}(x)u^{n+\theta}(x) - d^{n+\theta}(x)$$
$$= 0,$$

where $u^{n+\theta}(x)$ denotes the temporally discrete approximation to the value $u(x, (n + \theta)k)$ and $a_\ell^{n+\theta}(x)$ approximates $a_\ell(x, (n + \theta)k)$.

We must now select an appropriate representation for $u^{n+\theta}(x)$ on a spatial grid with nodes $x_i = ih$. If we assume Dirichlet boundary conditions, then the appropriate finite-element trial function will have the form

$$(4.4\text{-}2) \quad u^{n+\theta}(x) \simeq \hat{u}^{n+\theta}(x) = u_0\phi_0(x) + u_N\phi_N(x) + \sum_{i=1}^{N-1} u_i\phi_i(x).$$

For example, we might choose the functions ϕ_i to be the piecewise linear Lagrange interpolating basis. The terms $u_0\phi_0(x)$ and $u_N\phi_N(x)$ accommodate the Dirichlet boundary conditions by incorporating known boundary values for the coefficients u_0 and u_N.

Substitution of $\hat{u}^{n+\theta}(x)$ for $u^{n+\theta}(x)$ in Equation (4.4-1) yields a residual, that is,

$$(4.4\text{-}3) \quad \mathcal{L}(\hat{u}^{n+\theta}(x)) = R^{n+\theta}(x).$$

The Galerkin method, as described in Chapter Two, forces this residual to vanish in a weighted-average sense:

$$(4.4\text{-}4) \quad \int_0^1 R^{n+\theta}(x)\phi_j(x)dx = 0, \quad j = 1, \dots, N - 1.$$

Combining equations (4.4-1), (4.4-3) and (4.4-4) and suppressing dependence on x for simpler notation, we obtain

$$(4.4\text{-}5) \quad \int_0^1 \left(\frac{\hat{u}^{n+1} - \hat{u}^n}{k} - a_0^{n+\theta}\frac{\partial^2 \hat{u}^{n+\theta}}{\partial x^2} - a_1^{n+\theta}\frac{\partial \hat{u}^{n+\theta}}{\partial x}\right.$$
$$\left. - a_2^{n+\theta}\hat{u}^{n+\theta} - d^{n+\theta}\right)\phi_j dx = 0, \quad j = 1, \dots N - 1.$$

It is convenient at this point to apply integration by parts to the term containing the second derivative in Equation (4.4-5). One obtains

(4.4-6) $\displaystyle\int_0^1 \Bigg\{ \frac{\hat{u}^{n+1} - \hat{u}^n}{k}\phi_j + a_0^{n+\theta}\frac{\partial \hat{u}^{n+\theta}}{\partial x}\frac{\partial \phi_j}{\partial x}$

$$+ \left[\frac{\partial a_0^{n+\theta}}{\partial x}\frac{\partial \hat{u}^{n+\theta}}{\partial x} - a_1^{n+\theta}\frac{\partial \hat{u}^{n+\theta}}{\partial x} - a_2^{n+\theta}\hat{u}^{n+\theta} - d^{n+\theta} \right]\phi_j \Bigg\} dx$$

$$- a_0^{n+\theta}\frac{\partial \hat{u}}{\partial x}\phi_j \Bigg|_0^1 = 0.$$

Now define $\hat{u}^{n+\theta}$ to be $\theta\hat{u}^{n+1} + (1-\theta)\hat{u}^n$. Equation (4.4-6) becomes

(4.4-7) $\displaystyle\int_0^1 \Bigg[\left(\frac{1}{k} - \theta a_2^{n+\theta} \right)\hat{u}^{n+1}\phi_j + \theta a_0^{n+\theta}\frac{\partial \hat{u}^{n+1}}{\partial x}\frac{\partial \phi_j}{\partial x}$

$$+ \theta\left(\frac{\partial a_0^{n+\theta}}{\partial x} - a_1^{n+\theta} \right)\frac{\partial \hat{u}^{n+1}}{\partial x}\phi_j \Bigg] dx$$

$$= \int_0^1 \Bigg\{ \left[\frac{1}{k} + (1-\theta)a_2^{n+\theta} \right]\hat{u}^n\phi_j - (1-\theta)a_0^{n+\theta}\frac{\partial \hat{u}^n}{\partial x}\frac{\partial \phi_j}{\partial x}$$

$$- (1-\theta)\left(\frac{\partial a_0^{n+\theta}}{\partial x} - a_1^{n+\theta} \right)\frac{\partial \hat{u}^n}{\partial x}\phi_j + d^{n+\theta} \Bigg\} dx$$

$$+ a_0^{n+\theta}\frac{\partial \hat{u}}{\partial x}\phi_j \Bigg|_0^1 = 0, \qquad j = 1, \ldots, N-1.$$

Given either intial conditions or calculated values from the n-th time step, the right side of Equation (4.4-7) is known, since it contains values of $\hat{u}^n$ and known function values $a_0^{n+\theta}, a_1^{n+\theta}, a_2^{n+\theta}, d^{n+\theta}$. The boundary term $a_0^{n+\theta}(\partial \hat{u}/\partial x)\phi_j \mid_0^1$ vanishes, since in the case of Dirichlet boundary data each of the weighting functions $\phi_1, \ldots, \phi_{N-1}$ vanishes at both $x = 0$ and $x = 1$. Let us call the sum of the known terms $F_j^{n+\theta}$. Then, after substituting the trial function (4.4-2) into Equation (4.4-7) and neglecting the approximation error, we have

(4.4-8) $\displaystyle\sum_{i=0}^N u_i^{n+1} \int_0^1 \Bigg[\left(\frac{1}{k} - \theta a_2^{n+\theta} \right)\phi_i\phi_j + \theta a_0^{n+\theta}\frac{\partial \phi_i}{\partial x}\frac{\partial \phi_j}{\partial x}$

$$+ \theta\left(\frac{\partial a_0^{n+\theta}}{\partial x} - a_1^{n+\theta} \right)\frac{\partial \phi_i}{\partial x}\phi_j \Bigg] dx = F_j^{n+\theta},$$

$$j = 1, \ldots, N-1.$$

Before proceeding, let us pause to make a remark about boundary conditions. Our observation that the boundary term in Equation (4.4-7) vanishes pertains to Dirichlet boundary data only. If either endpoint $(x = 0$

or $x = 1$) of the domain were a Neumann or Robin boundary, matters would be different. For one thing, the nodal value of $\hat{u}$ at the boundary in question would remain unknown. To compensate for this extra unknown, the corresponding basis function ϕ_0 or ϕ_N would appear as a weighting function to supply the additional needed equation. In this case, however, the boundary term $a_0^{n+\theta}(\partial\hat{u}/\partial x)\phi_j|_0^1$ would not vanish for all values of j. Instead, we would need to use the Neumann or Robin data to evaluate this term. In either case, the result would again be a complete set of equations sufficient to determine all of the unknown nodal values of $\hat{u}$. We consider this procedure in a later subsection.

Now let us return to our development for Dirichlet conditions. Equation (4.4-8) yields an algebraic system having the tridiagonal matrix form

$$
\begin{bmatrix}
A_{11} & A_{12} & & & & \\
A_{21} & A_{22} & A_{23} & & & \\
& A_{32} & A_{33} & A_{34} & & \\
& & & \ddots & & A_{N-2,N-1} \\
& & & & A_{N-1,N-2} & A_{N-1,N-1}
\end{bmatrix}
\begin{bmatrix}
u_1^{n+1} \\
u_2^{n+1} \\
u_3^{n+1} \\
\vdots \\
u_{N-2}^{n+1} \\
u_{N-1}^{n+1}
\end{bmatrix}
$$

$$
=
\begin{bmatrix}
F_1^{n+\theta} \\
F_2^{n+\theta} \\
F_3^{n+\theta} \\
\vdots \\
F_{N-2}^{n+\theta} \\
F_{N-1}^{n+\theta}
\end{bmatrix}
-
\begin{bmatrix}
A_{10}u_0^{n+1} \\
0 \\
0 \\
\vdots \\
0 \\
A_{N-1,N}u_N^{n+1}
\end{bmatrix},
$$

where

$$
(4.4\text{-}9) \qquad A_{j,i} = \int_0^1 \left[\left(\frac{1}{k} - \theta a_2^{n+\theta} \right) \phi_i \phi_j + \theta a_0^{n+\theta} \frac{\partial \phi_i}{\partial x} \frac{\partial \phi_j}{\partial x} \right.
$$
$$
\left. + \theta \left(\frac{\partial a_0^{n+\theta}}{\partial x} - a_1^{n+\theta} \right) \frac{\partial \phi_i}{\partial x} \phi_j \right] dx, \quad j \neq 0, N .
$$

Notice that it is not actually necessary to integrate over the entire domain spatial $(0, 1)$, since each basis function $\phi_i(x)$ vanishes outside the two-element interval $[(i-1)h, (i+1)h]$. For the coefficients generated at the ends of the spatial domain, it is necessary to integrate over only one element, that is over $[0, h]$ for coefficients associated with the left side of $(0, 1)$ and over $[(N - 1)h, 1]$ for coefficients associated with the right side.

The integrals defined in Equation (4.4-9) are easily evaluated for the case of piecewise linear basis functions, provided we have established the functional form of the coefficients $a_0^{n+\theta}$, $a_1^{n+\theta}$, $a_2^{n+\theta}$. A common assumption is that these coefficients are piecewise constant, that is, that each coefficient is constant over each element $[ih, (i+1)h]$. Under this assumption the coefficients and their derivatives pass through the integral sign in any integral over just one element, and we have

$$A_{j,j-1} = \left(\frac{1}{k} - \theta a_2^{n+\theta}\right)\frac{h}{6} - \frac{1}{h}\left(\theta a_0^{n+\theta}\right) + \frac{1}{2}\left(\theta a_1^{n+\theta}\right),$$

$$A_{j,j} = \left(\frac{1}{k} - \theta a_2^{n+\theta}\right)\frac{2h}{3} + \frac{1}{h}\left(2\theta a_0^{n+\theta}\right),$$

$$A_{j,j+1} = \left(\frac{1}{h} - \theta a_2^{n+\theta}\right)\frac{h}{6} - \frac{1}{h}\left(\theta a_0^{n+\theta}\right) - \frac{1}{2}\left(\theta a_1^{n+\theta}\right).$$

Similar integrations must also be performed for the vector entries $F_j^{n+\theta}$. If we introduce the definition of A_{ij} into the matrix arising from Equation (4.4-8) and consider a typical j-th row, the discretized equation reads:

$$
\begin{aligned}
(4.4\text{-}10) \quad & \frac{h}{k}\left(\frac{u_{j-1}^{n+\theta}}{6} + \frac{2u_j^{n+\theta}}{3} + \frac{u_{j+1}^{n+\theta}}{6}\right) \\
& - \frac{\theta a_0^{n+\theta}}{h}\left(u_{j+1}^{n+\theta} - 2u_j^{n+\theta} + u_{j-1}^{n+\theta}\right) - \frac{\theta a_1^{n+\theta}}{2}\left(u_{j+1} - u_{j-1}\right) \\
& - a_2^{n+\theta}h\left(\frac{u_{j-1}^{n+\theta}}{6} + \frac{2u_j^{n+\theta}}{3} + \frac{u_{j+1}^{n+\theta}}{6}\right) = F_j^{n+\theta}.
\end{aligned}
$$

Equation (4.4-10) is similar to, yet different from, standard finite-difference forms. Specifically, after dividing through by h, the second and third terms in Equation (4.4.-10) represent standard finite-difference approximations. The first and last terms represent weighted averages of the values $u_{i-1,j}^{n+\theta}$, $u_{i,j}^{n+\theta}$, and $u_{i+1,j}^{n+\theta}$, where the weighting coefficients are precisely those used in Simpson's rule for numerical integration.

Galerkin approximations in time and space.

We now formulate a Galerkin scheme for Equation (4.3-6) using a finite-element approach in time. Let us assume that we wish to solve this PDE for all times from $t = 0$ up to a final value $t = t_f$. The approximating function now has the form

$$(4.4\text{-}11). \qquad u(x,t) \simeq \hat{u}(x,t) = \sum_{\ell=0}^{N} u_\ell \Phi_\ell(x,t)$$

The functions $\Phi_\ell(x,t)$ are now two-dimensional basis functions defined on the space-time domain $(0,1) \times (0, t_f)$ of the problem. We shall assume that these functions are nodal interpolating basis functions for a grid defined on $(0, 1) \times (0, t_f)$. From Galerkin's method we have

(4.4-12)
$$\int_0^1 \int_0^{t_f} \left[\frac{\partial \hat{u}(x, t)}{\partial t} - a_0(x, t) \frac{\partial^2 \hat{u}(x, t)}{\partial x^2} - a_1(x, t) \frac{\partial \hat{u}(x, t)}{(\partial x)} \right.$$

$$\left. - a_2(x, t) \hat{u}(x, t) - d(x, t) \right] \Phi_\ell(x, t) \, dx \, dt = 0,$$

where the index ℓ ranges over all non-Dirichlet nodes. Let us adopt the letters i and j as indices of spatial nodes and let n and m denote indices of temporal nodes. Also, let $N_t + 1$ be the number of nodes in the time direction and $N_x + 1$ be the number of nodes in the space direction. If we assume that the space-time basis functions are tensor products of spatial and temporal finite-element basis functions, we can write the basis function for node ℓ, located at (x_i, t^n), as $\Phi_\ell(x, t) = \phi_i(x) \phi^n(t)$. Thus Equation (4.4-12) becomes

(4.4-13)
$$\sum_{i=0}^{N_x} \sum_{n=0}^{N_t} u_i^n \left\{ \int_0^1 \int_0^{t_f} \left[\frac{\partial \phi^n(t)}{\partial t} \phi_i(x) - a_0(x, t) \frac{\partial^2 \phi_i(x)}{\partial x^2} \phi^n(t) \right. \right.$$

$$\left. - a_1(x, t) \frac{\partial \phi_i(x)}{\partial x} \phi^n(t) - a_2(x, t) \phi_i(x) \phi^n(t) \right] \phi_j(x) \phi^m(t) \, dx \, dt$$

$$\left. - \int_0^1 \int_0^{t_f} d(x, t) \phi_j(x) \phi^m(t) \, dx \, dt \right\} = 0 ,$$

where the indices (j, m) range over all non-Dirichlet nodes.

Now suppose for algebraic simplicity that a_0, a_1, a_2, and d are constant. Then we can write Equations (4.4-13) as follows:

(4.4-14)
$$\sum_{i=0}^{N_x} \sum_{n=0}^{N_t} \left\{ u_i^n \left[\int_0^{t_f} \frac{\partial \phi^n}{\partial t} \phi^m \, dt \int_0^1 \phi_i \phi_j \, dx \right] \right.$$

$$- \int_0^{t_f} \phi^n \phi^m \, dt \int_0^1 \left(a_0 \frac{\partial^2 \phi_i}{\partial x^2} + a_1 \frac{\partial \phi_i}{\partial x} + a_2 \phi_i \right) \phi_j \, dx$$

$$\left. - \int_0^{t_f} \int_0^1 d \, \phi_j \phi^m \, dx \, dt \right\} = 0.$$

In the simplest case, the basis functions $\phi_i(x)$ and $\phi^n(t)$ are piecewise Lagrange linear interpolating functions, and evaluating the integrals in Equation (4.4-14) for a typical space-time node $(ih, (n + 1)k)$ gives

(4.4-15)
$$\frac{h}{2}\left(\frac{u_{i-1}^{n+1}}{6} + \frac{2u_i^{n+1}}{3} + \frac{u_{i+1}^{n+1}}{6}\right)$$

$$- \frac{h}{2}\left(\frac{u_{i+1}^{n}}{6} + \frac{2u_i^{n}}{3} + \frac{u_{i+1}^{n}}{6}\right)$$

$$- \frac{k}{3}\left(a_0\frac{\delta_x^2}{h} + a_1\frac{\delta_x}{2} + a_2\right)u_i^{n+1}$$

$$- \frac{k}{6}\left(a_0\frac{\delta_x^2}{h} + a_1\frac{\delta_x}{2} + a_2\right)u_i^{n} - \frac{dk}{2} = 0.$$

Now compare Equation (4.4-15) with Equation (4.4-10), which we obtained using a finite-difference approximation in time. Division of Equation (4.4-15) by kh yields the time-differenced scheme (4.4-10) if we assume constant coefficients and $\theta = \frac{2}{3}$. This approach can be extended to higher-degree basis functions in time, but the results reported by Gray and Pinder (1974) for the diffusion equation are not very encouraging.

Solution pathology at sharp fronts.

Finite-element procedures, like finite-difference procedures, may exhibit pathologic behavior when the advection term in the transport equation dominates the diffusive term. This difficulty should not be surprising inasmuch as the advective term is approximated by the same discrete for-

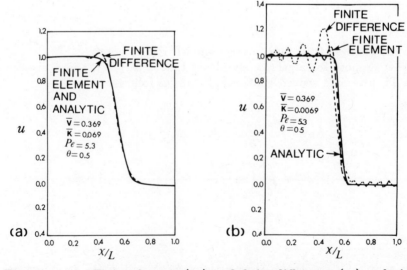

Figure 4-12. Finite-element (—) and finite-difference (...) solutions to the advection-diffusion transport equation for Peclet numbers of 5.3 and 53 (from Gray and Pinder, 1976).

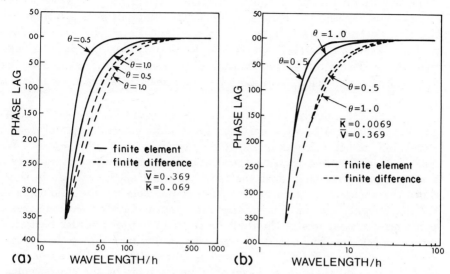

Figure 4-13. Phase lag Θ_m for finite-element and finite-difference methods (after Gray and Pinder, 1976).

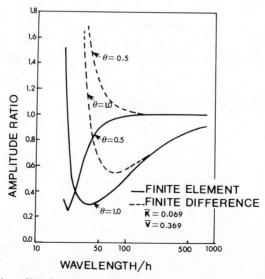

Figure 4-14. Amplitude ratio for finite-element and finite-difference methods (after Gray and Pinder, 1976).

mula in both the standard finite-difference method and piecewise linear finite-element method. Figure 4-12 illustrates finite-element solutions for two grid Peclet numbers Pe $\equiv \overline{V}/\overline{K} =$ 5.34 and Pe = 53.5. We also include the standard finite-difference solution for comparison. It is evident that, while the finite-element solution offers some improvement over the finite-difference solution, oscillatory behavior near the sharp front is still a problem.

The reason for this behavior can be seen from Figures 4-13 and 4-14. From Figure 4-13 we observe that the finite-element method does a better job, albeit marginally so, of propagating the shorter waves. We also see from Figure 4-13a that the Crank-Nicolson time weighting gives a smaller phase error than does the fully backward approximation in time. Since the smaller wavelengths are those responsible for the resolution of the sharp front, one can thus understand why the Crank-Nicolson scheme preserves the steepness of the front better than the lower-order backward difference scheme. But why do we see more oscillatory behavior in the case of the finite-difference method? Figure 4-14 provides insight into this question. Notice that the shorter waves are damped more by the finite-element method than by the finite-difference method. Thus, those Fourier components suffering phase errors also exhibit overdamping, so that the waves responsible for the spurious oscillations tend to disappear. One therefore expects to obtain a more oscillation-free solution with the finite-element method.

Figure 4-14 also indicates that backward difference schemes exert more numerical damping than the finite-element scheme. This fact is reflected in a more oscillation-free solution. However, with spatially backward finite differences, the numerical solution also normally exhibits increased numerical diffusion or smearing of the sharp front. We shall now examine nonstandard finite-element schemes that perform similarly.

Nonstandard finite-element methods.

The standard Galerkin finite-element approach can be modified to smooth the behavior of the numerical solution in the neighborhood of a sharp front. The most straightforward approach in the case of one-dimensional problems is to formulate a scheme that is analogous to the backward or upstream-weighted finite-difference approach. In this approach the advective term in Equation (4.4-10) is simply modified so that only the j-th and $(j-1)$-st nodes contribute to the discretization of the advection term when the fluid flows in the positive x-direction. The new formula reads

$$(4.4\text{-}16) \quad \frac{h}{k} \left(\frac{u_{j-1}^{n+\theta}}{6} + \frac{2u_j^{n+\theta}}{3} + \frac{u_{j+1}^{n+\theta}}{6} \right)$$

$$- \frac{\theta a_0^{n+\theta}}{h} \left(u_{j+1}^{n+\theta} - 2u_j^{n+\theta} + u_{j-1}^{n+\theta} \right) - \theta a_1^{n+\theta} \left(u_j^{n+\theta} - u_{j-1}^{n+\theta} \right)$$

$$- a_2^{n+\theta} h \left(\frac{u_{j-1}^{n+\theta}}{6} + \frac{2u_j^{n+\theta}}{3} + \frac{u_{j+1}^{n+\theta}}{6} \right) = F_j^{n+\theta} .$$

The solution obtained with the approximation (4.4-16) resembles that generated with an upstream-weighted finite-difference scheme. In particular, the solution is oscillation-free but exhibits substantial smearing near sharp concentration fronts, as shown in Figure 4-9.

Another approach to generating an oscillation-free solution is by using asymmetric weighting functions in the advection term. This method has the advantage of being readily extendable to two space dimensions. Because different weighting functions are used on different terms in the PDE, the method is not formally a Galerkin method. Our point of departure is Equation (4.4-6), which now reads

$$(4.4\text{-}17) \quad \int_0^1 \left[\frac{\hat{u}^{n+1} - \hat{u}^n}{k} \phi_j + a_0^{n+\theta} \frac{\partial \hat{u}}{\partial x} \frac{\partial \phi_j}{\partial x} \right.$$

$$+ \left(\frac{\partial a_1^{n+\theta}}{\partial x} \frac{\partial \hat{u}^{n+\theta}}{\partial x} - a_2^{n+\theta} \hat{u}^{n+\theta} - d^{n+\theta} \right) \phi_j$$

$$\left. - a_1^{n+\theta} \frac{\partial \hat{u}^{n+\theta}}{\partial x} \psi_j \right] dx - a_0^{n+\theta} \frac{\partial \hat{u}}{\partial x} \phi_j \Big|_0^1 = 0.$$

Here $\psi_j(x)$ denotes the asymmetric weighting function. Over the typical element system shown in Figure 4-15, this new basis function has the algebraic form

$$(4.4\text{-}18a) \qquad \psi_1(x) = 1 - \xi + 3\alpha(\xi^2 - \xi) , \quad 0 \le x \le h,$$

$$(4.4\text{-}18b) \qquad \psi_2(x) = \xi - 3\alpha(\xi^2 - \xi), \quad 0 \le x \le h,$$

where $\xi \equiv x/h$. The parameter α, specified by the analyst, dictates the degree of upsteam weighting in the final approximation. The shape of the weighting function $\psi_j(x)$ is illustrated in Figure 4-15. One can see readily from the figure why the approximation weights the upstream direction. Note that in this figure we have illustrated only one ψ_j for each element;

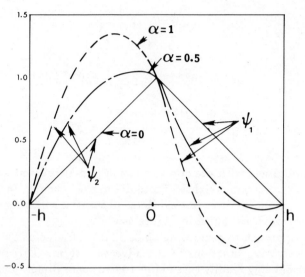

Figure 4-15. Asymmetric weighting functions ψ_j for $\alpha = 1, 0.5, 0$.

another function also has nonzero values on that element, as the definition (4.4-18) indicates.

Another nonstandard Galerkin procedure involves modification of the time derivative approximation. It has been observed, particularly in nonlinear equations, that **lumping** the time derivative coefficient typically generates a smoother numerical solution than that arising from the strict Galerkin, or **consistent**, formulation. In a lumped approximation, the first term in Equation (4.4-17) would be modified as follows:

$$\frac{h}{k}\left(\frac{u_{i-1}^{n+\theta}}{6} + \frac{2u_i^{n+\theta}}{3} + \frac{u_{i+1}^{n+\theta}}{6}\right)$$

$$\simeq \frac{h}{k}\left(\frac{u_i^{n+\theta}}{6} + \frac{2u_i^{n+\theta}}{3} + \frac{u_i^{n+\theta}}{6}\right) = \frac{h}{k}u_i^{n+\theta}.$$

Observe that this modification gives the standard finite-difference representation of the time derivative.

While asymmetric weighting functions provide the upstream weighting one often seeks for advection-dominated transport, an alternative algorithm based upon inexact Gauss quadrature is also possible. In this case we can compute the weighted-residual integrals using quadrature points displaced from the standard Gauss points, which are zeros of the associated Legendre polynomial as discussed in Chapter Two. This technique introduces artificial diffusion having the form $\alpha a_1 h/2$, where α measures the displacement of the quadrature point on the local-coordinate interval $[-1, 1]$. The artificial diffusion is very similar to that encountered in upstream weighted finite differences. For additional details on this approach see Hughes (1978) and Payre et al. (1982).

Galerkin finite-element method in two space dimensions.

In this subsection we extend the concepts of the finite-element method to accommodate two space dimensions. The extension is conceptually straightforward, although effective implementation requires some new concepts. We begin with a statement of our model problem,

$$(4.3\text{-}1) \qquad \frac{\partial u}{\partial t} + \nabla \cdot (\mathbf{v} u) - \nabla \cdot (\mathbf{K} \cdot \nabla u) + ru = 0 \ .$$

We have seen that there are two ways to treat the time derivative. One involves differencing the time derivative at the outset, leading to matrix equations in which the old time level acts essentially as a forcing function. The second approach incorporates the time dimension directly into the finite-element basis used to form the trial function. We now introduce a third option wherein each undetermined coefficient u_i of the trial function $\hat{u}$ is allowed to be a function of time. Thus, the trial function has the form

$$(4.4\text{-}19) \qquad u(x,y,t) \simeq \hat{u}(x,y,t) = \sum_{i=0}^{N_x} \sum_{j=0}^{N_y} u_{ij}(t)\phi_i(x)\phi_j(y)$$

$$= \sum_{i=0}^{N_x} \sum_{j=0}^{N_y} u_{ij}(t)\Phi_{ij}(x,y).$$

Here $\phi_i(x)$ and $\phi_j(y)$ are the one-dimensional basis functions discussed in Chapter Two. For simplicity, we shall assume that these basis functions are piecewise linear. $N_x + 1$ and $N_y + 1$ are the numbers of nodes in the x- and y-directions, respectively.

Following the Galerkin procedure, weighted averages of the residual, created by substituting $\hat{u}(x,y,t)$ for $u(x,y,t)$ in the PDE, are set equal to zero. Each weighting function in this case is a basis function $\phi_i(x)\phi_j(y) = \Phi_{ij}(x,y)$. Thus we have the following set of integral equations:

$$(4.4\text{-}20) \qquad \int_\Omega \left[\frac{\partial \hat{u}}{\partial t} + \nabla \cdot (\mathbf{v}\hat{u}) - \nabla \cdot (\mathbf{K} \cdot \nabla \hat{u}) + r\hat{u} \right] \Phi_{k\ell}(x,y)\,dx\,dy = 0 \ .$$

Here, as usual, Ω denotes the two-dimensional spatial domain of the problem, and the indices (k, ℓ) range over non-Dirichlet nodes.

Application of Green's theorem to the second-order term yields

$$(4.4\text{-}21) \qquad \int_\Omega \left\{ \left[\frac{\partial \hat{u}}{\partial t} + \nabla \cdot (\mathbf{v}\hat{u}) + r\hat{u} \right] \Phi_{k\ell} + (\mathbf{K} \cdot \nabla \hat{u}) \cdot \nabla \Phi_{k\ell} \right\} dx\, dy$$

$$- \oint_{\partial\Omega} (\mathbf{K} \cdot \nabla \hat{u}) \Phi_{k\ell} \cdot \mathbf{n}\, dx\, dy = 0.$$

The last term in Equation (4.4-21) is the boundary flux term. When this term appears in the equation, as it does for any equation in which node (k, ℓ) is a boundary node associated with an unknown, we can evaluate it using available boundary information. We shall address the question of boundary conditions more completely in a later subsection.

The introduction of the trial function (4.4-19) into Equations (4.4-21) yields

$$(4.4\text{-}22) \qquad \sum_{i=0}^{N_x} \sum_{j=0}^{N_y} \left(u_{ij} \int_\Omega \left\{ [\nabla \cdot (\mathbf{v}\Phi_{ij}) + r\Phi_{ij}]\Phi_{k\ell} \right.\right.$$

$$\left.\left. + (\mathbf{K} \cdot \nabla \Phi_{ij}) \cdot \nabla \Phi_{k\ell} \right\} dx\, dy + \frac{du_{ij}}{dt} \int_\Omega \Phi_{ij}\Phi_{k\ell}\, dx\, dy \right)$$

$$= \oint_{\partial\Omega} (\mathbf{K} \cdot \nabla \hat{u}) \Phi_{k\ell} \cdot \mathbf{n}\, dx\, dy.$$

We can write Equations (4.4-22) in matrix form as follows:

$$(4.4\text{-}23) \qquad\qquad\qquad \mathbf{A}\mathbf{u} + \mathbf{B}\frac{d\mathbf{u}}{dt} = \mathbf{f},$$

where, given the grid specified in Figure 4-16, the entries of the arrays are given by sums of elementwise integrals as follows:

$$A_{pq} = \sum_{e=1}^{E} \int_{\Omega_e} \left\{ [\nabla \cdot (\mathbf{v}\Phi_{ij}) + r\Phi_{ij}]\Phi_{k\ell} + (\mathbf{K} \cdot \nabla \Phi_{ij}) \cdot \nabla \Phi_{k\ell} \right\} dx\, dy,$$

$$B_{pq} = \sum_{e=1}^{E} \int_{\Omega_e} (\Phi_{ij}\Phi_{k\ell})\, dx\, dy,$$

$$f_p = \sum_{e=1}^{E_B} \int_{(\partial\Omega)_e} (\mathbf{K} \cdot \nabla \hat{u}) \Phi_{k\ell} \cdot \mathbf{n}\, dx\, dy.$$

In these definitions, $p = (k+1) + \ell(N_y + 1)$, $q = (i+1) + j(N_x + 1)$, and E and E_B are the number of elements and the number of elements intersecting the boundary, respectively. $(\partial\Omega)_e$ denotes the subset of the boundary $\partial\Omega$ belonging to element e. The evaluation of the integrals in the matrix equation (4.4-23) follows the method outlined earlier for the

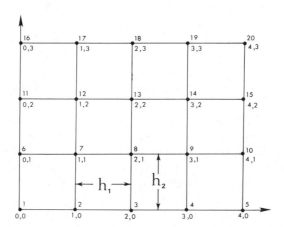

Figure 4-16. Finite-element grid for definition of elements in Equation (4.4-23).

evaluation of the integrals appearing in Equation (4.4-12). Provided the grid is rectangular, the integrations are easily done in closed form.

Having evaluated the coefficients A_{pq}, B_{pq}, and f_p, we are left with a set of ordinary differential equations. While numerous methods can be used to reduce this system to a set of algebraic equations, we shall use the simple finite-difference approach employed earlier. Thus we have

$$(4.4\text{-}24) \quad \theta\mathbf{A}^{n+\theta}\mathbf{u}^{n+1} + (1-\theta)\mathbf{A}^{n+\theta}\mathbf{u}^n + \frac{1}{k}\mathbf{B}^{n+\theta}(\mathbf{u}^{n+1} - \mathbf{u}^n) = \mathbf{f}^{n+\theta}.$$

Here we have allowed for the possibility of time-dependent coefficients and time-dependent boundary conditions. Rearranging Equation (4.4-24), we obtain

$$(4.4\text{-}25) \quad \mathbf{C}\mathbf{u}^{n+1} = \left(\theta\mathbf{A}^{n+\theta} + \frac{1}{k}\mathbf{B}^{n+\theta}\right)\mathbf{u}^{n+1}$$

$$= -\left[(1-\theta)\mathbf{A}^{n+\theta} - \frac{1}{k}\mathbf{B}^{n+\theta}\right]\mathbf{u}^n + \mathbf{f}^{n+\theta}.$$

The final task is to solve the matrix equation (4.4-25). The matrix $\mathbf{C}$ has nine nonzero diagonals, but the presence of the advective terms prevents it from being symmetric. Note also that, formally, we need to invert $\mathbf{C}$ only once, provided the time step k remains constant and the coefficients are time-independent. This observation can lead to significant computational savings, depending on the matrix solution method used.

In advection-dominated flows it may be necessary to use $\theta \neq \frac{1}{2}$ and to employ upstream weighting to avoid spurious oscillations in the numerical solution. Because we have used rectangular elements, the introduction of asymmetric weighting functions is straightforward. One simply multiplies the advective term by a weighting function that is asymmetric with respect to the appropriate coordinate direction. Thus, for the term $\partial(v_x u)/\partial x$, we would use an asymmetric weighting function $\psi_k(x)$. In practice, the degree of asymmetry imposed should be correlated with the magnitude of the grid Peclet number.

Extension of the foregoing analysis to problems that involve irregular spatial domains can be achieved in at least two different ways. In the first, we take advantage of the geometric flexibility available with triangular elements. In the second approach, we use tensor-product basis functions formulated in a rectangular local coordinate system on the square $[-1, 1] \times [-1, 1]$ and transform these functions to the irregular global coordinate system. In each of these approaches, it is notationally convenient to abandon the double indices (i, j) for nodes in favor of a single index. The trial function now appears as

$$\hat{u}(x, y, t) = \sum_{i=0}^{N} u_i(t)\phi_i(x, y) ,$$

where $\phi_i(x, y)$ is either a triangular basis function or a transformed tensor-product basis function. The Galerkin equations have the form

$$(4.4\text{-}26) \quad \sum_{i=0}^{N} u_i \int_{\Omega} \left[\nabla \cdot (\mathbf{v}\phi_i) + r\phi_i + (\mathbf{K} \cdot \nabla\phi_i) \cdot \nabla\phi_j \right] dx\, dy$$

$$+ \frac{du_i}{dt} \int \phi_i \phi_j dx\, dy = \oint_{\partial\Omega} (\mathbf{K} \cdot \nabla\hat{u})\phi_j \cdot \mathbf{n}\, dx\, dy,$$

where the index j ranges over non-Dirichlet nodes. While the algebraic changes we have introduced are minor, the technical ramifications involved in using this more general formulation are substantial. In the first place, the integrals appearing in the coefficient matrices are more difficult to evaluate, although this increased difficulty is relatively modest in the case of triangles. In the second place, the resulting matrix equations, while still banded, in general do not exhibit the regular structure associated with rectangular elements.

In the case of triangular elements, the implementation of Equations (4.4-26) is easily accomplished using the methodology presented in Section 2.12. First, we compute the coefficient matrices at the element level. Thus for piecewise planar basis functions one obtains a 3×3 matrix for each element. These element matrices are then assembled into a global coefficient matrix, as illustrated in Table 2-6. We then modify the global coefficient

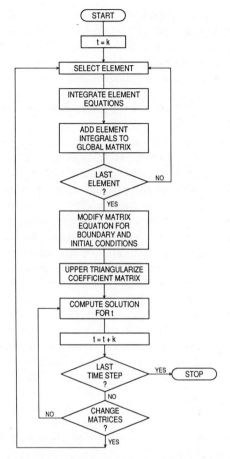

Figure 4-17. Flow chart for finite-element simulation of a diffusive process.

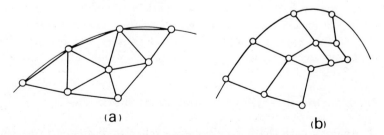

Figure 4-18. Representation of a segment of a region using triangular and isoparametric quadrilateral finite elements (after Botha and Pinder 1983).

243

matrix to accommodate Dirichlet boundary conditions. Conceptually, this process involves elimination of rows and columns associated with nodes where Dirichlet conditions are prescribed. However, to minimize storage in actual coding, one normally avoids from the outset the assembly of the matrix elements associated with Dirichlet nodes, opting rather to place the appropriate boundary information in the vector of known information directly. The resulting matrix equation is now solved for the unknown coefficients u_i^{n+1}. If we use a matrix solution routine based on some form of L-U decomposition, then we can take advantage of time-independent matrix entries by saving the upper triangularized matrix and using it repeatedly as we proceed stepwise through time. A flow chart summarizing the programming aspects of the procedure appears in Figure 4-17.

The use of isoparametric finite elements differs from the use of triangles only in the calculation of the Galerkin integrals and the structure of the coefficient matrix. Section 2.5 presents the concept of isoparametric transformations and applies this concept to an integral that typically arises in solving the transport equation. The procedure generally requires the use of numerical integration, which is described in Section 2.12. Figure 4-18 compares the representation of a small region using triangular and isoparametric elements. It is appropriate here to remark that, in using isoparametric elements, one must exercise care in locating mid-side nodes, particularly on a curved side. To guarantee a good approximation, the mid-side node should be located centrally along the side of the element, and the side should exhibit modest curvature. Remember that the curve that the analyst envisions through the side nodes is not necessarily the one recognized by the computer. The computer creates a polynomial representation of the side based on its nodal locations. This curve may or may not correspond to the analyst's conception and may exhibit undesirable wiggles owing to large curvatures and poorly placed mid-side nodes.

It is worth mentioning that upstream weighting can be employed in triangular and isoparametric quadrilateral elements as well as in simpler goemetries. When using isoparametric quadrilateral elements one simply uses the techniques introduced for rectangular elements, introducing the upstream weighting at the element level when integrations are being performed in the local coordinate system. In this coordinate system all deformed elements appear as squares of the form $[-1, 1] \times [-1, 1]$. We have already briefly described the use of asymmetric weighting functions on triangles in Section 3.16.

Alternating-direction finite-element methods.

By analogy with the finite-difference schemes discussed in Section 4.3, ADI finite-element methods can be formulated using either a matrix factoring strategy or the LOD approach. The factoring strategy is described for rectangular elements in Douglas and Dupont (1971). We shall employ the LOD procedure. We begin with the two-step form of the diffusion equation. To simplify the presentation we assume K_{xx} and K_{yy} to be spatially constant; however, the methodology also accommodates variable coefficients. We have

(4.4-27a)
$$\frac{1}{2}\frac{\partial u}{\partial t} = K_{xx}\frac{\partial u^2}{\partial x^2}$$

and

(4.4-27b)
$$\frac{1}{2}\frac{\partial u}{\partial t} = K_{yy}\frac{\partial u^2}{\partial y^2}.$$

Let us begin with a trial function having the form

$$\hat{u}(x,y,t) = \sum_{i=0}^{N_x}\sum_{j=0}^{N_y} u_{ij}(t)\phi_i(x)\phi_j(y),$$

where the nodal coordinates are $(x_i, y_j, t_n) = (ih_1, jh_2, nk)$, and $\phi_i(x)$ and $\phi_j(y)$ stand for piecewise linear Lagrange basis functions. Application of Galerkin's method to the pair (4.4-57) yields a set of integral equations having the forms

$$\int_\Omega \left(\frac{1}{2}\frac{\partial \hat{u}}{\partial t} - K_{xx}\frac{\partial^2 \hat{u}}{\partial x^2}\right)\phi_k(x)\phi_\ell(y)\,dx\,dy = 0\,,$$

$$\int_\Omega \left(\frac{1}{2}\frac{\partial \hat{u}}{\partial t} - K_{yy}\frac{\partial^2 \hat{u}}{\partial y^2}\right)\phi_k(x)\phi_\ell(y)\,dx\,dy = 0\,.$$

Now we apply Green's theorem to the second-order terms to obtain

(4.4-28a)
$$\int_\Omega \left(\frac{1}{2}\frac{\partial \hat{u}}{\partial t}\phi_k\phi_\ell + K_{xx}\frac{\partial \hat{u}}{\partial x}\frac{\partial \phi_k}{\partial x}\phi_k\phi_\ell\right)dx\,dy$$

$$-\oint_{\partial\Omega} K_{xx}\frac{\partial \hat{u}}{\partial x}\phi_k\phi_\ell\,dx\,dy = 0\,,$$

(4.4-28b)
$$\int_\Omega \left(\frac{1}{2}\frac{\partial \hat{u}}{\partial t}\phi_k\phi_\ell + K_{yy}\frac{\partial \hat{u}}{\partial y}\frac{\partial \phi_\ell}{\partial y}\phi_\ell\phi_k\right)dx\,dy$$

$$-\oint_{\partial\Omega} K_{yy}\frac{\partial \hat{u}}{\partial y}\phi_k\phi_\ell\,dx\,dy = 0\,.$$

Substituting the definition of $\hat{u}(x, y, t)$ into Equations (4.4-28) yields

$$(4.4\text{-}29a) \quad \sum_{i=0}^{N_x}\sum_{j=0}^{N_y}\left[\int_{\Omega}\left(\frac{1}{2}\frac{du_{ij}}{dt}\phi_i\phi_j\phi_k\phi_\ell + K_{xx}u_{ij}\frac{d\phi_i}{dx}\phi_j\frac{d\phi_k}{dx}\phi_\ell\right)dx\,dy\right.$$

$$\left. -u_{ij}\oint_{\partial\Omega}K_{xx}\frac{d\phi_i}{dx}\phi_j\phi_k\phi_\ell\,dx\,dy\right] = 0 ,$$

$$(4.4\text{-}29b) \quad \sum_{i=0}^{N_x}\sum_{j=0}^{N_y}\left[\int_{\Omega}\left(\frac{1}{2}\frac{du_{ij}}{dt}\phi_i\phi_j\phi_k\phi_\ell + K_{yy}u_{ij}\frac{d\phi_j}{dy}\phi_i\frac{d\phi_\ell}{dy}\phi_k\right)dx\,dy\right.$$

$$\left. -u_{ij}\oint_{\partial\Omega}K_{yy}\frac{d\phi_j}{dy}\phi_i\phi_k\phi_\ell\,dx\,dy\right] = 0 .$$

Assuming we have rectangular elements, Equations (4.4-29a) can be written as follows:

$$(4.4\text{-}30) \quad \sum_{i=0}^{N_x}\sum_{j=0}^{N_y}\left(\frac{1}{2}\frac{du_{ij}}{dt}\int_{\Omega_x}\phi_i\phi_k\,dx\int_{\Omega_y}\phi_j\phi_\ell\,dy\right.$$

$$+ K_{xx}u_{ij}\int_{\Omega_x}\frac{d\phi_i}{dx}\frac{d\phi_k}{dx}\,dx\int_{\Omega_y}\phi_j\phi_\ell\,dy$$

$$\left. - u_{ij}K_{xx}\int_{\partial\Omega_x}\frac{d\phi_i}{dx}\phi_k\,dx\int_{\partial\Omega_y}\phi_j\phi_\ell\,dy\right) = 0.$$

Here we have decomposed the two-dimensional spatial domain as $\Omega = \Omega_x \times \Omega_y$, taking advantage of our assumption that K_{xx} and K_{yy} are spatially uniform. In a rectangular grid aligned with the coordinate axes, we can write

$$(4.4\text{-}31) \quad \int_{\Omega_y}\phi_j\phi_\ell\,dy = \int_{\partial\Omega_y}\phi_j\phi_\ell\,dy.$$

Therefore we can write Equation (4.4-30) for each row j of the finite-element grid as follows:

$$(4.4\text{-}32) \quad \sum_{i=0}^{N_x}\left[\left(K_{xx}\int_{\Omega_x}\frac{d\phi_i}{dx}\frac{d\phi_k}{dx}\,dx\right)u_{ij} + \left(\frac{1}{2}\int_{\Omega_x}\phi_i\phi_k\,dx\right)\frac{du_{ij}}{dt}\right]$$

$$- \sum_{i=0}^{N_x}\left(K_{xx}\int_{\Omega_x}\frac{d\phi_i}{dx}\phi_k\,dx\right)u_{ij} = 0 .$$

Combining all equations of the form (4.4-32) into a matrix equation, we obtain the block structure

(4.4-33)

$$
\begin{bmatrix} \mathbf{A}_{00} & & & \\ & \mathbf{A}_{11} & & \\ & & \ddots & \\ & & & \mathbf{A}_{N_y N_y} \end{bmatrix} \begin{bmatrix} \mathbf{u}_0 \\ \mathbf{u}_1 \\ \vdots \\ \mathbf{u}_{N_y} \end{bmatrix}
$$

$$
+ \begin{bmatrix} \mathbf{B}_{00} & & & \\ & \mathbf{B}_{11} & & \\ & & \ddots & \\ & & & \mathbf{B}_{N_y N_y} \end{bmatrix} \begin{bmatrix} d\mathbf{u}_0/dt \\ d\mathbf{u}_1/dt \\ \vdots \\ d\mathbf{u}_{N_y}/dt \end{bmatrix} = \begin{bmatrix} \mathbf{F}_0 \\ \mathbf{F}_1 \\ \vdots \\ \mathbf{F}_{N_y} \end{bmatrix},
$$

where the blocks $\mathbf{A}_{\ell\ell}$ and $\mathbf{B}_{\ell\ell}$ have the tridiagonal forms

$$
\mathbf{A}_{\ell\ell} = \begin{bmatrix} A_{00} & A_{01} & & & \\ A_{10} & A_{11} & A_{12} & & \\ & & \ddots & & \\ & & & & A_{N_x-1,N_x} \\ & & A_{N_x,N_x-1} & A_{N_x,N_x} \end{bmatrix}_{y=\ell h_2}
$$

and

$$
\mathbf{B}_{\ell\ell} = \begin{bmatrix} B_{00} & B_{01} & & & \\ B_{10} & B_{11} & B_{12} & & \\ & & \ddots & & \\ & & & & B_{N_x-1,N_x} \\ & & B_{N_x,N_x-1} & B_{N_x,N_x} \end{bmatrix}_{y=\ell h_2}.
$$

The blocks $\mathbf{u}_\ell$ of unknowns and the forcing vector blocks $\mathbf{F}_\ell$ have the forms

$$
\mathbf{u}_\ell = \begin{bmatrix} u_{0\ell} \\ u_{1\ell} \\ \vdots \\ u_{N_x\ell} \end{bmatrix}, \quad \mathbf{F}_\ell = \begin{bmatrix} F_{0\ell} \\ F_{1\ell} \\ \vdots \\ F_{N_x\ell} \end{bmatrix}.
$$

The nonzero elements of these arrays are as follows:

$$
A_{ki} = K_{xx} \int_{\Omega_x} K_{xx} \frac{d\phi_i}{dx} \frac{d\phi_k}{dx} dx,
$$

$$
B_{ki} = \frac{1}{2} \int_{\Omega_x} \phi_i \phi_k \, dx,
$$

$$
F_k = \sum_{i=0}^{N_x} K_{xx} \left(\int_{\partial\Omega_x} \frac{d\phi_i}{dx} \phi_k dx \right) u_{ik}.
$$

When Dirichlet boundary conditions apply, we must modify the matrices $\mathbf{A}_{\ell\ell}$ by placing known information into the vectors $\mathbf{F}_\ell$. Equation (4.4-33) constitutes the **x-direction sweep**.

To illustrate the procedure for accommodating the time derivative, it is convenient to write Equation (4.4-33) in global matrix form as follows:

$$(4.4\text{-}34a) \qquad \mathbf{A}_x \mathbf{u} + \mathbf{B}_x \frac{d\mathbf{u}}{dt} = \mathbf{F}_x.$$

A development analogous to that presented for the x-direction sweep can now be formulated for the y-direction sweep. One simply follows the same steps outlined above, this time using Equation (4.4-29b) as the point of departure. The resulting set of equations will look like

$$(4.4\text{-}34b) \qquad \mathbf{A}_y \mathbf{u} + \mathbf{B}_y \frac{d\mathbf{u}}{dt} = \mathbf{F}_y.$$

We now employ Crank-Nicholson finite-difference approximations to the ordinary differential equations (4.3-34). If the time step is k, we get

$$(4.4\text{-}35a) \qquad \mathbf{A}_x \frac{1}{2}(\mathbf{u}^{n+1/2} + \mathbf{u}^n) + \mathbf{B}_x \frac{2}{k}(\mathbf{u}^{n+1/2} - \mathbf{u}^n)$$

$$= \frac{1}{2}(\mathbf{F}_x^{n+1/2} + \mathbf{F}_x^n)$$

and

$$(4.4\text{-}35b) \qquad \mathbf{A}_y \frac{1}{2}(\mathbf{u}^{n+1} - \mathbf{u}^{n+1/2}) + \mathbf{B}_y \frac{2}{k}(\mathbf{u}^{n+1} - \mathbf{u}^n)$$

$$= \frac{1}{2}(\mathbf{F}_y^{n+1} - \mathbf{F}_y^{n+1/2}).$$

Equations (4.4-35) constitute the final system of equations. They are solved sequentially, so that the vector $\mathbf{u}^{n+1/2}$ obtained from the x-direction sweep (4.4-35a) can be used to solve the y-direction sweep (4.4-35b).

Alternating-direction collocation method.

The collocation procedure is particularly attractive for the solution of nonlinear problems because, unlike the Galerkin finite-element method, it does not require any formal integration to generate coefficient matrices. When the method is applied in a standard manner, however, it requires a level of computational effort similar that associated with other numerical procedures, largely because of the cumbersome matrix structures that arise in more than one dimension. On the other hand, a very efficient alternating-direction formulation makes it possible to reduce certain problems to those involving the relatively convenient one-dimensional matrix structures. This approach is the topic of this subsection. As in the preceding subsection, we have two alternative strategies for develpoing the discrete equations: a matrix factoring approach and the LOD approach. Here, once again, we use the LOD procedure because it is algebraically more straightforward. We refer the reader to Celia and Pinder (1985) for a detailed discussion of the matrix factoring alternative.

Consider the two-dimensional advection-diffusion transport equation,

$$(4.4\text{-}36) \qquad \frac{\partial u}{\partial t} + \mathbf{v} \cdot \nabla u - K \nabla^2 u = 0,$$

where we assume, for simplicity, diffusion is isotropic and $\mathbf{v}$ and K are constant. For purposes of this subsection we consider discretizations involving rectangular elements only, refering the reader to Celia (1983) for a generalization.

We begin the development with the following LOD formulation:

$$(4.4\text{-}37a) \qquad \frac{1}{2}\frac{\partial u}{\partial t} + \mathcal{L}_x u = 0,$$

$$(4.4\text{-}37b) \qquad \frac{1}{2}\frac{\partial u}{\partial t} + \mathcal{L}_y u = 0,$$

where

$$\mathcal{L}_x(\cdot) \equiv v_x \frac{\partial}{\partial x}(\cdot) - K \frac{\partial^2}{\partial x^2}(\cdot),$$

and

$$\mathcal{L}_y(\cdot) \equiv v_y \frac{\partial}{\partial x}(\cdot) - K \frac{\partial^2}{\partial y^2}(\cdot).$$

Approximation of the time derivative using backward differences yields

$$(4.4\text{-}38a) \qquad \frac{u^{n+1/2} - u^n}{k} + \mathcal{L}_x u^{n+1/2} = 0,$$

$$(4.4\text{-}38b) \qquad \frac{u^{n+1} - u^{n+1/2}}{k} + \mathcal{L}_y u^{n+1/2} = 0,$$

centered at time level $t = (n + \frac{1}{2})k$. Substitution of Equation (4.4-38b) into (4.4-38a) gives the algebraic form of the overall time stepping scheme:

$$(4.4\text{-}39) \qquad \frac{u^{n+1} - u^n}{k} + \mathcal{L}_x u^{n+1} + \mathcal{L}_y u^{n+1} = -k \mathcal{L}_x \mathcal{L}_y u^{n+1}.$$

If we neglect the right side of Equation (4.4-39), we commit an error of magnitude $\mathcal{O}(k)$. As Celia (1983) explains, one can make a suitable correction for this error or use an iterative procedure to eliminate it. For now, we shall simply neglect it.

Because we are considering a PDE that is second-order in space, it is appropriate in a finite-element collocation formulation to use continuously differentiable Hermite polynomials as basis functions. Building on the development in Section 2.14, let us take as our trial function

$$(4.4\text{-}40) \quad \hat{u}(x,y,t) = \sum_{i=0}^{N_x} \sum_{j=0}^{N_y} \Big[u_{ij}(t) h_i^0(x) h_j^0(y) + u_{ij}^{(x)}(t) h_i^1(x) h_j^0(y)$$
$$+ u_{ij}^{(y)}(t) h_i^0(x) h_j^1(y) + u_{ij}^{(xy)} h_i^1(x) h_j^1(y) \Big] .$$

The functions h_i^0 and h_i^1 are the piecewise cubic Hermite polynomials on a uniform spatial grid of mesh h having N_x and N_y elements in the x- and y-directions, respectively.

In practice, we find it advantageous to introduce the local coordinate transformation

$$(4.4\text{-}41) \qquad\qquad \xi = 1 - 2\left(\frac{x_{i+1} - x}{h} \right) .$$

This transformation maps the x-interval $[x_i, x_{i+1}]$ to the ξ-interval $[-1, 1]$. In the ξ coordinate system, the cubic Hermites in the x-direction transform to the functions

$$(4.4 - 42a) \qquad \hat{h}_{-1}^0(\xi) = \frac{1}{4}(\xi^3 - 3\xi + 2),$$

$$(4.4 - 42b) \qquad \hat{h}_1^0(\xi) = \frac{1}{4}(\xi^3 - 3\xi - 2),$$

$$(4.4 - 42c) \qquad \hat{h}_{-1}^1(\xi) = \left(\frac{h}{2}\right)\left(\frac{1}{4}\right)(\xi^3 - \xi^2 - \xi + 1),$$

$$(4.4 - 42d) \qquad \hat{h}_1^1(\xi) = \left(\frac{h}{2}\right)\left(\frac{1}{4}\right)(\xi^3 + \xi^2 - \xi - 1).$$

We use a similar transformation to local η-coordinates in the y-direction. To employ these functions in the development that follows, it is necessary to use the chain rule to express x- and y-derivatives in the (ξ, η)-coordinate system. This procedure is analogous to that presented in Section 2.5 as part of our discussion of isoparametric transformations. For the remainder of this discussion, we shall continue to use the untransformed Hermite functions $h_i^j(x)$.

To construct an alternating-direction formulation, one first substitutes the trial function (4.4-40) into Equation (4.4-38a). When the terms in the resulting expression are grouped and rearranged, one obtains

$$
\sum_{j=0}^{N_y}\left\{h_j^0 \sum_{i=0}^{N_x}\left[u_{ij}^{n+1/2}\left(h_i^0 + k\mathcal{L}_x h_i^0\right) + (u_{ij}^{(x)})^{n+1/2}\left(h_i^1 + k\mathcal{L}_x h_i^1\right)\right]\right.
$$

$$
\left. + h_j^1 \sum_{i=0}^{N_x}\left[(u_{ij}^{(y)})^{n+1/2}\left(h_i^0 + k\mathcal{L}_x h_i^0\right) + (u_{ij}^{(xy)})^{n+1/2}\left(h_i^1 + k\mathcal{L}_x h_i^1\right)\right]\right\}
$$

$$
= \sum_{j=0}^{N_y}\left\{h_j^0 \sum_{i=0}^{N_x}\left[u_{ij}^n h_i^0 + (u_{ij}^{(x)})^n h_i^1\right] + h_j^1 \sum_{i=0}^{N_x}\left[(u_{ij}^{(y)})^n h_i^0 + (u_{ij}^{(xy)})^n h_i^1\right]\right\} .
$$

Inasmuch as the functions h_j^0 and h_j^1 are independent functions of y, and because the bracketed quantities are functions only of x, this last equation implies that

(4.4-43)
$$
\sum_{i=0}^{N_x}\left[u_{ij}^{n+1/2}\left(h_i^0 + k\mathcal{L}_x h_i^0\right) + (u_{ij}^{(x)})^{n+1/2}\left(h_i^1 + k\mathcal{L}_x h_i^1\right)\right]
$$

$$
= \sum_{i=0}^{N_x}\left[u_{ij}^n h_i^0 + (u_{ij}^{(x)})^n h_i^1\right] ,
$$

(4.4-44)
$$
\sum_{i=0}^{N_x}\left[(u_{ij}^{(y)})^{n+1/2}\left(h_i^0 + k\mathcal{L}_x h_i^0\right) + (u_{ij}^{(xy)})^{n+1/2}\left(h_i^1 + k\mathcal{L}_x h_i^1\right)\right]
$$

$$
= \sum_{i=0}^{N_x}\left[(u_{ij}^{(y)})^n h_i^0 + (u_{ij}^{(xy)})^n h_i^1\right] .
$$

These equations contain $4(N_x + 1)$ nodal parameters. We now discuss how to determine these values. To begin with, four equations will be generated by boundary conditions, as we shall illustrate shortly. To determine the remaining $4N_x$ parameters, we shall collocate Equations (4.4-43) and (4.4-44) at $4N_x$ interior points. These collocation points will correspond to the two-point Gauss points of numerical integration, as discussed in Section 2.15. In the local (ξ, η) coordinate system, these points are readily found from Table 2-7. As a conceptual aid, Figure 4-19 associates one collocation point with each undetermined parameter, a correspondence we shall use shortly.

Let us now write Equations (4.4-43) and (4.4-44) for each row of nodes, evaluating the terms in Equation (4.4-44) at the collocation points above

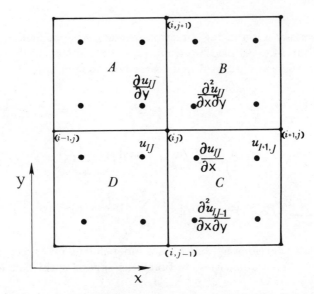

Figure 4-19. Diagrammatic representation of a finite-element collocation grid illustrating the nodes at the element corners and collocation points in the element interiors. The nodes carry the subscript (i,j), and the collocation points carry the subscripts (I,J). The elements are identified by the letters $(A-D)$.

the nodal row and those in Equation (4.4-43) at the collocation points below the nodal row. If we define the operator $\mathcal{A}_x(\cdot) \equiv (1 + k\mathcal{L}_x)(\cdot)$, we have for the x-direction sweep two sets of equations. The first set is as follows:

$$(4.4\text{-}45\text{a}) \qquad \sum_{i=0}^{N_x}\left[u_{ij}^{n+1/2}\mathcal{A}_x h_i^0 + (u_{ij}^{(x)})^{n+1/2}\mathcal{A}_x h_i^1 \right.$$
$$\left. - u_{ij}^n h_i^0 - (u_{ij}^{(x)})^n h_i^1\right]_{(x_\alpha,y_\alpha)} = 0.$$

Here (x_α,y_α) ranges over the collocation points assigned to nodal parameters of the form u_{IJ} and $u_{IJ}^{(x)}$ in Figure 4-19. The second set of equations is as follows:

$$(4.4\text{-}45\text{b}) \qquad \sum_{i=0}^{N_x}\left[(u_{ij}^{(y)})^{n+1/2}\mathcal{A}_x h_i^0 + (u_{ij}^{(xy)})^{n+1/2}\mathcal{A}_x h_i^1 \right.$$
$$\left. - (u_{ij}^{(y)})^n h_i^0 - (u_{ij}^{(xy)})^n h_i^1\right]_{(x_\alpha,y_\alpha)} = 0.$$

In this case, (x_α, y_α) is shorthand for the collocation points assigned to nodal parameters of the form $u_{IJ}^{(y)}$ and $u_{IJ}^{(xy)}$ in Figure 4-19. An analogous collection of equations arises for the y-direction sweep.

Before we can solve the system of equations (4.4-45) for each row, it is necessary to impose boundary conditions. Each equation of the form (4.4-45) contains certain known terms arising from forcing functions and nonzero boundary conditions. If we denote the vector containing these quantities as $\mathbf{F}_x$ and the counterpart associated with the y-direction equivalent of Equation (4.4-45) as $\mathbf{F}_y$, then one consistent representation of the boundary conditions (Celia and Pinder, 1985) has the matrix form

(4.4-46a) $\mathbf{F}_x = \mathbf{0}$,

(4.4-46b) $\mathbf{F}_y = [\mathbf{K}_x + k\mathbf{M}_x]^{-1}(k\mathbf{F} + \mathbf{B})$.

In these equations, $\mathbf{K}_x$ signifies the matrix associated with the approximation of spatial derivatives in the x-direction sweep, $\mathbf{M}_x$ is the matrix associated with the approximation of the time derivative in the x-direction sweep, $\mathbf{F}$ denotes the vector of forcing-function values, and $\mathbf{B}$ is the vector that accounts for nonzero boundary conditions. Note that when the coefficient matrices $\mathbf{K}_x$ and $\mathbf{M}_x$ are time-independent and the time step k is fixed, the matrices in Equations (4.4-46) need to be inverted only once.

Finite-element boundary conditions.

Up to now we have not explicitly shown how to treat boundary conditions in finite-element formulations of parabolic problems. For Dirichlet problems the treatment is straightforward: The prescribed boundary data permit us simply to assign values to certain nodal parameters, thus eliminating them from the list of unknowns. Similarly, with finite-element collocation using Hermite trial functions we can incorporate Neumann boundary data by assigning known values to nodal derivatives along the boundary, differentiating the given data tangentially along the boundary where appropriate. We shall devote the rest of this section to the treatment of Neumann conditions in a Galerkin setting.

When the governing equation is of the advection-diffusion transport type, application of Green's theorem to the Galerkin integrals will generate terms of the form

(4.4-47) $f_k = \sum_{e=1}^{E_B} \int_{(\partial\Omega)_e} [(\mathbf{A} \cdot \nabla \hat{u}) \cdot \mathbf{n} + B\hat{u}] \phi_k \, ds,$

where ds is the differential element of arclength, $\mathbf{A}$ and B are coefficients that can be functions of space and time, and $\mathbf{n}$ is the outward unit normal to one of the E_B boundary segments $(\partial\Omega)_e$. We can readily evaluate this term given Robin data prescribing the values of $(\mathbf{A} \cdot \nabla \hat{u}) \cdot \mathbf{n} + B\hat{u}$ on $\partial \Omega$. When B vanishes, the integrand of Equation (4.4-47) accommodates a Neumann condition prescribing boundary values of the flux $\mathbf{q} = (\mathbf{A} \cdot \nabla \hat{u}) \cdot \mathbf{n}$. Equation (4.4-47) can then be written as follows:

$$(4.4\text{-}48) \qquad f_k = \sum_{e=1}^{E_B} \int_{(\partial \Omega)_e} \mathbf{q}(s) \cdot \mathbf{n} \phi_k ds \equiv \sum_{e=1}^{E_B} I_k^e.$$

Consider first a piecewise linear boundary with the local ξ-coordinate system of Figure 4-20a. Equation (4.4-48) can be written for element e, using this local ξ coordinate system, as follows:

$$(4.4\text{-}49) \qquad I_k^e = \int_{-1}^{1} \mathbf{q}(s) \cdot \mathbf{n} \hat{\phi}_k(\xi) \frac{ds}{d\xi} d\xi.$$

Here $\hat{\phi}_{\hat{k}}$ is the transformed basis function, defined using local coordinates so that $\hat{\phi}_{\hat{k}}(\xi(s)) = \phi_k(s)$. For a piecewise linear element, $\hat{k} = \pm 1$, corresponding to the left and right end nodes at $\xi = \pm 1$. Using the side length ℓ_e explicitly to determine that $ds/d\xi = \ell_e/2$, we have

$$(4.4\text{-}50) \qquad I_k^e = \frac{\ell_e}{2} \int_{-1}^{1} \mathbf{q}(s) \cdot \mathbf{n} \hat{\phi}_{\hat{k}}(\xi) d\xi.$$

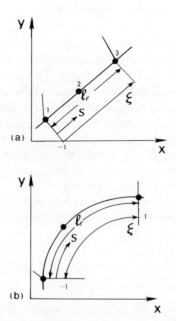

Figure 4-20. Quadratic element illustrating the dimensionless (ξ) coordinate system for a straight (a) and curved (b) element.

While we can approximate the normal flux $q_n \equiv \mathbf{q}(s) \cdot \mathbf{n}$ in many ways, it is convenient to use a representation $\hat{q}_n$ that is constant along each element side. Equation (4.4-50) then reduces to

$$(4.4\text{-}51) \qquad I_k^e = \frac{\ell_e q_n}{2} \int_{-1}^{1} \hat{\phi}_{\hat{k}}(\xi) ds = \frac{\ell_e q_n}{2}.$$

Thus, for our linear boundary segment, the specification of a Neumann boundary condition reduces to the allocation of half of the total flux across the element boundary to each node in the element. For the case of a quadratic boundary segment $(\partial\Omega)_e$, $\hat{k} = -1, 0, 1$, and we have

$$(4.4\text{-}52a) \qquad I_k^e = \frac{\ell_e}{2} q_n \int_{-1}^{1} \hat{\phi}_{-1}(\xi) d\xi = \frac{\ell_e q_n}{6},$$

when node k has local coordinate $\xi = -1$;

$$(4.4\text{-}52b) \qquad I_k^e = \frac{\ell_e}{2} q_n \int_{-1}^{1} \hat{\phi}_0(\xi) d\xi = \frac{2\ell_e q_n}{3},$$

when node k has local coordinate $\xi = 0$, and

$$(4.4\text{-}52c) \qquad I_k^e = \frac{\ell_e}{2} q_n \int_{-1}^{1} \hat{\phi}_1(\xi) d\xi = \frac{\ell_e q_n}{6},$$

when node k has local coordinate $\xi = 1$.

Occasionally it is advantageous to represent the normal flux q_n as a higher-degree polynomial over each element side. In this case the most convenient approach is to write, in local coordinates,

$$(4.4\text{-}53) \qquad q_n(\xi) = \sum_{\hat{k}} q_{n,\hat{k}} \hat{\phi}_{\hat{k}}(\xi),$$

where $q_{n,\hat{k}}$ denotes the flux assigned to the node whose local index is $\hat{k}$ and the sum ranges over all nodes along the element side defining the boundary. When each basis function $\hat{\phi}_{\hat{k}}(\xi)$ is piecewise linear, each boundary segment has two nodes having local coordinates $\xi = -1$ and $\xi = 1$, and we have

$$(4.4\text{-}54a) \quad I_k^e = \frac{\ell_e}{2} \sum_{\hat{k}=\pm 1} q_{n,\hat{k}} \int_{-1}^{1} \hat{\phi}_{\hat{k}}(\xi)\hat{\phi}_{-1}(\xi) d\xi = \frac{\ell_e}{2}\left(\frac{2q_{n,-1}}{3} + \frac{q_{n,1}}{3}\right),$$

for the node at $\xi = -1$, and

$$(4.4\text{-}54b) \quad I_k^e = \frac{\ell_e}{2} \sum_{\hat{k}=\pm 1} q_{n,\hat{k}} \int_{-1}^{1} \hat{\phi}_{\hat{k}}(\xi)\hat{\phi}_1(\xi) d\xi = \frac{\ell_e}{2}\left(\frac{2q_{n,-1}}{3} + \frac{q_{n,1}}{3}\right),$$

for the node at $\xi = 1$.

When the normal flux at the boundary is piecewise quadratic, we have three nodes on each boundary segment. The integral corresponding to the node whose local coordinate is $\xi = -1$ is as follows:

$$(4.4\text{-}55\text{a}) \quad I_k^e = \frac{\ell_e}{2} \sum_{\hat{k}=-1}^{1} q_{n,\hat{k}} \int_{-1}^{1} \hat{\phi}_{\hat{k}} \hat{\phi}_{-1} d\xi = \frac{\ell_e}{15} \left(2q_{n,-1} + q_{n,0} - \tfrac{1}{2} q_{n,1} \right) ;$$

the integral corresponding to the node at $\xi = 0$ is the following:

$$(4.4\text{-}55\text{b}) \quad I_k^e = \frac{\ell_e}{2} \sum_{\hat{k}=-1}^{1} q_{n,\hat{k}} \int_{-1}^{1} \hat{\phi}_{\hat{k}} \hat{\phi}_0 d\xi = \frac{\ell_e}{15} \left(q_{n,-1} + 8q_{n,0} + q_{n,1} \right) ,$$

and the integral at $\xi = 1$ is as follows:

$$(4.4\text{-}55\text{c}) \quad I_k^e = \frac{\ell_e}{2} \sum_{\hat{k}=-1}^{1} q_{n,\hat{k}} \int_{-1}^{1} \hat{\phi}_{\hat{k}} \hat{\phi}_1 d\xi = \frac{\ell_e}{15} \left(-\tfrac{1}{2} q_{n,-1} + q_{n,0} + 2q_{n,1} \right),$$

In the case of curved boundaries, such as those illustrated in Figure 4-19b, numerical integration is normally used. The general procedures outlined in Chapter Two can be applied, but the differential element ds of arclength must now be determined via the relationship

$$(4.4\text{-}56) \qquad ds = \frac{ds}{d\xi} d\xi = \sqrt{\left(\frac{dx}{d\xi}\right)^2 + \left(\frac{dy}{d\xi}\right)^2} \, d\xi.$$

Using this expression, the integral we seek becomes

$$(4.4\text{-}57) \qquad I_k^e = \int_{-1}^{1} q_n \phi_k(\xi) \sqrt{\left(\frac{dx}{d\xi}\right)^2 + \left(\frac{dy}{d\xi}\right)^2} \, d\xi,$$

which is amenable to numerical evaluation.

When the boundary integral in the approximating equation has the form (4.4-47) calling for Robin boundary data, then we can accommodate Robin data in essentially the same manner as in the Neumann case. Thus we compute the boundary term as follows:

$$(4.4\text{-}58) \qquad f_k = \sum_{e=1}^{E_B} \int_{(\partial\Omega)_e} q^* \phi_k ds ,$$

where q^* is the known boundary function . The procedures outlined above still are applicable in this case. However, when the boundary integral is of the form (4.4-48) calling for Neumann data the treatment of Robin data becomes more complicated. In this situation the Robin boundary condition

(4.4-59) $$q^*(s) = (\mathbf{A} \cdot \nabla \hat{u}) \cdot \mathbf{n} + B\hat{u}$$

gives rise to the integral term

(4.4-60) $$f_k = \sum_{e=1}^{E_B} \int_{(\partial\Omega)_e} (\mathbf{A} \cdot \nabla\hat{u}) \cdot \mathbf{n} \phi_k ds = \sum_{e=1}^{E_B} \int_{(\partial\Omega)_e} [q^*(s) - B\hat{u}]\phi_k ds .$$

Introduction of the definition of $\hat{u}$ yields

(4.4-61) $$f_k = \sum_{e=1}^{E_B} \int_{(\partial\Omega)_e} \left[q^*(s) - B \sum_{i=1}^{N} u_i \phi_i(x,y) \right] \phi_k(s) ds .$$

Thus the boundary integral term arising in this formulation generates a new coefficient associated with the unknown value u_i as well as known information. The coefficient multiplying u_i becomes part of the coefficient matrix, and the remainder of the information becomes part of the known vector on the right side of the matrix equation.

4.5. Problems for Chapter Four.

1. Consider a finite-difference discretization in one dimension such that (i) $x_2 - x_1 = \alpha(x_3 - x_2)$, where x_i are the coordinate locations of the finite difference nodes in space, and (ii) $t_2 - t_1 = \beta(t_3 - t_2)$, where t_i are the coordinate locations in time. Consider a three-point finite-difference approximation in space and a two-point finite-difference approximation in time of the equation

$$\frac{\partial u}{\partial t} - \frac{\partial^2 u}{\partial x^2} = 0.$$

Show that such an approximation is first-order accurate in space and time. Under what circumstances does the spatial approximation become second-order accurate?

2. Consider the alternating-direction algorithm in which the x-direction sweep is defined by

$$\frac{u_{ij}^* - u_{ij}^n}{k} = \frac{\delta_x(K_{xx}\delta_x u_{ij}^*)}{h_1^2} + \frac{\delta_y(K_{yy}\delta_y u_{ij}^n)}{h_2^2}$$

and the y sweep is defined by

$$\frac{u_{ij}^{n+1} - u_{ij}^n}{k} = \frac{\delta_x(K_{xx}\delta_x u_{ij}^*)}{h_1^2} + \frac{\delta_y(K_{yy}\delta_y u_{ij}^{n+1})}{h_2^2}$$

Demonstrate that this scheme is unconditionally stable.

3. Using a Taylor expansion, determine the order of accuracy of the alternating-direction formulation presented in Problem 2.

4. Consider the Galerkin finite-element approximation to the equation

$$\frac{\partial u}{\partial t} + v\frac{\partial u}{\partial x} - K\frac{\partial^2 u}{\partial x^2} = 0,$$

using piecewise quadratic basis functions in space. Show that the typical Galerkin equation centered at an interior mid-element node is equivalent to the following ordinary differential equation in the time domain:

$$\frac{1}{10}\left(\frac{du_{j-2}}{dt} + 8\frac{du_{j-1}}{dt} + \frac{du_j}{dt}\right)$$

$$+ v\left[\theta\frac{u_{j,k+1} - u_{j-2,k+1}}{2h} + (1-\theta)\frac{u_{j,k} - u_{j-2,k}}{2h}\right]$$

$$- K\left[\theta\frac{u_{j,k+1} - 2u_{j-k+1} + u_{j-2,k+1}}{h^2}\right.$$

$$\left. + (1-\theta)\frac{u_{j,k} - 2u_{j-1,k} + u_{j-2,k}}{h^2}\right] = 0.$$

5. Given the equation

$$\frac{\partial^2 u}{\partial x^2} + \frac{\partial^2 u}{\partial y^2} = 0$$

and a triangular linear finite-element with vertices $(0,0)$, $(1,1)$, and $(0,2)$, show that the element coefficient matrix generated by a Galerkin approximation to this equation is given by the following:

$$\begin{bmatrix} \left(\frac{1}{4}+\frac{1}{4}\right) & \left(-\frac{1}{2}+0\right) & \left(\frac{1}{4}-\frac{1}{4}\right) \\ \left(-\frac{1}{2}+0\right) & (1+0) & \left(-\frac{1}{2}+0\right) \\ \left(\frac{1}{4}-\frac{1}{4}\right) & \left(-\frac{1}{2}+0\right) & \left(\frac{1}{4}+\frac{1}{4}\right) \end{bmatrix}$$

(The material of Section 2.12 may be helpful in the solution to this problem.)

6. The finite-difference approximation to the equation

$$\frac{\partial u}{\partial t} + v\frac{\partial u}{\partial x} - K\frac{\partial^2 u}{\partial x^2} = 0$$

given by

$$\frac{u_{j,k+1} - u_{j,k}}{k} + v\frac{u_{j,k+1} - u_{j-1,k}}{h}$$

$$- K\frac{u_{j+1,k+1} - 2u_{j,k+1} + u_{j-1,k+1}}{h^2} = 0$$

is known to exhibit numerical smearing of sharp fronts. Show that, by accommodating the first term of the truncation error generated through the approximation of the first-order spatial derivative, one can reduce this numerical smearing. Hint: the procedure will result in a correction term to the difference approximation.

7. Assume the nodal spacing $x_3 - x_2 = \alpha(x_2 - x_1)$, where x_1, x_2 and x_3 are interior nodes (that is, none of these nodes is a boundary node). Show that the Galerkin finite-element method, formulated using piecewise linear basis functions to approximate the PDE

$$\frac{\partial u}{\partial t} - \frac{\partial^2 u}{\partial x^2} = 0,$$

is first-order accurate. Hint: consider carefully the interelement term arising from the application of integration by parts to the second-order spatial derivative.

4.6. References.

Aziz, K. and Settari, A., *Petroleum Reservoir Simulation*, London: Applied Science Publishers, Ltd., 1979.

Boris, J.P. and Book, D.L., "Flux-corrected transport I. SHASTA, a fluid transport algorithm that works," *J. Comput. Phys.* **11** (1973), 38-69.

Celia, M.A., "Collocation on Deformed Finite Elements and Alternating Direction Collocation Methods," Ph.D. Dissertation, Department of Civil Engineering, Princeton University, Princeton, New Jersey, 1983.

Celia, M. A. and Pinder, G. F., "An analysis of alternating-direction methods for parabolic equations," *Num. Meth. Partial Differential Equations,* *1* (1985), 57-70.

Chaudhari, N.M., "An improved numerical technique for solving multidimensional miscible displacement equations," *Soc. Pet. Eng. J.* (1971), 277-284.

Douglas, J. Jr. and Dupont, T., "Alternating-direction Galerkin methods on rectangles," in B. Hubbard, Ed., *Numerical Solution of Partial Differential Equations 2* (SYNSPADE 1970), New York: Academic Press, 1971, pp. 133-214.

D'Yakonov, E.G., "Difference schemes with split operators for multidimensional unsteady problems," *U.S.S.R. Comp. Math., 4:2,* 92-110, 1963.

Forsythe, G.E. and Wasow, W.R., *Finite-Difference Methods for Partial Differential Equations,* New York: Wiley, 1960.

Gray, W.G. and Pinder, G.F., "An analysis of the numerical solution of the transport equation," *Water Resour. Res., 12:3* (1976), 547-555.

Gray, W.G., and Pinder, G.F., "Galerkin approximation of the time derivative in the finite element analysis of groundwater flow," *Water Resour. Res., 10:4* (1974), 821-828.

Hughes, T.J.R., "A simple scheme for developing upwind finite elements," *Int. J. Num. Methods Eng., 12* (1978), 1359-1365.

Lax, P.D. and Richtmyer, R.D., "Survey of the stability of linear finite difference equations," *Comm. Pure Appl Math. 9* (1956), 267-291.

McKee, S. and Mitchell, A.R. "Alternating direction methods for parabolic equations in two space dimensions with a mixed derivative," *Comput. J., 13* (1970), 81-86.

Mitchell, A.R. and Fairweather, G., "Improved forms of the alternating direction methods of Douglas, Peaceman, and Rachford for solving parabolic and elliptic equations," *Numerische Mathematik, 6* (1964), 285-292.

Mitchell, A.R., and Griffiths, D.F., *The Finite Difference Method in Partial Differential Equations*, Chichester, U.K.: Wiley, 1980.

Morrow, R. and Cram, L.E, ."Flux-corrected transport and diffusion on a non-uniform mesh," *J. Comput. Phys., 57* (1985), 129-136.

Noye, J.,"Finite difference methods for partial differential equations," in J. Noye, ed., *Numerical Solution of Partial Differential Equations*, Amsterdam: North Holland, 1982.

Paolucci, S. and Chenoweth, D.R., "Stability of the explicit finite differenced transport equation," *Jour. Comp. Phys., 47* (1982), 489-496.

Payre, G., deBroissa, M., and Baxinet, J. "An upwind finite element method via numerical integration," *Int. J. Num. Methods Eng., 18* (1982), 381-396.

Pinder, G.F. and Gray, W.G., *Finite Element Simulation in Surface and Subsurface Hydrology*, New York: Academic Press, 1977.

Samarskii, A.A., "Economical difference schemes for parabolic equations with mixed derivatives," *Z. Vycisl. Mat. Mat. Fiz., 4* (1964), 753-759.

Yanenko, N.N., *The Method of Fractional Steps: The Solution of Problems of Mathematical Physics in Several Variables* (English translation by M. Holt), New York: Springer-Verlag, 1971.

Zauderer, E., *Partial Differential Equations of Applied Mathematics*, New York: Wiley, 1983.

CHAPTER FIVE
NONDISSIPATIVE SYSTEMS

5.1. Introduction.

Hyperbolic partial differential equations typically describe evolutionary phenomena in the absence of dissipative influences such as viscosity and diffusion. Generally, first-order hyperbolic equations model transport phenomena, while second-order equations or systems of first-order equations govern the dynamic behavior of such continuous mechanical systems as elastic solids or inviscid fluids. To motivate the theory developed in this chapter, we focus attention on several physical examples. These include the transport of a solute dissolved in a fluid, the motion of a compressible inviscid fluid, and the self-advected motion of a compressible viscous fluid governed by Burgers' equation.

Solute transport.

In Sections 1.5 and 4.1 we investigated the species mass balance for a substance dissolved in a moving fluid. If ρ^S is the concentration of solute, measured as mass per unit volume, then when the fluid flow is isochoric, $\nabla \cdot \mathbf{v} = 0$. In this case, the advection-diffusion transport equation (1.5-4) reduces to

$$(5.1\text{-}1) \qquad \frac{\partial \rho^S}{\partial t} + \mathbf{v} \cdot \nabla \rho^S = \nabla \cdot (K^S \nabla \rho^S),$$

where K^S denotes the diffusion coefficient and $\mathbf{v}$ is the barycentric velocity of the fluid. In flows where the advective term $\mathbf{v} \cdot \nabla \rho^S$ dominates the transport process, we frequently neglect the diffusion term, getting

$$(5.1\text{-}2) \qquad \frac{\partial \rho^S}{\partial t} + \mathbf{v} \cdot \nabla \rho^S = 0.$$

If the velocity $\mathbf{v}$ is known, we may regard Equation (5.1-2) as a first-order PDE for the concentration $\rho^S(\mathbf{x}, t)$. This is the **advective transport equation**.

In deriving Equation (1.5-4), we assumed that the solute undergoes no chemical reactions. Suppose instead that the solute is subject to a reaction such as biodegradation or nuclear decay. In this case we must include an exchange term r^S, getting

$$(5.1\text{-}3) \qquad \frac{\partial \rho^S}{\partial t} + \mathbf{v} \cdot \nabla \rho^S = r^S.$$

The right side of this equation represents the rate of increase in $\rho^S(\mathbf{x}, t)$ attributable to homogeneous chemical reactions. Frequently, one considers such reactions to proceed at rates proportional to the amount of solute present. In these cases the exchange term is a linear function of concentration, that is, $r^S = -r\rho^S$, where the case $r > 0$ corresponds to a first-order decay reaction. This assumption alters Equation (5.1-3) to read

$$(5.1\text{-}4) \qquad \frac{\partial \rho^S}{\partial t} + \mathbf{v} \cdot \nabla \rho^S + r\rho^S = 0.$$

One might call this equation the **advection-reaction** equation.

Inviscid fluid motion.

As another example, consider the motion of a Newtonian fluid. To get an equation governing the flow, we begin with Cauchy's first law, Equation (1.3-9). If body forces are absent, $\mathbf{b} = \mathbf{0}$, and we get

$$(5.1\text{-}5) \qquad \rho \frac{\partial \mathbf{v}}{\partial t} + \rho \mathbf{v} \cdot \nabla \mathbf{v} - \nabla \cdot \mathbf{t} = \mathbf{0}.$$

Now let us recall from Section 1.4 that, for a Newtonian viscous fluid, the stress tensor $\mathbf{t}$ depends linearly on the deformation rate $\mathbf{D}$ according to the constitutive law

$$(5.1\text{-}6) \qquad \mathbf{t} = -[p + \lambda \operatorname{tr}(\mathbf{D})]\mathbf{1} + 2\mu\mathbf{D},$$

where $\mathbf{1}$ denotes the unit tensor and

$$(5.1\text{-}7) \qquad \mathbf{D} = \tfrac{1}{2}[\nabla \mathbf{v} + (\nabla \mathbf{v})^{\mathsf{T}}].$$

In general, we must complement Equation (5.1-5) by the mass balance equation (1.3-7), which we rewrite as follows:

$$(5.1\text{-}8) \qquad \frac{1}{\rho}\frac{D\rho}{Dt} = -\nabla \cdot \mathbf{v}.$$

Let us consider two special cases. First, when the fluid is incompressible, the density is a constant, and Equation (5.1-8) implies

(5.1-9) $\text{tr}(\mathbf{D}) = \nabla \cdot \mathbf{v} = 0.$

Thus, equation (5.1-5) becomes

(5.1-10) $\dfrac{\partial \mathbf{v}}{\partial t} + \mathbf{v} \cdot \nabla \mathbf{v} + \dfrac{1}{\rho}\nabla p = \mu \nabla^2 \mathbf{v}.$

Equations (5.1-9) and (5.1-10) are the Navier-Stokes equations for incompressible fluids. Observe that Equation (5.1-10) is equivalent to Equation (1.4-10).

In the second case we suppose that the fluid is compressible, so we allow $\nabla \cdot \mathbf{v} \neq 0$. However, we assume that dissipative influences such as viscosity and heat conduction are negligble, so that in particular the fluid is inviscid and $\lambda = \mu = 0$. Thus the stress tensor reduces to $\mathbf{t} = -p\mathbf{1}$. Our balance laws, assuming body forces and heat supply are absent, are the following:

$$\text{(Mass)} \quad \frac{1}{\rho}\frac{D\rho}{Dt} = -\nabla \cdot \mathbf{v}$$

(5.1 − 11) $\text{(Momentum)} \quad \dfrac{\partial \mathbf{v}}{\partial t} + \mathbf{v} \cdot \nabla \mathbf{v} + \dfrac{1}{\rho}\nabla p = \mathbf{0}$

(5.1 − 12) $\text{(Energy)} \quad \rho\dfrac{DE}{Dt} + p\nabla \cdot \mathbf{v} = 0.$

We further assume that the fluid flows in a state of local thermodynamic equilibrium. This assumption implies (Eringen, 1980, Chapter Four) that there exists an equation of state

$$p = p(\rho, S),$$

where S is the fluid's entropy, and further that the Clausius-Duhem inequality (1.3-16) becomes an entropy *balance*, that is, an equation. Using our assumption that $\mathbf{t} = -p\mathbf{1}$ and $\mathbf{q} = 0$, we can rewrite this equation as follows:

$$\frac{DS}{Dt} - \frac{1}{T}\left(\frac{DE}{Dt} + \frac{p}{\rho}\nabla \cdot \mathbf{v}\right) = 0.$$

Notice that the term in parentheses vanishes by the energy balance (5.1-12), so we can conclude that $DS/Dt = 0$. This equation shows that inviscid compressible flows subject to an equation of state are **isentropic**; in other words, the entropy following a particle path remains constant.

Since the fluid obeys an equation of state $p = p(\rho, S)$ and S remains constant along particle paths, we can eliminate the pressure from the momentum balance. Let us define $a^2 = \partial p/\partial \rho$, which will be positive for any thermodynamically stable process. Then Equation (5.1-11) reduces to

$$(5.1\text{-}13) \qquad \frac{\partial \mathbf{v}}{\partial t} + \mathbf{v} \cdot \nabla \mathbf{v} + \frac{a^2}{\rho} \nabla \rho = 0.$$

As we shall discuss shortly, the variable $a = \sqrt{\partial p/\partial \rho}$ represents the **speed of sound** in the fluid.

Burgers' equation (self-advection).

For one-dimensional flows, the condition of incompressibility is too restrictive to yield PDEs of mathematical interest. To see why, suppose the fluid moves in the direction of the only spatial coordinate involved in the problem. Let x be that coordinate, and denote by u the nonvanishing component of velocity. Then Equation (5.1-9) implies $\partial u/\partial x = 0$, that is u is a function of t only. To avoid this rather trivial case, let us suppose that the specific volume $(1/\rho)$ can change without producing a change in pressure, and that the pressure is independent of position. Then Equation (5.1-10) reduces to

$$(5.1\text{-}14) \qquad \frac{\partial u}{\partial t} + u\frac{\partial u}{\partial x} - \mu\frac{\partial^2 u}{\partial x^2} = 0.$$

This PDE, called Burgers' equation, serves as a one-dimensional analog of the Navier-Stokes equation (1.4-10). It is a parabolic PDE. However, many commonly encountered fluids, such as air, have small viscosity, and when μ is small it is appropriate to consider Equation (5.1-14) to be of "mixed type". Indeed, as $\mu \to 0$ the solutions of Equation (5.1-14) behave like those of the hyperbolic equation

$$(5.1\text{-}15) \qquad \frac{\partial u}{\partial t} + u\frac{\partial u}{\partial x} = 0$$

in a sense that we shall make clearer in this chapter. Observe that Equation (5.1-15) resembles a one-dimensional version of the transport equation (5.1-3); however, it is nonlinear since the coefficient v in this case is just the unknown u.

The wave equation.

For one-dimensional, compressible, inviscid flows in which the nonvanishing component u of the velocity points in the x-direction, Equations (5.1-8) and (5.1-13) collapse to give

$$(5.1\text{-}16a) \qquad \frac{\partial \rho}{\partial t} + \frac{\partial}{\partial x}(\rho u) = 0,$$

$$(5.1\text{-}16\text{b}) \qquad \frac{\partial u}{\partial t} + u\frac{\partial u}{\partial x} + \frac{a^2}{\rho}\frac{\partial \rho}{\partial x} = 0.$$

These equations, though one-dimensional in space, are still nonlinear. Let us consider the special case when the velocity u and the density fluctuations, together with the derivatives of both quantities, are small. If we decompose $\rho = \rho_0 + \tilde{\rho}$, where ρ_0 is a uniform, stationary value and $\tilde{\rho} << \rho$ denotes the small fluctuations, then Equations (5.1-16) admit considerable simplification. Since ρ_0 is stationary and uniform, $\partial\rho_0/\partial t = \partial\rho_0/\partial x = 0$, and consequently $\partial\rho/\partial t = \partial\tilde{\rho}/\partial t$ and $\partial\rho/\partial x = \partial\tilde{\rho}/\partial x$. Thus Equations (5.1-16) yield

$$\frac{\partial \rho}{\partial t} + \rho\frac{\partial u}{\partial x} + u\frac{\partial \rho}{\partial x} = 0,$$

$$\frac{\partial u}{\partial t} + u\frac{\partial u}{\partial x} + \frac{a^2}{\rho}\frac{\partial \rho}{\partial x} = 0.$$

We shall retain all terms, such as $\rho_0\partial u/\partial x$, in which only one small factor appears. However, the terms $u\partial\rho/\partial x$ and $u\partial u/\partial x$ contain products of small quantities, and, these being much smaller than the other terms, we shall neglect them. Thus we obtain

$$(5.1\text{-}17\text{a}) \qquad \frac{\partial \rho}{\partial t} + \rho\frac{\partial u}{\partial x} = 0,$$

$$(5.1\text{-}17\text{b}) \qquad \rho\frac{\partial u}{\partial t} + a^2\frac{\partial \rho}{\partial x} = 0.$$

These equations are the linearized versions of our original Equations (5.1-16). Notice that, upon differentiating Equation (5.1-17a) with respect to t and Equation (5.1-17b) with respect to x, one can eliminate $\partial^2 u/\partial x\partial t$ to conclude that the density ρ satisfies the second-order equation

$$(5.1\text{-}18) \qquad \frac{\partial^2 \rho}{\partial t^2} - a^2\frac{\partial^2 \rho}{\partial x^2} = 0.$$

Similarly, by differentiating Equation (5.1-17a) with respect to x and Equation (5.1-17b) with respect to t, we find a PDE for the velocity u:

$$(5.1\text{-}19) \qquad \frac{\partial^2 u}{\partial t^2} - a^2\frac{\partial^2 u}{\partial x^2} = 0.$$

Both of these second-order equations are instances of the classical wave equation. The parameter a is the speed at which disturbances in the value of ρ or u propagate in the fluid, and for this reason we identify it as the speed of sound. Recall that a similar equation arose in Chapter One when we investigated the equations governing longitudinal displacements in an elastic solid. There, we derived Equation (1.4-13), which is another version of the standard one-dimensional wave equation.

5.2. Well Posed Problems.

For hyperbolic equations, most well posed problems are initial or initial-boundary value problems. We shall examine typical well posed problems for the two linear model equations discussed in the previous section. Our discussion is based on the notion of characteristic curves, which we shall define shortly. For the remainder of this section let us denote the unknown function in any PDE by u.

Transport.

Consider the transport equation (5.1-4) in one space dimension,

$$(5.2\text{-}1) \qquad \frac{\partial u}{\partial t} + v\frac{\partial u}{\partial x} = -ru.$$

We can construct a set of curves in the (x, t)-plane along which this PDE reduces to a set of ordinary differential equations as follows. Consider a general curve $(x(s), t(s))$, parametrized by a variable s such as arclength. According to the chain rule, the rate of change of the unknown $u(x, t)$ along such a curve is as follows:

$$(5.2\text{-}2) \qquad \frac{dt}{ds}\frac{\partial u}{\partial t} + \frac{dx}{ds}\frac{\partial u}{\partial x} = \frac{du}{ds}.$$

For this equation to be consistent with Equation (5.2-1), u must obey the first-order ordinary differential equation

$$(5.2\text{-}3) \qquad \frac{du}{ds} = -ru$$

along all curves $(x(s), t(s))$ satisfying the differential equations

$$(5.2\text{-}4) \qquad \frac{dt}{ds} = 1, \quad \frac{dx}{ds} = v,$$

that is, along the curves $dx/dt = v$ in the (x, t)-plane. Thus we have found a family of curves $(x(s), t(s))$ along which the original PDE reduces to an ordinary differential equation. The general solution to Equation (5.2-3) has the form

$$u(s) = U_o\text{exp}\int_o^s r(x(\sigma), t(\sigma))d\sigma,$$

where U_o is an arbitrary constant.

The curves $(x(s), t(s))$ defined by the differential equations (5.2-4) are called **characteristic curves**. They are curves along which the PDE (5.2-1) says nothing more than the chain rule. The ordinary differential equation for $u(x(s), t(s))$ along these curves, given in Equation (5.2-3), is called the **characteristic equation**. When the original PDE is sufficiently simple, determining the characteristic curves and solving the characteristic equation can form the basis for an exact solution scheme. To illustrate this, consider Equation (5.2-1), and assume the transport velocity v and the coefficient r to be constant. Suppose the solution $u(x, t)$ has initial values $u(x, 0) = u_0(x)$ along the x-axis. We can view these initial conditions as defining a curve in (x, t, u)-space, parametrized as

$$(5.2\text{-}5) \qquad t = 0, \quad x = \ell, \quad u = u_0(\ell),$$

where ℓ measures progress along the initial curve $t = 0$ in some fashion. From Equations (5.2-3) and (5.2-4), the characteristic curves and characteristic equations are

$$(5.2\text{-}6) \qquad \frac{dt}{ds} = 1, \quad \frac{dx}{ds} = v, \quad \frac{du}{ds} = -ru.$$

Solving the system (5.2-6) of ordinary differential equations subject to the initial conditions (5.2-5), we find

$$(5.2\text{-}7) \qquad t = s, \quad x = vs + \ell, \quad u = u_0(\ell)e^{-rs}.$$

Geometrically, Equations (5.2-7) indicate that, if the variable s parametrizes progress in time, then along the characteristic curves $x = vs + \ell$ the solution $u(x, t)$ decays exponentially. The family of solution curves $u = u_0(\ell)e^{-rs}$, indexed by values of the continuous initial parameter ℓ, forms a surface over the (x, t)-plane. This surface is precisely the graph of the solution $u(x, t)$ for the given initial data. Algebraically, we compute the function $u(x, t)$ by eliminating the parameter s from Equations (5.2-7) to get

$$(5.2\text{-}8) \qquad x = vt + \ell, \quad u = u_0(\ell)e^{-rt}.$$

Since we know $\ell = x - vt$, we can write

$$(5.2\text{-}9) \qquad u(x, t) = u_0(x - vt)e^{-rt}.$$

From a physical standpoint, Equation (5.2-9) states that the concentration $u(x, t)$ undergoes two types of modification as time increases. First, the initial concentration profile advects downstream with velocity v. Indeed, if the decay rate $r = 0$, then

$$(5.2\text{-}10) \qquad\qquad u(x, t) = u_0(x - vt),$$

implying that the concentration at time t is exactly the initial concentration that occurred at a point located upstream by a distance vt. Figure 5-1 illustrates the solution domain for this case. Second, the initial concentration undergoes exponential decay. In fact, if $v = 0$, no advection occurs, and Equation (5.2-1) is a pure ordinary differential equation describing decay via first-order reaction with solution $u = u_0 e^{-rt}$.

The problem we have just solved is a typical initial-value problem for a first-order PDE. Well posed initial-boundary-value problems also occur in applications. For example, in Equation (5.2-1) assume that we have an initial distribution $u_0(x), 0 < x < 1$, and that the concentration of solute entering the region for $t > 0$ is $u(0, t) = u_B(t)$. To solve this problem, we can extend the definition of the initial function u_0 via Equation (5.2-10). In the case when $r = 0$, this extension yields

$$(5.2\text{-}11) \qquad\qquad u_B(t) = u_0(-vt), \quad t > 0.$$

This fact leads us to define

$$(5.2\text{-}12) \qquad\qquad u_0(x) \equiv u_B(-x/v), \quad x < 0.$$

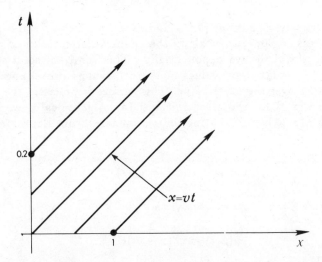

Figure 5-1. Solution domain and characteristic curves for the advection problem.

Accordingly, this initial-boundary value problem has a unique solution in the half of the space-time plane for which $x - vt \leq 1$. Indeed, the solution is given by Equation (5.2-10) in the region $0 \leq x - vt \leq 1$ and by

(5.2-13) $$u(x, t) = u_B(t - x/v)$$

in the region $x - vt < 0$.

For the parabolic heat equation, we have seen in Chapter Four that we need a second boundary condition (for example, at $x = 1$) to define a well posed problem. However, this is not the case for the transport equation. In fact, if we were to impose the additional boundary condition $u(1, t) = u_R(t)$ with $t > 0$, then a solution would exist only if this condition were consistent with the solution values arriving at $x = 1$ from upstream locations:

(5.2-14a) $$u_R(t) = u_0(1 - vt); \quad 0 < t \leq v^{-1}$$

and

(5.2-14b) $$u_R(t) = u_B(1 - vt); \quad v^{-1} < t.$$

Thus, it is not possible to prescribe arbitrary values of u at the downstream point $x = 1$.

On the other hand, if we perturb the transport equation (5.2-1) by adding a second-order term (a "diffusion term"), no matter how small, then the downstream boundary condition can be satisfied. Thus, instead of Equation (5.2-1), we consider the second-order PDE

(5.2-15) $$\frac{\partial u}{\partial t} - \epsilon \frac{\partial^2 u}{\partial x^2} + v \frac{\partial u}{\partial x} + ru = 0.$$

When $\epsilon > 0$, Equation (5.2-15) governs an advection-diffusion process. If the ratio ϵ/v is very small, the equation is advection-dominated. Since the order of the equation has formally changed as a result of adding the term $-\epsilon \partial^2 u/\partial x^2$, we can now accommodate additional boundary conditions. In this sense we see an abrupt, qualitative change in the mathematical properties of the system as $\epsilon \to 0$ continuously. Because of this qualitative change in behavior, the perturbation is said to be **singular**. As a rule, such singular perturbations give rise to **boundary layers**, which are thin regions of the (x, t)-plane where the presence of the perturbed term modifies the solution drastically. Away from the boundary layer, the solution remains almost unchanged from that of the pure first-order equation.

As an illustration, consider the steady-state version of Equation (5.2-15), in which $\partial u/\partial t = 0$. Assume $r = 0$, and impose the boundary conditions

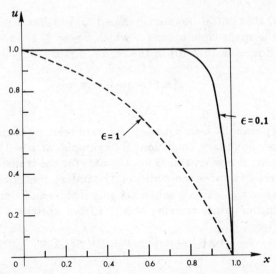

Figure 5-2. Solutions to the singularly perturbed advection equation for different values of the perturbation parameter ϵ.

(5.2-16) $u(0) = 1, \quad u(1) = 0.$

Set $v = 2$ and write $\epsilon = \delta^{-1}$. The solution is as follows:

(5.2-17) $u(x) = e^{\delta x}\dfrac{\sinh[\delta(1 - x)]}{\sinh(\delta)}.$

When $\epsilon = 0$, the steady state satisfies a first-order differential equation $(du/dx = 0)$, and only one of the boundary conditions can be satisfied. The results for different nonzero values of ϵ appear in Figure 5-2. As $\epsilon \to 0$, the function $u(x)$ approaches the solution of the transport equation (5.2-1) subject to the boundary condition $u(0) = 1$, except at a boundary layer located next to $x = 1$. However, the solution to Equation (5.2-1) satisfying $u(1) = 0$ is $u = 0$. For this example, then, the solution satisfying the boundary condition at the left endpoint of the interval is the only one which is physically meaningful.

Wave propagation.

In one space dimension, when the wave speed $a > 0$ is constant, the wave equation (5.1-19) can be brought into the canonical form

(5.2-18) $\dfrac{\partial^2 u}{\partial \vartheta^2} - \dfrac{\partial^2 u}{\partial x^2} = 0$

by the change of variables $at = \vartheta$. Furthermore, it is relatively easy to show that the change of variables

(5.2-19) $$\xi = x + \vartheta, \qquad \eta = x - \vartheta$$

yields an alternative canonical form of this equation, namely,

(5.2-20) $$\frac{\partial^2 u}{\partial \xi \partial \eta} = 0.$$

This last form serves as a basis for a solution to the wave equation. Notice that we can view Equation (5.2-20) as an ordinary differential equation for $\partial u/\partial \xi$. In this interpretation, Equation (5.2-20) asserts that $\partial u/\partial \xi$ does not depend on η, so we can write $\partial u/\partial \xi = \phi'(\xi)$ for some function ϕ. But this equation implies $\partial(u - \phi)/\partial \xi = 0$, which in turn shows that $u - \phi$ does not depend on ξ. Therefore $u(\xi) - \phi(\xi) = \psi(\eta)$ for some function ψ, and we see that

(5.2-21) $$u = \phi(\xi) + \psi(\eta),$$

or, in expanded form,

(5.2-22) $$u = \phi(x - \vartheta) + \psi(x + \vartheta).$$

Equation (5.2-22) yields the general solution of Equation (5.2-18) in terms of two functions ϕ and ψ that so far remain arbitrary but differentiable. More precisely, a function $u(x, t)$ is a solution of the wave equation (5.2-18) if and only if it can be written in the form (5.2-22).

When we substitute for ϑ in terms of the original time variable t, Equation (5.2-22) becomes

(5.2-23) $$u = \phi(x - at) + \psi(x + at).$$

Observe that $\phi(x - at)$ remains constant when $x - at$ is constant (see Figure 5-3). Thus $\phi(x - at)$ represents a waveform propagating with unchanged shape toward the right – that is, an **advancing** wave – with speed a. Similarly, $\psi(x + at)$ is a wave propagating towards the left – that is, a **receding** wave – with the same speed. We conclude from this discussion that any solution of the wave equation must be the superposition of two waves, one propagating in the positive x-direction and the other propagating in the negative x-direction. Both waves have the same speed a.

Let us apply these results to the propagation of sound in an inviscid but compressible fluid in a pipe. For the time being, assume that the pipe is infinitely long and that the fluid moves axially, that is, only in directions parallel to the axis of the pipe. From Equations (5.1-18) and (5.1-19) it follows that the velocity and density (as well as the pressure) of the fluid

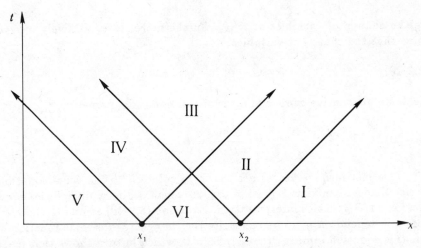

Figure 5-3. Solution domain for the second-order wave equation in one space dimension.

obey the wave equation. From physical considerations, it is natural to suspect that an initial-value problem prescribing the initial displacement u and the initial velocity $\partial u/\partial t$ will be well posed. This is indeed so, as we shall see. Such a problem is known as a **Cauchy problem** for the wave equation.

Let the initial conditions for the wave equation (5.1-19) be as follows:

$$(5.2\text{-}24) \qquad u(x,0) = p(x), \qquad \frac{\partial u}{\partial t}(x,0) = q(x), \qquad -\infty < x < \infty.$$

For convenience, let $Q(x)$ be any antiderivative of $q(x)$. Using Equation (5.2-23), we see that the conditions (5.2-24) imply

$$(5.2\text{-}25a) \qquad\qquad \psi(x) + \phi(x) = p(x)$$
$$(5.2\text{-}25b) \qquad\qquad \psi'(x) - \phi'(x) = a^{-1}Q'(x).$$

Differentiating Equation (5.2-25a) and solving the system (5.2-25) for ψ', we get

$$\psi'(x) = \frac{p'(x) + a^{-1}Q'(x)}{2}.$$

Integrating this equation and using the result to solve for ϕ in Equation (5.2-25a), we obtain

$$\psi(x) = \frac{p(x) + a^{-1}Q(x)}{2} + C_0$$
$$\phi(x) = \frac{p(x) - a^{-1}Q(x)}{2} - C_0,$$

where C_0 is the constant of integration. Writing Q as an explicit integral of q, we see that the solution of the Cauchy problem has the form

$$(5.2\text{-}26) \qquad u(x,t) = \tfrac{1}{2}\left[p(x+at) + p(x-at) + a^{-1}\int_{x-at}^{x+at} q(\tau)\,d\tau\right].$$

Let us interpret this result. Figure 5-3 shows the (x,t)-plane divided into six regions. Observe that

- regions I and V are not affected by the initial conditions (the "perturbation") in the open interval (x_1, x_2);

- regions II, III, IV, and VI are affected by the initial conditions in (x_1, x_2);

- however, regions II, III, and IV would be affected if the initial conditions were modified outside (x_1, x_2), while region VI would not be affected under these circumstances.

In view of these facts, we say that the **domain of influence** of the open interval (x_1, x_2) is the union of regions II, III, IV, and VI, while region VI is the **domain of determinacy** of (x_1, x_2). Finally, taking into account the form of the solution in Equation (5.2-26), given any point (x, t), we say that the closed interval $[x - at, x + at]$ is the **interval of dependence** of (x, t).

We now see that the initial values of u and $\partial u/\partial t$ along the pipe determine the solution $u(x, t)$ at every point of the (x, t)-plane only if they are prescribed along the whole real line $(-\infty < x < \infty)$, that is, when the pipe is infinitely long. When the pipe is finite, as will be the case in any actual physical problem, it is necessary to supplement the initial conditions with suitable boundary conditions to guarantee a well posed problem. Such conditions may be Dirichlet, Neumann, or Robin boundary conditions.

It is possible to construct the solution of an initial-boundary-value problem for the wave equation using an extension of the method of characteristics introduced above for the transport equation. For example, consider the wave equation (5.2-18) in the open x-interval $(-1, 1)$ subject to the initial conditions

$$(5.2\text{-}27) \qquad u(x,0) = p(x), \qquad \frac{\partial u}{\partial t}(x,0) = q(x), \qquad -1 < x < 1,$$

and Dirichlet boundary conditions

$$(5.2\text{-}28) \qquad u(-1,t) = u(1,t) = 0, \qquad t \geq 0.$$

The functions $p(x)$ and $q(x)$ are defined only for $x \in (-1, 1)$.

To solve this problem, observe that the solution (5.2-26) of the initial-value problem can be written as

(5.2-29) $$u(x,t) = \phi(t-x) + \psi(t+x),$$

where, for any σ along the real line, the advancing and receding components have the forms

(5.2-30) $$\phi(\sigma) = \frac{p(-\sigma) - Q(-\sigma)}{2}, \qquad \psi(\sigma) = \frac{p(\sigma) + Q(\sigma)}{2},$$

as discussed above. Let us define the initial functions $\phi_0(x)$ and $\psi_0(x)$ for $-1 \le x \le 1$ by Equations (5.2-30), setting $\sigma = x$. The boundary conditions (5.2-28) imply that

$$\phi(t-1) + \psi(t+1) = 0, \qquad \phi(t+1) + \psi(t-1) = 0.$$

Replacing $t - 1$ by σ in these equations, one gets

(5.2-31) $$\psi(\sigma + 2) = -\phi(\sigma), \qquad \phi(\sigma + 2) = -\psi(\sigma).$$

Repeated application of these identities yields a recurrence relation. Thus we see that the receding component ψ is periodic, obeying the equations

(5.2-32a) $$\psi(\sigma) = -\phi_0(\sigma - 4n + 2), \quad 4n - 3 \le \sigma \le 4n - 1,$$
$$n = 1, 2, \ldots,$$

(5.2-32b) $$\psi(\sigma) = -\psi_0(\sigma - 4n), \quad 4n - 1 \le \sigma \le 4n + 1,$$
$$n = 1, 2, \ldots.$$

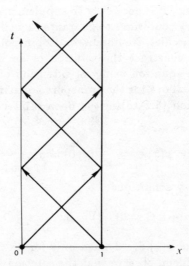

Figure 5-4. Reflecting characteristic curves for the second-order wave equation on a bounded spatial domain.

Similarly, the advancing component obeys the equations

(5.2-33a) $\phi(\sigma) = -\psi_0(\sigma - 4n + 2), \quad 4n - 3 \leq \sigma \leq 4n - 1,$
$n = 1, 2, \ldots,$

(5.2-33b) $\phi(\sigma) = -\phi_0(\sigma - 4n), \quad 4n - 3 \leq \sigma \leq 4n - 1,$
$n = 1, 2, \ldots.$

With these definitions of ϕ and ψ, Equation (5.2-29) again gives the solution of the initial-boundary value problem. One can interpret these results in terms of successive reflections, as illustrated in Figure 5-4.

5.3. General Properties of Nonlinear Equations.

There are many applications where the coefficients v and r in Equation (5.2-1) may be functions of the sought solution u. In these cases, the equation is no longer linear. However, so long as v and r depend at most on x, t, and u and not on derivatives of u, we say that the equation is **quasilinear**. In these cases the solutions to the PDE may still be fairly accessible, despite the nonlinearity. This is the case, for instance, with Burgers' equation. We now review the theory of first-order quasilinear equations, again basing our discussion on the method of characteristics.

First-order quasilinear equations.

The general form of a first-order quasilinear equation in the independent variables x and t is as follows:

(5.3-1) $$a(x, t, u)\frac{\partial u}{\partial x} + b(x, t, u)\frac{\partial u}{\partial t} + c(x, t, u) = 0.$$

Let $(x(s), t(s))$ be the coordinates of an arbitrary point on a smooth curve that is parametrized by s in the (x, t)-plane, as shown in Figure 5-5. By analogy with our discussion of Equation (5.2-4) for the transport equation, such a curve is a characteristic curve for the PDE (5.3-1) when it is a solution curve to the ordinary differential equations

(5.3-2) $$\frac{dx}{ds} = a, \qquad \frac{dt}{ds} = b.$$

Now, however, the functions a and b depend on the unknown solution $u(x, t)$.

By the chain rule, the rate of change of u along any curve $(x(s), t(s))$ is the following:

(5.3-3) $$\frac{du}{ds} = \frac{\partial u}{\partial x}\frac{dx}{ds} + \frac{\partial u}{\partial t}\frac{dt}{ds}.$$

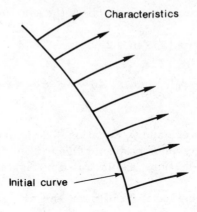

Figure 5-5. Initial and characteristic curves for a first-order quasilinear PDE.

When $u(x, t)$ is a solution of the quasilinear PDE (5.3-1) and the curve $(x(s), t(s))$ is a characteristic, we can substitute Equations (5.3-2) into Equation (5.3-3) to deduce that

$$(5.3\text{-}4) \qquad \frac{du}{ds} = a\frac{\partial u}{\partial x} + b\frac{\partial u}{\partial t} = -c.$$

Hence, any solution $u(x, t)$ to Equation (5.3-1) satisfies the system of ordinary differential equations

$$(5.3\text{-}5) \qquad \frac{dx}{ds} = a(x, t, u), \quad \frac{dt}{ds} = b(x, t, u), \quad \frac{du}{ds} = -c(x, t, u).$$

The solutions of the system (5.3-5) define curves in (x, t, u)-space. Under suitable regularity assumptions for the coefficients a, b and c, through any point (x_0, t_0, u_0) of this three-dimensional space there passes exactly one curve satisfying Equations (5.3-5). If $u(x, t)$ satisfies the quasilinear PDE (5.3-1), and if the point (x_0, t_0, u_0) lies on the surface $(x, t, u(x, t))$, then the corresponding solution curve to Equations (5.3-5) lies entirely on the surface $(x, t, u(x, t))$. In other words, the graph of any solution $u(x, t)$ to Equation (5.3-1) is composed of solution curves to Equation (5.3-5).

Let us now construct the solution surface by means of characteristics. The procedure we shall follow provides a methodology for computing the solution of initial value problems or Cauchy problems for Equation (5.3-1). We wish to find a surface $(x, t, u(x, t))$ passing through a prescribed initial curve in (x, t, u)-space, as shown in Figure 5-5. Let us parametrize this initial curve by

$$(5.3\text{-}6) \qquad x = x_0(\ell), \quad t = t_0(\ell), \quad u = u_0(\ell).$$

For every fixed value of the parameter ℓ, let

(5.3-7) $$x = x(s, \ell), \quad t = t(s, \ell), \quad u = U(s, \ell)$$

be the coordinates of the curve satisfying Equations (5.3-5) and passing through the point with coordinates given by Equation (5.3-6). Thus, if the parameter s has its origin on the initial curve (x_0, t_0, u_0), then

(5.3-8) $$x(0, \ell) = x_0(\ell), \quad t(0, \ell) = t_0(\ell), \quad U(0, \ell) = u_0(\ell).$$

Assume that, in some region of the (x, t)-plane containing the projection $(x_0(\ell), t_0(\ell))$ of the initial curve, the first two equations in (5.3-7) can be inverted uniquely. This allows us to solve for the parameters s and ℓ in terms of x and t, giving functional relationships

(5.3-9) $$s = s(x, t), \quad \ell = \ell(x, t).$$

Then, knowing $U(x, \ell)$, we can compute the solution u as follows:

(5.3-10) $$u(x, t) = U(s(x, t), \ell(x, t)).$$

It then follows from the chain rule and Equations (5.3-5) that

(5.3-11) $$a\frac{\partial u}{\partial t} + b\frac{\partial u}{\partial t} = \frac{\partial U}{\partial s} = -c.$$

This construction hinges on our ability to invert the transformation $(x, t) = (x(s, \ell), t(s, \ell))$. By the inverse function theorem (Williamson et al., 1972, Chapter Four), we can do this in some neighborhood of the projection $(x_0(\ell), t_0(\ell))$ of the initial curve (5.3-6) in the (x, t)-plane, provided the Jacobian determinant of the transformation does not vanish there. This requires

$$\det \begin{bmatrix} \partial x/\partial s & \partial x/\partial \ell \\ \partial t/\partial s & \partial t/\partial \ell \end{bmatrix} = \frac{\partial x}{\partial s}\frac{\partial t}{\partial \ell} - \frac{\partial t}{\partial s}\frac{\partial x}{\partial \ell} \neq 0.$$

On the initial curve, however, $x = x_0(\ell)$ and $t = t_0(\ell)$, so $\partial x/\partial \ell = dx_0/d\ell$ and $\partial t/\partial \ell = dt_0/d\ell$. Also, since $\partial x/\partial s$ and $\partial t/\partial s$ are derivatives of x and t in the direction of the characteristic curves, $\partial x/\partial s = a$ and $\partial t/\partial s = b$. Thus the determinant condition reduces to

(5.3-12) $$a\frac{dt_0}{d\ell} - b\frac{dx_0}{d\ell} = \left(\frac{dx_0}{d\ell}, \frac{dt_0}{d\ell}\right) \cdot (-b, a) \neq 0.$$

Since $(-b, a)$ is a vector normal to the characteristic curves defined by the equations $(dx/ds, dt/ds) = (a, b)$, Equation (5.3-12) states that the vector $(dx_0/d\ell, dt_0/d\ell)$ tangent to the (x, t)-projection of the initial curve must have a nonzero component normal to any characteristic curve. In other words, the local existence and uniqueness of the construction yielding

Equation (5.3-10) are guaranteed only when the projection of the initial curve is nowhere tangent to a characteristic curve.

Even if the condition given in Equation (5.3-12) holds, the inverse function theorem guarantees the validity of our construction (5.3-10), and hence a unique solution $u(x, t)$, only in some neighborhood of the initial curve. In this sense, existence and uniqueness of solutions to the quasilinear equation (5.3-1) are indeed local properties. In general, we cannot guarantee that continuous solutions exist in the large, and we may be forced to admit discontinuous functions $u(x, t)$ to get any solutions at all.

This last observation, although mathematically problematic, corresponds with physical reality in many instances. Discontinuous solutions typically arise when the smoothing effects of dissipative influences are physically negligible. As an example, let us examine Equation (5.1-15), which we obtained by setting the viscosity coefficient $\mu = 0$ in Burgers' equation. The characteristic curves for this equation satisfy

$$(5.3\text{-}13) \qquad \frac{dx}{ds} = u, \quad \frac{dt}{ds} = 1, \quad \frac{du}{ds} = 0.$$

The curves are clearly straight lines in this case, since u is constant along them. Consider the Cauchy problem defined by the following parametrized initial data:

$$(5.3\text{-}14) \qquad x = \ell, \quad t = 0, \quad u = u_0(\ell).$$

Solving the ordinary differential equations (5.3-13) subject to these initial data yields

$$(5.3\text{-}15) \qquad x(\ell, s) = \ell + u_0(\ell)s, \quad t(\ell, s) = s,$$

and

$$(5.3\text{-}16) \qquad u(x, t) = u_0(\ell(x, t)),$$

provided it is possible to invert the system (5.3-15) to get $\ell(x, t)$.

As an illustration, let

$$(5.3\text{-}17) \qquad u_0(\ell) = \begin{cases} -\alpha, & \ell < -1 \\ \alpha\ell, & -1 \le \ell \le 1 \\ \alpha, & \ell > 1, \end{cases}$$

where α is a constant. We wish to invert Equations (5.3-15) to get the functions $\ell(x, t)$ and $s(x, t)$. When $\alpha \ge 0$, the inverse is well defined for every $t \ge 0$. In this case, the upper half of the (x, t)-plane can be divided into three regions as shown in Figure 5-6a. The inverse functions are

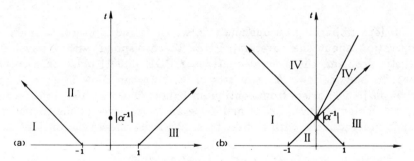

Figure 5-6. Solution domain for the inviscid Burgers' equation.

$$(5.3\text{-}18) \quad s(x,t) = t; \quad \ell(x,t) = \begin{cases} x + \alpha t, & (x,t) \quad \text{in Region I,} \\ x/(1 + \alpha t), & (x,t) \quad \text{in Region II,} \\ x - \alpha t, & (x,t) \quad \text{in Region III,} \end{cases}$$

where the regions are given by

$$(5.3\text{-}19) \qquad \begin{array}{ll} \text{I}: & x < -1 - \alpha t, \\ \text{II}: & -1 - \alpha t < x < 1 + \alpha t, \\ \text{III}: & 1 + \alpha t < x. \end{array}$$

If $\alpha < 0$ there is still a unique inverse function $\ell(x,t)$ in Regions I, II, and III of Figure 5-6b. In fact, the explicit definition of $\ell(x,t)$ is again given by Equation (5.3-18). However, in this case Regions I, II, and III do not cover the entire upper half of the (x,t)-plane. Indeed, Region IV of that figure is not covered.

Shocks.

As the previous example illustrates, a nonlinear hyperbolic PDE may not possess a continuous solution in the large. To model the behavior of actual physical systems in the absence of continuous solutions it is necessary to admit discontinuous solutions. By analogy with the discontinuities that arise in high-speed gas dynamics, jumps in the solutions to hyperbolic PDEs are called **shocks**.

To illustrate this phenomenon, let us consider the viscous and inviscid forms of Burgers' equation, Equations (5.1-14) and (5.1-15). The hyperbolic PDE in this case is Equation (5.1-15), which is an approximation of the parabolic PDE (5.1-14) in the limit of very small viscosity μ. Thus, to be physically realistic, a solution to Equation (5.1-15) must be the limit of the solutions to the viscous equation (5.1-14). The main objective of the methodology presented below is to supply short cuts for deriving such solutions without carrying out the actual limit process, which would require far more effort.

Before embarking on our analysis, we shall find it useful to make an observation about the governing PDE. To correspond with a physically realistic problem, the general quasilinear PDE (5.3-1) must be a special case of a balance law. As a matter of fact, we can show in general how Equation (5.3-1) arises from continuum balance laws without reference to any particular physics. Let A and B be antiderivatives of the coefficients a and b, respectively, with respect to u. We write these relationships as

$$A(x, t, u) = \int^u a(x, t, \xi)\, d\xi$$

and

$$B(x, t, u) = \int^u b(x, t, \xi)\, d\xi.$$

Then Equation (5.3-1) is equivalent to

(5.3-20) $$\frac{\partial B}{\partial t} + \frac{\partial}{\partial x}(Bv) + \frac{\partial}{\partial x}(A - Bv) - (\partial_1 A + \partial_2 B)u + c = 0,$$

for any differentiable function v. Here, $\partial B/\partial t$ denotes the derivative with respect to t of the function of $B(x, t, u(x, t))$, while we use the notation $\partial_i B$ to signify the partial derivative of B with respect to its i-th argument, all other arguments being held fixed. Equation (5.3-20) is an instance of the general balance law (1.3-5), with $\rho \Psi = B$, $\tau = Bv - A$, and $\rho g = (\partial_1 A + \partial_2 B)u - c$. (Notice that the generalized flux τ is a scalar, since we have only one space dimension.) When the arbitrary function v is taken to be zero, Equation (5.3-20) reduces to the following:

(5.3-21a) $$\frac{\partial B}{\partial t} + \frac{\partial A}{\partial x} - (\partial_1 A + \partial_2 B)u + c = 0.$$

By analogy with Equation (1.3-6), the corresponding jump condition is as follows:

(5.3-21b) $$\lfloor A - Bv_\Sigma \rfloor = 0.$$

Here, as in Chapter One, v_Σ denotes the shock velocity.

With this observation in mind, let us continue our investigation of Burgers' equation (5.1-15). We can write this in the conservation form of Equation (5.3-21a) by identifying

(5.3-22) $$\rho \Psi = u, \quad \tau = \tfrac{1}{2}u^2; \quad g = 0.$$

The associated jump condition is

(5.3-23) $$\lfloor \tfrac{1}{2}u^2 - uv_\Sigma \rfloor = 0,$$

that is,

$$2\lfloor u \rfloor v_\Sigma = \lfloor u^2 \rfloor.$$

Suppose we denote the values of u on either side of the shock Σ by u_- and u_+. If we call $\bar{u} = (u_+ + u_-)/2$, then it is easy to show that

(5.3-24) $$\lfloor u^2 \rfloor = 2\,\bar{u}\,\lfloor u \rfloor.$$

If we now assume that $\lfloor u \rfloor \neq 0$, we find that

(5.3-25) $$v_\Sigma = \bar{u}.$$

In other words, the shock velocity equals the average of the values of the fluid velocity u on either side of the shock. If the locus of the shock is $x = \Sigma(t)$, then Equation (5.3-25) is equivalent to the differential equation

(5.3-26) $$\Sigma'(t) = \bar{u}(\Sigma(t), t).$$

Now we are ready to complete the analysis of the problem given by the initial conditions (5.3-17) for the case $\alpha < 0$. The shock in the (x, t)-plane will start at the point $(0, |\alpha^{-1}|)$, as shown in Figure 5-6b. All of the characteristics that start in the closed interval $[-1,1]$ of the x-axis pass through the point $(0, |\alpha^{-1}|)$, and therefore through the shock, before entering Region IV. Thus, only characteristics that start in the intervals $(-\infty, -1)$ and $(1, +\infty)$ have to be considered in Region IV. Hence,

(5.3-27) $$\bar{u}(x, t) \equiv \frac{\alpha - \alpha}{2} = 0.$$

in Region IV. It follows from Equation (5.3-26) that $\Sigma'(t) \equiv 0$, and the shock is a standing shock, namely, that part of the t-axis for which $t \geq |\alpha^{-1}|$.

To get a moving shock, let us consider instead of Equation (5.3-17) the initial conditions

$$u_0(\ell) = \begin{cases} -\alpha & , & \ell < -1 \\ \alpha\ell & , & -1 \leq \ell \leq 0 \\ 0 & , & 0 < \ell, \end{cases}$$

where $\alpha < 0$. Then the region where the inverse function $\ell(x, t)$ is not unique is Region IV' in Figure 5-6b. From the Equation (5.3-26) giving the shock speed, we find

(5.3-28) $$\Sigma'(t) = \frac{|\alpha|}{2}$$

as the equation for the shock. This yields a straight line $\Sigma(t)$ starting at the point $(0, |\alpha^{-1}|)$, as seen illustrated in Figure 5-6b for the case $|\alpha| = 1$.

These examples can give insight into some of the questions regarding well-posedness of first-order hyperbolic PDEs. Consider the problem of finding the steady state of the inviscid Burgers' equation (5.1-15) in the interval $-1 < x < 1$, subject to the boundary conditions $u(-1) = 1$ and $u(1) = -1$. By setting $\partial u / \partial t = 0$ in Equation (5.1-15), we see that the steady state must satisfy $\partial u / \partial x = 0$. Thus u must be a constant, and such a function cannot satisfy the boundary conditions. However, if we admit discontinuous solutions we can allow u to be *piecewise* constant, thus making it possible to satisfy the boundary conditions. However, once we admit discontinuous solutions the boundary conditions no longer suffice to guarantee uniqueness. Indeed, for a time-independent solution the shock speed $v_\Sigma = 0$, and Equation (5.3-25) therefore reduces to $\bar{u} = 0$. There are many time-independent functions that satisfy the boundary conditions and this jump condition. Indeed, for every $x_\Sigma \in (-1, 1)$, the function

$$(5.3\text{-}29) \qquad u(x) = \begin{cases} 1, & -1 \le x < x_\Sigma \\ -1, & x_\Sigma \le x \le 1 \end{cases}$$

is a mathematically acceptable solution. Thus, there is a lack of uniqueness for this problem.

This nonuniqueness is connected with the fact that the steady state depends on the initial conditions of the problem. For example, if the boundary conditions given above are supplemented by the initial conditions

$$(5.3\text{-}30) \qquad u(x, 0) = -x, \quad -1 \le x \le 1,$$

then the solution is essentially the steady shock depicted in Figure 5-6b. Indeed, we can decompose the region in the (x, t)-plane where $-1 \le x \le 1$, $t > 0$ into the three regions I, II, and III depicted in Figure 5-6b. The solution is then

$$(5.3\text{-}31) \qquad u(x, t) = \begin{cases} 1, & \text{in Region I} \\ x/(t-1), & \text{in Region II} \\ -1, & \text{in Region III.} \end{cases}$$

For the initial conditions (5.3-30), the steady state attained as $t \to \infty$ corresponds to the case $x_\Sigma = 0$ in Equation (5.3-29). For other initial conditions, the same boundary conditions will yield different values of x_Σ. We should note, however, that not all steady states are sensitive to the initial conditions. For example, if we impose the boundary conditions $u(-1) = 1$ and $u(1) = 0$, then the jump condition is $\bar{u} = \frac{1}{2}$. In this case the only admissible steady state is $u(x) = 1$ for $-1 \le x \le 1$, regardless of the initial conditions.

There is an important observation that must be made regarding uniqueness of the conservation form of a PDE. We wrote Burgers' equation in the conservation form (5.3-21a) by adopting the definitions in Equation (5.3-22). This choice is consistent with the balance or the conservation laws used to derive Burgers' equation in Section 5.1. Other choices of conservation form are also feasible, but they can be misleading because their associated jump conditions may be incorrect. For example, Burgers' equation can also be brought into the conservation form (5.3-21a) by adopting the definitions

$$(5.3\text{-}32) \qquad \rho\Psi = \tfrac{3}{2}u^2; \quad \tau = u^3; \quad g = 0$$

instead of those in Equation (5.3-22). If we were to use this form of the PDE, the jump condition (5.3-21b) would be the following

$$(5.3\text{-}33) \qquad 3\lfloor u^2 \rfloor v_\Sigma = 2\lfloor u^3 \rfloor.$$

To see that this equation differs from our previous jump condition, observe that Equation (5.3-24) and the relationship

$$(5.3\text{-}34) \qquad \lfloor u^3 \rfloor = \lfloor u \rfloor [\overline{u^2} + 2(\overline{u})^2],$$

taken together, imply

$$(5.3\text{-}35) \qquad 3\overline{u}v_\Sigma = \overline{u^2} + 2(\overline{u})^2$$

This condition on v_Σ contradicts that given in Equation (5.3-25). In other words, while the definitions (5.3-32) lead to a PDE that is mathematically equivalent to the inviscid Burgers' equation, the solutions consistent with these definitions will typically fail to be nondissipative limits to solutions of the momentum balance when shocks are present. Thus, to obtain physically meaningful solutions it is wise to use a form of the PDE that is directly derivable from the fundamental balance laws associated with the physics being modeled.

5.4. Finite-Difference Methods for Linear Problems.

Given a hyperbolic PDE, one can approximate it by finite differences in many different ways. The most direct approach would be to approximate each one of the derivatives occurring in the equation by one of the standard difference schemes discussed in Chapter Two. For example, in the transport equation, we could approximate $\partial u/\partial x$ by central, forward, or backward differences. However, not all of the algorithms generated in this way are efficient or even useful. To develop effective algorithms in a systematic fashion requires some analysis.

TABLE 5.1. Finite-difference forms for the advection equation.
(Source: Lapidus and Pinder, 1982)

	Finite-Difference Form	Symbolic Representation[a]	Stability	Explicit or Implicit
(5.4-3)	$\dfrac{u_{r+1,s} - u_{r-1,s}}{2k} + b\dfrac{u_{r,s+1} - u_{r,s-1}}{2h} = 0$	$C - C$	$\dfrac{bk}{h} < 1$	Explicit
(5.4-4)	$\dfrac{u_{r+1,s} - u_{r,s}}{k} + b\dfrac{u_{r,s+1} - u_{r,s-1}}{2h} = 0$	$F - C$	Unstable	Explicit
(5.4-5)	$\dfrac{u_{r+1,s} - u_{r-1,s}}{2k} + b\dfrac{u_{r,s} - u_{r,s-1}}{h} = 0$	$C - B$	Unstable	Explicit
(5.4-6)	$\dfrac{u_{r+1,s} - u_{r,s}}{k} + b\dfrac{u_{r,s} - u_{r,s-1}}{h} = 0$	$F - B$	$\dfrac{bk}{h} < 1$	Explicit

TABLE 5.1 (continued).

(5.4-7)	$\dfrac{u_{r+1,s} - u_{r-1,s}}{2k} + b\dfrac{u_{r+1,s+1} - u_{r+1,s-1}}{2h} = 0$	$C - C_{r+1}$	Stable	Implicit
(5.4-8)	$\dfrac{u_{r+1,s} - u_{r,s}}{k} + b\dfrac{u_{r+1,s+1} - u_{r+1,s-1}}{2h} = 0$	$F - C_{r+1}$	Stable	Implicit
(5.4-9)	$\dfrac{u_{r+1,s} - u_{r,s}}{k} + b\dfrac{u_{r+1,s} - u_{r+1,s-1}}{h} = 0$	$F - B_{r+1}$	Stable	Explicit / Implicit

[a]Symbols have the following meanings: C, centered difference; F, forward difference; B, backward difference. "F-C" means "forward-in-time, centered-in-space."

In particular, knowledge of a scheme's truncation error, consistency, and stability in their usual senses is not generally sufficient for the discretization of hyperbolic PDEs. Information about the ability of an algorithm to propagate waves in a physically realistic manner is also necessary. For this purpose Fourier analysis furnishes a valuable tool in studying numerical schemes for hyperbolic equations. Fourier analysis also offers the benefit of yielding a great deal of information with relatively little effort. Although rigorously applicable only to linear PDEs with constant coefficients and to approximations on a uniform grid, this methodology furnishes inferences that have a far wider range of relevance.

As a simple model equation, let us investigate the linear transport or advection equation

$$(5.4\text{-}1) \qquad\qquad \frac{\partial u}{\partial t} + b \frac{\partial u}{\partial x} = 0,$$

subject to the initial condition

$$(5.4\text{-}2) \qquad\qquad u(x,0) = u_0(x), \quad -\infty < x < \infty.$$

We assume $b > 0$ is constant. We might alternatively impose an initial condition restricted to a ray in the x-axis (for example, the set $0 \leq x < \infty$), but in that case the initial data would have to be supplemented by a boundary condition as discussed in Section 5.2. We shall introduce the methodology for Fourier analysis in connection with this simple example, leaving extensions to the wave equation for later discussion.

Difference algorithms.

Table 5.1 lists examples of standard finite-difference analogs for Equation (5.4-1). We number these approximations (5.4-3) through (5.4-9). For example, using central differences for both the time and space derivatives, one gets the difference analog

$$(5.4\text{-}3) \qquad \frac{u_{r+1,s} - u_{r-1,s}}{2k} + b \frac{u_{r,s+1} - u_{r,s-1}}{2h} = 0.$$

This approximation is frequently called the **leapfrog** formula since it hops from time level $n-1$ to time level $n+1$. The symbols h and k denote the meshes of the spatial and temporal grids, respectively, and we shall assume these grids to be uniform. The parameter

$$(5.4\text{-}10) \qquad\qquad \overline{C} = \frac{bk}{h}$$

is the **Courant number**; using it allows us to rewrite Equation (5.4-3) as

$$u_{r+1,s} = u_{r-1,s} - \overline{C}(u_{r,s+1} - u_{r,s-1})$$

The other analogs appearing in Table 5.1 arise similarly through the use of various combinations of central, forward and backward finite differences around node (r, s) for the first-order derivatives occurring in Equation (5.4-1).

Truncation error, consistency, and stability.

Let us denote the order of accuracy of a difference scheme by $\mathcal{O}(h^p + k^q)$. From the constructions of Equations (5.4-3) through (5.4-9) it is easy to establish their orders of accuracy using the approximation-theoretic approach described in Chapter Two. For example, the central-central formula used in Equation (5.4-3) is accurate to $\mathcal{O}(h^2 + k^2)$. To be consistent, we need only guarantee that both of the exponents p and q are greater than zero. Since this is true for all of the schemes presented in Table 5.1, each of them is consistent.

The Lax Equivalence Theorem (see Richtmyer and Morton, 1967, Chapter Three) states that, given a well posed initial value problem and a consistent finite-difference approximation to it, stability is a necessary and sufficient condition for convergence. For explicit algorithms, we shall show via Fourier analysis that the inequality

(5.4-11) $$\overline{C} = \frac{bk}{h} \leq 1$$

is a necessary condition for stability. Inequality (5.4-11) is called the **Courant-Friedrichs-Lewy (CFL) condition**.

To show how the condition (5.4-11) arises, we rely on the von Neumann method introduced in Section 2.9. Consider a finite-difference solution whose Fourier modes have the typical form $\xi^r e^{i\beta s h}$. Here ξ^r is the amplitude of the mode at time level r, and βh is the spatial wavelength of the mode at spatial node s. For a linear finite-difference scheme to be stable, it suffices to guarantee that no Fourier mode in the numerical solution will grow without bound as $t \to \infty$ or, equivalently, as $r \to \infty$. This criterion requires that $|\xi| \leq 1$ for every real β. For example, substituting a typical Fourier mode $\xi^r e^{i\beta s h}$ into Equation (5.4-3), we see that

$$\xi^{r+1} e^{i\beta s h} = \xi^{r-1} e^{i\beta s h} - \overline{C} \xi^r [e^{i\beta(s+1)h} - e^{i\beta(s-1)h}].$$

Dividing through by $\xi^r e^{i\beta s h}$ and identifying $\sin \beta h = (e^{i\beta h} - e^{-i\beta h})/2\hat{\imath}$, we obtain

$$\xi = \xi^{-1} - 2\hat{\imath}\overline{C} \sin \beta h.$$

Solving for the amplification factor ξ yields two roots:

$$(5.4\text{-}12) \qquad \xi = -i\overline{C}\sin\beta h \pm \sqrt{1 - \overline{C}^2\sin^2\beta h}.$$

When $\overline{C}^2\sin^2\beta h > 1$, the magnitude of one of the roots exceeds unity. However, if $\overline{C}^2\sin^2\beta h \leq 1$, then both roots satisfy the condition $|\xi| \leq 1$. Therefore, the central-central algorithm of Equation (5.4-3) is stable provided $\overline{C}^2\sin^2\beta h \leq 1$ for every real β. Since $|\sin\beta k| \leq 1$ for all real arguments βh, this condition is tantamount to $\overline{C} \leq 1$, which imposes a limit on the time step:

$$(5.4\text{-}13) \qquad k \leq \frac{h}{b}.$$

Thus in this case the CFL condition is not only necessary but also sufficient for stability.

It is worth remarking that the CFL condition is not so severe as stability conditions typical of parabolic schemes. There the time-step limitation is typically proportional to the *square* of the spatial mesh, as in the condition

$$(5.4\text{-}14) \qquad k \leq \alpha h^2,$$

where α is some positive constant. Therefore, while many people prefer unconditionally stable implicit schemes for parabolic equations, explicit schemes still find frequent use in the discretization of hyperbolic equations.

Next, consider the scheme (5.4-4), for which substitution of a typical Fourier mode and simplification leads to

$$(5.4\text{-}15) \qquad \xi = 1 - 2i\overline{C}\sin\beta h$$

Multiplying each side of the equation by its complex conjugate gives

$$(5.4\text{-}16) \qquad |\xi|^2 = 1 + 4\overline{C}^2\sin^2\beta h.$$

Given any $\overline{C} \neq 0$, there are values of the real parameter β for which $|\xi| > 1$, and hence the algorithm is unconditionally unstable. A similar argument shows that the scheme (5.4-5) is also unconditionally unstable. Substituting the typical Fourier mode into the difference equation yields

$$(5.4\text{-}17) \qquad -2\overline{C}(1 - e^{-i\beta h}) = \xi - \frac{1}{\xi}.$$

Choosing β so that $\beta h = \pi$, we find

$$(5.4\text{-}18) \qquad \xi^2 + 4\overline{C}\xi - 1 = 0,$$

and hence

$$\xi = -2\overline{C} \pm \sqrt{4\overline{C}^2 + 1}.$$

For this choice of β, then, $|\xi| > 1$ for one of the roots, no matter how we select $\overline{C}$. It follows that the scheme (5.4-5) is unstable.

Peter Lax proposed a way to overcome the instability of the forward-central scheme (5.4-4). His modification consists of replacing the term $u_{r,s}$ by the average of u at the spatial nodes $s+1$ and $s-1$. This tactic exploits the approximation

(5.4-19) $$u_{r,s} = \tfrac{1}{2}(u_{r,s+1} + u_{r,s-1}) + \mathcal{O}(k^2)$$

to produce the difference scheme

(5.4-20) $$u_{r+1,s} = \tfrac{1}{2}(1 - \overline{C})u_{r,s+1} + \tfrac{1}{2}(1 + \overline{C})u_{r,s-1}.$$

Using the fact that the approximation (5.4-5) is accurate to $\mathcal{O}(k + h^2)$ and taking into account Equation (5.4-19), we can see that Equation (5.4-20) is also accurate to $\mathcal{O}(k + h^2)$. However, the latter equation is conditionally stable. To demonstrate this, we note that Fourier analysis of Equation (5.4-20) reveals the amplification factor

$$\xi = \tfrac{1}{2}(e^{i\beta h} + e^{-i\beta h}) - \tfrac{1}{2}\overline{C}(e^{i\beta h} - e^{-i\beta h}) = \cos\beta h - i\overline{C}\sin\beta h.$$

Consequently,

$$|\xi|^2 = \cos^2\beta h + \overline{C}^2 \sin^2\beta h = 1 - (1 - \overline{C}^2)\sin^2\beta h.$$

From this identity it is clear that $|\xi| \leq 1$ for every real β when $\overline{C} \leq 1$. Thus we have shown that Lax's method is explicit, accurate to $\mathcal{O}(k + h^2)$, and conditionally stable.

By adopting a slightly different point of view, we can give an enlightening interpretation to the procedure used to derive the Lax method from Equation (5.4-4). In this procedure we start from the equation

(5.4-4) $$\frac{u_{r+1,s} - u_{r,s}}{k} + b\frac{u_{r,s+1} - u_{r,s-1}}{2h} = 0$$

and replace the term $u_{r,s}$ by $\tfrac{1}{2}(u_{r,s+1} + u_{r,s-1})$. This yields

(5.4-22) $$\frac{u_{r+1,s} - \tfrac{1}{2}(u_{r,s+1} + u_{r,s-1})}{k} + b\frac{u_{r,s+1} - u_{r,s-1}}{2h} = 0.$$

By rearranging terms, we can write this equation equivalently as a perturbed version of the forward-central scheme, namely,

(5.4-23) $$\frac{u_{r+1,s} - u_{r,s}}{k} + b\frac{u_{r,s+1} - u_{r,s-1}}{2h} = \frac{h^2}{2k}\left(\frac{u_{r,s+1} + u_{r,s-1} - 2u_{r,s}}{h^2}\right).$$

But this difference scheme is formally an analog of the PDE

$$(5.4\text{-}24) \qquad \frac{\partial u}{\partial t} + b\frac{\partial u}{\partial x} = \frac{h^2}{2k}\frac{\partial^2 u}{\partial x^2},$$

in which $\partial u/\partial x$ and $\partial^2 u/\partial x^2$ are approximated by central differences while $\partial u/\partial t$ is approximated by forward differences. Observe that the difference equation (5.4-23) is still consistent with the first-order PDE, since the coefficient of the difference analog to $\partial^2 u/\partial x^2$ vanishes as $h^2/2k \to 0$. However, the second-order error term on the right side of Equation (5.4-23) mimics a term that would be present if we had included the effects of diffusion or viscosity in the original PDE. Thus, by adding a vanishing dissipative term we have transformed an unstable approximation to a stable one.

This observation yields an immediate generalization. When constructing finite-difference approximations to Equation (5.4-1), one can just as well consider a parabolic extension

$$(5.4\text{-}25) \qquad \frac{\partial u}{\partial t} + b\frac{\partial u}{\partial x} = \alpha\frac{\partial^2 u}{\partial x^2},$$

where the dissipation coefficient or "viscosity" α depends on the grid meshes h and k. Clearly, if the resulting difference analog is to be consistent with Equation (5.4-1), it is necessary that $\alpha \to 0$ when both $h, k \to 0$.

Perhaps the most popular method for accommodating this **artificial viscosity** approach is the **Lax-Wendroff scheme**,

$$(5.4\text{-}26) \quad u_{r+1,s} = (1 - \overline{C}^2)u_{r,s} - \frac{\overline{C}}{2}C(1 - \overline{C})u_{r,s+1} + \frac{\overline{C}}{2}(1 + \overline{C})u_{r,s-1}.$$

This scheme arises when we take the viscosity parameter as $\alpha = b^2 k/2$ in Equation (5.4-25), that is, when we replace $h^2/2k$ by $b^2 k/2$ in Equation (5.4-23). We have as a result

$$(5.4\text{-}27) \quad \frac{u_{r+1,s} - u_{r,s}}{k} + \frac{b(u_{r,s+1} - u_{r,s-1})}{2h} = \frac{b^2 k}{2h^2}(u_{r,s+1} - 2u_{r,s} + u_{r,s-1}).$$

This scheme has the advantage that, unlike Equation (5.4-23), its artificial dissipative term has a magnitude that adapts to the strength of the advective term through the parameter b.

Dissipation and dispersion.

Classically, one uses Fourier analysis to analyze the stability of finite-difference schemes. In Chapter Four, we reviewed a somewhat newer application of Fourier analysis, namely, assessing the dissipation and dispersion

of Fourier wave components by numerical approximations. This application also proves fruitful in the setting of hyperbolic PDEs. To demonstrate
the methodology, we shall continue to use the advection equation (5.4-1),
subject to the initial conditions (5.4-2), as our model.

Let us begin by representing the initial data as a Fourier integral,

$$(5.4\text{-}28) \qquad u_0(x) = \int\limits_{-\infty}^{\infty} U_0(\beta)e^{i\beta x}\,d\beta.$$

This representation exhibits the initial values as a continuous superposition
of waves of the form $e^{i\beta x}$. The parameter β denotes the wave number of
the wave $e^{i\beta x}$, and U_0 is the amplitude with which this wave contributes
to u_0. When the initial conditions simply specify $u_0(x) = e^{i\beta x}$, it is easy
to see that the exact solution is $u(x,t) = e^{i\beta(x-bt)}$. Thus, given Equation
(5.4-28), the exact solution is just the continuous superposition

$$(5.4\text{-}29) \qquad u(x,t) = \int\limits_{-\infty}^{\infty} U_0(\beta)e^{i\beta(x-bt)}\,d\beta,$$

since the PDE (5.4-1) is linear. This observation suggests that the analysis
of difference schemes using the initial condition $u_0(x) = e^{i\beta x}$ will reveal
much about their wave-propagating properties.

Using our previous notation for Fourier modes, we can write a typical component of the numerical solution supplied by a given algorithm as
$\xi^r e^{i\beta sh}$. We can also write the amplification factor ξ in polar form as

$$(5.4\text{-}30) \qquad \xi = |\xi|e^{-i\beta b'k},$$

where ξ and b' depend on the wave number β. On the other hand, the
exact solution $u(x,t)$ at time level r has a different Fourier decomposition,
whose typical component we shall write as $\xi_e^r e^{i\beta sh}$, with

$$(5.4\text{-}31) \qquad \xi_e = e^{-i\beta bk}.$$

Since $|\xi_e| = 1$, the exact solution consists of Fourier modes propagating,
without amplification, with velocity b depending on the wave number β.
In contrast, the numerical solution is composed of waves propagating with
velocity b' and growing or decaying in amplitude by the factor $|\xi|$ at each
time level. The numerical and exact solution coincide if and only if $\xi/\xi_e =
1$. Let us define the ratio

$$(5.4\text{-}32) \qquad R = \frac{\xi}{\xi_e} = |\xi|e^{-i\beta(b'-b)k}.$$

TABLE 5.2. Dissipation or damping in the F-B and Lax-Wendroff approximations.

βh	FB [from (5.4-6)]			Lax-Wendroff [from (5.4-22)]		
	$\overline{C} = 0.05$	$\overline{C} = 0.25$	$\overline{C} = 0.50$	$\overline{C} = 0.05$	$\overline{C} = 0.25$	$\overline{C} = 0.50$
18°	0.9977	0.9908	0.9877	1.0000	0.9999	0.9998
30°	0.9936	0.9746	0.9659	1.0000	0.9995	0.9983
45°	0.9860	0.9435	0.9239	0.9999	0.9975	0.9919
60°	0.9760	0.9014	0.8660	0.9997	0.9926	0.9763
90°	0.9513	0.7906	0.7071	0.9988	0.9703	0.9014
120°	0.9260	0.6614	0.5000	0.9972	0.9318	0.7603
180°	0.9000	0.5000	0.0000	0.9950	0.8750	0.5000

Then $|R|$ measures the error in the amplification between time levels, while the number

$$(5.4\text{-}33) \qquad\qquad b' - b = -\frac{1}{\beta k}\tan^{-1}\frac{\Im(R)}{\Re(R)}$$

measures the error in the wave velocity. [Here, $\Re(z)$ and $\Im(z)$ denote the real and imaginary parts, respectively, of a complex number z.] Generally, when $|R| < 1$ we say that the scheme exhibits **numerical dissipation** or damping, while when $b' - b \neq 0$ we say that the scheme exhibits **numerical dispersion**. Notice that the degree of dissipation or dispersion may vary with the wave number.

As an illustration, by selecting a series of values for βh, Roberts and Weiss (1966) developed the data in Table 5.2 for the advection equation with Courant numbers $\overline{C} = 0.05, 0.25$ and 0.5. These data indicate the numerical dissipation or damping for one time step in the forward-backward (F-B) approximation (5.4-6) of Table 5.1 and in the Lax-Wendroff scheme reviewed earlier. It is obvious from the table that the Lax-Wendroff approximation introduces much less numerical damping than the F-B approximation. In fact, if we choose $\overline{C} = \frac{1}{2}$ then the F-B approximation damps the $\beta = 2\pi/6h$ component by about 50 percent in five time steps. Only when $\overline{C} \to 0$ does this damping decrease. However, there is a limit to how far one can reduce the Courant number while maintaining practical levels of computational efficiency.

5.5. Finite-Difference Methods for Nonlinear Problems.

Nonlinear hyperbolic PDEs present many of the difficulties that we have encountered in the linear case. However, the nonlinearities often make formal analysis more difficult, and as a result much of our reasoning is limited to analogies via linearized versions of the equations to be solved. In this section we shall discuss two classes of algorithms in common use for the solution of nonlinear hyperbolic equations. The first class comprises methods based on the direct approximation of primitive equations arising from balance laws. The second class, which we shall treat more briefly, consists of techniques derived from Riemann's methods for initial-value problems. For simplicity, we restrict attention to conservation laws in one space dimension.

Conservative schemes.

In view of the relationship between balance laws and the conservation form (5.3-21) discussed in Section 5.3, we shall write any system of conservation laws in the form

$$(5.5\text{-}1a) \qquad\qquad \frac{\partial \boldsymbol{\psi}}{\partial t} + \frac{\partial \mathbf{A}(\boldsymbol{\psi})}{\partial x} = \mathbf{0}$$

with jump conditions across a shock having velocity v_Σ given by

(5.5-1b) $$\lfloor \mathbf{A} - \boldsymbol{\psi} v_\Sigma \rceil = \mathbf{0}.$$

Here, $\mathbf{A}(\boldsymbol{\psi})$ is a known function of $\boldsymbol{\psi}$, which in turn may be a scalar unknown or a vector composed of M unknown component functions.

Equations like (5.5-1a) and (5.5-1b) are local versions of global balance laws having the form

$$\int_V \left[\frac{\partial \boldsymbol{\psi}}{\partial t} + \frac{\partial \mathbf{A}(\boldsymbol{\psi})}{\partial x} \right] dx = \mathbf{0}$$

over some region V. By extension, let us consider the integral form of the balance law taken over a region in both the space and time variables:

$$\int_\Omega \left[\frac{\partial \boldsymbol{\psi}}{\partial t} + \frac{\partial \mathbf{A}(\boldsymbol{\psi})}{\partial x} \right] dx\, dt = \mathbf{0}.$$

Here, Ω is a subset of the (x, t)-plane, as drawn in Figure 5-7. Applying the divergence theorem in this two-dimensional (space-time) setting, we can reduce the integral over Ω to obtain

(5.5-2) $$\int_{\partial \Omega} [\boldsymbol{\psi} n_t + \mathbf{A}(\boldsymbol{\psi}) n_x] dx\, dt = \mathbf{0}.$$

Here, n_t and n_x stand for the unit outward normal vectors in the time and space directions, respectively, that is, for ± 1, depending on which edge of Ω one is considering.

Now let us introduce a grid in the (x, t)-plane, as shown in Figure 5-8. We assume the grid is uniform in each coordinate direction, having spatial and temporal meshes h and k, respectively. If Ω is the single open rectangle $(x_{s-1/2}, x_{s+1/2}) \times (t_r, t_{r+1})$, then the integral form (5.5-2) yields

(5.5-3) $$\int_{(s-1/2)h}^{(s+1/2)h} \boldsymbol{\psi}(x, t_{r+1}) dx - \int_{(s-1/2)h}^{(s+1/2)h} \boldsymbol{\psi}(x, t_r) dx$$
$$+ \int_{rk}^{(r+1)k} \mathbf{A}(\boldsymbol{\psi}(x_{s+1/2}, t)) dt - \int_{rk}^{(r+1)k} \mathbf{A}(\boldsymbol{\psi}(x_{s-1/2}, t)) dt = \mathbf{0}.$$

This integrated version of the original PDE suggests a rather simple idea for transforming Equation (5.5-1) into a finite-difference equation. Indeed, we can write Equation (5.5-3) as

(5.5-4) $$\frac{\mathbf{U}_{r+1,s} - \mathbf{U}_{r,s}}{k} + \frac{\mathbf{F}_{r+1/2,s+1/2} - \mathbf{F}_{r+1/2,s-1/2}}{h} = \mathbf{0},$$

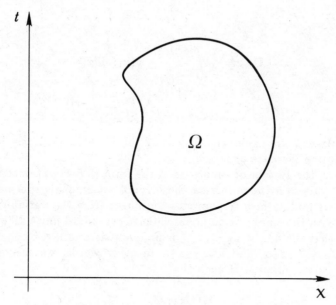

Figure 5-7. Subset Ω of the (x, t)-plane over which the integral form of Equation (5.5-1a) holds.

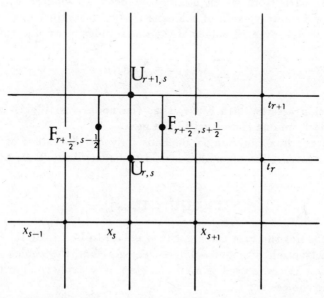

Figure 5-8. Finite-difference grid for the integrated scheme in Equation (5.5-3).

where

(5.5-5)
$$\mathbf{U}_{r,s} = \frac{1}{h} \int_{(s-1/2)h}^{(s+1/2)h} \boldsymbol{\psi}(x, t_r)\, dx;$$

$$\mathbf{F}_{r+1/2, s+1/2} = \frac{1}{k} \int_{rk}^{(r+1)k} \mathbf{A}(\boldsymbol{\psi}(x_{s+1/2}, t))\, dt.$$

The integrals $\mathbf{F}_{r,s}$ are called the **fluxes**. Equation (5.5-4) clearly has the form of a finite-difference equation.

However, the system of equations of the form (5.5-4) obtained by letting r and s vary over the numerical mesh is not yet uniquely specified. We still need to stipulate how to compute the fluxes from the variables $\mathbf{U}_{r,s}$. In keeping with the emphasis on nodal values in standard finite differences, we assume that the flux $\mathbf{F}_{r+1/2, s+1/2}$ is uniquely determined by the values of $\mathbf{U}_{r,s}, \mathbf{U}_{r,s+1}, \mathbf{U}_{r+1,s}$, and $\mathbf{U}_{r+1,s+1}$. In other words, we assume that there exists some function $\mathbf{H}$ such that

(5.5-6) $$\mathbf{F}_{r+1/2, s+1/2} = \mathbf{H}(\mathbf{U}_{r,s}, \mathbf{U}_{r,s+1}, \mathbf{U}_{r+1,s}, \mathbf{U}_{r+1,s+1})$$

for all nodal indices r and s. When the function $\mathbf{H}$ is specified, Equation (5.5-4) furnishes a well defined numerical scheme. Moreover, as a consequence of our integral formulation, this scheme reflects the conservation form of the PDE from which it arose. Indeed, so long as $\mathbf{F}_{r,s} \to \mathbf{0}$ as $s \to \pm\infty$, a discrete version of the conservation law still holds. Summing Equations (5.5-4) over all values of the nodal index s, we find that

(5.5-7)
$$\sum_s \mathbf{U}_{r+1,s} = \sum_s \mathbf{U}_{r,s}.$$

This equation asserts, in a sense, that the nodal variable $\mathbf{U}$ is globally "conserved" from one time level to the next.

When the flux function $\mathbf{H}$ depends only on the values of $\mathbf{U}$ at the earlier time level r, the algorithm is said to be **explicit**. Thus a scheme is explicit when

(5.5-8) $$\mathbf{H} = \mathbf{H}(\mathbf{U}_{r,s}, \mathbf{U}_{r,s+1}).$$

When this is not the case the algorithm is said to be **implicit**. Of course, in the construction of a finite-difference algorithm, the choice of the actual function $\mathbf{H}$ to be used is a crucial step. For explicit algorithms, two conceivable choices are

(5.5-9a) $$\mathbf{H}(\mathbf{U}_1, \mathbf{U}_2) = \tfrac{1}{2}[\mathbf{A}(\mathbf{U}_1) + \mathbf{A}(\mathbf{U}_2)]$$

and

$$(5.5\text{-}10a) \qquad \mathbf{H}(\mathbf{U}_1, \mathbf{U}_2) = \mathbf{A}\left(\frac{\mathbf{U}_1 + \mathbf{U}_2}{2}\right).$$

(We use the symbols $\mathbf{U}_1$, $\mathbf{U}_2$, etc., merely as dummy variables to indicate functional dependencies.) We shall eventually show that neither of these two schemes is computationally attractive. Corresponding to these choices are analogous ones for fully implicit schemes, namely,

$$(5.5\text{-}11) \qquad \mathbf{H}(\mathbf{U}_1, \mathbf{U}_2, \mathbf{U}_3, \mathbf{U}_4) = \tfrac{1}{2}[\mathbf{A}(\mathbf{U}_3) + \mathbf{A}(\mathbf{U}_4)]$$

and

$$(5.5\text{-}12) \qquad \mathbf{H}(\mathbf{U}_1, \mathbf{U}_2, \mathbf{U}_3, \mathbf{U}_4) = \mathbf{A}\left(\frac{\mathbf{U}_3 + \mathbf{U}_4}{2}\right).$$

Some remarks about the algorithmic aspects of these schemes are in order. Consider first the explicit case. Equation (5.5-4), written in the form

$$(5.5\text{-}13) \qquad \mathbf{U}_{r+1,s} = \mathbf{U}_{r,s} - \sigma(\mathbf{F}_{r+1/2,s+1/2} - \mathbf{F}_{r+1/2,s-1/2}),$$

$$(5.5\text{-}14) \qquad \sigma = \frac{k}{h},$$

supplies the values of $\mathbf{U}$ at the new time level $(r+1)$ in terms of quantities known at time level r. However, for implicit algorithms such as those obtained from Equations (5.5-11) and (5.5-12), Equations (5.5-13) constitute a system of coupled nonlinear algebraic equations for the values of $\mathbf{U}$ at the new time level. To solve such a system numerically, it is generally necessary to consider linearized versions of the equations or to use an iterative solution scheme such as Newton's method. For example, to linearize Equation (5.5-13), one might replace the definition (5.5-11) by a linear projection to the new time level. If $\mathbf{A}'$ denotes the derivative of $\mathbf{A}$, we would get

$$(5.5\text{-}15) \qquad \mathbf{H}(\mathbf{U}_1, \mathbf{U}_2, \mathbf{U}_3, \mathbf{U}_4) = \frac{\mathbf{A}(\mathbf{U}_1) + \mathbf{A}(\mathbf{U}_2)}{2}$$
$$+ \frac{\alpha}{2}[\mathbf{A}'(\mathbf{U}_1)(\mathbf{U}_3 - \mathbf{U}_1) + \mathbf{A}'(\mathbf{U}_2)(\mathbf{U}_4 - \mathbf{U}_2)].$$

Similarly, one would replace Equation (5.5-12) by

$$(5.5\text{-}16) \qquad \mathbf{H}(\mathbf{U}_1, \mathbf{U}_2, \mathbf{U}_3, \mathbf{U}_4) = \mathbf{A}\left(\frac{\mathbf{U}_1 + \mathbf{U}_2}{2}\right)$$
$$+ \frac{\alpha}{2}\mathbf{A}'\left(\frac{\mathbf{U}_1 + \mathbf{U}_2}{2}\right)(\mathbf{U}_3 + \mathbf{U}_4 - \mathbf{U}_1 - \mathbf{U}_2).$$

In these equations, $\alpha \geq 0$ is a parameter to be chosen. In the case of coupled conservation laws, when the unknown function ψ, and hence its integrated form $\mathbf{U}$, are vectors, $\mathbf{A}'$ denotes the Jacobian matrix of $\mathbf{A}$ as a function of ψ. We shall discuss a variety of linearizations based on Newton's method in Chapter Six.

The modified equation approach.

As in the case of the linear equations discussed in Section 5.4, the procedures just explained only offer hints about schemes that might be tried. The actual performance of any particular scheme must be subjected to further analysis before we accept it as feasible. One technique for this analysis is the **modified equation** approach. Although it is a powerful technique, we should remark that this approach is purely heuristic in nature. The basic idea is to analyze the differential equation that a solution of Equations (5.5-13) and (5.5-6) actually satisfies. For this purpose, given a function $\mathbf{U}(x,t)$ satisfying the discrete equation (5.5-13) on a uniform mesh, we compute $\partial \mathbf{U}/\partial t + \partial \mathbf{A}(\mathbf{U})/\partial x$. If $\psi = \mathbf{U}$ were the exact solution of Equation (5.5-1a), this quantity would vanish identically. However, Equation (5.5-13) only gives a discrete analog of Equation (5.5-1a), and typically there will be a nonzero residual. Therefore substitution of $\mathbf{U}$ into the exact differential operator yields the following **modified equation**:

$$(5.5\text{-}17) \qquad \frac{\partial \mathbf{U}}{\partial t} + \frac{\partial}{\partial x}\mathbf{A}(\mathbf{U}) = h^N \mathbf{Q}(\mathbf{U}) + \mathcal{O}\left(h^{N+1}\right).$$

Here $\mathbf{Q}$ stands for some nonlinear spatial differential operator. The value of the exponent N will depend on the particular scheme considered.

The computations required to obtain the modified equation (5.5-17) admit some simplification through the introduction of an auxiliary function

$$
\begin{aligned}
(5.5\text{-}18) \quad \mathbf{e}(\mathbf{U}, h, \sigma, x, t) &\equiv \{[\mathbf{U}(x, t+\sigma h) - \mathbf{U}(x,t)]/\sigma \\
&+ \mathbf{H}(\mathbf{U}(x,t), \mathbf{U}(x+h,t), \mathbf{U}(x,t+\sigma h), \mathbf{U}(x+h,t+\sigma h)) \\
&- \mathbf{H}(\mathbf{U}(x-h,t), \mathbf{U}(x,t), \mathbf{U}(x-h,t+\sigma h), \mathbf{U}(x,t+\sigma h))\}/h.
\end{aligned}
$$

Here, as before, $\sigma = k/h$ is the ratio of the time step to the spatial grid mesh. Enforcing (5.5-13) together with (5.5-6) is tantamount to requiring that $\mathbf{e}$ vanish at each one of the nodes in the space-time grid. However, away from the nodes, $\mathbf{e} \neq 0$.

Observe that increments of $\mathbf{U}$ and $\mathbf{H}$ occur in the expression for $\mathbf{e}$ given by Equation (5.5-18). We can evaluate these increments using a Taylor expansion around the point (x,t), truncating the series at the second-order term. We get

$$\mathbf{U}(x, t+\sigma h) - \mathbf{U}(x,t) = \frac{\partial \mathbf{U}}{\partial t}(x,t)\sigma h + \frac{1}{2}\frac{\partial^2 \mathbf{U}}{\partial t^2}(x,t)\sigma^2 h^2 + \mathcal{O}\left(h^3\right),$$

for the time increment in $\mathbf{U}$;

$$
\begin{aligned}
&\mathbf{H}(\mathbf{U}(x,t),\mathbf{U}(x+h,t),\mathbf{U}(x,t+\sigma h),\mathbf{U}(x+h,t+\sigma h)) \\
&\quad - \mathbf{H}(\mathbf{U}(x,t),\mathbf{U}(x,t),\mathbf{U}(x,t),\mathbf{U}(x,t)) \\
&= h\left[\frac{\partial \mathbf{U}}{\partial x}\partial_2\mathbf{H} + \sigma\frac{\partial \mathbf{U}}{\partial t}\partial_3\mathbf{H} + \left(\frac{\partial \mathbf{U}}{\partial x} + \sigma\frac{\partial \mathbf{U}}{\partial t}\right)\partial_4\mathbf{H}\right] \\
&\quad + \frac{h^2}{2}\left\{\frac{\partial^2\mathbf{U}}{\partial x^2}\partial_2\mathbf{H} + \frac{\partial \mathbf{U}}{\partial x}\left[\frac{\partial \mathbf{U}}{\partial x}\partial_{22}\mathbf{H} + \sigma\frac{\partial \mathbf{U}}{\partial t}\partial_{32}\mathbf{H}\right.\right. \\
&\quad + \left.\left(\frac{\partial \mathbf{U}}{\partial x} + \sigma\frac{\partial \mathbf{U}}{\partial t}\right)\partial_{42}\mathbf{H}\right] + \sigma^2\frac{\partial^2\mathbf{U}}{\partial t^2}\partial_3\mathbf{H} \\
&\quad + \sigma\frac{\partial \mathbf{U}}{\partial t}\left[\frac{\partial \mathbf{U}}{\partial x}\partial_{23}\mathbf{H} + \sigma\frac{\partial \mathbf{U}}{\partial t}\partial_{33}\mathbf{H} + \left(\frac{\partial \mathbf{U}}{\partial x} + \sigma\frac{\partial \mathbf{U}}{\partial t}\right)\partial_{43}\mathbf{H}\right] \\
&\quad + \left(\frac{\partial^2\mathbf{U}}{\partial x^2} + 2\sigma\frac{\partial^2\mathbf{U}}{\partial x\partial t} + \sigma^2\frac{\partial^2\mathbf{U}}{\partial t^2}\right)\partial_4\mathbf{H} + \left(\frac{\partial \mathbf{U}}{\partial x} + \sigma\frac{\partial \mathbf{U}}{\partial t}\right) \\
&\quad \cdot\left.\left[\frac{\partial \mathbf{U}}{\partial x}\partial_{24}\mathbf{H} + \sigma\frac{\partial \mathbf{U}}{\partial t}\partial_{34}\mathbf{H} + \left(\frac{\partial \mathbf{U}}{\partial x} + \sigma\frac{\partial \mathbf{U}}{\partial t}\right)\partial_{44}\mathbf{H}\right]\right\} + \mathcal{O}(h^3),
\end{aligned}
$$

for the time increment in $\mathbf{H}$, and

$$
\begin{aligned}
&\mathbf{H}(\mathbf{U}(x,t),\mathbf{U}(x,t),\mathbf{U}(x,t),\mathbf{U}(x,t)) \\
&\quad - \mathbf{H}(\mathbf{U}(x-h,t),\mathbf{U}(x,t),\mathbf{U}(x-h,t+\sigma h),\mathbf{U}(x,t+\sigma h)) \\
&= h\left[\frac{\partial \mathbf{U}}{\partial x}\partial_1\mathbf{H} + \left(\frac{\partial \mathbf{U}}{\partial x} - \sigma\frac{\partial \mathbf{U}}{\partial t}\right)\partial_3\mathbf{H} - \sigma\frac{\partial \mathbf{U}}{\partial t}\partial_4\mathbf{H}\right] \\
&\quad - \frac{h^2}{2}\left\{\frac{\partial^2\mathbf{U}}{\partial x^2}\partial_1\mathbf{H} + \frac{\partial \mathbf{U}}{\partial x}\left[\frac{\partial \mathbf{U}}{\partial x}\partial_{11}\mathbf{H} + \left(\frac{\partial \mathbf{U}}{\partial x} - \sigma\frac{\partial \mathbf{U}}{\partial t}\right)\partial_{31}\mathbf{H}\right.\right. \\
&\quad - \left.\sigma\frac{\partial \mathbf{U}}{\partial t}\partial_{41}\mathbf{H}\right] + \left(\frac{\partial^2\mathbf{U}}{\partial x^2} - 2\sigma\frac{\partial^2\mathbf{U}}{\partial x\partial t} + \sigma^2\frac{\partial^2\mathbf{U}}{\partial t^2}\right)\partial_3\mathbf{H} \\
&\quad + \left(\frac{\partial \mathbf{U}}{\partial x} - \sigma\frac{\partial \mathbf{U}}{\partial t}\right)\left[\frac{\partial \mathbf{U}}{\partial x}\partial_{13}\mathbf{H} + \left(\frac{\partial \mathbf{U}}{\partial x} - \sigma\frac{\partial \mathbf{U}}{\partial t}\right)\partial_{33}\mathbf{H} - \sigma\frac{\partial \mathbf{U}}{\partial t}\partial_{43}\mathbf{H}\right] \\
&\quad + \sigma^2\frac{\partial^2\mathbf{U}}{\partial t^2}\partial_4\mathbf{H} - \sigma\frac{\partial \mathbf{U}}{\partial t}\left[\frac{\partial \mathbf{U}}{\partial x}\partial_{14}\mathbf{H} + \left(\frac{\partial \mathbf{U}}{\partial x} - \sigma\frac{\partial \mathbf{U}}{\partial t}\right)\partial_{34}\mathbf{H}\right. \\
&\quad - \left.\left.\sigma\frac{\partial \mathbf{U}}{\partial t}\partial_{44}\mathbf{H}\right]\right\} + \mathcal{O}(h^3),
\end{aligned}
$$

for the space increment in $\mathbf{H}$. Here we denote the derivative of the function $\mathbf{H}(\cdot,\cdot,\cdot,\cdot)$ with respect to its i-th argument, $i = 1,\ldots,4$, by $\partial_i\mathbf{H}$. Observe that when $\mathbf{U}$ is a vector, then $\partial_i\mathbf{H}$ is a matrix. Similarly, $\partial_{ji}\mathbf{H}$, the derivative of $\partial_i\mathbf{H}$ with respect to its j-th argument, $j = 1,\ldots,4$, is an array with three indices. Whenever an algebraic expression above contains a product of two entities, we assume the product appropriate for the arrays being multiplied. In the right side of these equations the arguments of the derivatives $\partial_i \mathbf{H}$ and $\partial_{ji} \mathbf{H}$ have been omitted; it must be understood that all these arguments are taken to be equal to $\mathbf{U}(x, t)$.

Combining these expansions yields the following expression for $\mathbf{e}$:

$$(5.5\text{-}19) \quad \mathbf{e} = \frac{\partial \mathbf{U}}{\partial t} + \frac{\partial \mathbf{U}}{\partial x} \cdot \sum_{j=1}^{4} \partial_j \mathbf{H}$$

$$+ \left\{ \frac{1}{2}\sigma \frac{\partial^2 \mathbf{U}}{\partial t^2} + \frac{1}{2}\frac{\partial^2 \mathbf{U}}{\partial x^2} \sum_{j=1}^{4}(-1)^j \partial_j \mathbf{H} + \sigma \frac{\partial^2 \mathbf{U}}{\partial x \partial t}\left(\partial_3 \mathbf{H} + \partial_4 \mathbf{H}\right) \right.$$

$$+ \frac{\partial \mathbf{U}}{\partial x}\frac{\partial \mathbf{U}}{\partial x}\left[\partial_{24}\mathbf{H} - \partial_{13}\mathbf{H} + \frac{1}{2}\sum_{j=1}^{4}(-1)^j \partial_{jj}\mathbf{H}\right]$$

$$+ \sigma\frac{\partial \mathbf{U}}{\partial t}\left[\frac{\partial \mathbf{U}}{\partial x}\left(\partial_{33}\mathbf{H} + \partial_{44}\mathbf{H} + \partial_{14}\mathbf{H}\right.\right.$$

$$\left.\left.\left. + \partial_{23}\mathbf{H} + \partial_{24}\mathbf{H} + 2\partial_{34}\mathbf{H}\right)\right]\right\}h + \mathcal{O}\left(h^2\right).$$

As noted previously, $\mathbf{e}$ vanishes at the nodes, so that

$$(5.5\text{-}20) \qquad \frac{\partial \mathbf{U}}{\partial t} + \frac{\partial \mathbf{U}}{\partial x}\sum_{j=1}^{4}\partial_j \mathbf{H} = h\mathbf{Q}(\mathbf{U}) + \mathcal{O}\left(h^2\right),$$

where $\mathbf{Q}(\mathbf{U})$ is the negative of the expression appearing between braces in the right side of Equation (5.5-19). Comparing (5.5-20) with (5.5-17), we see that when

$$(5.5\text{-}21) \qquad\qquad \mathbf{A}'(\mathbf{U}) = \sum_{j=1}^{4}\partial_j \mathbf{H},$$

the first of these equations reduces to the second one with $N = 1$. Thus, Equation (5.5-21) is a consistency condition. Observe that $\mathbf{H}(\mathbf{U}, \mathbf{U}, \mathbf{U}, \mathbf{U})$ is a function of $\mathbf{U}$ whose derivative is precisely $\sum_{j=1}^{4}\partial_j \mathbf{H}$. Taking this into account, we see that the consistency condition (5.5-21) implies

$$(5.5\text{-}22) \qquad \mathbf{H}(\mathbf{U}, \mathbf{U}, \mathbf{U}, \mathbf{U}) = \mathbf{A}(\mathbf{U}) + \text{ constant of integration.}$$

For consistent schemes further simplications are possible. Indeed, for such schemes

$$(5.5\text{-}23) \qquad\qquad \frac{\partial \mathbf{U}}{\partial t} + \frac{\partial \mathbf{A}}{\partial x}(\mathbf{U}) = h\mathbf{Q}(\mathbf{U}) + \mathcal{O}\left(h^2\right).$$

From this fact we deduce that

$$\frac{\partial^2 \mathbf{U}}{\partial t^2} + \mathcal{O}(h) = -\frac{\partial^2 \mathbf{A}}{\partial x \partial t} = -\frac{\partial}{\partial x}\left(\mathbf{A}'\frac{\partial \mathbf{U}}{\partial t}\right)$$

$$= -\frac{\partial}{\partial x}\left[(\mathbf{A}')^2\frac{\partial \mathbf{U}}{\partial x}\right],$$

$$\frac{\partial^2 \mathbf{U}}{\partial x \partial t} + \mathcal{O}(h) = -\frac{\partial}{\partial x}\left(\mathbf{A}'\frac{\partial \mathbf{U}}{\partial x}\right).$$

These observations allow us to write

$$\mathbf{Q}(\mathbf{U}) = -\frac{1}{2}\frac{\partial}{\partial x}\left[(\mathbf{A}')^2\frac{\partial \mathbf{U}}{\partial x}\right]$$

$$-\frac{1}{2}\left(-\partial_1\mathbf{H} + \partial_2\mathbf{H} - \partial_3\mathbf{H} + \partial_4\mathbf{H}\right)\frac{\partial^2 \mathbf{U}}{\partial x^2}$$

$$+\sigma\left(\partial_3\mathbf{H} + \partial_4\mathbf{H}\right)\frac{\partial}{\partial x}\left(\mathbf{A}'\frac{\partial \mathbf{U}}{\partial x}\right)$$

$$-\frac{1}{2}\Bigg[\left(-\partial_{11}\mathbf{H} + \partial_{22}\mathbf{H} - \partial_{33}\mathbf{H}\right.$$

$$\left.+ \partial_{44}\mathbf{H} - \partial_{13}\mathbf{H} + \partial_{24}\mathbf{H}\right)\frac{\partial \mathbf{U}}{\partial x}\Bigg]\frac{\partial \mathbf{U}}{\partial x}$$

$$+\sigma\bigg(\partial_{33}\mathbf{H} + \partial_{44}\mathbf{H} + \partial_{13}\mathbf{H} + \partial_{14}\mathbf{H}$$

$$+ \partial_{23}\mathbf{H} + \partial_{24}\mathbf{H} + 2\partial_{34}\mathbf{H}\bigg) + \mathcal{O}(h).$$

Let us define the matrix

$$(5.5\text{-}24) \qquad \mathbf{D}(\mathbf{U}) \equiv \sigma\left(\partial_3\mathbf{H} + \partial_4\mathbf{H} - \frac{1}{2}\mathbf{A}'\right)\mathbf{A}' - \frac{1}{2}\sum_{j=1}^{4}(-1)^j\partial_j\mathbf{H}.$$

The reader should verify by direct computation that

$$\frac{\partial}{\partial x}\left[\mathbf{D}(\mathbf{U})\cdot\frac{\partial \mathbf{U}}{\partial x}\right] = \mathbf{Q}(\mathbf{U}) + \mathcal{O}(h).$$

Hence, we can write Equation (5.5-23) as follows:

$$(5.5\text{-}25) \qquad \frac{\partial \mathbf{U}}{\partial t} + \frac{\partial \mathbf{A}(\mathbf{U})}{\partial x} = h\frac{\partial}{\partial x}\left[\mathbf{D}(\mathbf{U})\cdot\frac{\partial \mathbf{U}}{\partial x}\right] + \mathcal{O}(h^2).$$

Equation (5.5-25) exhibits the lowest-order error term as a form of artificial dissipation with coefficient $\mathbf{D}(\mathbf{U})$. By analogy with physical dissipative influences such as viscosity, the numerical scheme (5.5-20) will be stable if all eigenvalues of $\mathbf{D}(\mathbf{U})$ have positive real parts (Brenier and Hennart, 1985). When $\mathbf{U}$ is a scalar function, so is $\mathbf{D}(\mathbf{U})$, and the stability condition reduces to the scalar inequality $\mathbf{D}(\mathbf{U}) > 0$.

Analysis of performance.

In view of Equations (5.5-22) and (5.5-8), for explicit schemes where $\mathbf{H} = \mathbf{H}(\mathbf{U}_1, \mathbf{U}_2)$ the consistency condition will be satisfied if the following condition holds:

$$(5.5\text{-}26) \qquad\qquad \partial_1 \mathbf{H} + \partial_2 \mathbf{H} = \mathbf{A}'.$$

Thus the matrix function $\mathbf{D}$ defined in Equation (5.5-24) reduces to the somewhat simpler form,

$$(5.5\text{-}27) \qquad \mathbf{D}(\mathbf{U}) = -\frac{\sigma}{2}(\mathbf{A}')^2 + \tfrac{1}{2}(\partial_1 \mathbf{H} - \partial_2 \mathbf{H}).$$

Observe that the matrix $(\mathbf{A}')^2$ will have eigenvalues with positive real parts provided the vector function $\mathbf{A}(\psi)$ has real coordinate functions. Let us say that an explicit scheme is **symmetric** if

$$(5.5\text{-}28) \qquad\qquad \mathbf{H}(\mathbf{U}_1, \mathbf{U}_2) = \mathbf{H}(\mathbf{U}_2, \mathbf{U}_1).$$

This implies that the derivatives of $\mathbf{H}$ satisfy the condition

$$(5.5\text{-}29) \qquad\qquad \partial_1 \mathbf{H} = \partial_2 \mathbf{H},$$

and, as a consequence,

$$(5.5\text{-}30) \qquad\qquad \mathbf{D}(\mathbf{U}) = -\frac{\sigma}{2}(\mathbf{A}')^2.$$

Thus, $\mathbf{D}$ has eigenvalues with negative real parts. This argument shows that symmetric schemes are unconditionally unstable unless $\mathbf{A}' = \mathbf{0}$. As an immediate application, we conclude that the algorithms proposed in Equations (5.5-9a) and (5.5-10a), being clearly symmetric, are unconditionally unstable. However, Equation (5.5-27) shows that the difficulty can be overcome if $\partial_1 \mathbf{H} \neq \partial_2 \mathbf{H}$. Following this idea, let us add a term to the definitions (5.5-9a) and (5.5-10a) to make them asymmetric:

$$(5.5\text{-}9\text{b}) \qquad \mathbf{H}(\mathbf{U}_1, \mathbf{U}_2) = \tfrac{1}{2}[\mathbf{A}(\mathbf{U}_1) + \mathbf{A}(\mathbf{U}_2)] + \beta(\mathbf{U}_1 - \mathbf{U}_2),$$

$$(5.5\text{-}10\text{b}) \qquad \mathbf{H}(\mathbf{U}_1, \mathbf{U}_2) = \mathbf{A}\left(\frac{\mathbf{U}_1 + \mathbf{U}_2}{2}\right) + \beta(\mathbf{U}_1 - \mathbf{U}_2).$$

Here, β is a parameter whose value we shall choose to guarantee stability. Observe first that the consistency condition remains valid despite the addition of the new terms, since $\beta(\mathbf{U}_1 - \mathbf{U}_2)$ vanishes when $\mathbf{U}_1 = \mathbf{U}_2 = \mathbf{U}$. Denoting by $\mathbf{I}$ the identity matrix and using Equation (5.5-27), we see that

$$(5.5\text{-}31) \qquad\qquad \mathbf{D}(\mathbf{U}) = \beta \mathbf{I} - \frac{\sigma}{2}(\mathbf{A}')^2.$$

This equation shows that the scheme is stable whenever we select β so that

$$(5.5\text{-}32) \qquad\qquad \frac{2\beta}{\sigma} > \rho^2(\mathbf{A}'),$$

where $\rho(\mathbf{A}')$ stands for the spectral radius of the matrix $\mathbf{A}'$.

Equation (5.5-24) can also be used to analyze implicit algorithms. As an illustration, let us apply it to the schemes defined by Equations (5.5-15) and (5.5-16). These schemes are easily verified to be consistent. Observe also that, for both of these algorithms,

$$(5.5\text{-}33) \qquad \partial_3\mathbf{H} + \partial_4\mathbf{H} = \alpha\mathbf{A}', \quad \partial_1\mathbf{H} + \partial_3\mathbf{H} = \partial_2\mathbf{H} + \partial_4\mathbf{H},$$

and therefore

$$(5.5\text{-}34) \qquad\qquad \mathbf{D}(\mathbf{U}) = \sigma(\alpha - \tfrac{1}{2})(\mathbf{A}')^2.$$

Hence, stability is guaranteed when $\alpha > \tfrac{1}{2}$.

As an application of the modified equation approach, we briefly treat the nonlinear wave equation (5.1-16). One possible conservation form is the coupled pair

$$(5.5\text{-}35) \qquad\qquad \frac{\partial \rho}{\partial t} + \frac{\partial}{\partial x}(\rho u) = 0,$$

$$\frac{\partial(\rho u)}{\partial t} + \frac{\partial}{\partial x}(\rho u^2 + p) = 0.$$

To cast this system into the general form (5.5-1a), it is necessary to express all the variables involved in terms of only two of them, say, the density ρ and the momentum density $\Phi = \rho u$. For isentropic flows p is a function of ρ only, so that the presence of this function in system (5.5-35) poses no problem. On the other hand, u can be eliminated by means of the relationship $u = \Phi/\rho$. Then, the system (5.5-1a) with

$$(5.5\text{-}36) \qquad \psi = \begin{bmatrix} \rho \\ \Phi \end{bmatrix}; \quad \mathbf{A}(\psi) = \begin{bmatrix} \Phi \\ p + \Phi^2/\rho \end{bmatrix} \equiv \begin{bmatrix} A_1(\rho, \Phi) \\ A_2(\rho, \Phi) \end{bmatrix}$$

is identical to the system (5.5-35). The derivative of $\mathbf{A}$ in this case is just the matrix

$$(5.5\text{-}37) \qquad \mathbf{A}'(\boldsymbol{\psi}) = \begin{bmatrix} \partial A_1/\partial \rho & \partial A_1/\partial \Phi \\ \partial A_2/\partial \rho & \partial A_2/\partial \Phi \end{bmatrix} = \begin{bmatrix} 0 & 1 \\ a^2 - u^2 & 2u \end{bmatrix},$$

where $a^2 = \partial p/\partial \rho$. The matrix $\mathbf{A}'$ has real eigenvalues $u \pm a$. Using Equation (5.5-32), we see that the explicit schemes (5.5-9b) and (5.5-10b) will be stable [that is, $\mathbf{D}(\mathbf{U})$ will have positive eigenvalues] if and only if

$$(5.5\text{-}38) \qquad \qquad \beta > \frac{\sigma}{2}(u + a).$$

Riemann's method.

Riemann's method is a technique for solving initial-value problems having piecewise constant initial data. As such, it furnishes the basis for several numerical schemes using piecewise constant approximations to the true initial conditions. We introduce the method for constant initial data possessing a single jump discontinuity, then we generalize the results to data with any number of jump discontinuites.

First consider the one-dimensional initial-value problem

$$(5.5\text{-}39a) \qquad \qquad \frac{\partial \boldsymbol{\psi}}{\partial t} + \frac{\partial \mathbf{A}(\boldsymbol{\psi})}{\partial x} = 0,$$

with initial conditions

$$(5.5\text{-}39b) \qquad \qquad \boldsymbol{\psi}(x,0) = \begin{cases} \boldsymbol{\psi}_L & \text{if} \quad x < \bar{x}, \\ \boldsymbol{\psi}_R & \text{if} \quad x \geq \bar{x}. \end{cases}$$

Here $\bar{x}$, $\boldsymbol{\psi}_L$ and $\boldsymbol{\psi}_R$ are prescribed constant values. By the chain rule, we can rewrite the conservation form (5.5-39a) of the PDE as

$$\frac{\partial \boldsymbol{\psi}}{\partial t} + \mathbf{A}'(\boldsymbol{\psi}) \frac{\partial \boldsymbol{\psi}}{\partial x} = 0.$$

If $\boldsymbol{\psi}$ is a scalar function, then $\mathbf{A}'(\boldsymbol{\psi})$ is a scalar factor that plays a role analogous to that of the fluid velocity in the first-order transport equation (5.2-1). In the extension to vector conservation laws, $\mathbf{A}'(\boldsymbol{\psi})$ plays a similar role. However, if $\boldsymbol{\psi}$ and $\mathbf{A}$ are vector functions, then $\mathbf{A}'(\boldsymbol{\psi})$ is a matrix, and the speeds of propagation of waves in $\boldsymbol{\psi}$ are the eigenvalues of the matrix $\mathbf{A}'(\boldsymbol{\psi})$. Since these speeds can be no greater in magnitude than the spectral radius $\rho(\mathbf{A}'(\boldsymbol{\psi})) = R$, we conclude that

$$(5.5\text{-}40) \qquad \qquad \boldsymbol{\psi}(x,t) = \begin{cases} \boldsymbol{\psi}_L & \text{if} \quad x < \bar{x} - Rt \\ \boldsymbol{\psi}_R & \text{if} \quad x > \bar{x} + Rt, \end{cases}$$

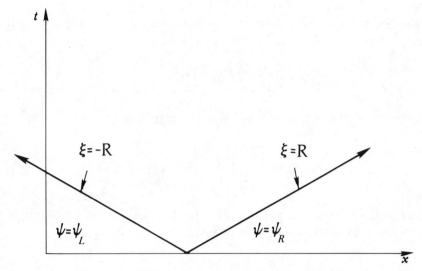

Figure 5-9. Solution domain for the problem defined in Equations (5.5-39).

as depicted in Figure 5-9. Moreover, the homogeneity of the first-order derivatives in Equation (5.5-39a) implies that the solution $\psi(x, t)$ is a function of x and t through the variable $\xi = (x - \bar{x})/t$ only. Hence, there is a function $\mathbf{W}(\psi_L, \psi_R, \xi)$ such that

(5.5-41) $$\psi(x, t) = \mathbf{W}(\psi_L, \psi_R, \xi).$$

As a simple example consider the following linearized version of system (5.5-35) that holds for perturbations of a fluid initially at rest:

$$\frac{\partial \rho}{\partial t} + \rho_0 \frac{\partial u}{\partial x} = 0,$$
$$\frac{\partial u}{\partial t} + \frac{a^2}{\rho_0} \frac{\partial \rho}{\partial x} = 0.$$

This system is just the first-order form of the standard second-order wave equation. In this case $\psi = (\rho, u)^\mathsf{T}$, and $\mathbf{A}$ is the linear function defined by

$$\mathbf{A}(\psi) = \begin{bmatrix} 0 & \rho_0 \\ a^2/\rho_0 & 0 \end{bmatrix} \begin{bmatrix} \rho \\ u \end{bmatrix} = \mathbf{A}'\psi.$$

Thus the derivative of $\mathbf{A}$ is just the matrix $\mathbf{A}'$, whose eigenvalues are $\pm a$.

Therefore the spectral radius of $\mathbf{A}'$ is $R = a$. If we set $\boldsymbol{\psi}_L = (1,0)^\top, \boldsymbol{\psi}_R = (1,1)^\top$, and $\bar{x} = 0$, then the initial conditions are

$$\boldsymbol{\psi}(0, s) = \begin{bmatrix} \rho(0, x) \\ u(0, x) \end{bmatrix} = \begin{cases} (1,0)^\top, & \text{if } x < 0 \\ (1,1)^\top, & \text{if } x > 0, \end{cases}$$

and $\xi = x/t$. The solution at any time $t > 0$ is then

$$\boldsymbol{\psi}(x, t) = \begin{cases} (1,0)^\top, & \text{if } x < -at \\ (s, \tfrac{1}{2})^\top, & \text{if } -at \le x \le at \\ (1,1)^\top, & \text{if } at < x. \end{cases}$$

[One method of verifying the solution in this rather simple example is to solve the equivalent second-order system

$$\frac{\partial^2 u}{\partial t^2} = a^2 \frac{\partial^2 u}{\partial x^2}, \quad -\infty < x < \infty,$$

$$u(x, 0) = u_0(x) = \begin{cases} 0, & \text{if } x < 0 \\ 1, & \text{if } x > 0, \end{cases}$$

$$\frac{\partial u}{\partial t}(x, 0) = 0,$$

whose solution is $u(x, t) = \tfrac{1}{2} u_0(x - at) + \tfrac{1}{2} u_0(x + at)$.] Therefore, for this simple problem we can exhibit the function $\mathbf{W}$ explicity as follows:

$$\mathbf{W}(\boldsymbol{\psi}_L, \boldsymbol{\psi}_R, \xi) = \begin{cases} \boldsymbol{\psi}_L, & \text{if } \xi < -a \\ (\boldsymbol{\psi}_L + \boldsymbol{\psi}_R)/2, & \text{if } -a \le \xi \le a \\ \boldsymbol{\psi}_R, & \text{if } a < \xi. \end{cases}$$

It is only fair to mention that the solution of the Riemann problem for nonlinear equations is a somewhat involved task; for an example we refer the reader to Isaacson (to appear).

Now consider a slightly more elaborate version of the initial-value problem defined in Equations (5.5-39). In this version, we replace the initial conditions (5.5-39b) by the piecewise constant condition

(5.5-39c) $\boldsymbol{\psi}(x, 0) = \boldsymbol{\psi}_i, \quad x_i < x < x_{i+1}, \quad i = 0, \pm 1, \pm 2, \ldots,$

$\boldsymbol{\psi}_i$ being constant for each integer i. Choose a time increment k such that

(5.5-42) $2Rk < \inf_i \{h_i\},$

where $h_i = x_{i+1} - x_i$ and inf denotes the infimum or greatest lower bound

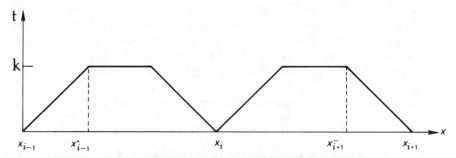

Figure 5-10. "Transported" endpoints of the finite-difference grid used to accommodate the initial conditions (5.5-39c).

taken over all intervals in the grid. Then, reasoning as for Equations (5.5-41) on each interval (x_i, x_{i+1}), we find

(5.5-43) $$\psi(x,t) = \mathbf{W}(\psi_{i-1}, \psi_i, (x - x_i)/t),$$

$$x_{i-1}^+ < x < x_{i+1}^-, \quad 0 \le t \le k.$$

Here, for every integer i, the numbers x_i^+ and x_i^- denote the "transported" endpoints,

(5.5-44) $$x_i^+ = x_i + Rk; \quad x_i^- = x_i - Rk$$

as shown in Figure 5-10. Clearly, the solution defined in Equation (5.5-43) will cover the entire band $-\infty < x < \infty$, $0 \le t \le k$ of the (x,t)-plane when i ranges over all integers. Thus we know how to construct the solution of the initial-value problem (5.5-39a, c) in that band whenever the function $\mathbf{W}$ is known.

Gudonov schemes and Glimm's scheme.

Observe that the function $\psi(x, k)$, obtained by setting $t = k$ in Equation (5.5-43), is not generally piecewise constant and so cannot be written in the form (5.5-39c). This fact is inconvenient because, if $\psi(x, k)$ were piecewise constant, then by repeated application of Riemann's method it would be possible to construct ψ successively in the bands of the (x,t)-plane whose time limits are $(0, k), (k, 2k), \ldots$. In other words, we could regard the solution $\psi(x, k)$ as the piecewise constant initial condition for a problem solved on $(k, 2k)$, and so forth, to advance ψ in time.

A natural way of overcoming this difficulty is to approximate the function $\psi(x, k)$ by a piecewise constant function. When this is done, the Riemann's procedure for solving initial-value problems can be repeated in each time level interval $(nk, (n+1)k)$. This basic idea underlies several well known schemes, of which Gudonov's scheme and Glimm's scheme are exemplary.

To introduce these schemes, let us replace the initial data (5.5-39c) by

$$(5.5\text{-}45) \qquad \psi(x, t^n) = U_i^n; \quad x_i < x < x_{i+1}; \quad i = 0, \pm 1, \pm 2 \dots.$$

Here, the superscript n indicates the n-th time level. If we integrate the PDE (5.5-39a) over the rectangular region illustrated in Figure 5-11 we obtain the following:

$$(5.5\text{-}46) \quad U_i^{n+1} = U_1^n h_i + A\big(W(U_{i-1}^n, U_i^n, 0)k\big) - A\big(W(U_i^n, U_{i+1}^n, 0)k\big),$$

where we define the variables U_i^{n+1} by the equation

$$(5.5\text{-}47) \qquad 4U_i^{n+1} = \frac{1}{h_i} \int_{x_i}^{x_{i+1}} \psi(x, t^{n+1})dx$$

and use the function W defined in our discussion of Riemann's problem. Equation (5.5-46) is **Gudonov's (explicit) scheme**.

Now consider the initial value problem defined by Equations (5.5-39a) and (5.5-45), and assume that we have a uniform grid, that is, one for which $h_i = h$, independent of i. Define $x_{i+1/2} = x_i + h/2$. To develop Glimm's scheme, we introduce in addition a half-step in time, specifying a constant value $U_i^{n+1/2}$ at the half-time-step level in the shifted space interval $x_{i-1/2} < x < x_{i+1/2}$, drawn in Figure 5-11. When the time increment k satisfies the restriction in Equation (5.5-42), we have

$$(5.5\text{-}48) \qquad \psi(x, t^{n+1/2}) = W(U_{i-1}^n, U_i^n, (x - x_i)/(t^{n+1/2} - t^n)),$$

$$x_{i-1/2} < x < x_{i+1/2}.$$

For any value of x in the interval $(x_{i-1/2}, x_{i+1/2})$, the value of $\psi(x, t^{n+1/2})$

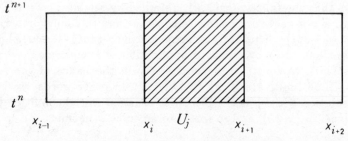

Figure 5-11. Rectangular region of integration for the PDE (5.5-39a).

serves as a prediction of $U_{n-1/2}^{n+1/2}$. In Glimm's method one chooses x randomly in the interval $(x_{i-1/2}, x_{i+1/2})$, getting a prediction of the form

$$(5.5\text{-}49) \qquad U_{i-1/2}^{n+1/2} = W\left(U_{i-1/2}^n, U_i^n, (\xi_n - \tfrac{1}{2})\frac{h}{k}\right),$$

where ξ_n signifies a random number between 0 and 1. The shift in the time interval is corrected by taking a second half-step; thus, we obtain

$$(5.5\text{-}50) \qquad U_i^{n+1} = W\left(U_{i-1/2}^{n+1/2}, U_{i+1/2}^{n+1/2}, (\xi_n - \tfrac{1}{2})\frac{h}{k}\right).$$

For further details, we refer the reader to Peyret and Taylor (1983).

5.6. Finite Elements for Hyperbolic Equations.

The historical motivation for finite-element methods was the solution of elliptic problems. The methods extend to parabolic problems without too much difficulty, as we discussed in Chapter Four. However, hyperbolic problems pose more of a challenge for finite-element methods, because these problems lie quite far in spirit from those problems for which the methods were first constructed. The kinds of numerical difficulties encountered are essentially the same as those we have discussed in Section 4.3 in connection with solution pathologies at sharp fronts. In species transport settings, such fronts typically occur in advection-dominated flows, for which the Peclet number is much larger than unity. Since hyperbolic systems are nondissipative, they represent the extreme case for which the Peclet number or some analog of it is infinite.

In this section we shall consider methods using a fixed spatial grid. Various moving-grid and adaptive grid refinement methods have been developed in the last few years. While some of these techniques are closely related to methods we shall study, major algorithmic difficulties must be overcome before these methods become established for systems of equations or for equations in several dimensions.

As in Section 4.4, we base our development on the general principles presented in Section 3.16. Let us start with the first-order advection equation

$$(5.6\text{-}1) \qquad \frac{\partial u}{\partial t} + b\frac{\partial u}{\partial x} = 0,$$

where b is a positive constant, together with the boundary and initial conditions

$$(5.6\text{-}2) \qquad u(0,t) = f(t),$$
$$(5.6\text{-}3) \qquad u(x,0) = u_0(x).$$

Galerkin approximation.

There is no particular difficulty in formally applying the standard Galerkin approach to Equation (5.6-1). We begin by defining the trial function $\hat{u}$ in terms of a set of interpolating basis functions $\phi_0(x), \ldots, \phi_N(x)$:

$$(5.6\text{-}4) \qquad u(x,t) \simeq \hat{u}(x,t) = \sum_{j=0}^{N} v_j(t)\phi_j(x)$$

Observe that we accommodate time dependence by allowing the coefficients $v_j(t)$ to be transient. Substituting $\hat{u}$ into Equation (5.6-1) yields a residual,

$$(5.6\text{-}5) \qquad R(x,t) = \frac{\partial \hat{u}}{\partial t}(x,t) + b\frac{\partial \hat{u}}{\partial x}(x,t).$$

If, for example, the functions $\phi_j(x)$ are Lagrange basis functions on a grid $x_0 < x_1 < \cdots < x_N$, requiring $\hat{u}$ to satisfy the boundary condition (5.6-2) forces $v_0(t) = f(t)$. The method of weighted residuals is now applied using the basis functions ϕ_i, $i = 1, \ldots, N$, as test functions. Thus, we force the residual to be orthogonal to each of the functions ϕ_i, $i = 1, \ldots, N$, yielding integral equations having the form

$$(5.6\text{-}6) \qquad \int_{\Omega} R(x,t)\phi_i(x)dx = 0, \qquad i = 1, \ldots, N,$$

where the integral is taken over the entire spatial domain Ω of the problem. Substitution of the residual (5.6-5) into the Galerkin equations (5.6-6) yields the integral equations

$$(5.6\text{-}7) \qquad \int_{\Omega} \left(\frac{\partial \hat{u}}{\partial t} + b\frac{\partial \hat{u}}{\partial x} \right) \phi_i dx = 0, \qquad i = 1, \ldots, N.$$

This set of equations is equivalent to a system of ordinary differential equations whose matrix form is as follows:

$$(5.6\text{-}8) \qquad \mathbf{A}\frac{d\mathbf{u}}{dt} + \mathbf{C}\mathbf{u} = \mathbf{0}.$$

The matrices $\mathbf{A}$ and $\mathbf{C}$ have entries

$$a_{ij} = \int_{\Omega} \phi_j\phi_i dx$$

$$c_{ij} = b\int_{\Omega} \frac{d\phi_j}{dx}\phi_i dx,$$

respectively, and the vector $\mathbf{u}$ has coordinates v_j, $j = 1, \ldots, N$. There are many schemes available for solving systems of the form (5.6-8). However, for problems whose solutions have sharp fronts, one commonly observes that the numerical solutions exhibit undesirable oscillatory behavior near the fronts. Much research has been devoted to finite-element procedures capable of overcoming this type of error.

Before proceeding with the discussion of oscillatory behavior near sharp fronts, we should point out that there is an alternative formulation of the Galerkin method that we could have employed. In the formulation we have just presented, the basis functions are functions of the space coordinates alone, and hence only the space variable was discretized. Such a procedure is usually referred to as **semidiscretization**. As another option, we might consider two-dimensional basis functions having the form $\phi_i = \phi_i(x, t)$. In this case, we replace the trial function defined in Equation (5.6-4) by the following:

$$(5.6\text{-}9) \qquad \hat{u}(x, t) = \sum_{j=0}^{N} v_j \phi_j(x, t),$$

where now the nodes represent points in space-time and the nodal coefficients v_j are constant. Substitution of this expression into the Galerkin equations (5.6-7) yields integral equations of the form

$$(5.6\text{-}10) \qquad \sum_{j=1}^{N} v_j \int_{\Omega} \left(\frac{\partial \phi_j}{\partial t} + b \frac{\partial \phi_j}{\partial x} \right) \phi_i \, dx \, dt = 0.$$

We can write this set of equations in matrix form as follows:

$$(5.6\text{-}11) \qquad \mathbf{Au} = \mathbf{0},$$

where now the matrix $\mathbf{A}$ has entries

$$a_{ij} = \int_{\Omega} \left(\frac{\partial \phi_j}{\partial t} + b \frac{\partial \phi_j}{\partial x} \right) \phi_i \, dx \, dt,$$

and the vector $\mathbf{u}$ has coordinates v_i. Thus we have replaced the set (5.6-8) of ordinary differential equations by a set of linear algebraic equations.

Let us briefly comment on the relationship between the scheme (5.6-10) and standard finite-difference approximations. If suitable finite-difference schemes are used to approximate the time derivative appearing in the differential system (5.6-8), it is possible to reconcile that equation set with the algebraic system (5.6-11) by choosing the functions $\phi_i(x, t)$ appropriately.

We leave this possibility for the reader to investigate. In this sense procedures based on space-time basis functions are more general than semidiscretization procedures.

To demonstrate the difficulties encountered in the solution of Equation (5.6-1) using standard Galerkin methods, let us formulate a problem using piecewise linear Lagrange basis functions, as discussed in Section 2.4. These functions are defined as follows:

$$(5.6\text{-}12) \qquad \phi_i(x) = \begin{cases} (x - x_{i-1})/(x_i - x_{i-1}), & x_{i-1} \leq x \leq x_i \\ (x_{i+1} - x)/(x_{i+1} - x_i), & x_i \leq x \leq x_{i+1} \\ 0, & \text{otherwise.} \end{cases}$$

When we introduce this choice of basis into Equations (5.6-8), we obtain for a uniform spatial grid the general expression

$$(5.6\text{-}13) \qquad \frac{1}{6}\left(\frac{dv_{i-1}}{dt} + 4\frac{dv_i}{dt} + \frac{dv_{i+1}}{dt}\right) + b\left(\frac{v_{i+1} - v_{i-1}}{2h}\right) = 0,$$

where h stands for the mesh of the spatial grid. This expression must be modified, of course, for boundary nodes. Observe that the space derivative has been replaced by a central difference approximation, and the time derivative is represented by a weighted average of three adjacent nodal values. From our experience with advection-dominated transport in Chapter Four, we should anticipate that this central difference approximation of the advection term will lead to numerical difficulties in some circumstances.

To elucidate the nature of such difficulties and to compare the finite-element technique with finite-difference approximations, we adopt a trick introduced by Vichnevetsky and Peiffer (1975). First, consider a more general form of the Galerkin equation (5.6-13) in which we allow variable weighting of the time derivatives:

$$(5.6\text{-}14) \quad \alpha\frac{dv_i}{dt} + \left(\frac{1-\alpha}{2}\right)\left(\frac{dv_{i-1}}{dt} + \frac{dv_{i+1}}{dt}\right) + b\left(\frac{v_{i+1} - v_{i-1}}{2h}\right) = 0.$$

The standard finite-difference scheme corresponds to the choice $\alpha = 1$, while for $\alpha = \frac{2}{3}$ we retrieve the finite-element form (5.6-13). Let us relabel the values of v_i associated with odd indices i as V_i. We can now rewrite Equation (5.6-14) as follows:

$$(5.6\text{-}15a) \quad \alpha\frac{dv_i}{dt} + \left(\frac{1-\alpha}{2}\right)\left(\frac{dV_{i-1}}{dt} + \frac{dV_{i+1}}{dt}\right) + b\left(\frac{V_{i+1} - V_{i-1}}{2h}\right) = 0,$$

for even values of i, and

$$(5.6\text{-}15b) \quad \left(\frac{1-\alpha}{2}\right)\left(\frac{dv_{i-1}}{dt} + \frac{dv_{i+1}}{dt}\right) + \alpha\frac{dV_i}{dt} + b\left(\frac{v_{i+1} - v_{i-1}}{2h}\right) = 0,$$

for odd values of i. This system of equations is consistent with the system of PDEs

(5.6-16a)
$$\alpha\frac{\partial U}{\partial t} + (1-\alpha)\frac{\partial V}{\partial t} + b\frac{\partial V}{\partial x} = 0,$$

(5.6-16b)
$$(1-\alpha)\frac{\partial U}{\partial t} + \alpha\frac{\partial V}{\partial t} + b\frac{\partial U}{\partial x} = 0,$$

where we consider values v_i and V_i to be nodal values of smooth functions $U(x,t)$ and $V(x,t)$, respectively. We now take the arithmetic mean of Equations (5.6-16) to give

(5.6-17a)
$$\frac{\partial}{\partial t}\left(\frac{U+V}{2}\right) + b\frac{\partial}{\partial x}\left(\frac{U+V}{2}\right) = 0.$$

By subtraction, the same equations yield

(5.6-17b)
$$\frac{\partial}{\partial t}\left(\frac{U-V}{2}\right) + \frac{b}{1-2\alpha}\frac{\partial}{\partial x}\left(\frac{U-V}{2}\right) = 0.$$

The function $\frac{1}{2}(U+V)$ in Equation (5.6-17a) represents the smooth part of the numerical solution; it travels with velocity b, as it should in accordance with the original equation (5.6-1). However, the function $\frac{1}{2}(U-V)$ represents the node-to-node oscillations in the numerical solution. These oscillations would vanish if the nodal values v_i and V_i were both point realizations of the exact solution u. Thus $\frac{1}{2}(U-V)$ represents a spurious oscillatory part of the numerical solution traveling with speed $b/(1-2\alpha)$ in accordance with Equation (5.6-17b). Vichnevetsky and Peiffer (1975) perform a rigorous frequency analysis supporting this interpretation and showing that the oscillations associated with $\frac{1}{2}(U-V)$ have wavelength $2h$.

Nonstandard Galerkin methods.

In Section 4.4 we found that the standard Galerkin finite-element approach can be modified to smooth the behavior of the numerical solution in the neighborhood of a sharp front. Similar procedures can be used to eliminate spurious oscillations in the numerical solution of our model first-order hyperbolic equation. We begin, as in the previous case, by selecting a trial function based on piecewise linear Lagrange basis functions:

(5.6-18)
$$u(x,t) \simeq \hat{u}(x,t) = \sum_{j=0}^{N} v_j(t)\phi_j(x).$$

From the method of weighted residuals, we require

$$\int_\Omega R(x,t)w_i(x)dx = 0, \qquad i = 1,\ldots,N,$$

where the functions $w_1(x),\ldots,w_N(x)$ are weighting functions and

(5.6-19) $$R(x,t) = \frac{\partial \hat{u}}{\partial t}(x,t) + b\frac{\partial \hat{u}}{\partial x}(x,t).$$

Therefore, the weighted residual equations have the form

(5.6-20) $$\int_\Omega \left(\frac{\partial \hat{u}}{\partial t} + b\frac{\partial \hat{u}}{\partial x}\right) w_i dx = 0.$$

If we choose the basis functions ϕ_i as the weighting functions w_i, we arrive at the standard Galerkin formulation just discussed. However, we shall instead select different weighting functions $w_i(x) = \psi_i(x)$, where $\psi_i(x)$ denotes the asymmetric weighting function introduced in Section 4.4. Over the typical element system shown in Figure 4-15, these weighting functions have the algebraic form

(5.6-21a) $\qquad \psi_1(x) = 1 - \xi + 3\alpha(\xi^2 - \xi), \qquad 0 \le \xi \le h,$

(5.6-21) $\qquad \psi_2(x) = \xi - 3\alpha(\xi^2 - \xi), \qquad 0 \le \xi \le h,$

where $\xi \equiv x/h$ represents a local element coordinate. The parameter α, assigned by the analyst, dictates the degree of upstream weighting in the final approximation, as illustrated in Figure 4-15.

We now generate a set of ordinary differential equations in time by substituting Equations (5.6-21) and (5.6-18) into the Galerkin-like equation (5.6-20) and then integrating over the spatial domain. The resulting expression for a typical internal node i is the following:

(5.6-22) $$h\left(\frac{1}{6} + \frac{\alpha}{4}\right)\frac{dv_{i-1}}{dt} + \frac{2h}{3}\frac{dv_i}{dt} + h\left(\frac{1}{6} - \frac{\alpha}{4}\right)\frac{dv_{i+1}}{dt}$$
$$= \frac{b}{2}(1+\alpha)v_{i-1} - \alpha b v_i - \frac{b}{2}(1-\alpha)v_{i+1}.$$

It is apparent from the definition (5.6-21) of the weighting functions that, when $\alpha = 0$, the asymmetric weighting function ψ_i reduces to the standard basis function ϕ_i, and thus we recover the Galerkin approximation. On the other hand, if we choose $\alpha = 1$, the right side of Equation (5.6-22) becomes upstream weighted. In other words, the spatial approximation is biased in the direction opposite to the sense of b. The scheme becomes

(5.6-23) $$\frac{5h}{12}\frac{dv_{i-1}}{dt} + \frac{2h}{3}\frac{dv_i}{dt} - \frac{h}{12}\frac{dv_{i+1}}{dt} = b(v_{i-1} - v_i).$$

This scheme, unfortunately, is not convergent.

To obtain a convergent asymmetric scheme, we apply the asymmetric weighting functions $\psi_i(x)$ to the spatial or advective term only, using the symmetric functions $\phi_i(x)$ as weighting functions in the terms arising from the time derivative. Thus we obtain Galerkin-like equations of the form

$$(5.6\text{-}24) \qquad \int_\Omega \left(\frac{\partial \hat{u}}{\partial t} \phi_i + b \frac{\partial \hat{u}}{\partial x} \psi_i \right) dx = 0.$$

Observe that, strictly speaking, this particular scheme does not fall under the rubric of the method of weighted residuals, since it does not use a consistent weighting function throughout the integrand. We shall rely on Fourier analysis to investigate its numerical properties.

Substituting the trial function (5.6-18) into Equation (5.6-24) and performing the indicated integrations, we obtain, for a typical interior node i,

$$(5.6\text{-}25) \qquad \frac{h}{6} \frac{dv_{i-1}}{dt} + \frac{2h}{3} \frac{dv_i}{dt} + \frac{h}{6} \frac{dv_{i+1}}{dt}$$
$$= \frac{b}{2}(1+\alpha)v_{i-1} - \alpha b v_i - \frac{b}{2}(1-\alpha)v_{i+1}.$$

Let us introduce a finite-difference aproximation for dv_i/dt, so that the ordinary differential equation (5.6-25) becomes

$$(5.6\text{-}26) \qquad \tfrac{1}{6}\left(v_{i-1}^{r+1} - v_{i-1}^r\right) + \tfrac{2}{3}\left(v_i^{r+1} - v_i^r\right) + \tfrac{1}{6}\left(v_{i+1}^{r+1} - v_{i+1}^r\right)$$
$$= \overline{C}\theta \left[\frac{(1+\alpha)}{2} v_{i-1}^{r+1} - \alpha v_i^{r+1} - \frac{(1-\alpha)}{2} v_{i+1}^{r+1} \right]$$
$$+ \overline{C}(1-\theta) \left[\frac{(1+\alpha)}{2} v_{i-1}^{r} - \alpha v_i^{r} - \frac{(1-\alpha)}{2} v_{i+1}^{r} \right].$$

Here $\overline{C} = bk/h$, where k denotes the time step. This approximation becomes a backward difference, Crank-Nicolson or forward difference scheme when $\theta = 1$, $\theta = \frac{1}{2}$, or $\theta = 0$, respectively.

To demonstrate the effect of asymmetry on the numerical solution to (5.6-1), we consider the propagation of a step discontinuity over the range $0 \le x \le 1$. The following system of equations applies:

$$\frac{\partial u}{\partial t} + b \frac{\partial u}{\partial x} = 0,$$
$$u(x,0) = 0, \qquad 0 < x < 1,$$
$$u(0,t) = u_0 = 1, \qquad 0 \le t.$$

The solutions obtained using the scheme (5.6-26) with $\alpha = 0$ (the standard Galerkin method) and $\alpha = 0.25$ (upstream weighting) appear in Figure 5-12

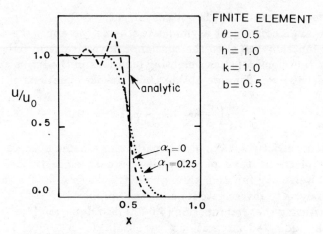

Figure 5-12. Solutions to the advection equation using the asymmetric Galerkin scheme with upstream weighting parameter $\alpha = 0$ and $\alpha = 0.25$.

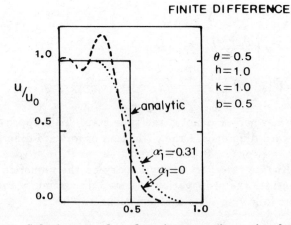

Figure 5-13. Solutions to the advection equation using lumped versions of the asymmetric Galerkin scheme.

(after Lapidus and Pinder, 1982). A Crank-Nicolson time approximation was used in each case.

It is appropriate to ask how the upstream-weighted finite-element approximation compares with upstream-weighted finite differences. One way to convert Equation (5.6-26) to a finite-difference equation is by **lumping** the terms associated with the time derivative. This procedure replaces the weighted average

$$\tfrac{1}{6}\left(v_{i-1}^{r+1} - v_{i-1}^r\right) + \tfrac{2}{3}\left(v_i^{r+1} - v_i^r\right) + \tfrac{1}{2}\left(v_{i+1}^{r+1} - v_{i+1}^r\right)$$

by the expression $v_i^{r+1} - v_i^r$, yielding the finite-difference approximation

$$(5.6\text{-}27) \qquad v_i^{r+1} - v_i^r = \overline{C}\theta \left(\frac{1+\alpha}{2}v_{i-1}^{r+1} - \alpha v_i^{r+1} - \frac{1-\alpha}{2}v_{i+1}^{r+1}\right)$$
$$+ \overline{C}(1-\theta)\left(\frac{1+\alpha}{2}v_{i-1}^r - \alpha v_i^r - \frac{1-\alpha}{2}v_{i+1}^r\right).$$

Solutions to this equation for the same parameters used in the finite-element scheme appear in Figure 5-13. Thus we can compare upstream-weighted finite-difference and finite-element schemes by examining Figures 5-12 and 5-13. In each case a value of α is just large enough to suppress numerically induced oscillations. The calculations indicate that the upstream-weighted schemes suppress spurious oscillations, which we attribute to numerical dispersion or phase error, at the expense of artificial numerical dissipation. This dissipation smears the sharp front and generates a solution having features more appropriate to a second-order parabolic equation than to a first-order hyperbolic equation. In other words, the solution exhibits a diffusive appearance that is consistent with what one would expect from the advective-diffusive transport equation. However, in this case physical diffusion is absent, and the smearing we see is purely *numerical* diffusion.

Orthogonal collocation formulation.

The formulation of orthogonal collocation for hyperbolic problems is essentially the same as for parabolic problems. We simply require

$$(5.6\text{-}28) \qquad \left(\frac{\partial \hat{u}}{\partial t} + b\frac{\partial \hat{u}}{\partial x}\right)\Bigg|_{\bar{x}_i} = 0, \qquad i = 1, 2, \ldots, M,$$

where M is the number of collocation points $\bar{x}_i$. Using Hermite cubic basis functions, the trial function on an N-element spatial grid has the form

$$(5.6\text{-}29) \qquad \hat{u}(x,t) = \sum_{j=0}^{N}\left[v_j(t)h_j^0(x) + v_j'(t)h_j^1(x)\right],$$

where v_j and v'_j stand for the value and slope, respectively, of $\hat{u}$ at the node x_j. Recall from Chapter Two that, in global coordinates, the Hermite cubic basis functions are defined on each element $[x_j, x_{j+1}]$ in a uniform grid as

$$(5.6\text{-}30) \qquad h_j^0(x) = \frac{1}{h^3} \left(x - x_j - h \right)^2 \left[2(x - x_j) + h \right],$$

$$h_{j+1}^0(x) = -\frac{1}{h^3} \left(x - x_j \right)^2 \left[2(x - x_j) - 3h \right],$$

$$h_j^1(x) = \frac{1}{h^2} \left(x - x_j \right) \left(x - x_j - h \right)^2,$$

$$h_{j+1}^1(x) = \frac{1}{h^2} \left(x - x_j - h \right) \left(x - x_j \right)^2,$$

where $h = x_{j+1} - x_j$ is the grid mesh.

The treatment of boundary conditions using this trial function deserves some explanation. Typical boundary conditions appropriate for the one-dimensional pure advection equation have the form

$$u(0, t) = f(t).$$

As we discussed earlier in this chapter, a two-point boundary-value problem, while appropriate for parabolic PDEs, generally overdetermines the solution to a first-order hyperbolic PDE. Thus, strictly speaking, we have at most one boundary condition. Taken together with the usual $2N$ collocation equations associated with the two-point Gauss quadrature ordinates $x_{j1} = [j + \frac{1}{2}(1 - 1/\sqrt{3})]/h$ and $x_{j2} = [j + \frac{1}{2}(1 - 1/\sqrt{3})]/h$, the boundary data fail to provide enough conditions to determine the $2N + 2$ nodal coefficients in the expansion (5.6-26). To get the needed extra equation, we reason that the hyperbolic PDE is actually an approximation to a parabolic PDE in which we have neglected dissipative influences. On the strength of this reasoning it seems appropriate to impose a "zero dissipation" condition at the other end, say $x = L$, of the spatial domain:

$$\frac{\partial u}{\partial x}(L, t) = 0.$$

Thus we can set $v_0(t) = f(t)$ and $v'_N(t) = 0$, and the $2N$ collocation equations suffice to determine the remaining nodal coefficients. For the case $b = 1$, combining Equations (5.6-28) and (5.6-29) then yields

$$(5.6\text{-}31) \qquad \left[\sum_{j=0}^{N} \left(\frac{dv_j}{dt} h_j^0 + \frac{dv'_j}{dt} h_j^1 + v_j \frac{dh_j^0}{dx} + v'_j \frac{dh_j^1}{dx} \right) \right]_{x_{j\alpha}} = 0,$$

$$j = 1, 2, \ldots, M; \quad \alpha = 1, 2.$$

In each element $[x_j, x_{j+1}]$ there are two collocation points $x_{j\alpha}$. In standard formulations these points are located at $x_{j1} = [j + \frac{1}{2}(1 - 1/\sqrt{3})]h$ and $x_{j2} = [j + \frac{1}{2}(1 + 1/\sqrt{3})]h$, which are the two-point Gauss quadrature points for that interval. Thus $M = 2N$, and if Equations (5.6-31) are supplemented with two boundary conditions the total number of conditions is $2N + 2$. Therefore our collocation equations together with the boundary conditions now suffice to determine the unknowns v_i and v_i' for $i = 0, 1, \dots, N$.

For the two collocation points located in a typical element $[x_i, x_{i+1}]$, only two terms in the sum occurring in Equation (5.6-31) remain nonzero, and the corresponding collocation equations for the element reduce to

$$(5.6\text{-}32) \quad \left[\frac{dv_i}{dt}h_i^0 + \frac{dv_i'}{dt}h_i^1 + v_i\frac{dh_i^0}{dx} + v_i'\frac{dh_i^1}{dx} + \frac{dv_{i+1}}{dt}h_{i+1}^0\right.$$

$$\left. + \frac{dv_{i+1}'}{dt}h_{i+1}^1 + v_{i+1}\frac{dh_{i+1}^0}{dx} + v_{i+1}'\frac{dh_{i+1}^1}{dx}\right]_{x_{i\alpha}} = 0, \quad \alpha = 1, 2.$$

Adopting the notation $a = (1 + 1/\sqrt{3})/2$ and $c = (1 - 1/\sqrt{3})/2$, we can write the collocation points as $x_{i1} = (i + a)h$ and $x_{i2} = (i + c)h$. Evaluating the polynomials h_i^0, h_i^1, dh_i^0/dx, dh_i^1/dx, and so on, at the collocation points permits us to transform Equations (5.6-32) into the equations

$$(5.6\text{-}33a) \quad -a^2(2a - 3)\frac{dv_i}{dt} + a^2ch\frac{dv_i'}{dt} - \frac{6ca}{h}v_i - a(3c - 1)v_i'$$

$$+ c^2(2a + 1)\frac{dv_{i+1}}{dt} - c^2ah\frac{dv_{i+1}'}{dt} + \frac{6ac}{h}v_{i+1} - c(3a - 1)v_{i+1}'$$

$$= 0,$$

$$(5.6\text{-}33b) \quad c^2(2a + 1)\frac{dv_i}{dt} + c^2ah\frac{dv_i'}{dt} - \frac{6ca}{h}v_i - c(3a - 1)v_i'$$

$$- a^2(2a - 3)\frac{dv_{i+1}}{dt} - a^2ch\frac{dv_{i+1}'}{dt} + \frac{6ac}{h}v_{i+1} - a(3c - 1)v_{i+1}'$$

$$= 0.$$

In matrix form, Equations (5.6-33) become

$$\begin{bmatrix} -a^2(2a - 3) & a^2ch & c^2(2a + 1) & -c^2ah \\ c^2(2a + 1) & c^2ah & -a^2(2a - 3) & -a^2ch \end{bmatrix} \begin{bmatrix} dv_i/dt \\ dv_i'/dt \\ dv_{i+1}/dt \\ dv_{i+1}'/dt \end{bmatrix}$$

$$+ \begin{bmatrix} -6ca/h & -a(3c - 1) & 6ca/h & -c(3a - 1) \\ -6ca/h & -c(3a - 1) & 6ca/h & -a(3c - 1) \end{bmatrix} \begin{bmatrix} v_i \\ v_i' \\ v_{i+1} \\ v_{i+1}' \end{bmatrix} = 0.$$

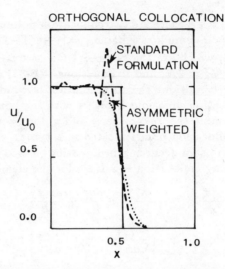

Figure 5-14. Solution to an advected step-wave problem using standard orthogonal collocation.

If we assemble all the element matrix equations of this form into a single global matrix equation for the entire grid, we get an equation of the form

$$(5.6\text{-}34) \qquad \mathbf{A}\frac{d\mathbf{u}}{dt} + \mathbf{B}\mathbf{u} = \mathbf{0}.$$

This system of ordinary differential equations can be solved, for example, by using a finite-difference approximation for $d\mathbf{u}/dt$. Employing a general time-weighted scheme, Equation (5.6-34) becomes

$$(5.6\text{-}35) \qquad \mathbf{A}\left(\frac{\mathbf{u}^{r+1} - \mathbf{u}^{r}}{k}\right) + \theta\mathbf{B}\mathbf{u}^{r+1} + (1-\theta)\mathbf{B}\mathbf{u}^{r} = \mathbf{0},$$

where k denotes the time step and superscripts index the time level. Figure 5-14 presents a solution to the step-wave propagation problem using the Crank-Nicolson scheme that results when $\theta = 0.5$. The oscillatory behavior in this numerical solution is quite unsatisfactory.

Orthogonal collocation with upstream bias.

To eliminate spurious oscillations we can use either one of two distinct techniques. The first is to develop asymmetric basis functions analogous to the upstream weighting schemes described earlier. In this case the desired asymmetry can be achieved by using the modified Hermite polynomials introduced in Chapter Four. The new functions have the forms

(5.6-36a)　　　　　$h_j^{0,a}(x) = h_j^0(x) + \alpha_0 w(x),$

(5.6-36b)　　　　　$h_{j+1}^{0,a}(x) = h_{j+1}^0(x) + \alpha_0 w(x),$

(5.6-36c)　　　　　$h_j^{1,a}(x) = h_j^1(x) + \alpha_1 w(x) h,$

(5.6-36d)　　　　　$h_{j+1}^{1,a}(x) = h_{j+1}^1(x) + \alpha_1 w(x) h.$

Here, the function w generating the asymmetry is a piecewise quartic polynomial that vanishes except on the element $[x_j, x_{j+1}]$, where it is given by

(5.6-37)　　　　　$w(x) = \dfrac{16}{h^4}(x - x_j)^2 (x - x_j - h)^2.$

The undetermined parameters α_0 and α_1 occurring in (5.6-36) have to be chosen depending on the numerical behavior of the system. When $\alpha_0 = \alpha_1 = 0$, we recover the standard orthogonal collocation.

　　　The second approach involves moving the collocation points. To impose upstream bias on the standard collocation formulation, we simply shift the collocation points away from the Gauss points in an upstream direction. In particular, if the coefficient b in the advective term is positive and we denote by $x_{i\alpha}$ a two-point Gauss quadrature point in a typical element $[x_i, x_i + h]$, then the corresponding upstream collocation point has the form $x_{i\alpha}^* = x_{i\alpha} - \varsigma h$. Here $\varsigma > 0$ is a parameter chosen so that $x_{i\alpha}^*$ still lies inside the element $[x_i, x_i + h]$. As with all of the other asymmetric schemes discussed in this section, it is important to use the upstream bias in the advective term only, yielding collocation equations of the form

$$\frac{\partial \hat{u}}{\partial t}(x_{i\alpha}) + b\frac{\partial \hat{u}}{\partial x}(x_{i\alpha}^*) = 0, \qquad j = 1, \ldots, M.$$

　　　It is easy to see how this scheme causes numerical diffusion. If we expand the piecewise quadratic function $\partial \hat{u}/\partial x$ in a Taylor series about $x_{i\alpha}$, we get

$$\frac{\partial \hat{u}}{\partial x}(x_{i\alpha}^*) = \frac{\partial \hat{u}}{\partial x}(x_{i\alpha}) - \varsigma h\frac{\partial^2 \hat{u}}{\partial x^2}(x_i\alpha) + \mathcal{O}(h^2).$$

Thus, our upstream collocation equation becomes

$$\frac{\partial \hat{u}}{\partial t}(x_{i\alpha}) + b\frac{\partial \hat{u}}{\partial x}(x_{i\alpha}) - \varsigma h b\frac{\partial^2 \hat{u}}{\partial x^2}(x_{i\alpha}) = \mathcal{O}(h^2).$$

The third term on the left side of this equation clearly mimics the effects of diffusion, the effective numerical diffusion coefficient $\varsigma h b$ vanishing as $\varsigma \to 0$ or $h \to 0$. Allen and Pinder (1983) and Allen (1983) examine this scheme in more detail.

Dissipation and dispersion.

In Section 5.4 we introduced the concepts of dissipation and dispersion in connection with the finite-difference solution of hyperbolic PDEs. The extension of this concept to finite-element methods is not difficult in concept. In practice, however, finite-element analysis is generally more algebraically complicated. We now undertake an analysis of dissipation and dispersion for these methods, since no other approach provides the same degree of insight into the behavior of numerical solutions to hyperbolic equations.

Recall from Equation (5.4-32) that the ratio

$$(5.6\text{-}38) \qquad\qquad R = \frac{\xi}{\xi_e}$$

provides a measure of the error in the propagation of a wave component in a numerical solution. In particular, $|R|$ measures the error in the amplification between time levels. Generally, $|R| < 1$ implies numerical dissipation or damping, while for a wave component $e^{i\beta x}$ the quantity $b' - b = -(1/\beta h)\tan^{-1}[\Im(R)/\Re(R)]$ measures numerical dispersion. The numerical phase change of the wave component $e^{i\beta x}$ after one time step is the following:

$$(5.6\text{-}39) \qquad\qquad \theta_n(\beta) = \tan^{-1}\frac{\Im(\xi)}{\Re(\xi)}.$$

Let $L(\beta)$ be the wavelength of the wave component $e^{i\beta x}$, so that

$$(5.6\text{-}40) \qquad\qquad N(\beta) = \frac{L(\beta)}{bk}$$

represents the number of time steps of length k necessary for this component to propagate through one wavelength. After this component has propagated a distance $L(\beta)$, the value of the numerical phase angle will be $N\theta_n$, while the correct phase angle of the component in the exact solution is 2π. Thus, by analogy with Equation (4.3-26), the phase-lag error is as follows:

$$(5.6\text{-}41) \qquad\qquad \Theta(\beta) = N\theta_n - 2\pi.$$

Let us apply this analysis to the Galerkin finite-element method. We shall develop the theory for the asymmetric approach since we can easily recover the standard Galerkin scheme by setting $\alpha = 0$. Recall Equation (5.6-22), which says

$$(5.6\text{-}22) \quad h\left(\frac{1}{6}+\frac{\alpha}{4}\right)\frac{dv_{i-1}}{dt} + \frac{2h}{3}\frac{dv_i}{dt} + h\left(\frac{1}{6}-\frac{\alpha}{4}\right)\frac{dv_{i+1}}{dt}$$

$$= \frac{b}{2}(1+\alpha)v_{i-1} - \alpha b v_i - \frac{b}{2}(1-\alpha)v_{i+1}.$$

Using a weighted finite-difference approximation for the time derivatives yields

$$(5.6\text{-}42) \quad \left[\left(\frac{1}{6}+\frac{\alpha}{4}\right) - \theta\overline{C}\frac{1+\alpha}{2}\right]v_{i-1}^{r+1} + \left[\tfrac{2}{3}+\theta\alpha\overline{C}\right]v_i^{r+1}$$

$$+ \left[\left(\frac{1}{6}-\frac{\alpha}{4}\right) + \theta\overline{C}\frac{1-\alpha}{2}\right]v_{i+1}^{r+1}$$

$$= \left[\left(\frac{1}{6}+\frac{\alpha}{4}\right) + (1-\theta)\overline{C}\frac{1+\alpha}{2}\right]v_{i-1}^{r}$$

$$+ \left[\tfrac{2}{3} - (1-\theta)\alpha\overline{C}\right]v_i^{r} + \left[\left(\frac{1}{6}-\frac{\alpha}{4}\right) - (1-\theta)\overline{C}\frac{1-\alpha}{2}\right]v_{i+1}^{r},$$

where $\overline{C} = bk/h$. Consider a typical wave component of the numerical solution at a time level r, having the form

$$(5.6\text{-}43) \qquad\qquad v_s^r(\beta) = e^{i\beta sh}.$$

At the next time level, we have

$$(5.6\text{-}44) \qquad\qquad v_s^{r+1}(\beta) = \xi(\beta)e^{i\beta sh},$$

where $\xi(\beta)$ denotes the amplification factor for the wave number β. Substitution of the typical wave components (5.6-43) and (5.6-44) into the discrete equation (5.6-42) yields

$$(5.6\text{-}45) \quad \xi\left[\frac{2}{3} + \theta\alpha\overline{C} + e^{-i\beta h}\left(\frac{1}{6}+\frac{\alpha}{4} - \theta\overline{C}\frac{1+\alpha}{2}\right)\right.$$

$$\left. + e^{i\beta h}\left(\frac{1}{6}-\frac{\alpha}{4}+\theta\overline{C}\frac{1-\alpha}{2}\right)\right]$$

$$= \frac{2}{3} - (1-\theta)\alpha\overline{C} + e^{-i\beta h}\left[\frac{1}{6}+\frac{\alpha}{4}+(1-\theta)\overline{C}\frac{1+\alpha}{2}\right]$$

$$+ e^{i\beta h}\left[\frac{1}{6}-\frac{\alpha}{4}-(1-\theta)\overline{C}\frac{1-\alpha}{2}\right].$$

This equation is easily solved for the amplification factor $\xi(\beta)$. For the standard Galerkin formulation ($\alpha = 0$), we obtain

$$(5.6\text{-}46) \qquad \xi_G = \frac{\frac{2}{3} + e^{-i\beta h}\left[\frac{1}{6} + \frac{1}{2}\overline{C}(1-\theta)\right] + e^{i\beta h}\left[\frac{1}{6} - \frac{1}{2}\overline{C}(1-\theta)\right]}{\frac{2}{3} + e^{-i\beta h}\left(\frac{1}{6} - \frac{1}{2}\overline{C}\theta\right) + e^{i\beta h}\left(\frac{1}{6} + \frac{1}{2}\overline{C}\theta\right)}.$$

Plots of typical values of ξ_G appear in Figure 5-15.

According to Figure 5-15a, all wave components of the numerical solution propagate with the same amplitude as the corresponding components of the exact solution. This feature of the standard Galerkin approach may appear to be an asset. However, we shall see that in some instances numerical damping can, in fact, be an advantage. To understand how, consider the

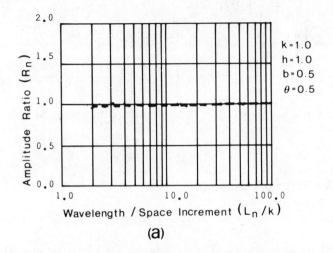

(a)

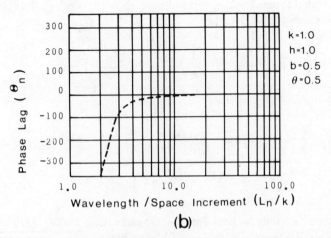

(b)

Figure 5-15 Amplification factor for the standard Galerkin formulation of the advection equation.

phase-lag diagram in Figure 5-15b. Waves longer than $10h$ propagate with very little phase-lag error, and for such waves propagation with the correct amplitude is indeed desirable. Shorter waves, however, exhibit significant phase-lag error. In fact, the wave having length $2h$ does not propagate at all.

The shorter-wavelength components that get out of phase are precisely the components responsible for the oscillatory behavior observed in the numerical solution to the steep front problem, depicted in Figure 5-12. The case shown in this figure is particularly sensitive to phase error, because the accurate propagation of a sharp front requires the high-frequency resolution associated with small-wavelength components. Thus, if the shorter wave components could be propagated more accurately, the solution quality would improve. Figure 5-12 suggests that this improvement can be achieved by reducing the spatial increment h. This strategy obviously increases the computational effort required to solve the problem numerically. Another possibility is the use of higher-order methods, that is, methods whose error shrinks faster as $h \rightarrow 0$. In particular, the use of higher-degree basis functions will generally enhance the accuracy of the scheme, but, as Gray and Pinder (1976) demonstrate, the improvement in sharp-front representation will not be as marked as one might anticipate. Higher-order methods also have the disadvantage that they require an increase in computational effort, typically by yielding matrix equations whose structure is less sparse than for low-order methods.

A third possible course of action is to eliminate, through numerical damping, those wave components that exhibit deleterious phase-lag error. In this case, the numerical front, devoid of these small-wavelength components, will no longer be sharp but rather will be artificially smoothed or smeared. This smearing is precisely the effect of numerical diffusion. In the standard Galerkin case numerical dissipation is easily introduced through the approximation of the time derivative. By setting $\theta = 0.5$, we have used the centered-in-time approximation in this example. This temporal weighting generates no numerical dissipation. As we increase θ from 0.5 to 1.0, however, numerical dissipation begins to appear. One can see this by comparing the magnitudes of the numerator and denominator of the amplification factor in Equation (5.6-46).

Now let us examine the wave propagating characteristics of the modified Galerkin scheme employing asymmetric weighting functions. As mentioned previously, the use of asymmetric weighting functions throughout the weighted-residual equations, as implemented in Equation (5.6-22), typically lead to nonconvergent schemes. Instead, we can obtain a convergent scheme such as Equation (5.6-26) by employing symmetric weighting on the temporal term and asymmetric weighting on the spatial term in the PDE. Substitution of the typical wave components (5.6-43) and (5.6-44) into this latter scheme yields, after some manipulation,

$$(5.6\text{-}47) \quad \xi_A = \left\{ \frac{2}{3} - (1-\theta)\overline{C}\alpha + e^{-i\beta h}\left[\frac{1}{6} + (1-\theta)\overline{C}\frac{1+\alpha}{2}\right] \right.$$

$$+ e^{i\beta h}\left[\frac{1}{6} - (1-\theta)\overline{C}\frac{1-\alpha}{2}\right]\Bigg\}$$

$$\cdot \left[\frac{2}{3} + \theta\overline{C}\alpha + e^{-i\beta h}\left(\frac{1}{6} - \theta\overline{C}\frac{1+\alpha}{2}\right)\right.$$

$$\left. + e^{i\beta h}\left(\frac{1}{6} + \theta\overline{C}\frac{1-\alpha}{2}\right)\right]^{-1}.$$

Figure 5-12 illustrates the effect of the asymmetric weighting functions. Observe that, as α increases from zero, the spurious oscillations in the numerical solution decrease in amplitude. Simultaneously, the sharp front becomes artificially smeared, owing to numerical diffusion. The trade-off between numerical diffusion and spurious oscillations is evident. Since both types of error are qualitatively incorrect, the decision as to the optimal value of α is a largely subjective one. In some nonlinear problems, oscillations can lead to nonconvergent numerical solutions and are therefore intolerable. More often, however, the choice between smearing and oscillations is simply a question of esthetics.

To understand the effect of asymmetric weighting on wave components of the solution, it is necesary to compare Figures 5-16 and 5-15. This comparison reveals that the phase-lag error of the finite-element scheme is essentially unaffected by the introduction of upstream weighting in the approximation to the spatial derivative. The amplitude ratio, however, experiences marked change. In particular, for the upstream weighted scheme a broad spectrum of components undergo severe damping. In fact, those components that exhibit the most significant phase-lag error are also those that experience the most numerical damping. Thus one can conclude that the dissipative process eliminates the spurious oscillations through selective damping of specific wave components in the numerical solution. Unfortunately, this damping also destroys those wave components needed to resolve the sharp front. The solution now exhibits the smeared appearance characteristic of numerical diffusion, as shown in Figure 5-12.

Finally, let us present the results reported in Lapidus and Pinder (1982) for the analysis of the collocation schemes used to generate Figure 5-14. For symmetric basis functions and a centered difference approximation in time, we obtain the amplitude ratios and phase-lag errors plotted in Figure 5-17. As expected, this standard scheme imposes no numerical damping on the numerical solution. The phase-lag error for this scheme is smaller for standard collocation than for the other schemes we have considered, even when we account for the fact that the Hermite cubic trial function requires two coefficients per spatial node.

For collocation using asymmetric weighting functions, the results for $\alpha_0 = 0.16$ and $\alpha_1 = 0.09$ appear in Figure 5-18. The amplitude ratio

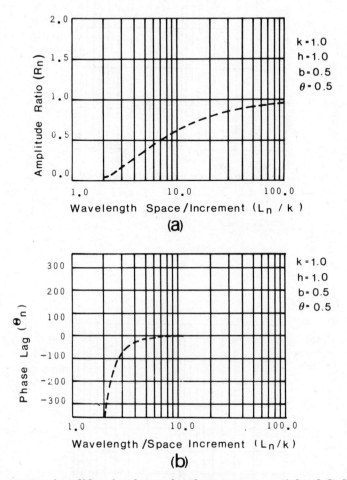

Figure 5-16. Amplification factor for the upstream-weighted Galerkin formulation of the advection equation.

is modified, as expected, by the introduction of asymmetry. The small-wavelength components are numerically damped. Also, for the first time in our discussion of finite-element methods, we see a phase acceleration. We do not observe spurious oscillations ahead of the front, however, because the wave components that would give rise to these oscillations are numerically damped.

In summary, we conclude that phase-lag errors of small-wavelength components lead to spurious oscillations in both finite-difference and finite-element schemes. The troublesome wave components can be suppressed by introducing numerical dissipation. One way of achieving this is by using asymmetric weighting functions or, in the case of collocation schemes, by

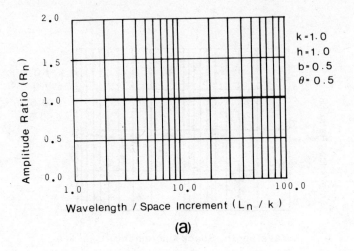

(a)

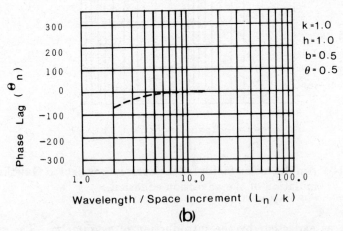

(b)

Figure 5-17. Amplitude ratios and phase lag errors for the standard collocation formulation of the advection equation.

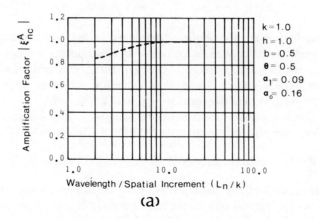

(a)

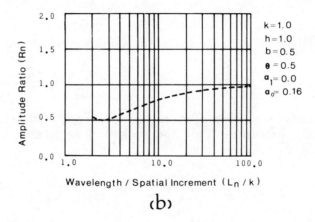

(b)

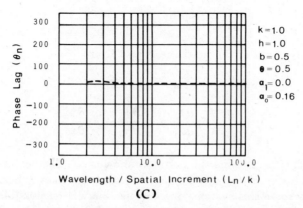

(C)

Figure 5-18. Amplitude ratios and phase lag errors for the upstream-weighted collocation formulation of the advection equation.

using asymmetric basis functions or displaced collocation points. However, the wave components that are suppressed in this manner are precisely those required to resolve sharp fronts. Consequently, the smoothed solutions are also artificially smeared and thus resemble solutions appropriate for problems where physical dissipation is present. For problems in which the numerical solutions are representable by long-wavelength components, the spurious short-wavelength oscillations should be less troublesome.

The finite-element method in several dimensions.

The extension of the procedures discussed above to several space dimensions can be carried out along lines very similar to those presented in Section 4.4. We shall illustrate this extension briefly, using as a model equation the following PDE in three space dimensions:

$$(5.6\text{-}48) \qquad \frac{\partial u}{\partial t} + b_1 \frac{\partial u}{\partial x_1} + b_2 \frac{\partial u}{\partial x_2} + b_3 \frac{\partial u}{\partial x_3} = 0.$$

To obtain a Galerkin approximation, we define the trial function

$$(5.6\text{-}49) \qquad u(t, x_1, x_2, x_3) \simeq \hat{u}(t, x_1, x_2, x_3) = \sum_{j=1}^{N} v_j(t) \phi_j(x_1, x_2, x_3),$$

accommodating time dependence in the nodal coefficients, as done above in one space dimension. Then the method of weighted residuals yields Galerkin equations of the form

$$(5.6\text{-}50) \qquad \sum_{j=1}^{N} \int_{\Omega} \left[\frac{dv_j}{dt} + v_j \left(b_1 \frac{\partial \phi_j}{\partial x_1} + b_2 \frac{\partial \phi_j}{\partial x_2} + \frac{\partial \phi_j}{\partial x_3} \right) \right] \phi_i \, dx = 0.$$

In matrix form, these Galerkin equations can be written as a system of ordinary differential equations,

$$(5.6\text{-}51) \qquad \mathbf{A} \frac{d\mathbf{u}}{dt} + \mathbf{C}\mathbf{u} = \mathbf{f}.$$

The typical entries of the matrices $\mathbf{A}$ and $\mathbf{C}$ are

$$(5.6\text{-}52\text{a}) \qquad a_{ij} = \int_{\Omega} \phi_j \phi_i \, dx,$$

$$(5.6\text{-}52\text{b}) \qquad c_{ij} = \int_{\Omega} \left(b_1 \frac{\partial \phi_j}{\partial x_1} + b_2 \frac{\partial \phi_j}{\partial x_2} + b_3 \frac{\partial \phi_j}{\partial x_3} \right) \phi_i \, dx,$$

and the vector $\mathbf{u}$ contains the nodal unknowns v_i. The vector $\mathbf{f}$ appearing on the right side of the system (5.6-51) contains known boundary information. To solve this system, one can use any of the possible approaches mentioned earlier. In fact, much of the material presented in Section 4.4 is directly applicable to multidimensional hyperbolic equations.

5.7. Problems for Chapter Five.

1. Show that, in an incompressible fluid, the equation governing solute transport with linear decay is as follows:

$$\frac{\partial \rho^S}{\partial t} + \mathbf{v} \cdot \nabla \rho^S + (\nabla \cdot \mathbf{v} + \gamma) \rho^S = 0.$$

Consider the case when two reacting solutes, of concentrations ρ^1 and ρ^2, have the same velocity. We have

$$\frac{\partial \rho^\alpha}{\partial t} + \mathbf{v} \cdot \nabla \rho^\alpha + (\nabla \cdot \mathbf{v}) \rho^\alpha + \gamma^\alpha = 0, \quad \alpha = 1, 2.$$

The most general linear relation for the reaction terms γ^α is $\boldsymbol{\gamma} = \mathbf{G}\boldsymbol{\rho}$, where $\mathbf{G}$ is a 2×2 matrix, while $\boldsymbol{\gamma}$ and $\boldsymbol{\rho}$ denote the vectors (γ^1, γ^2) and (ρ^1, ρ^2), respectively. Prove that total mass of the solutes is conserved if and only if the entries of $\mathbf{G}$ satisfy the relationship

$$g_{11} + g_{21} = g_{21} + g_{22} = 0.$$

There are applications in which the transport velocity may not be the same for both solutes. Thus, the transport velocities could be $\mathbf{v}^1 \neq \mathbf{v}^2$. Prove that in this case, too, the relationship just stated is a necessary and sufficient condition for mass conservation.

2. Consider the one-dimensional version of Problem 1. Assume that there is conservation of mass, that $v^1 = v^2$ and both are constant, and that $\gamma_{11} = \gamma_{22} = 1$. Prove that the total concentration of solute satisfies Equation (5.1-2). Use this fact to simplify the method of solution, and write a computer program, using any suitable numerical procedure, to solve this problem with the following initial and boundary conditions:

$$\rho^1(x, 0) = \rho^2(x, 0) = 0, \quad \rho^1(0, t) = .08, \quad \rho^2(0, t) = .01.$$

Consider the same problem with $v^2 = 1.2v^1$. Show that, in this case, the method of solution cannot be simplified by use of equation (5.1-2). Supply a numerical formulation of this problem anyway.

3. This problem consists in the study of the motion produced by pistons on a compressible fluid (air, for example), contained in a tube of uniform cross section. A complete treatment of this problem is not simple, since the governing equation is nonlinear and shocks may form. However, if the motion of the piston is sufficiently slow and viscous effects can be neglected, the motion of the fluid is essentially horizontal. In this case the velocity u satisfies the linear wave equation (5.1-19). Consider a tube extending from

$x = -1$ to $x = 1$ containing a homogeneous fluid, so that $a^2 = \text{constant}$. Obtain the solution to this linear problem by defining the functions $\phi(\sigma)$ and $\psi(\sigma)$ for $-1 \leq \sigma \leq \infty$ so that (5.2-29) holds. Assume that the fluid in the tube is initially at rest and that the velocities of two pistons moving at the end points $x = -1$ and $x = 1$ are given by $g_1(t)$ and $g_2(t)$, $t \geq 0$, respectively.

4. Determine the truncation error and carry out a stability analysis for each one of the finite-difference schemes shown in Table 5.1.

5. Use central difference approximations to transform the leapfrog formula (5.4-3) into one which involves only the values $u_{r-1,s-1}$, $u_{r-1,s+1}$, $u_{r+1,s-1}$, and $u_{r+1,s+1}$. Show that, by replacing $2k$ with k and $2h$ with h and making the corresponding adjustments in the subscripts, one obtains the scheme

$$u_{r+1,s+1} = u_{r,s} + \epsilon(u_{r,s+1} - u_{r+1,s}),$$

where

$$\epsilon = \frac{1-C}{1+C}.$$

This is a two-level scheme. Show that it is unconditionally stable. Construct the corresponding dispersion curves for different values of ϵ (including $\epsilon = 0$), and compare these curves with those obtained using other algorithms.

6. Solve the step propagation problem

$$\frac{\partial u}{\partial t} + \frac{\partial u}{\partial x} = 0, \quad (x,t) \in (0,\infty) \times (0,\infty)$$

$$u(0,t) = 1, \quad t \geq 0,$$

$$u(x,0) = 0, \quad x > 0,$$

using the scheme developed in Problem 5. Try several values of ϵ, and compare the results with those obtained using $\epsilon = 1$.

7. Use Equation (5.5-36) to show that the jump conditions for the nonlinear wave equation are as follows:

$$\lfloor \rho \rfloor v_\Sigma = \lfloor \rho u \rfloor = \bar{\rho} \lfloor u \rfloor + \bar{u} \lfloor \rho \rfloor$$

$$\lfloor \rho u \rfloor v_\Sigma = \lfloor p + \rho u^2 \rfloor = \lfloor p \rfloor + \bar{\rho} u \lfloor u \rfloor + \bar{u} \lfloor \rho u \rfloor,$$

where the bar stands for the average across the discontinuity of the left and right limits of the corresponding function. Thus, for example

$$\bar{u} = \tfrac{1}{2}(u_+ + u_-).$$

Show that, for perturbations from a continuous initial state, these jump conditions reduce to the following:

$$\lfloor \rho \rfloor v_\Sigma = \rho_0 \lfloor u \rfloor + u_0 \lfloor \rho \rfloor$$
$$(\rho_0 \lfloor u \rfloor + u_0 \lfloor \rho \rfloor) \, v_\Sigma = a_0^2 \lfloor \rho \rfloor + u_0^2 \lfloor \rho \rfloor + 2\rho_0 u_0 \lfloor u \rfloor.$$

Prove that these two equations imply $v_\Sigma = u_0 \pm a$. This equation was used when we obtained the solution of Riemann's problem for the linearized version of (5.5-35), with $u_0 = 0$.

8. In some sense, the simplest functional relationship between p and ρ is a linear one, as follows:

$$p = p_0 + a^2 \rho,$$

where a^2 is a constant. Assuming this form, solve Riemann's problem for the nonlinear wave equation (5.5-35), using the jump conditions Developed in Problem 7. Develop the solution given $a^2 = 1$, $\rho_L = .9$, $u_L = .3$, $\rho_R = 1.1$, and $u_R = .2$.

9. Equation (5.6-1) is first-order. Because of this fact, for a solution to be admissible, it is only required that it be continuous. Using this property one can relax the continuity conditions imposed in the derivation of the orthogonal collocation solution. Indeed, one can replace Equation (5.6-29) by Equation (5.6-4), with $\phi_1(x)$ given by Equation (5.6-12). For this choice only one collocation point at each subinterval of the partition can be used. For this collocation point to be a Gauss quadrature point, it must be the midpoint in each subinterval. Prove that this choice yields the scheme (5.7-4).

10. Develop a computer program to solve the advection equation using the asymmetric Galerkin scheme for piecewise linear basis functions, leaving the weighting parameter α free. Verify the results shown in Figure 5-12 for the step front problem.

5.8. References.

Allen, M.B. and Pinder, G.F., "Collocation simulation of multiphase porous-medium flow," *Soc. Pet. Eng. J.* (1983), 135-142.

Allen, M.B., "How upstream collocation works," *Int. J. Num. Meth. Eng. 19:12* (1983), 1753-1763.

Barret, J.W. and Morton, K.W., "Approximate symmetrization and Petrov-Galerkin methods for diffusion-convection problems," *Comput. Methods Appl. Mech. Eng., 45* (1984), 97-122.

Brenier, Y. and Hennart, J.P., "Introduction to Numerical Hyperbolic Equations," *Communicaciones Tecnicas IIMAS-UNAM, 84* (1985).

Eringen, A.C., *Mechanics of Continua*, 2nd ed., Huntington, New York: Krieger, 1980.

Fletcher, G.A.J., *Computational Galerkin Methods*, New York: Springer-Verlag, 1984.

Gray, W.G. and Pinder, G.F., "An analysis of the numerical solution of the transport equation," *Water Resour. Res., 12:3* (1976), 547-555.

Holt, M., *Numerical Methods in Fluid Dynamics*, New York: Springer-Verlag, 1984.

Isaacson, E.L., "Global solution of a Riemann problem for a non-strictly hyperbolic system of conservation laws arising in enhanced oil recovery", *J. Comp. Phys.*, (to be published).

Lapidus, L. and Pinder, G.F., *Numerical Solution of Partial Differential Equations in Science and Engineering*, New York: Wiley, 1982.

Meyer, R.E., *Introduction to Mathematical Fluid Dynamics*, New York: Wiley, 1971.

Morton, K.W., "Generalized Galerkin methods for hyperbolic problems," *Comput. Methods Appl. Mech. Eng., 52* (1985), 847-871.

Peyret, R. and Taylor T.D., *Computational Methods for Fluid Flow*, New York: Springer-Verlag, 1983.

Richtmyer, R.D. and Morton, K.W., *Difference Methods for Initial-Value Problems*, 2nd ed., New York: Wiley, 1967.

Roberts, K.V. and Weiss, N.O., "Convective difference schemes," *Math. Comp., 20* (1966), 272-285.

Vichnevetsky, R. and Bowles, J.B., *Fourier Analysis of Numerical Approximations of Hyperbolic Equations*, Philadelphia: Society for Industrial and Applied Mathematics, 1982.

Vichnevetsky, R. and Peiffer, B., "Error waves in finite element and finite difference methods for hyperbolic equations," in *Advances in Computer Methods for Partial Differential Equations,* R. Vichnevetsky, Ed., Bethlehem, Pennsylvania: AICA, 1975, pp. 53-72.

Warming, R.F. and Hyett, B.J., "The modified equation approach to the stability and accuracy analysis of finite difference methods," *J. Comput. Phys., 14* (1974), 159-179.

Williamson, R.E, Crowell, R.H., and Trotter, H.F., *Calculus of Vector Functions*, 3rd ed., Englewood Cliffs, New Jersey: Prentice-Hall, 1972.

CHAPTER SIX
HIGH-ORDER, NONLINEAR,
AND COUPLED SYSTEMS

6.1. Introduction.

So far in our exposition we have focused primarily on numerical methods for solving first- or second-order linear PDEs. While this class of equations is important and quite broad, in many problems of interest in science and engineering the appropriate PDEs may fall outside its boundaries. For example, some PDEs arising in structural mechanics involve derivatives of order higher than two. The biharmonic equation developed in Section 1.4 is one of the important paradigms in this regard. Also, many of the PDEs occurring in applications are nonlinear. While we have devoted some attention to nonlinear hyperbolic equations in Chapter Five, we have yet to examine the numerical treatment of nonlinear PDEs of elliptic or parabolic type. Finally, a plethora of applications require the solution of simultaneous, coupled PDEs. These applications are so numerous that we cannot hope to give a systematic treatment of them all here. We shall, however, give an overview of two important coupled systems, namely, the problem of a deforming solid and the problem of simultaneous flow of oil, gas, and water in a petroleum reservoir.

6.2. The Biharmonic Equation.

Let us begin by considering the biharmonic equation,

$$(6.2\text{-}1) \qquad\qquad \nabla^4 u = 0,$$

on a two-dimensional domain Ω, where $\nabla^4 = \partial^4/\partial x^4 + 2\partial^4/\partial x^2 \partial y^2 + \partial^4/\partial y^4$ in Cartesian coordinates. As Section 1.4 explains, this equation governs the Airy stress function for plane strain in an elastic solid. The equation arises in related contexts as well. For example, it governs the transverse deflection of a two-dimensional elastic plate under a transverse load. Two sets of boundary conditions commonly occur for this equation. The first set is

$$(6.2\text{-}2) \qquad \begin{aligned} u(\mathbf{x}) &= \alpha(\mathbf{x}) \;, \\ \nabla^2 u(\mathbf{x}) &= \beta(\mathbf{x}) \;, \end{aligned}$$

for $\mathbf{x} \in \partial\Omega$. When $\alpha(\mathbf{x}) \equiv \beta(\mathbf{x}) \equiv 0$ in flat-plate applications, these boundary conditions model a plate with simply supported edges. The second set is

$$(6.2\text{-}3) \qquad u(\mathbf{x}) = \frac{\partial u}{\partial n}(\mathbf{x}) = 0 \;, \quad \mathbf{x} \in \partial\Omega \;,$$

where $\partial u/\partial n = \nabla u \cdot \mathbf{n}$ stands for the outward normal derivative of u on $\partial\Omega$. These boundary conditions represent a plate with clamped edges.

There is a noteworthy difference between boundary conditions (6.2-2) and (6.2-3). When we impose the simply-supported plate conditions (6.2-2), the boundary-value problem for Equation (6.2-1) admits a factored form

$$(6.2\text{-}4) \qquad \begin{aligned} \nabla^2 u(\mathbf{x}) &= v(\mathbf{x}) \;, & \mathbf{x} &\in \Omega \;, \\ u(\mathbf{x}) &= \alpha(\mathbf{x}) \;, & \mathbf{x} &\in \partial\Omega \;; \\ \nabla^2 v(\mathbf{x}) &= 0 \;, & \mathbf{x} &\in \Omega \;, \\ v(\mathbf{x}) &= \beta(\mathbf{x}) \;, & \mathbf{x} &\in \partial\Omega \;. \end{aligned}$$

Thus a single fourth-order problem reduces to a coupled pair of second-order problems. The boundary-value problem for Equation (6.2-1) with clamped-edge conditions admits no such factorization. As we shall see, this observation has significant implications for discrete solution techniques.

Finite-difference approximation: single-equation approach.

We can derive finite-difference approximations to Equation (6.2-1) through straightforward application of the central difference operators defined in Section 2.6. Let us distinguish these operators acting in the x- and y-directions by denoting them as δ_x and δ_y. Assuming that the domain Ω is a rectangle $(a, b) \times (c, d)$, we can establish a two-dimensional grid $\Delta = \Delta_x \times \Delta_y$, where $\Delta_x : (a =)x_o < \cdots < x_n(= b)$ and $\Delta_y : (c =)y_o < \cdots < y_m(= d)$ as in Section 2.5. For convenience, let Δ be uniform in each coordinate direction, so that $x_i - x_{i-1} = h$ and $y_j - y_{j-1} = k$. Thus we can approximate the operator $\nabla^4 = \partial^4/\partial x^4 + 2\partial^4/\partial x^2 \partial y^2 + \partial^4/\partial y^4$ by writing difference analogs for each term as follows:

$$\left.\frac{\partial^4 u}{\partial x^4}\right|_{(x_i,y_j)} = \frac{1}{h^4}\delta_x^2(\delta_x^2 u_{i,j}) + \mathcal{O}(h^2)$$

$$= \frac{1}{h^4}\delta_x^2(u_{i-1,j} - 2u_{i,j} + u_{i+1,j}) + \mathcal{O}(h^2)$$

$$= \frac{1}{h^4}(u_{i-2,j} - 4u_{i-1,j} + 6u_{i,j} - 4u_{i+1,j}$$
$$+ u_{i+2,j}) + \mathcal{O}(h^2) .$$

Similarly,

$$\left.\frac{\partial^4 u}{\partial y^4}\right|_{(x_i,y_j)} = \frac{1}{k^4}(u_{i,j-2} - 4u_{i,j-1} + 6u_{i,j}$$
$$- 4u_{i,j+1} + u_{i,j+2}) + \mathcal{O}(k^2) .$$

Finally,

$$\left.\frac{\partial^4 u}{\partial x^2 \partial y^2}\right|_{(x_i,y_j)} = \frac{1}{h^2}\delta_x^2\left(\frac{1}{k^2}\delta_y^2 u_{i,j}\right) + \mathcal{O}(h^2 + k^2)$$

$$= \frac{1}{h^2 k^2}\Big[u_{i+1,j+1} + u_{i-1,j+1} + u_{i-1,j-1} + u_{i+1,j-1}$$
$$- 2\left(u_{i+1,j} + u_{i,j+1} + u_{i-1,j} + u_{i,j-1}\right) + 4u_{i,j}\Big]$$
$$+ \mathcal{O}(h^2 + k^2) .$$

Combining these approximations, we arrive at a second-order difference analog for $\nabla^4 u = 0$, each equation of which couples 13 unknown nodal values of u. Figure 6-1 illustrates a typical difference molecule for the grid point (x_i, y_j), along with the weights corresponding to each nearby node in the difference approximation.

It is clear from this diagram that a typical row in the matrix arising from this discretization will contain 13 nonzero entries.

A difference molecule of this size leads to rather special considerations for treating boundary conditions, since boundary data will affect all difference equations centered no more than one node away from $\partial\Omega$. To accommodate the coupling over as many as five adjacent nodes, as shown in Figure 6-1, let us establish a layer of fictitious nodes around the boundary, as depicted for a representative boundary segment in Figure 6-2. We must now establish a one-to-one correspondence between finite-difference equations and nodal unknowns, using the boundary conditions. Consider first the simply-supported plate conditions (6.2-2). At a typical boundary

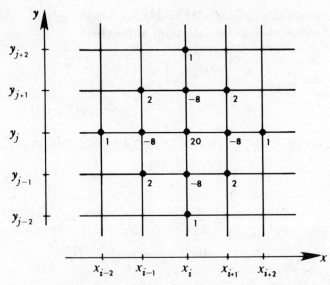

Figure 6-1. Thirteen-point difference molecule for the biharmonic operator ∇^4 in two dimensions, with numerical weights assigned to each node.

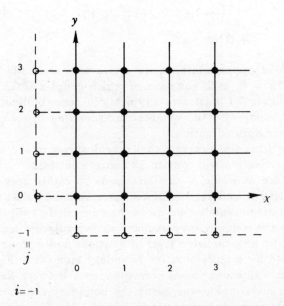

Figure 6-2. Arrangement of fictitious nodes near the boundary for the finite-difference analog of the biharmonic equation.

node like (x_0, y_0) or (x_0, y_2) the Dirichlet boundary condition $u(\mathbf{x}) = \alpha(\mathbf{x})$ gives explicit equations for the corresponding nodal value $u_{0,0}$ or $u_{0,2}$. The difficulty arises when we try to write the 13-point difference approximation along the interior layer of nodes, such as (x_1, y_1), (x_2, y_1), (x_1, y_2), and so forth, that lie adjacent to the boundary. Here we need to use the fictitious nodes to accommodate the fact that the difference equation calls for values of u at nodes lying outside the actual boundary $\partial\Omega$ of the domain. For example, when the grid meshes h and k in the x- and y-directions are equal, the difference equation associated with the unknown value $u_{1,2}$ at the node (x_1, y_2) requires

$$
\begin{aligned}
&20u_{1,2} - 8(u_{1,3} + u_{2,2} + u_{1,1} + u_{0,2}) \\
&\quad + 2(u_{0,3} + u_{2,3} + u_{2,1} + u_{0,1}) \\
&\quad + u_{1,4} + u_{3,2} + u_{1,0} + u_{-1,2} = 0 \ .
\end{aligned}
$$

(6.2-5)

Having added these fictitious nodes, we now need to ensure that they do not spoil the balance between equations and unknowns. Since there are difference equations centered at the nodal unknown $u_{1,3}$, $u_{2,2}$, $u_{1,1}$, $u_{2,3}$, $u_{2,1}$, $u_{1,4}$, and $u_{3,2}$ and Dirichlet boundary conditions assigning values to $u_{0,1}$, $u_{0,2}$, $u_{0,3}$, and $u_{0,1}$, we only need to eliminate the fictitious unknown $u_{-1,2}$ from Equation (6.2-5) to guarantee a one-to-one correspondence between equations and unknowns. Applying the difference approximation

$$
\frac{1}{h^2}\left(u_{-1,2} + u_{0,3} + u_{1,2} + u_{0,1} - 4u_{0,2}\right)
$$

$$
\simeq \nabla^2 u(x_0, y_2) = \beta(x_0, y_0)
$$

to the second-order boundary condition, we obtain

$$
u_{-1,2} = h^2 \beta(x_0, y_0) + 4u_{0,2} - u_{0,3} - u_{1,2} - u_{0,1} \ .
$$

Substituting this identity into Equation (6.2-5) yields an equation involving only interior nodal values and known boundary values. Keeping the unknowns on the left, this equation becomes

$$
\begin{aligned}
&19u_{1,2} - 8(u_{1,3} + u_{2,2} + u_{1,1}) + 2(u_{2,3} + u_{2,1}) + u_{1,4} + u_{3,2} \\
&= 4\alpha(x_0, y_2) - \alpha(x_0, y_3) - \alpha(x_0, y_1) - h^2\beta(x_0, y_0) \ .
\end{aligned}
$$

This same technique applies to the elimination of fictitious nodal values from equations centered at other nodes along the interior layer.

A similar strategy works in the case of clamped-edge conditions (6.2-3). To eliminate the fictitious variable $u_{-1,2}$ from Equation (6.2-4) in this instance, we invoke a second-order difference approximation to the normal-derivative condition,

$$\frac{1}{2h}(u_{1,2} - u_{-1,2}) \simeq \left.\frac{\partial u}{\partial n}\right|_{(x_0,y_2)} = 0 \; .$$

This yields $u_{-1,2} = u_{1,2}$, so after accounting for the homogeneous Dirichlet boundary data we can rewrite Equation (6.2-5) as

$$21u_{1,2} - 8(u_{1,3} + u_{2,2} + u_{1,1}) + 2(u_{2,3} + u_{2,1}) + u_{1,4} + u_{3,2} = 0 \; ,$$

an equation involving no fictitious nodes.

The discretizations based on the 13-point difference molecule just described yield large, sparse sets of linear equations having the matrix form $\mathbf{A}\mathbf{u} = \mathbf{b}$, where $\mathbf{A}$ is the coefficient matrix, $\mathbf{u}$ is the vector of unknown nodal values, and $\mathbf{b}$ is a vector of known boundary information. The construction of this matrix equation is no different in principle from analogous constructions presented earlier, for example, in Chapter Three. However, in practical computations the sensitivity of $\mathbf{u}$ to perturbations in the matrix entries or in the forcing vector $\mathbf{b}$ can be a crucial consideration. In Chapter Three we encountered the important notion that, when the condition number $\text{cond}(\mathbf{A}) = \|\mathbf{A}\|\|\mathbf{A}^{-1}\|$ is very large, small errors in the entries of $\mathbf{A}$ or $\mathbf{b}$ can produce large errors in the computed values of $\mathbf{u}$. This notion is the basis for an even more important observation regarding the numerical solution of the biharmonic equation.

Consider first, for simplicity's sake, the usual finite-difference solution of Poisson's equation $\nabla^2 u = f$ on a rectangle using a two-dimension-al grid of uniform mesh. As we saw in Chapter Three, this problem yields a sparse linear system $\mathbf{A}_1\mathbf{u} = \mathbf{b}_1$ in which the matrix $\mathbf{A}_1$ is symmetric. For such matrices, as we have seen in Section 2.9, the Euclidean norm is $\|\mathbf{A}_1\|_2 = |\lambda_{\max}|$, the magnitude of the eigenvalue of $\mathbf{A}_1$ lying furthest from the origin. Since the eigenvalues of $\mathbf{A}_1^{-1}$ are the reciprocals of those of $\mathbf{A}_1$, it follows that $\|\mathbf{A}_1^{-1}\|_2 = 1/|\lambda_{\min}|$, which is finite provided $\mathbf{A}_1$ is nonsingular. Thus the condition number for difference analogs to Poisson's equation is $\text{cond}(\mathbf{A}_1) = |\lambda_{\max}/\lambda_{\min}|$. One can show that, for Poisson problems on uniform grids, $\lambda_{\max} = \mathcal{O}(h^{-2})$ and $\lambda_{\min} = \mathcal{O}(1)$ as $h \to 0$. Thus $\text{cond}(\mathbf{A}_1) = \mathcal{O}(h^{-2})$ as $h \to 0$, and as a result finite-difference approximations on fine grids tend to yield poorly conditioned matrix equations. In practice, overcoming this difficulty requires preconditioning, iterative improvement, or high-precision machine arithmetic, all of which imply greater computational expense.

In the case of the biharmonic equation the probem is even worse. In this case the 13-point discretization of $\nabla^4 u = \nabla^2(\nabla^2 u)$ typically yields a symmetric matrix $\mathbf{A}$ whose eigenvalues are *squares* of the eigenvalues for the matrix $\mathbf{A}_1$ of Poisson's problem on the same grid (Birkhoff and Lynch, 1984, Section 3.8). Thus for the biharmonic equation we can expect

$$\text{cond}(\mathbf{A}) = \frac{|\lambda_{\max}|}{|\lambda_{\min}|} = \frac{\mathcal{O}(h^{-4})}{\mathcal{O}(1)} = \mathcal{O}(h^{-4}).$$

This implies an even faster increase in condition number for fine grids than with second-order problems and a concomitantly faster increase in computational expense.

Finite-difference approximation: coupled-equation approach.

The reasoning just given suggests that it may be more efficient to solve the factored form of the biharmonic equation whenever this is possible. The strategy here is to replace a single set of nodal equations having condition number $\mathcal{O}(h^{-4})$ by a possibly sparser set of roughly twice as many equations having a much smaller condition number $\mathcal{O}(h^{-2})$. To do this, let us return to Equations (6.2-4), which we rewrite as follows:

$$\nabla^2 \begin{bmatrix} u \\ v \end{bmatrix} = \begin{bmatrix} v \\ 0 \end{bmatrix} \quad \text{on} \quad \Omega,$$

$$\begin{bmatrix} u \\ v \end{bmatrix} = \begin{bmatrix} \alpha \\ \beta \end{bmatrix} \quad \text{on} \quad \partial\Omega .$$

Now we can simply approximate $\nabla^2 u$ and $\nabla^2 v$ using the standard second-order difference analogs on the same grid Δ defined for the unfactored form treated above. Thus

$$\nabla^2 u = \frac{u_{i-1,j} - 2u_{i,j} + u_{i+1,j}}{h^2}$$
$$+ \frac{u_{i,j-1} - 2u_{i,j} + u_{i,j+1}}{k^2} + \mathcal{O}(h^2 + k^2),$$

and similarly for v. The resulting difference equations furnish a collection of algebraic equations having the form

$$\frac{1}{h^2}(u_{i-1,j} - 2u_{i,j} + u_{i+1,j}) + \frac{1}{k^2}(u_{i,j-1} - 2u_{i,j} + u_{i,j+1}) = v_{i,j},$$

$$\frac{1}{h^2}(v_{i-1,j} - 2v_{i,j} + v_{i+1,j}) + \frac{1}{k^2}(v_{i,j-1} - 2v_{i,j} + v_{i,j+1}) = 0,$$

at each interior node (x_i, y_j). While this approach requires the solution of twice as many equations as the 13-point difference formulations, the advantages in matrix sparseness and slower degradation in matrix condition number as $h \to 0$ often outweigh the disadvantages associated with increases in matrix order.

Finite-element approximations: single-equation approach.

With finite-element aproximations we face a similar choice between solving Equation (6.2-1) as one fourth-order PDE or as a coupled system of two second-order PDEs. Let us begin by examining the first option, using both the simply supported plate conditions (6.2-2) and the clamped-plate boundary conditions (6.2-3).

Before selecting a particular basis for the trial function $\hat{u}(\mathbf{x})$, let us examine the Galerkin formalism to see what requirements $\hat{u}$ will have to satisfy. Assume $\hat{u}$ has an expansion of the form

$$(6.2\text{-}6) \qquad \hat{u}(\mathbf{x}) = u_\partial(\mathbf{x}) + \sum_{i=1}^{M} u_i \phi_i(\mathbf{x}).$$

Here, the basis functions $\phi_i(\mathbf{x})$, $i = 1, \dots, M$, satisfy homogeneous boundary conditions, that is, $\phi_i(\mathbf{x}) = 0$ for $\mathbf{x} \in \partial\Omega$. We shall see in a moment what boundary values to impose on the function $u_\partial(\mathbf{x})$. Given such a trial function, the Galerkin integral equations are as follows:

$$\int_\Omega \nabla^4 \hat{u}(\mathbf{x}) \phi_j(\mathbf{x}) \, d\mathbf{x} = 0, \quad j = 1, \dots, M.$$

To reduce the formal smoothness constraints on $\hat{u}$ implied in these equations, we apply Green's theorem to shift differentiation from the trial function to the weighting functions $\phi_j(\mathbf{x})$. One application of Green's theorem yields

$$-\int_\Omega \nabla[\nabla^2 \hat{u}(\mathbf{x})] \cdot \nabla\phi_j(\mathbf{x}) \, d\mathbf{x} + \oint_{\partial\Omega} \phi_j(\mathbf{x}) \nabla[\nabla^2 \hat{u}(\mathbf{x})] \cdot \mathbf{n} \, d\mathbf{x} = 0,$$

and another gives

$$(6.2\text{-}7) \qquad \int_\Omega \nabla^2 \hat{u}(\mathbf{x}) \, \nabla^2 \phi_j(\mathbf{x}) \, d\mathbf{x} + \oint_{\partial\Omega} \phi_j(\mathbf{x}) \nabla[\nabla^2 \hat{u}(\mathbf{x})] \cdot \mathbf{n} \, dx$$

$$- \oint_{\partial\Omega} \nabla^2 \hat{u}(\mathbf{x}) \nabla\phi_j(\mathbf{x}) \cdot \mathbf{n} \, dx = 0.$$

Since the trial function $\hat{u}$ and the weighting functions ϕ_j possess the same degree of smoothness, there is no point in shifting differentiation between the two any further.

Before examining the treatment of boundary conditions, let us digress for a moment to make an observation that has important implications for coding. Equation (6.2-7) implies that the second derivatives $\nabla^2 \phi_j(\mathbf{x})$ and $\nabla^2 \hat{u}(\mathbf{x})$ must have, at worst, jump discontinuities between finite elements Ω_e if the first integral is to admit decomposition into a sum of integrals over individual elements, as in

$$\int_\Omega \nabla^2 \hat{u} \, \nabla^2 \phi_j \, d\mathbf{x} = \sum_e \int_{\Omega_e} \nabla^2 \hat{u} \, \nabla^2 \phi_j \, d\mathbf{x}.$$

Such a decomposition is highly desirable, since it allows elementwise computation of the matrix entries in a computer code. Therefore, to guarantee the validity of this decomposition, we must choose basis functions $\{\phi_j\}_{j=1}^N$ that have continuous gradients, that is, that belong to $C^1(\Omega)$. In general, basis functions that allow elementwise decomposition of the Galerkin volume integrals are called **conforming elements**.

Now we return to the issue of boundary conditions. The presence of the boundary integrals in the Galerkin equations (6.2-7) implies that any boundary conditions having the forms

$$\nabla^2 u(\mathbf{x}) = \beta(\mathbf{x}) \quad \text{on} \quad \partial\Omega,$$
$$\nabla[\nabla^2 u(\mathbf{x})] \cdot \mathbf{n} = \gamma(\mathbf{x}) \quad \text{on} \quad \partial\Omega,$$

can be accommodated via straightforward substitution of the functions β and γ in the boundary terms. Therefore these are natural boundary conditions. However, no such mechanism exists for imposing the lower-order boundary conditions

$$u(\mathbf{x}) = \alpha(\mathbf{x}) \quad \text{on} \quad \partial\Omega,$$
$$\nabla u(\mathbf{x}) \cdot \mathbf{n} = \xi(\mathbf{x}) \quad \text{on} \quad \partial\Omega,$$

and our only recourse is to impose them a priori in the construction of the trial function $\hat{u}$. Therefore, these are essential boundary conditions.

In view of these observations, the Hermite bicubic interpolating functions constitute a feasible choice of basis functions for Galerkin approximations to the biharmonic equation. With this choice the trial function will look like

$$\hat{u}(\mathbf{x}) = u_\partial(\mathbf{x}) + \sum_{i=1}^N [u_i \phi_{00i}(\mathbf{x}) + u_i^{(x)} \phi_{10i}(\mathbf{x})$$
$$+ u_i^{(y)} \phi_{01i}(\mathbf{x}) + u_i^{(xy)} \phi_{11i}(\mathbf{x})].$$

In this representation the functions $\phi_{jki}(\mathbf{x})$ are tensor products of the usual one-dimensional Hermite cubics described in Section 2.4; specifically,

$\phi_{jki}(x, y) = h_i^j(x)h_i^k(y)$. The unknown coefficients $u_i, u_i^{(x)}, u_i^{(y)}$, and $u_i^{(xy)}$ stand for the values of $\hat{u}, \partial\hat{u}/\partial x, \partial\hat{u}/\partial y$, and $\partial^2\hat{u}/\partial x\partial y$, respectively, at the node $\mathbf{x}_i$. Recall from Chapter Two that the interpolation error associated with this type of trial function is $\mathcal{O}(h^4 + k^4)$.

For example, for simply supported plate conditions, we accommodate essential boundary conditions in the trial function of Equation (6.2-6) by defining the boundary function

$$u_\partial(\mathbf{x}) = \sum_\partial u_i\,\phi_{00i}(\mathbf{x}).$$

Here, the notation $\sum_\partial$ indicates the sum over boundary nodes $\mathbf{x}_i$. The coefficients u_i in this sum are known boundary values given by $u_i = \alpha(\mathbf{x}_i)$. The remaining condition $\nabla^2 u = \beta(\mathbf{x})$ on $\partial\Omega$ is a natural boundary condition. Thus we need not explicitly incorporate it into the definition of $\hat{u}$, since we can use it to compute boundary integrals.

For clamped-edge conditions the construction of $u_\partial(\mathbf{x})$ involves only slightly more complication, depending on the shape of the domain Ω. If Ω is a rectangle $(a, b) \times (c, d)$ as above, then $\partial\Omega$ consists of line segments parallel to the x- and y-axes. If $\sum_{\partial_x}$ and $\sum_{\partial_y}$ stand for sums over boundary nodes along the segments parallel to the x- and y- axes, respectively, then

$$u_\partial(\mathbf{x}) = \sum_\partial u_i\phi_{00i}(\mathbf{x}) + \sum_{\partial_x} u_i^{(y)}\phi_{01i}(\mathbf{x})$$
$$+ \sum_{\partial_y} u_i^{(x)}\phi_{10i}(\mathbf{x}) + \sum_{\text{corners}} u_i^{(xy)}\phi_{11i}(\mathbf{x}).$$

Since the clamped-edge conditions force $u(\mathbf{x}) = \nabla u(\mathbf{x}) \cdot \mathbf{n} = 0$ along $\partial\Omega$, each of the boundary nodal values u_i and the nodal normal derivatives $u_i^{(y)}, u_i^{(x)}$ appearing in these sums vanishes. Moreover, differentiating $\nabla u(\mathbf{x}) \cdot \mathbf{n} = 0$ tangentially along $\partial\Omega$ shows that $\partial^2 u/\partial x\partial y = 0$ on $\partial\Omega$, so that the nodal cross-derivatives $u_i^{(xy)}$ also vanish on the boundaries. Thus, in the clamped-edge case, $u_\partial(\mathbf{x}) = 0$, and the remainder of the trial function $\hat{u}$ involves only the interior nodal values u_i and the unknown tangential derivatives $u_i^{(x)}$ or $u_i^{(y)}$ along $\partial\Omega$.

It is worth mentioning that other continuously differentiable finite-element bases for the biharmonic equation have appeared in the literature. In particular there are at least three approaches using triangular elements (Strang and Fix, 1973, Section 1.9). The first and most straightforward of these is to use continuously differentiable piecewise quintic polynomials defined on triangles. Such quintics have 21 degrees of freedom. Of these degrees of freedom, 18 must determine the values of $\hat{u}, \partial\hat{u}/\partial x, \partial\hat{u}/\partial y, \partial^2\hat{u}/\partial x^2, \partial^2\hat{u}/\partial x\partial y$, and $\partial^2\hat{u}/\partial y^2$ at the vertices of each triangle to force continuous differentiability at the vertices. The remaining three degrees of freedom determine the mid-side values of the normal derivative $\partial \hat{u}/\partial n$ to guarantee continuous differentiability across the edges of the triangle. Figure 6-3a illustrates this element.

For the full quintic element the interpolation error is $\mathcal{O}(h^6)$, where, as above, h signifies the maximum dimension of a triangle in the grid.

The second approach using triangles is to eliminate the three off-vertex degrees of freedom in the quintic triangle by forcing the quintic to reduce to a cubic along the edges. These reduced quintics involve fewer unknowns per element than the standard quintic, 18 instead of 21, as shown in Figure 6-3b. Reduced quintics therefore required somewhat less computational effort for a given triangulation than full quintics. The penalty paid for this gain is a slight decrease in accuracy: The reduced quintic element has an interpolation error that is $\mathcal{O}(h^5)$.

The third continuously differentiable triangular element that we shall mention is the **Clough-Tocher element** (Clough and Tocher, 1966). The idea here is to divide each triangle into three "daughter" triangles, as shown in Figure 6-3c We then impose internal continuity constraints leaving only 12 values, defining $\hat{u}, \partial \hat{u}/\partial x, \partial \hat{u}/\partial y$ at the vertices and $\partial \hat{u}/\partial n$ at the mid-side nodes, as elemental degrees of freedom. To accomplish this, we allow each daughter triangle to be cubic in x and y, thus allowing for 10 initial degrees of freedom per daughter triangle and therefore 30 initial degrees of freedom for the "parent" triangle. By imposing common values of $\hat{u}, \partial \hat{u}/\partial x$, and $\partial \hat{u}/\partial y$ at each parent vertex and at the vertex common to the three daughter triangles, we arrive at constraints sufficient to eliminate 16 of the 30 degrees of freedom. Then, by requiring common values of $\partial \hat{u}/\partial n$ along the internal edges of the daughter triangles, we eliminate three additional degrees of freedom, leaving 12 degrees of freedom undetermined. For details of the algebra involved in this elimination we refer the reader to the original paper by Clough and Tocher (1966).

There is one final class of approaches worth mentioning in the single-equation formulation of Equation (6.2-1). The motivation for this approach is the desire to abandon the smoothness constraints on $\hat{u}(\mathbf{x})$ implied by equation (6.2-6). As we have seen, these constraints lead to elements requiring many degrees of freedom and thus to Galerkin matrices having undesirably large bandwidths. Finite-element basis funcitons that violate these smoothness constraints are called **nonconforming** elements. One example of a nonconforming element used in connection with the biharmonic equation is the **Morley element**, which is a quadratic polynomial defined over a triangular region. Such a polynomial has six degrees of freedom, which define the three nodal values at the vertices and the three normal derivatives at mid-side nodes. This smaller number of unknowns per element allows great computational efficiency compared to conforming

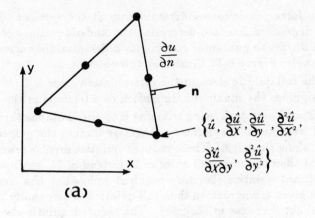

$$\frac{\partial u}{\partial n}$$

n

$$\left\{\hat{u}, \frac{\partial \hat{u}}{\partial x}, \frac{\partial \hat{u}}{\partial y}, \frac{\partial^2 \hat{u}}{\partial x^2},\right.$$
$$\left.\frac{\partial^2 \hat{u}}{\partial x \partial y}, \frac{\partial^2 \hat{u}}{\partial y^2}\right\}$$

(a)

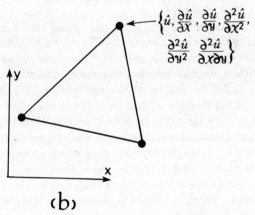

$$\left\{\hat{u}, \frac{\partial \hat{u}}{\partial x}, \frac{\partial \hat{u}}{\partial y}, \frac{\partial^2 \hat{u}}{\partial x^2},\right.$$
$$\left.\frac{\partial^2 \hat{u}}{\partial y^2}, \frac{\partial^2 \hat{u}}{\partial x \partial y}\right\}$$

(b)

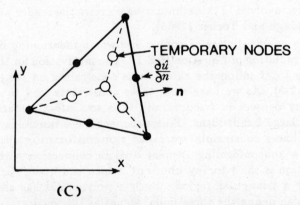

TEMPORARY NODES

$$\frac{\partial \hat{u}}{\partial n}$$

n

(C)

Figure 6-3. Continuously differentiable triangular finite elements: (a) the complete C^1 quintic element, (b) the reduced C^1 quintic element, (c) the Clough-Tocher element (after Lapidus and Pinder, 1982, p. 460.)

346

elements. However, far from being continuously differentiable, trial functions using bases of Morley elements are not even continuous across the edges of triangles. Therefore, elementwise calculation of the Galerkin integrals is no longer formally valid. Nevertheless, Galerkin approximations using this element still yield convergent approximations to Equation (6.2-1). The analysis of nonconforming finite-element methods lies beyond the scope of this book, and we refer the reader to Griffiths and Mitchell (1984) for a review of the subject.

Finite-element approximations: coupled-equation approach.

We close our discussion of the biharmonic equation with an introduction to finite-element approximations for the coupled equations (6.2-4). In the interest of slightly more generality, let us consider a nonhomogeneous version of the biharmonic equation, $\nabla^4 u = f$. In this case the coupled second-order PDEs become

$$\nabla^2 \begin{bmatrix} u \\ v \end{bmatrix} = \begin{bmatrix} v \\ f \end{bmatrix} \quad \text{on} \quad \Omega.$$

Thus we seek two trial functions $\hat{u}(\mathbf{x})$ and $\hat{v}(\mathbf{x})$. It will turn out that, compared with the single-equation approach, this two-equation formulation relaxes the smoothness constraints on the finite-element bases for $\hat{u}$ and $\hat{v}$.

To see how, let us consider trial functions of the form

$$\hat{u}(\mathbf{x}) = u_\partial(\mathbf{x}) + \sum_{i=1}^{N} u_i \phi_i(\mathbf{x}),$$

$$\hat{v}(\mathbf{x}) = v_\partial(\mathbf{x}) + \sum_{j=1}^{M} v_j \psi_j(\mathbf{x}).$$

Notice that we have allowed for different bases $\{\phi_i\}_{i=1}^{N}$ and $\{\psi_j\}_{j=1}^{M}$ in the construction of $\hat{u}$ and $\hat{v}$. Such finite-element formulations of factored PDE's using separate expansions of the unknown u and one of its derivatives (in this case $\nabla^2 u$) are called **mixed methods**. As with Galerkin formulations for single equations, each of the interior basis functions $\phi_i(\mathbf{x})$, $\psi_j(\mathbf{x})$ must satisfy homogeneous boundary conditions. This means $\phi_i(\mathbf{x}) = 0$ and $\psi_j(\mathbf{x}) = 0$, whenever $\mathbf{x} \in \partial\Omega$. The Galerkin integral equations for these trial functions become

$$\int_\Omega (\nabla^2 \hat{u} - \hat{v}) \psi_k \, d\mathbf{x} = 0, \quad k = 1, \ldots, M,$$

$$\int_\Omega (\nabla^2 \hat{v} - f) \phi_\ell \, d\mathbf{x} = 0, \quad \ell = 1, \ldots, N.$$

(The particular arrangement of weighting functions in these equations arises from the variational formulation of the problem; for details we refer the reader to Carey and Oden, 1983, Section 3.3.) Making use of Green's theorem, we get

(6.2-8a)
$$-\int_\Omega \nabla \hat{u} \cdot \nabla \psi_k \, d\mathbf{x} - \int_\Omega \hat{v} \psi_k \, d\mathbf{x}$$

$$+ \oint_{\partial\Omega} \psi_k \nabla \hat{u} \cdot \mathbf{n} \, d\mathbf{x} = 0, \quad k = 1, \dots, M,$$

$$-\int_\Omega \nabla \hat{v} \cdot \nabla \phi_\ell \, d\mathbf{x} - \int_\Omega f \phi_\ell \, d\mathbf{x}$$

$$+ \oint_{\partial\Omega} \phi_\ell \nabla \hat{v} \cdot \mathbf{n} \, d\mathbf{x} = 0, \quad \ell = 1, \dots, N.$$

Now we can see how the mixed formulation allows us to use trial functions satisfying less stringent smoothness constraints. The integrals over Ω in Equations (6.2-8a) will admit elementwise decompositions of the form

$$\int_\Omega (\cdot) \, d\mathbf{x} = \sum_e \int_{\Omega_e} (\cdot) \, d\mathbf{x}$$

provided the gradients of the basis functions are at worst jump-discontinuous at element boundaries. Therefore the trial functions $\hat{u}$ and $\hat{v}$ need only belong to $C^0(\Omega)$ for the finite-element method to be conforming. In particular, we can choose tensor-product Lagrange bases or piecewise-planar triangles to form our trial functions. Thus the mixed formulation enjoys less restrictive smoothness requirements than the fourth-order formulation.

Equations (6.2-8a) also show that boundary conditions specifying values of $\partial u / \partial n = \nabla u \cdot \mathbf{n}$ and $\partial v / \partial n = \nabla(\nabla^2 u) \cdot \mathbf{n}$ along $\partial \Omega$ will be natural boundary conditions for the coupled-equation approach. On the other hand, boundary conditions specifying u and $v = \nabla^2 u$ on $\partial \Omega$ will be essential boundary conditions, which we must incorporate into the definitions of $\hat{u}(\mathbf{x})$ and $\hat{v}(\mathbf{x})$ through the boundary terms $u_\partial(\mathbf{x})$ and $v_\partial(\mathbf{x})$.

Equations (6.2-8a) give rise to matrix equations having an interesting structure. Assume for simplicity that the boundary conditions are homogeneous, so that

$$u = 0 \quad \text{on} \quad \partial\Omega,$$

$$\nabla^2 u = v = 0 \quad \text{on} \quad \partial\Omega,$$

and hence $u_\partial(\mathbf{x}) \equiv v_\partial(\mathbf{x}) \equiv 0$ in the trial functions. The Galerkin equations (6.2-7) then expand to give

(6.2-8b)
$$\sum_{i=1}^{N} u_i \int_{\Omega} \nabla \phi_i \cdot \nabla \psi_k \, dx + \sum_{j=1}^{M} v_j \int_{\Omega} \psi_j \psi_k \, dx$$

$$= \oint_{\partial \Omega} \psi_k \nabla \hat{u} \cdot \mathbf{n} \, dx, \quad k = 1, \dots, N \, ,$$

(6.2-8c)
$$\sum_{j=1}^{M} v_j \int_{\Omega} \nabla \psi_j \cdot \nabla \phi_\ell \, dx = - \int_{\Omega} f \phi_\ell \, dx$$

$$+ \oint_{\partial \Omega} \phi_\ell \nabla \hat{v} \cdot \mathbf{n} \, dx, \quad \ell = 1, \dots, M \, .$$

Notice that the boundary integrals in each of these sets of equations all vanish, since the interior basis functions $\phi_\ell(\mathbf{x})$ and $\phi_k(\mathbf{x})$ vanish on $\partial \Omega$. Now denote by $\mathbf{K}$ the $M \times N$ matrix whose (i, k)-th entry is $\int_{\Omega} \nabla \phi_i \cdot \nabla \psi_k \, dx$, by $\mathbf{M}$ the $M \times M$ matrix whose (j, k)-th entry is $\int_{\Omega} \psi_j \psi_k \, dx$, by $\mathbf{u}$ and $\mathbf{v}$ the column vectors containing the nodal values $u_1, \dots, u_N$ and $v_1, \dots, v_M$, respectively, and by $\mathbf{f}$ the M-dimensional vector whose ℓ-th entry is $\int_{\Omega} f \phi_\ell \, dx$. Then Equations (6.2-8) assume the form

$$\begin{bmatrix} \mathbf{K} & \mathbf{M} \\ \mathbf{0} & \mathbf{K}^{\mathsf{T}} \end{bmatrix} \begin{bmatrix} \mathbf{u} \\ \mathbf{v} \end{bmatrix} = \begin{bmatrix} \mathbf{0} \\ -\mathbf{f} \end{bmatrix}.$$

This matrix structure has the advantage that it is already "almost" upper triangular. This property, together with the fact that the matrices $\mathbf{K}$ and $\mathbf{M}$ are typically sparse, makes it possible to envision highly efficient algorithms for solving the matrix equations arising from mixed methods. These observations serve to mitigate further the disadvantages associated with solving for two unknowns instead of one at every grid point.

6.3. Nonlinear Problems.

Many of the PDEs arising in practical problems are nonlinear. Typically, the nonlinearity owes its existence to dependencies of material properties or forcing functions on the unknown in the problem. This section reviews several approaches to the numerical solution of nonlinear PDEs, relying on the physically motivated examples described below. Several exercises at the end of this chapter introduce still other approaches to the discretization of nonlinear PDEs.

Steady, nonlinear heat flow.

To begin with, let us examine two types of nonlinearity that commonly occur in steady-state problems. We start with the energy balance for heat flow developed in Sections 1.4 and 4.1. When heat sources are present, Equation (4.1-6) becomes

$$(6.3\text{-}1) \qquad\qquad \rho c_v \frac{\partial T}{\partial t} = \nabla \cdot (k_H \nabla T) + \rho h,$$

where T is the unknown temperature, ρ is the mass density, c_v is the heat capacity of the material, and h represents the external supply of heat. In a steady state, $\partial T/\partial t = 0$, and we can write Equation (6.3-1) as

$$(6.3\text{-}2) \qquad\qquad \nabla \cdot (k_H \nabla T) = -\rho h.$$

In simple cases when k_H is constant, the heat supply term may still depend on the temperature T. Such a dependence occurs, for example, in systems governed by thermostats or in materials whose density ρ exhibits significant dependence on temperature. Under these assumptions, Equation (6.3-2) becomes

$$(6.3\text{-}3) \qquad\qquad \nabla^2 T = -\rho(T)h(T)/k_H,$$

which is a nonlinear version of Poisson's equation.

If, in addition, the heat flux coefficient k_H depends on T, then the terms involving spatial derivatives in Equation (6.3-2) also become nonlinear, and we are left with the elliptic equation

$$(6.3\text{-}4) \qquad\qquad \nabla \cdot [k_H(T)\nabla T] = -\rho(T)h(T).$$

Thus in steady-state heat-flow problems nonlinearity can occur either in the forcing term modeling heat supplies or through the material property governing the rate of heat flux.

Nonlinear diffusion.

Next, consider a transient problem involving the diffusion of a solute in a fluid, as examined in Section 1.5, when we relax the assumption that the solute transport is passive. In the absence of advection and chemical reactions, the species mass balance equation for the dissolved solute is

$$(6.3\text{-}5) \qquad\qquad \frac{\partial}{\partial t}\left(\rho \omega^S\right) + \nabla \cdot \mathbf{j}^S = 0.$$

Let us assume, as in Section 1.5, that the diffusive flux $\mathbf{j}^S$ obeys Fick's law, only now we shall allow the diffusion coefficient K^S to depend on the unknown mass fraction ω^S. Thus, we shall assume

$$\mathbf{j}^S = -K^S(\omega^S)\nabla(\rho\omega^S); \quad K^S > 0.$$

Also, since we are abandoning the assumption that the solute transport is passive, we may as well allow the overall mixture density ρ to vary with the amount of solute present by letting $\rho = \rho(\omega^S)$. However, we shall assume that gradients in density arising from this dependence are small compared with gradients in ω^S. Using these constitutive assumptions, we find that Equation (6.3-5) reduces to the nonlinear parabolic equation

$$\frac{\partial}{\partial t}\left[\rho(\omega^S)\omega^S\right] - \nabla \cdot \left[\rho(\omega^S)K^S(\omega^S)\nabla\omega^S\right] = 0.$$

If we now recognize that the combination $\rho(\omega^S)\omega^S$ is just the solute density $\rho^S(\omega^S)$ and call $\rho(\omega^S)K^S(\omega^S) = k_D(\omega^S)$, then we arrive at the slightly simpler form,

(6.3-6) $$\frac{\partial}{\partial t}\left[\rho^S(\omega^S)\right] - \nabla \cdot \left[k_D(\omega^S)\nabla\omega^S\right] = 0.$$

This equation exhibits nonlinearity both in the accumulation term and in the flux term.

The remainder of this section is devoted to the discussion of methods for solving each of the types of problems just cited. There are three such problems: the **nonlinear Poisson equation** generalized from Equation (6.3-3),

(6.3-7) $$\nabla^2 u = f(u);$$

the **nonlinear steady heat flow equation** generalized from Equation (6.3-4),

(6.3-8) $$\nabla \cdot [K(u)\nabla u] = f(u);$$

and the **nonlinear diffusion equation** generalized from Equation (6.3-6),

(6.3-9) $$\frac{\partial}{\partial t}[c(u)] - \nabla \cdot [K(u)\nabla u] = 0.$$

While these equations are representative of many nonlinear forms that occur in science and engineering, they by no means exhaust all of the possibilities. Notice in particular that Equations (6.3-7) through (6.3-9) do not include nonlinear equations applicable to nondissipative systems. In this regard, we have already discussed nonlinear hyperbolic PDEs, such as Burgers' equation, in Chapter Five. In Section 6.5 we shall see how considerations similar to those discussed in Chapter Five arise in a nonlinear coupled system governing oil-reservoir flows.

The nonlinear Poisson equation.

The nonlinear Poisson problem is in many ways the simplest of the four examples we shall discuss, so let us attack it first. Consider, for example, a Galerkin finite-element approximation to the homogeneous boundary-value problem

$$(6.3\text{-}10) \qquad \nabla^2 u = f(u) \quad \text{in } \Omega,$$

$$u \equiv 0 \quad \text{on } \Omega.$$

If we replace the unknown function $u(\mathbf{x})$ by a trial function

$$(6.3\text{-}11) \qquad \hat{u}(\mathbf{x}) = \sum_{i=1}^{N} u_i \, \phi_i(\mathbf{x}),$$

then applying the method of weighted residuals with the basis functions $\phi_1, \ldots, \phi_N$ serving as weighting functions yields the Galerkin integral equations

$$(6.3\text{-}12) \qquad \int_\Omega \left[\nabla^2 \hat{u} - f(\hat{u}) \right] \phi_j \, d\mathbf{x} = 0, \quad j = 1, \ldots, N.$$

Using Green's theorem and observing that the boundary contributions vanish since $\hat{u} = 0$ on $\partial\Omega$, we obtain

$$(6.3\text{-}13) \qquad \sum_{i=1}^{N} u_i \int_\Omega \nabla\phi_i \cdot \nabla\phi_j \, d\mathbf{x} + \int_\Omega f\left(\sum_{i=1}^{N} u_i \phi_i \right) \phi_j \, d\mathbf{x} = 0,$$

$$j = 1, \ldots, N.$$

These Galerkin equations have an equivalent matrix form

$$(6.3\text{-}14) \qquad \mathbf{A}\mathbf{u} = -\mathbf{f}(\mathbf{u}),$$

where the (j, i)-th entry of the $N \times N$ matrix $\mathbf{A}$ is $\int_\Omega \nabla\phi_i \cdot \nabla\phi_j d\mathbf{x}$, the j-th entry of the vector $\mathbf{f}$ is $\int_\Omega f(\hat{u})\phi_j d\mathbf{x}$, and the i-th entry of $\mathbf{u}$ is the unknown coefficient u_i. Because the right side of Equation (6.3-14) depends on the unknown vector $\mathbf{u}$, we cannot solve this matrix equation by a single pass through any of the linear matrix solution algorithms described in Sections 3.11 through 3.15. However, we can make use of these computationally attractive linear methods by adopting an *iterative* strategy in which each iteration requires the solution of a linear approximation to Equation (6.3-14).

As a simple example, suppose we begin with an initial guess $\mathbf{u}^{(0)}$ for the solution vector $\mathbf{u}$. We can use this guess to evaluate the forcing vector

$\mathbf{f}$, giving a tentative value $\mathbf{f}(\mathbf{u}^{(0)})$ in the right side of Equation (6.3-14). Given this right side, we can solve the equation $\mathbf{A}\mathbf{u}^{(1)} = -\mathbf{f}(\mathbf{u}^{(0)})$ for a new iterative value $\mathbf{u}^{(1)}$ that, we hope, gives a better approximation to the true solution $\mathbf{u}$. We can proceed in this fashion, at each iteration using the most recent known iterative value $\mathbf{u}^{(m)}$ to solve for a new value $\mathbf{u}^{(m+1)}$ via the linear matrix equation

$$(6.3\text{-}15) \qquad\qquad \mathbf{A}\mathbf{u}^{(m+1)} = -\mathbf{f}(\mathbf{u}^{(m)}).$$

This iterative scheme is an example of the method of **successive substitution**.

In practice, we can never expect $\mathbf{u}^{(m+1)}$ to equal the exact finite-element solution $\mathbf{u}$. But, assuming that the scheme produces sufficiently better approximations $\mathbf{u}^{(m+1)}$ at each iteration, we can run the iterations a finite number of times until the approximate solution satisfies some **convergence criterion**. For example, we might compute the **residual** $\mathbf{R}^{(m+1)}$ $= \mathbf{A}\mathbf{u}^{(m+1)} + \mathbf{f}(\mathbf{u}^{(m+1)})$ at each iteration, stopping the iterations as soon as $\|\mathbf{R}^{(m+1)}\|$ falls below a prescribed tolerance in some norm. As an alternative, if we can estimate the **error** $\mathbf{e}^{(m+1)} = \mathbf{u}^{(m+1)} - \mathbf{u}$, we can stop iterating as soon as $\|\mathbf{e}^{(m+1)}\|$ is small enough.

Two questions immediately arise. First, when can we expect the iterative scheme to converge? That is, for which problems can we expect Equation (6.3-15) to yield an iterative sequence $\{\mathbf{u}^{(m)}\}_{m=0}^{\infty}$ such that $\mathbf{u}^{(m)} \to \mathbf{u}$ as $m \to \infty$? Second, if the iterative scheme does converge, at what rate does the error decrease? For many nonlinear problems in practice, these questions are quite difficult to answer with any precision. We can, however, outline a framework for the analysis of particular problems.

Consider first the question whether a scheme converges. Let us begin by writing Equation (6.3-15) in the equivalent form $\mathbf{u}^{(m+1)} = -\mathbf{A}^{-1}\mathbf{f}(\mathbf{u}^{(m)})$. This equation is a special case of the more general iterative scheme

$$(6.3\text{-}16) \qquad\qquad \mathbf{u}^{(m+1)} = \mathbf{g}(\mathbf{u}^{(m)}),$$

where $\mathbf{g}$ is some function mapping N-vectors to N-vectors. By definition, such a function satisfies a **Lipschitz condition** of order L in a region $\mathcal{U}$ of N-space if there exists a positive constant L such that, whenever the vectors $\mathbf{v}$ and $\mathbf{w}$ lie in $\mathcal{U}$,

$$\|\mathbf{g}(\mathbf{v}) - \mathbf{g}(\mathbf{w})\| \leq L\|\mathbf{v} - \mathbf{w}\|,$$

in some norm. When the Lipschitz constant $L < 1$, the function $\mathbf{g}$ is called a **strict contraction**. To see how these concepts relate to the issue of convergence, observe that, according to Equation (6.3-14), the true solution to the finite-element discretization of the Poisson problem satisfies $\mathbf{u} = \mathbf{g}(\mathbf{u})$. In other words, $\mathbf{u}$ is a **fixed point** of the function $\mathbf{g}$. The following theorem, proved in Ortega and Rheinboldt (1970, p. 120), establishes the crucial connection between strict contractions and convergence of their associated iterative schemes under successive substitution:

Contraction Mapping Theorem. *Given a function* $\mathbf{g}$ *that maps N-vectors from a region $\mathcal{U}$ of N-space into $\mathcal{U}$, if $\mathbf{g}$ is a strict contraction on $\mathcal{U}$, then the iterative scheme (6.3-16) converges to a unique fixed point $\mathbf{u} \in \mathcal{U}$ for any initial guess $\mathbf{u}^{(0)} \in \mathcal{U}$.*

Therefore, to guarantee that the sucessive substitution scheme (6.3-16) converges, it suffices to find a region $\mathcal{U}$ of N-dimensional Euclidean space on which the function $\mathbf{g}$ is a strict contraction. For highly complex problems, this may be quite difficult to do rigorously, and it may be necessary to rely on experimental calculations to estimate where $\mathbf{g}$ will be a strict contraction.

In problems where we can estimate the Lipschitz constant L, we can also estimate the error at each iteration. Ortega and Rheinboldt (1970, p. 385) prove that, given a strict contraction as in the theorem just stated, the error $\mathbf{e}^{(m)} = \mathbf{u}^{(m)} - \mathbf{u}$ at the m-th iteration obeys the bound

$$\|\mathbf{e}^{(m)}\| \leq \frac{L}{L-1}\|\mathbf{u}^{(m)} - \mathbf{u}^{(m-1)}\|.$$

The right side of this inequality is computable at any iteration $m \geq 1$, so all we need to get error bounds in a computer code is an estimate of L. Whenever $\mathbf{g}$ is a differentiable function, it is possible to show (using the mean value theorem) that L is related to the Jacobian matrix $\mathbf{g}'(\mathbf{u})$ of $\mathbf{g}$ by the identity

$$L = \sup_{\mathbf{v} \in \mathcal{U}} \|\mathbf{g}'(\mathbf{v})\|,$$

that is, L is the least upper bound of $\|\mathbf{g}'(\mathbf{v})\|$ taken over all $\mathbf{v} \in \mathcal{U}$.

We can also estimate how fast the iterative scheme converges when $\mathbf{g}$ is differentiable. Since the true finite-element solution $\mathbf{u}$ is a fixed point of $\mathbf{g}$,

$$\mathbf{e}^{(m+1)} = \mathbf{u}^{(m+1)} - \mathbf{u} = \mathbf{g}(\mathbf{u}^{(m)}) - \mathbf{g}(\mathbf{u})$$
$$= \mathbf{g}(\mathbf{u} + \mathbf{e}^{(m)}) - \mathbf{g}(\mathbf{u}).$$

Therefore, in any norm,

$$\|\mathbf{e}^{(m+1)}\| = \|\mathbf{e}^{(m)}\| \frac{\|\mathbf{g}(\mathbf{u} + \mathbf{e}^{(m)}) - \mathbf{g}(\mathbf{u})\|}{\|\mathbf{e}^{(m)}\|},$$

and we find that the ratio of successive error norms obeys

$$\frac{\|\mathbf{e}^{(m+1)}\|}{\|\mathbf{e}^{(m)}\|} = \frac{\|\mathbf{g}(\mathbf{u} + \mathbf{e}^{(m)}) - \mathbf{g}(\mathbf{u})\|}{\|\mathbf{e}^{(m)}\|}.$$

If the iterative scheme converges, then $\mathbf{e}^{(m)} \to \mathbf{0}$ as $m \to \infty$. Given the hypothesis that $\mathbf{g}$ is a differentiable function, we see that the right side of the last equation approaches the norm of the Jacobian matrix of $\mathbf{g}$, evaluated at the solution $\mathbf{u}$, as $m \to \infty$:

$$(6.3\text{-}17) \qquad \lim_{m \to \infty} \frac{\|\mathbf{e}^{(m+1)}\|}{\|\mathbf{e}^{(m)}\|} = \|\mathbf{g}'(\mathbf{u})\|.$$

Thus, at each iteration after the first few, we can expect the error to decrease roughly by a factor of $\|\mathbf{g}'(\mathbf{u})\|$.

Error-norm ratios such as the one estimated in Equation (6.3-17) provide the basis for the most common method of measuring the speed with which iterative schemes converge. We say that a scheme has **order of convergence** (or **asymptotic convergence rate**) α if there exists some constant C such that

$$\lim_{m \to \infty} \frac{\|\mathbf{e}^{(m+1)}\|}{\|\mathbf{e}^{(m)}\|^\alpha} \le C.$$

We have just demonstrated that the method of successive substitution has order of convergence $\alpha = 1$, a fact that we commonly describe by saying that successive substitution converges linearly.

Nonlinear steady heat flow via successive substitution.

A similar approach using successive substitution applies to the equation

$$\nabla \cdot [K(u)\,\nabla u] = f(u),$$

only now the presence of a nonlinear coefficient $K(u)$ in the flux term on the left leads to some special considerations. Consider again a Galerkin finite-element formulation for a homogeneous Dirichlet problem forcing $u = 0$ on $\partial\Omega$. Employing the same trial function as for the nonlinear Poisson problem, we can derive the following Galerkin integral equations:

$$\int_\Omega [K(\hat{u})\,\nabla\hat{u}\cdot\nabla\phi_j + f(\hat{u})\phi_j]\,d\mathbf{x} = 0, \quad j = 1,\ldots,N.$$

Substituting the trial function given in Equation (6.3-11), we get

$$\sum_{i=1}^N \int_\Omega K(\hat{u})\,\nabla\phi_i\cdot\nabla\phi_j\,d\mathbf{x} + \int_\Omega f(\hat{u})\phi_j\,d\mathbf{x} = 0,$$

$$j = 1,\ldots,N.$$

This set of equations has an equivalent matrix form,

$$(6.3\text{-}18) \qquad \mathbf{A}(\mathbf{u})\mathbf{u} = -\mathbf{f}(\mathbf{u}),$$

where the entries of the matrix $\mathbf{A}$ and the vector $\mathbf{f}$ are as follows:

(6.3-19a)
$$A_{ji} = \int_{\Omega} K(\hat{u}) \, \nabla \phi_i \cdot \nabla \phi_j \, d\mathbf{x},$$

(6.3-19b)
$$f_j = \int_{\Omega} f(\hat{u}) \phi_j \, d\mathbf{x}.$$

Notice that, in contrast to the matrix equation (6.3-14) for the non-linear Poisson problem, the forcing function *and* the matrix $\mathbf{A}$ in Equation (6.3-8) depend nonlinearly on the unknown solution $\mathbf{u}$. Given our previous development, this fact poses no real conceptual difficulty in formulating an iterative method. We can construct a successive substitution scheme simply by lagging the evaluation of $\mathbf{u}$ by an iteration whenever it appears as the argument of a function:

$$\mathbf{A}(\mathbf{u}^{(m)})\mathbf{u}^{(m+1)} = -\mathbf{f}(\mathbf{u}^{(m)}).$$

Each stage in this iterative procedure requires the solution of a linear matrix equation, and we can investigate the performance of the scheme as before if we examine the iterated function $\mathbf{g}(\mathbf{u}) = -\mathbf{A}^{-1}(\mathbf{u})f(\mathbf{u})$.

As a practical matter, however, the nonlinearities in the flux term now interfere with the computation of the matrix entries in $\mathbf{A}$. At the m-th iteration, the matrix entries given in Equation (6.3-19a) will look like

$$A_{ji}^{(m)} = \int_{\Omega} K(\hat{u}^{(m)}) \, \nabla \phi_i \cdot \nabla \phi_j \, d\mathbf{x},$$

where $\hat{u}^{(m)}$ denotes the function obtained by substituting the coefficients stored in the most recently computed vector $\mathbf{u}^{(m)}$ into the trial function (6.3-11). These integrals may be difficult or impossible to compute exactly if the material property K has a complicated functional form.

One way to circumvent this problem is to use numerical quadrature, as discussed in Section 2.12. With this tactic, we would use the trial function $\hat{u}^{(m)}(\mathbf{x}) = \sum_{i=1}^{N} u_i^{(m)} \phi_i(\mathbf{x})$ to evaluate $K(\hat{u}^{(m)}(\mathbf{x}))$ at the requisite sampling points, multiply these values by the corresponding values of $\nabla \phi_i(\mathbf{x}) \cdot \nabla \phi_j(\mathbf{x})$, and add the results together in a weighted sum to compute an approximate value of each integral. Observe that this process requires a fair amount of calculation at every iteration, thereby contributing significantly to the cost of the computations.

A common alternative to this straightforward approach is to use functional coefficients, as introduced in Section 3.6. Thus, we might adopt a finite-element representation for K having the form

$$K(\hat{u}^{(m)}(\mathbf{x})) \simeq \hat{K}^{(m)}(\mathbf{x}) = \sum_{\ell=1}^{M} K_{\ell}^{(m)} \psi_{\ell}(\mathbf{x}).$$

Two observations are in order here. First, the basis functions ψ_ℓ used in this representation of K may be different from those used in the trial function $\hat{u}$. This flexibility may allow some computational savings in some formulations. For example, if $\hat{u}$ has a piecewise Hermite cubic expansion, it may be much simpler to compute a piecewise Lagrange linear expansion for K than to construct another Hermite cubic expansion for it. Second, the coefficients $K_\ell^{(m)}$ are simply the values of the function K at the values $\hat{u}^{(m)}(\mathbf{x}_\ell)$ corresponding to the appropriate spatial location in the expansion of $\hat{K}^{(m)}(\mathbf{x})$. If the nodes of the interpolating functions ψ_ℓ are also nodes of the interpolating functions ϕ_i, then we need not explicitly interpolate $\hat{u}^{(m)}$ to compute $\hat{K}^{(m)}$. Rather, we can simply set

$$K_\ell^{(m)} = K(u_\ell^{(m)}),$$

where $u_\ell^{(m)}$ signifies the nodal value corresponding to the node at which ψ_ℓ is centered.

To see how such an approximate scheme works computationally, notice that it yields approximate matrix entries having the form

$$(6.3-20) \quad A_{ji}^{(m)} \simeq \int_\Omega \hat{K}^{(m)}(\mathbf{x})\, \nabla\phi_i(\mathbf{x}) \cdot \nabla\phi_j(\mathbf{x})\, d\mathbf{x}$$

$$= \int_\Omega \left[\sum_{\ell=1}^{M} K(u_\ell^{(m)} \psi_\ell(\mathbf{x}) \right] \nabla\phi_i(\mathbf{x}) \cdot \nabla\phi_j(\mathbf{x})\, d\mathbf{x}$$

$$= \sum_{\ell=1}^{M} K(u_\ell^{(m)}) \int_\Omega \psi_\ell \nabla\phi_i \cdot \nabla\phi_j\, d\mathbf{x}.$$

In light of the fact that each of the basis functions ψ_ℓ, ϕ_i, and ϕ_j vanishes over most of the finite-element grid, we see that very few of the integrals appearing in this sum will be nonzero. The sum will therefore be very easy to compute once we have calculated the appropriate coefficients $K(u_\ell^{(m)})$. More important, these integrals are independent of $\hat{u}$, and we can compute them once and for all in advance of starting the iterative procedure. The calculation of the nonlinear flux coefficient can now be limited to the nodes, and the evaluation of matrix entries amounts to the computation of small linear combinations of these nodal values. As mentioned in Section 3.16, these linear combinations will be especially easy to compute if we simply choose piecewise constant basis functions ψ_ℓ for the finite-element representation of K.

Finally, we can adopt a similar functional representation to compute the forcing vector $\mathbf{f}(\mathbf{u})$ in the matrix equation (6.3-18). Using the same basis functions $\{\psi_\ell\}_{\ell=1}^{M}$ as for the coefficient K, we get

$$f(\hat{u}^{(m)}(\mathbf{x})) \simeq \hat{f}^{(m)}(\mathbf{x}) = \sum_{\ell=1}^{M} f_{\ell}^{(m)} \psi_{\ell}(\mathbf{x}),$$

where $f_{\ell}^{(m)} = f(u_{\ell}^{(m)})$. Thus we can approximate the j-th entry of the vector $\mathbf{f}$ as follows:

$$(6.3\text{-}21) \qquad f_j \simeq \int_{\Omega} \left[\sum_{\ell=1}^{M} f(u_{\ell}^{(m)}) \psi_{\ell}(\mathbf{x}) \right] \phi_j(\mathbf{x})\, d\mathbf{x}$$

$$= \sum_{\ell=1}^{M} f(u_{\ell}^{(m)}) \int_{\Omega} \psi_{\ell}(\mathbf{x}) \phi_j(\mathbf{x})\, d\mathbf{x}.$$

Once again, only a few of the integrals $\int_{\Omega} \psi_{\ell}\phi_j\, d\mathbf{x}$ will be nonzero, and those that are can be computed in advance of the iterations. The computation of the forcing vector thus also reduces to the calculation of small linear combinations of function values at each iteration.

Nonlinear steady heat flow via Newton's method.

Another class of iterative approaches for discretized nonlinear PDEs uses **Newton's method** as its point of departure. If we write the discrete equation (6.3-18) for the coefficient vector $\mathbf{u} = (u_1, \ldots, u_N)$ in the form

$$(6.3\text{-}22) \qquad \mathbf{F}(\mathbf{u}) = \mathbf{A}(\mathbf{u})\mathbf{u} + \mathbf{f}(\mathbf{u}) = \mathbf{0},$$

then it becomes apparent that solving the finite-element equations is equivalent to finding a root of the nonlinear vector function $\mathbf{F}$. Newton's method for an equation of the form $\mathbf{F}(\mathbf{u}) = \mathbf{0}$ is an iterative scheme based on the following observation. If $\mathbf{F}$ is twice continuously differentiable as a function of the vector $\mathbf{u}$, then it has a Taylor expansion about any known iterative value $\mathbf{u}^{(m)}$:

$$(6.3\text{-}23) \qquad \mathbf{F}(\mathbf{u}^{(m)} + \Delta\mathbf{u}) = \mathbf{F}(\mathbf{u}^{(m)}) + \mathbf{F}'(\mathbf{u}^{(m)})\Delta\mathbf{u} + \mathcal{O}(\|\Delta\mathbf{u}\|_2^2).$$

Here, $\Delta\mathbf{u}$ is an increment in the value of $\mathbf{u}^{(m)}$. $\mathbf{F}'(\mathbf{u})$ denotes the Jacobian matrix of $\mathbf{F}(\mathbf{u})$, whose (j, i)-th entry is $\partial F_j/\partial u_i$, F_j being the j-th component function of $\mathbf{F}(\mathbf{u}) = (F_1(\mathbf{u}), \ldots, F_N(\mathbf{u}))$.

Given an iterative value $\mathbf{u}^{(m)}$, if we could find an increment $\Delta\mathbf{u}$ such that $\mathbf{F}(\mathbf{u}^{(m)} + \Delta\mathbf{u}) = \mathbf{0}$, then we would be able to solve Equation (6.3-22) exactly. However, the nonlinearity of $\mathbf{F}$ usually prevents this. Let us settle instead on neglecting the term $\mathcal{O}(\|\Delta\mathbf{u}\|_2^2)$ in Equation (6.3-23) and forcing the resulting *linear* approximation $\mathbf{F}(\mathbf{u}^{(m)}) + \mathbf{F}'(\mathbf{u}^{(m)})\Delta\mathbf{u}$ to vanish:

$$(6.3\text{-}24a) \qquad \mathbf{F}'(\mathbf{u}^{(m)})\Delta\mathbf{u} = -\mathbf{F}(\mathbf{u}^{(m)}).$$

Since $\mathbf{u}^{(m)}$ is a known vector, the entries of the matrix $\mathbf{F}'(\mathbf{u}^{(m)})$ and the vector $\mathbf{F}(\mathbf{u}^{(m)})$ can be computed to give a linear matrix equation for the increment vector $\Delta\mathbf{u}$. We then form a new iterate $\mathbf{u}^{(m+1)}$ by setting

$$(6.3\text{-}24\text{b}) \qquad\qquad\qquad \mathbf{u}^{(m+1)} = \mathbf{u}^{(m)} + \Delta\mathbf{u}.$$

While this new iterate will generally fail to solve the nonlinear problem exactly, we expect that it will furnish a better approximation to the true solution than the previous iterate. Having computed the new iterate, we can go on to compute yet another new iterate, proceeding in this fashion until the residual $\mathbf{F}(\mathbf{u}^{(m+1)})$ is small enough in some norm. Equations (6.3-24) constitute Newton's method for solving $\mathbf{F}(\mathbf{u}) = \mathbf{0}$ iteratively.

Before examining the application of this scheme to our discretized PDE, let us see what to expect for its convergence rate. Our argument will rest on the assumption that the Jacobian matrix $\mathbf{F}'$ is nonsingular throughout some neighborhood of the exact root $\mathbf{u}$. Provided $\mathbf{F}$ is sufficiently smooth, we can use Taylor's theorem to expand $\mathbf{F}$ at its exact root $\mathbf{u}$ about the known iterate $\mathbf{u}^{(m)}$ as follows:

$$\mathbf{0} = \mathbf{F}(\mathbf{u}) = \mathbf{F}(\mathbf{u}^{(m)}) + \mathbf{F}'(\mathbf{u}^{(m)})(\mathbf{u} - \mathbf{u}^{(m)}) + \mathbf{R}.$$

In this expansion the remainder $\mathbf{R} = \mathcal{O}(\|\mathbf{u} - \mathbf{u}^{(m)}\|^2)$, meaning that $\|\mathbf{R}\|/\|\mathbf{u} - \mathbf{u}^{(m)}\|^2 \leq C$ for some constant C as $m \to \infty$. Now notice that $\mathbf{u} - \mathbf{u}^{(m)} = -\mathbf{e}^{(m)}$, where $\mathbf{e}^{(m)}$ is the error at the m-th iteration, and that $\mathbf{F}(\mathbf{u}^{(m)}) = -\mathbf{F}'(\mathbf{u}^{(m)})(\mathbf{u}^{(m+1)} - \mathbf{u}^{(m)})$ by the definition of Newton's method. Making these substitutions, we find

$$\mathbf{F}'(\mathbf{u}^{(m+1)} - \mathbf{u}) = \mathbf{R},$$

and

$$\frac{\|\mathbf{R}\|}{\|\mathbf{e}^{(m)}\|^2} \to C \quad \text{as} \quad m \to \infty.$$

When Newton's method converges, $\mathbf{F}'(\mathbf{u}^{(m)}) \to \mathbf{F}'(\mathbf{u})$ as $m \to \infty$, and, since $\mathbf{u}^{(m+1)} - \mathbf{u} = \mathbf{e}^{(m+1)}$, taking norms yields

$$\frac{\|\mathbf{e}^{(m+1)}\|}{\|\mathbf{e}^{(m)}\|^2} \leq \frac{\|\mathbf{R}\|}{\|\mathbf{e}^{(m)}\|^2} \frac{1}{\|\mathbf{F}'(\mathbf{u}^{(m)})\|} \leq \frac{C}{\|\mathbf{F}'(\mathbf{u})\|}$$

$$\text{as} \quad m \to \infty.$$

Since the right side of this limit is a constant, we see that the order of Newton's method is two provided our assumptions about $\mathbf{F}$ hold. In other words, Newton's method converges quadratically.

Now let us apply the method to Equation (6.3-22). To make the development both simple and realistic, consider the numerical approximation to this equation that results when we adopt functional representations of the forms (6.3-20) and (6.3-21) for the entries of $\mathbf{A}$ and $\mathbf{f}$. Thus

$$A_{jn}(\mathbf{u}) \simeq \sum_{\ell=1}^{M} \alpha_{nj\ell} K(u_\ell),$$

$$f_j(\mathbf{u}) \simeq \sum_{\ell=1}^{M} \beta_{j\ell} f(u_\ell),$$

where $\alpha_{nj\ell} = \int_\Omega \psi_\ell \nabla \phi_n \cdot \nabla \phi_j \, d\mathbf{x}$ and $\beta_{j\ell} = \int_\Omega \psi_\ell \phi_j \, d\mathbf{x}$ are constants and u_ℓ stands for an entry in the sought solution vector $\mathbf{u}$. Therefore the j-th component of the function $\mathbf{F}$ takes the form

$$F_j(\mathbf{u}) = \sum_{n=1}^{N} A_{jn} u_n + f_j(\mathbf{u})$$

$$\simeq \sum_{n=1}^{N} \left[\sum_{\ell=1}^{M} \alpha_{nj\ell} K(u_\ell) \right] u_n + \sum_{\ell=1}^{M} \beta_{j\ell} f(u_\ell).$$

We can thus compute the (j,i)-th entry of the Jacobian matrix for $\mathbf{F}$ approximately as follows:

(6.3-25) $$\frac{\partial F_j}{\partial u_i}(\mathbf{u}^{(m)}) \simeq \sum_{n=1}^{N} \alpha_{nji} K'(u_i^{(m)}) u_n^{(m)}$$

$$+ \sum_{\ell=1}^{M} \alpha_{ij\ell} K(u_\ell^{(m)}) + \beta_{ji} f'(u_i^{(m)}).$$

Observe that, even though each of these entries is a relatively small linear combination, each requires the evaluation of the nonlinear functions $K(u)$, $K'(u)$, and $f'(u)$ at several values of their arguments. What is more, we must recompute each entry in the $N \times N$ Jacobian matrix $\mathbf{F}'$ and the forcing vector $-\mathbf{F}$ at every iteration. These calculations can contribute significantly to the computational cost of Newton's method, especially if, as often happens, the partial derivatives of $\mathbf{F}$ require more effort to compute than $\mathbf{F}$ itself.

There are many variants on Newton's method aimed at reducing this computational cost. One variant that is quite simple is to use the initial Jacobian matrix throughout the iterations. Thus, instead of Equations (6.3-24), we use

$$\mathbf{F}'(\mathbf{u}^{(0)}) \, \Delta \mathbf{u} = -\mathbf{F}(\mathbf{u}^{(m)}),$$

$$\mathbf{u}^{(m+1)} = \mathbf{u}^{(m)} + \Delta \mathbf{u}.$$

This scheme, called the **modified Newton's method**, clearly relies on the assumption that the Jacobian matrix $\mathbf{F}'$ varies slowly as a function of $\mathbf{u}$ in some neighborhood of the exact solution. The method typically exhibits slower convergence, but in many problems the savings per iteration gained by avoiding the construction and inversion of a new Jacobian matrix can outweigh the cost of the extra iterations required.

Another class of variants on Newton's method bypasses the computation of derivatives altogether. In many practical problems the functions K and f may not be amenable to differentiation in closed form. In these cases Equation (6.3-25) does not provide a computable expression for $\partial F_i/\partial u_i$. One alternative is to replace this partial derivative by a finite-difference approximation:

$$\frac{\partial F_j}{\partial u_i}(\mathbf{u}^{(m)}) \simeq \frac{1}{h_i^{(m)}} \left[F_j(\mathbf{u}^{(m)} + h_i^{(m)}\mathbf{e_i}) - F_j(\mathbf{u}^{(m)}) \right],$$

where $\mathbf{e}_i$ denotes the i-th unit basis vector whose j-th entry is the Kronecker symbol δ_{ij} and $h_i^{(m)}$ signifies an increment chosen in some systematic way. We shall review two ways of choosing this increment, referring the reader to Ortega and Rheinboldt (1970) for a thorough discussion of the various possibilities as well as for proofs of the convergence rates stated below.

One seemingly natural choice for the increment $h_i^{(m)}$ is $h_i^{(m)} = u_i^{(m)} - u_i^{(m-1)}$, whose magnitude presumably decreases as $m \to \infty$. Obviously, this choice makes sense only if $u_i^{(m)} \neq u_i^{(m-1)}$. The resulting scheme, known as the **secant method**, requires $N^2 + 1$ evaluations of the component functions F_j at each iteration, and its order of convergence is $(1 + \sqrt{5})/2 \simeq 1.618$. A somewhat more sophisticated approach is **Steffensen's method**, in which we choose $h_i^{(m)} = F_i(\mathbf{u}^{(m)})$, which should also decrease in magnitude as $k \to \infty$. When discrepancies in the units of $h_i^{(m)}$ and $F_i(\mathbf{u}^{(m)})$ arise, or when $F_i(\mathbf{u}^{(m)})$ has an inappropriate magnitude, we can multiply the latter by a suitable scaling factor. Steffensen's method offers the advantage of retaining the quadratic convergence rate associated with Newton's method while obviating the computation of derivatives.

One final Newton-like method that we shall examine is based on the original PDE rather than its discrete analog. Let us write the steady heat-flow problem as follows:

$$\mathcal{F}(u) \equiv \nabla \cdot [K(u)\nabla u] - f(u) = 0.$$

We can derive a linearized form of this equation by evaluating the function $K(u)$ at a known iterative level m and linearly projecting the remaining occurrences of $u(\mathbf{x})$ to the next iterative level:

$$\nabla \cdot [K(u^{(m)})\nabla u^{(m+1)}] - f(u^{(m+1)}) = 0.$$

Now we rewrite this equation so that the increment $\delta u = u^{(m+1)} - u^{(m)}$ appears as the unknown. Replacing $u^{(m+1)}$ by $u^{(m)} + \delta u$, adopting the approximation $f(u^{(m+1)}) \simeq f(u^{(m)}) + f'(u^{(m)})\delta u$, and rearranging, we get

(6.3-26) $$\left\{ \nabla \cdot [K(u^{(m)})\nabla] - f'(u^{(m)}) \right\} \delta u$$

$$= -\left\{ \nabla \cdot [K(u^{(m)})\nabla u^{(m)}] - f(u^{(m)}) \right\} = -\mathcal{F}(u^{(m)}).$$

This scheme furnishes a linear operator equation that we can solve for the iterative increment δu, after which we set $u^{(m+1)} = u^{(m)} + \delta u$ to begin a new iteration. Notice the formal similarity between Equation (6.3-26) and the matrix equation (6.3-24a) defining Newton's method: An operator evaluated at the most recently known iterative level acts on an unknown increment to yield the negative of the most recent residual.

To discretize Equation (6.3-26), we can establish, for example, a finite-element representation

(6.3-27) $$\hat{u}^{(m)}(\mathbf{x}) = u_\partial(\mathbf{x}) + \sum_{i=1}^{N} u_i^{(m)} \phi_i(\mathbf{x}),$$

where $u_\partial(\mathbf{x})$ satisfies the known essential boundary conditions and the basis functions $\phi_i(\mathbf{x})$ satisfy homogeneous boundary conditions. We then use an analogous trial function for the unknown increment δu:

$$\delta\hat{u}(\mathbf{x}) = \sum_{i=1}^{N} \delta u_i\, \phi_i(\mathbf{x}).$$

If we substitute these expansions into the operator equation (6.3-27) and apply some version of the method of weighted residuals, we shall produce a matrix analog of Equation (6.3-26) that is linear in the vector $(\delta u_1, \ldots, \delta u_N)$ of unknown nodal increments. We then solve this matrix equation at each iteration, using the new increment vectors to update the coefficients in Equation (6.3-27) according to the relationship $u_i^{(m+1)} = u_i^{(m)} + \delta u_i$ at each iterative level. As with Newton's method, we can stop iterating as soon as the residual $\mathcal{F}(\hat{u}^{(m)})$ is small enough in some norm.

The nonlinear diffusion equation.

We now turn our attention to the nonlinear diffusion equation

(6.3-9) $$\frac{\partial}{\partial t}[c(u)] - \nabla \cdot [K(u)\nabla u] = 0.$$

Two features distinguish this equation from the nonlinear steady heat-flow problem just considered. First, we must discretize the equation in both space and time, so the issue of time-stepping algorithms becomes crucial.

Second, Equation (6.3-9) exhibits nonlinearity in the accumulation term $\partial[c(u)]/\partial t$. In the context of the species mass balance from which we derived Equation (6.3-9), the nonlinearity in the accumulation term has a common physical interpretation in terms of compositional density effects. Indeed, if we apply the chain rule to the time derivative of the solute density $c(u)$, we find

$$(6.3\text{-}28) \qquad \frac{\partial}{\partial t}[c(u)] = \frac{dc}{du}(u)\frac{\partial u}{\partial t}.$$

The function dc/du represents the rate of change of density with respect to solute mass fraction u, which we might identify as a "compressibility" due to the effects of mixture composition.

It is quite common to discretize Equation (6.3-9) in time by first applying the chain rule as in Equation (6.3-28) and then approximating the quantity $\partial u/\partial t$ by finite differences or some other method. Thus we obtain a **semidiscrete** equation whose form depends on which time level we choose for the evaluations of the remaining occurrences of u in the PDE. If the index n signifies the most recent known time level, then evaluating the spatial terms and the coefficient dc/du at the next unknown time level $n + 1$ produces an implicit equation,

$$\frac{1}{k}\frac{dc}{du}(u^{n+1})(u^{n+1} - u^n) - \nabla \cdot \left[K(u^{n+1})\nabla u^{n+1}\right] = 0,$$

where k denotes the time step.

At this point we can choose any one of several schemes to discretize the spatial variations. Let us examine the application of finite-element collocation, as discussed in Section 2.14. For simplicity, we shall treat the one-dimensional analog,

$$(6.3\text{-}29) \qquad \frac{1}{k}\frac{dc}{du}(u^{n+1} - u^n) - \frac{\partial}{\partial x}\left[K(u^{n+1})\frac{\partial u^{n+1}}{\partial x}\right] = 0,$$

on a spatial interval $(0, L)$. Let us consider this equation subject to the mixed, constant boundary conditions

$$u(0, t) = \bar{u}_0, \quad t > 0,$$

$$\frac{\partial u}{\partial x}(L, t) = \bar{u}_L', \quad t > 0,$$

and initial conditions

$$u(x, 0) = u_I(x), \quad x \in (0, L).$$

For the finite-element basis functions we select the piecewise Hermite cubic interpolating polynomials $\{h_i^0(x), h_i^1(x)\}_{i=0}^N$ on a uniform grid $\Delta : 0 = x_0 < x_1 < \cdots < x_N = L$ having mesh $x_i - x_{i-1} = h$. For the given boundary data, our trial function at time level n will have the form

$$
(6.3\text{-}30) \qquad \hat{u}^n(x) = \bar{u}_0 h_0^0(x) + (u_0')^n h_0^1(x)
$$

$$
+ \sum_{i=1}^{N-1} \left[(u_i)^n h_i^0(x) + (u_i')^n h_i^1(x) \right]
$$

$$
+ (u_n)^n h_N^0(x) + \bar{u}_L' h_N^1(x),
$$

where the coefficients $(u_1)^n, \ldots, (u_N)^n$ stand for unknown values of $\hat{u}^n(x)$ at the nodes $x_1, \ldots, x_N$, and the coefficients $(u_0')^n, \ldots, (u_{N-1}')^n$ stand for unknown values of $d\hat{u}^n/dx$ at the nodes $x_0, \ldots, x_{N-1}$. Thus at each time level $t = nk$ we must solve for $2N$ unknown coefficients defining the trial function $\hat{u}^n \in C^1((0, L))$.

To get the necessary $2N$ equations, we collocate Equation (6.3-29) at the $2N$ Gauss points, as discussed in Section 2.14. Let us denote these points as $\bar{x}_\ell$, $\ell = 1, \ldots, 2N$. Recall that these points, when viewed in the local coordinates ξ mapping a typical element $[x_{i-1}, x_i]$ onto $[-1, 1]$, lie at $\xi(\bar{x}_\ell) = \pm 1/\sqrt{3} \simeq \pm 0.57735$. Using these points, we arrive at the equations

$$
\frac{1}{k} \frac{dc}{du} (\hat{u}^{n+1}(\bar{x}_\ell)) \left[\hat{u}^{n+1}(\bar{x}_\ell) - \hat{u}^n(\bar{x}_\ell) \right]
$$

$$
- \frac{\partial}{\partial x} \left[K(\hat{u}^{n+1}(\bar{x}_\ell)) \frac{\partial \hat{u}^{n+1}}{\partial x}(\bar{x}_\ell) \right] = 0, \quad \ell = 1, \ldots, 2N,
$$

or, equivalently,

$$
(6.3\text{-}31) \quad \frac{1}{k} \frac{dc}{du} (\hat{u}^{n+1}(\bar{x}_\ell)) \left[\hat{u}^{n+1}(\bar{x}_\ell) - \hat{u}^n(\bar{x}_\ell) \right]
$$

$$
- \frac{\partial K}{\partial x} (\hat{u}^{n+1}(\bar{x}_\ell)) \frac{\partial \hat{u}^{n+1}}{\partial x}(\bar{x}_\ell) - K(\hat{u}^{n+1}(\bar{x}_\ell)) \frac{\partial^2 \hat{u}^{n+1}}{\partial x^2}(\bar{x}_\ell)
$$

$$
= 0, \qquad \ell = 1, \ldots, 2N.
$$

We now confront the fact that the nonlinear functions dc/du, K, and $\partial K/\partial x$ make the discrete equations impossible to solve, except perhaps in very simple cases. To circumvent this difficulty, we must devise convenient approximations for these coefficients and implement an iterative scheme to accommodate their dependence on the unknown $\hat{u}^{n+1}$. For the first task, let us approximate dc/du and K by functional representations using piecewise Lagrange linear expansions:

$$\frac{dc}{du}(\hat{u}^{n+1}) \simeq \left(\widehat{\frac{dc}{du}}\right)^{n+1}(x) = \sum_{i=0}^{N}\left(\frac{dc}{du}\right)^{n+1}_i \ell_i(x),$$

$$K(\hat{u}^{n+1}) \simeq \hat{K}^{n+1}(x) = \sum_{i=0}^{N} K_i^{n+1}\ell_i(x).$$

In these equations, $\ell_i(x)$ denotes the piecewise Lagrange linear basis function associated with the node x_i, and the nodal values $(dc/du)_i^{n+1}$ and K_i^{n+1} are just the values of the functions dc/du and K at the nodal values of $\hat{u}^{n+1}$:

$$\left(\frac{dc}{du}\right)_i^{n+1} = \frac{dc}{du}((u_i)^{n+1});$$

$$K_i^{n+1} = K((u_i)^{n+1}).$$

Observe that this representation for K suggests a natural representation for $\partial K/\partial x$, namely,

$$\frac{\partial K}{\partial x}(\hat{u}^{n+1}) \simeq \frac{\partial \hat{K}}{\partial x}(x) = \sum_{i=0}^{N} K_i^{n+1}\frac{d\ell_i}{dx}(x).$$

We are left with the job of constructing an iterative scheme to handle the nonlinearities. Following our discussion of the nonlinear steady heat-flow equation, we shall employ a Newton-like scheme in which, given a known iterative value $\hat{u}^{n+1,m}(x)$ for the new time level $n+1$, we solve for an increment

$$\delta\hat{u}(x) = \sum_{i=1}^{N}\left[\delta u_i\, h_i^0(x) + \delta u_i'\, h_i^1(x)\right]$$

and compute the coefficients of the new iterate $\hat{u}^{n+1,m+1}(x)$ according to the updating rules

$$(u_i)^{n+1,m+1} = (u_i)^{n+1,m} + \delta u_i,$$

$$(u_i')^{n+1,m+1} = (u_i')^{n+1,m} + \delta u_i'.$$

Observe that the boundary values $(u_0)^{n+1} = \bar{u}_0$ and $(u_N')^{n+1} = \bar{u}_L'$ are known, so $\delta u_0 = \delta u_N' = 0$ at every stage.

In solving for these increments, we can lag all nonlinear coefficients by an iteration, projecting linear occurrences of the unknown $\hat{u}^{n+1}$ forward to the next unknown iterative level. We thus arrive at a linearized set of collocation equations,

$$\frac{1}{k}\left(\widehat{\frac{dc}{du}}\right)^{n+1,m}(\bar{x}_\ell)\left[\hat{u}^{n+1,m}(\bar{x}_\ell) + \delta\hat{u}(\bar{x}_\ell) - \hat{u}^n(\bar{x}_\ell)\right]$$

$$-\frac{\partial \hat{K}^{n+1,m}}{\partial x}(\bar{x}_\ell)\left[\frac{\partial \hat{u}^{n+1,m}}{\partial x}(\bar{x}_\ell) + \frac{\partial \delta\hat{u}}{\partial x}(\bar{x}_\ell)\right]$$

$$-\hat{K}^{n+1,m}(\bar{x}_\ell)\left[\frac{\partial^2 \hat{u}^{n+1,m}}{\partial x^2}(\bar{x}_\ell) + \frac{\partial^2 \delta\hat{u}}{\partial x^2}(\bar{x}_\ell)\right] = 0,$$

$$\ell = 1,\ldots,2N.$$

If we move all terms known at the m-th iterative level to the right side of this equation and keep the terms involving the unknown coefficients of $\delta\hat{u}$ on the left, we get

$$(6.3\text{-}32)\quad \left[\frac{1}{k}\left(\widehat{\frac{dc}{du}}\right)^{n+1,m}(\bar{x}_\ell) - \frac{\partial \hat{K}^{n+1,m}}{\partial x}(\bar{x}_\ell)\frac{\partial}{\partial x} - \hat{K}^{n+1,m}(\bar{x}_\ell)\frac{\partial^2}{\partial x^2}\right]\delta\hat{u}(\bar{x}_\ell)$$

$$= -\left\{\frac{1}{k}\left(\widehat{\frac{dc}{du}}\right)^{n+1,m}(\bar{x}_\ell)\left[\hat{u}^{n+1,m}(\bar{x}_\ell) - \hat{u}^n(\bar{x}_\ell)\right]\right.$$

$$\left. -\frac{\partial \hat{K}^{n+1,m}}{\partial x}(\bar{x}_\ell)\frac{\partial \hat{u}^{n+1,m}}{\partial x}(\bar{x}_\ell) - \hat{K}^{n+1,m}(\bar{x}_\ell)\frac{\partial^2 \hat{u}^{n+1,m}}{\partial x^2}(\bar{x}_\ell)\right\}$$

$$\equiv -\hat{R}^{n+1,m}(\bar{x}_\ell),\quad \ell = 1,\ldots,2N.$$

We terminate the iterative calculations for each new time level $n+1$ as soon as the residual norm $\|\hat{R}^{n+1,m+1}\|_\infty$ falls below some prescribed tolerance,

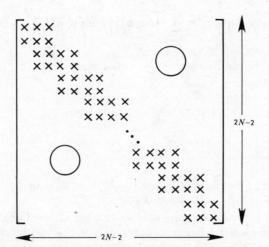

Figure 6-4. Structure of the matrix arising at each iteration in the finite-element collocation scheme for solving the nonlinear diffusion equation.

setting $\hat{u}^{n+1} = \hat{u}^{n+1,m+1}$. Given the computed value $\hat{u}^n(x)$ at any time level, we can begin the iterations for the next time level by setting $\hat{u}^{n+1,0} = \hat{u}^n$. We leave to the reader the task of showing that each iteration involves the solution of a set of linear equations whose matrix has the structure drawn in Figure 6-4.

Although the discrete equations (6.3-32) are consistent with the original PDE (6.3-12), they tend to produce approximate solutions having global mass balance errors. More precisely, unless the time step k is very small or the coefficient dc/du varies slowly as a function of the unknown u, the mass error over a single time step can be significant. The integral

$$\int_0^L \left[c(\hat{u}^{n+1}) - c(\hat{u}^n) - k\frac{\partial}{\partial x}\left(\hat{K}^{n+1}\frac{\partial \hat{u}^{n+1}}{\partial x} \right) \right]\, dx$$

furnishes a computable measure of this error. The problem with Equations (6.3-32) is their treatment of the accumulation term: The value of the coefficient $(\widehat{dc/du})$ at the level $(n+1, m)$ may fail to be representative of its value over the entire time step from $t = nk$ to $t = (n+1)k$. To avoid the thorny issue of how to evaluate this coefficient more accurately, it is sometimes best to abandon the convenience of the chain-rule factorization in Equation (6.3-28) in favor of a *direct* discretization of the accumulation term. Thus we need a functional representation for the constitutive function $c(u)$, which, for consistency with the Hermite cubic representation used to compute the spatial flux terms, we can take to be

$$\hat{c}^{n+1} = \sum_{i=0}^{N} \left[c((u_i)^{n+1})h_i^0(x) + \frac{dc}{du}((u_i)^{n+1})(u_i')^{n+1}h_i^1(x) \right].$$

Notice that we have used the chain rule to develop a representation for the slope $\partial \hat{c}/\partial x$ at each spatial node.

Using this approach, one can discretize the accumulation term in the iterative time-stepping scheme as follows:

$$\frac{\partial \hat{c}}{\partial t}\bigg|^{n+1,m} \simeq \frac{1}{k}\left[\hat{c}^{n+1,m} + \left(\frac{dc}{du}\right)^{n+1,m}\delta \hat{u} - \hat{c}^n \right],$$

where we use the piecewise linear representation for $(\widehat{dc/du})$ as before to avoid polynomial approximations having large degree. The collocation equations that result are similar to those in Equation (6.3-32), the only exception being that

$$\hat{R}^{n+1,m}(\bar{x}_\ell) = \frac{1}{k}\left[\hat{c}^{n+1,m}(\bar{x}_\ell) - \hat{c}^n(\bar{x}_\ell) \right]$$

$$- \frac{\partial \hat{K}^{n+1,m}}{\partial x}(\bar{x}_\ell)\frac{\partial \hat{u}^{n+1,m}}{\partial x}(\bar{x}_\ell) - \hat{K}^{n+1,m}(\bar{x}_\ell)\frac{\partial^2 \hat{u}^{n+1,m}}{\partial x^2}(\bar{x}_\ell).$$

Now, when $\|\hat{R}^{n+1,m+1}\|_\infty$ is small, the integrated mass balance error over the time step will be small as well. For an application of this technique to a problem with strong nonlinearities in the accumulation term, we refer the reader to Allen and Murphy (1985). A subsequent paper (Allen and Murphy, 1986) demonstrates the extension of this approach to two space dimensions.

6.4. The Simulation of Solid Deformation.

As our first example of a coupled system of PDEs, we shall consider the problem of a fluid-saturated porous medium that undergoes deformation as a result of fluid withdrawal. This system has great importance in areas where groundwater withdrawal leads to land subsidence. The numerical method of choice for the simulation of solid deformation is the finite-element method. The reason for this is the relative ease with which the method accommodates the moving boundaries of the porous medium as it deforms. Our development in this section follows that presented in Safai (1977).

Governing equations.

The governing equations describing soil deformation are obtained by substituting suitable constitutive relationships into the mixture balance laws appropriate for a fluid-saturated porous medium. Let us consider first the momentum balance for the fluid-solid mixture, recalling the framework for deriving multiphase balance laws presented in Section 1.5. If we denote the rock phase by the index R and the fluid phase by the index F, then summing the momentum balances for the two phases R and F yields

$$\rho \frac{D\mathbf{v}}{Dt} - \nabla \cdot (\mathbf{t}^R + \mathbf{t}^F) - \rho \mathbf{b} = \mathbf{0}.$$

Here, ρ stands for the overall mixture density, $\mathbf{v}$ is the barycentric velocity of the mixture, and $\mathbf{t}^R$ and $\mathbf{t}^F$ stand for the stress tensors in the solid and fluid, respectively. The symbol $\mathbf{b}$ signifies the total body force acting on the mixture, which we assume to be accounted for by the influence of gravity. In the most sophisticated models of solid displacement, every term in this equation may contribute to the net motion of the mixture. However, in many applications the motion resulting from changes in the stress term dominates the effects of the mixture's inertia and the influence of gravity. If we neglect these latter terms, we are left with

$$\nabla \cdot (\mathbf{t}^R + \mathbf{t}^F) = \mathbf{0}.$$

In Section 1.5 we argued that the stress in the fluid phase can be approximated by its isotropic part $-p^F\mathbf{1}$, where p^F is the mechanical pressure in the fluid. Adopting this approximation, we see that the overall momentum balance reduces to the following:

(6.4-1)
$$\nabla \cdot \mathbf{t}^R - \nabla \cdot (p^F \mathbf{1}) = 0.$$

Hydrologists working with solid deformation problems often call the quantities $\mathbf{t}^R$ and p^F the **effective stress** and the **excess pore-water pressure**, respectively.

Now let us examine the mass balances for the two phases. Before writing these equations, let us establish some assumptions and terminology. In the following, we shall assume that the fluid and solid are incompressible in the sense that their intrinsic mass densities ρ^F and ρ^R have negligible rates of change in time and space, even though changes in the volume fractions $\phi^F \equiv \phi$ and $\phi^R = 1 - \phi$ may cause the *overall* mixture density to vary. We shall also refer to the fluid and solid displacements $\mathbf{U}^F$ and $\mathbf{U}^R$. We define each of these quantities, as for the solid displacement discussed in Section 1.4, as the distance between a material point in the appropriate phase and its location in the initial configuration of the mixture. Thus we may write the velocity in each phase α as $\mathbf{v}^\alpha = \partial \mathbf{U}^\alpha / \partial t$. Finally, recall that in Section 1.5 we derived Darcy's law for the velocity of a fluid in a porous medium by assuming the rock matrix to be stationary. In the problem of solid deformation we must abandon that assumption, since the solid phase will be moving. However, we can easily accommodate solid motions by referring to the **relative velocity** $\mathbf{v}^F - \mathbf{v}^R$ in the formulation of the Stokes drag of Section 1.5. The result is the following generalization of Darcy's law:

$$\mathbf{v}^F - \mathbf{v}^R \equiv \frac{\partial \mathbf{U}^F}{\partial t} - \frac{\partial \mathbf{U}^R}{\partial t} = -\frac{K}{\phi \rho^F g} \nabla p^F.$$

Observe that we have assumed gravitational forces to be negligible and have adopted the hydrologists' notation in using the hydraulic conductivity K, defined in Section 3.1. We use g to denote the magnitude of gravitational acceleration.

With these remarks in mind, consider the fluid mass balance,

$$\frac{\partial}{\partial t}(\phi \rho^F) + \nabla \cdot (\phi \rho^F \mathbf{v}^F) = 0.$$

Let us expand the time derivative and use the **fluid compressibility**, defined as $\beta = (1/\rho^F)d\rho^F/dp^F$, to rewrite the time derivative of the fluid density. We get

$$\phi \beta \rho^F \frac{\partial p^F}{\partial t} + \rho^F \frac{\partial \phi}{\partial t} + \nabla \cdot (\phi \rho^F \mathbf{v}^F) = 0.$$

Next, notice that, according to Darcy's law, we can replace the fluid velocity in this last equation by using the identity

$$\mathbf{v}^F = -\frac{K}{\phi \rho^F g}\nabla p^F + \mathbf{v}^R.$$

Then if we use the assumption that gradients of ρ^F are negligible and further assume that the porous medium has uniform hydraulic conductivity, we get the following flow equation for the fluid:

$$\phi\beta\frac{\partial p^F}{\partial t} + \frac{\partial \phi}{\partial t} - \frac{K}{\rho^F g}\nabla^2 p^F + \nabla \cdot (\phi\mathbf{v}^R) = 0.$$

Turning to the mass balance for the solid matrix, we have

$$\frac{\partial}{\partial t}[(1-\phi)\rho^R] + \nabla \cdot [(1-\phi)\rho^R\mathbf{v}^R] = 0.$$

Now we expand the time derivative as before, this time assuming that the solid grains of the porous matrix are incompressible so that $\partial\rho^R/\partial t = 0$. We have as a result

$$-\frac{\partial \phi}{\partial t} = -\nabla \cdot [(1-\phi)\mathbf{v}^R]$$
$$= -\nabla \cdot \mathbf{v}^R + \nabla \cdot (\phi\mathbf{v}^R)$$
$$= -\nabla \cdot \left(\frac{\partial \mathbf{U}^R}{\partial t}\right) + \nabla \cdot (\phi\mathbf{v}^R),$$

where we have taken advantage of the spatial uniformity of ρ^R. Adding this equation to the flow equation for the fluid derived above and interchanging spatial and temporal differential operators, we obtain the final form of the flow equation,

$$(6.4\text{-}2) \qquad \phi\beta\frac{\partial p^F}{\partial t} - \frac{K}{\rho^F g}\nabla^2 p^F + \frac{\partial}{\partial t}(\nabla \cdot \mathbf{U}^R) = 0.$$

In what follows, we shall consider the solid matrix to be a linear elastic solid. Thus we can rewrite the term $\nabla \cdot \mathbf{t}^R$ in Equation (6.4-1) using Hooke's law, Equation (1.4-12), which states

$$(6.4\text{-}3) \qquad \mathbf{t}^R = 2\mu\mathbf{e}^R + \lambda\mathrm{tr}(\mathbf{e})\mathbf{1},$$

where λ and μ are the Lamé constants of the rock matrix and $\mathbf{e} = [\nabla\mathbf{U}^R + (\nabla\mathbf{U}^R)^\top]$ is the infinitesimal Eulerian strain. Substituting this constitutive relationship into Equation (6.4-1) yields

$$(6.4\text{-}4) \qquad \mu\nabla^2\mathbf{U}^R + (\lambda+\mu)\nabla(\nabla \cdot \mathbf{U}^R) - \nabla \cdot (p^F\mathbf{1}) = \mathbf{0}.$$

To keep the notation from becoming too cumbersome, in the remainder of this section we shall drop the phase indices F and R from the governing equations.

Galerkin formulation.

Let us discretize Equations (6.4-1) [or (6.4-4)] and (6.4-2), treating the displacement vector $\mathbf{U}$ and the excess pore-water pressure p as the dependent variables. Let the approximating trial functions for these variables be defined as

(6.4-5a) $$\hat{\mathbf{U}}(\mathbf{x}, t) = \sum_{i=0}^{N} \mathbf{U}_i(t)\phi_i(\mathbf{x})$$

and

(6.4-5b) $$\hat{p}(\mathbf{x}, t) = \sum_{i=0}^{N} p_i(t)\phi_i(\mathbf{x}).$$

For convenience, we shall also keep track of the effective stress by adopting a finite-element representation for $\mathbf{t}$:

(6.4-5c) $$\hat{\mathbf{t}}(\mathbf{x}, t) = \sum_{i=0}^{N} \mathbf{t}_i(t)\phi_i(\mathbf{x}).$$

In these expansions, $\mathbf{U}_i$, p_i, and $\mathbf{t}_i$ are coefficients to be determined. The reader should be careful not to confuse the unindexed symbol ϕ, signifying the porosity, with the basis functions $\phi_i(\mathbf{x})$.

The substitution of the trial functions (6.4-5) into the governing equations yields a residual, which Galerkin's method forces to be orthogonal to each basis function ϕ_i associated with a non-Dirichlet node. Thus we obtain a set of equations of the form

(6.4-6a) $$\int_{\Omega} [\nabla \cdot \hat{\mathbf{t}} - \nabla \cdot (\hat{p}\mathbf{1})]\phi_j(\mathbf{x}) \, d\mathbf{x} = \mathbf{0},$$

(6.4-6b) $$\int_{\Omega} \left[\frac{K}{\rho g}\nabla^2\hat{p} - \phi\beta\frac{\partial\hat{p}}{\partial t} - \frac{\partial}{\partial t}(\nabla \cdot \hat{\mathbf{U}}) \right] \phi_j(\mathbf{x}) \, d\mathbf{x} = 0,$$

where Ω denotes the spatial domain of the problem. Application of Green's theorem to the first term in Equations (6.4-6a) and (6.4-6b) and introduction of the constitutive relationship expressed in Equation (6.4-4) yields

(6.4-7a) $$\int_{\Omega} \left[\lambda(\nabla \cdot \hat{\mathbf{U}})\mathbf{1} \cdot \nabla\phi_j + \mu(\nabla\hat{\mathbf{U}} \cdot \nabla\phi_j + \hat{\mathbf{U}}\nabla \cdot \nabla\phi_j) + (\nabla\hat{p})\phi_j \right] d\mathbf{x}$$

$$= \oint_{\partial\Omega} (\mathbf{n} \cdot \hat{\mathbf{t}})\phi_j \, d\mathbf{x}$$

and

(6.4-7b)
$$\int_\Omega \left[\frac{K}{\rho g} \nabla \hat{p} \cdot \nabla \phi_j + \phi \beta \frac{\partial \hat{p}}{\partial t} \phi_j + \frac{\partial}{\partial t} (\nabla \cdot \hat{\mathbf{U}}) \phi_j \right] dx$$

$$= \oint_{\partial \Omega} \frac{K}{\rho g} \mathbf{n} \cdot (\phi_j \nabla \hat{p}) \, dx.$$

Consider now the case when Ω is two-dimensional, with x and z being the two Cartesian coordinate directions. Substitution of the definitions of $\hat{\mathbf{U}}$ and $\hat{p}$ from Equations (6.4-5) into the integral equations (6.4-7) yields three sets of equations. For the first component of Equation (6.4-7a) we get

(6.4-8a)
$$\sum_{i=0}^N \int_\Omega \left[\left(\lambda \frac{\partial \phi_i}{\partial x} \frac{\partial \phi_j}{\partial x} + 2\mu \frac{\partial \phi_i}{\partial x} \frac{\partial \phi_j}{\partial x} + \mu \frac{\partial \phi_i}{\partial z} \frac{\partial \phi_j}{\partial z} \right) (U_x)_i \right.$$

$$\left. + \left(\lambda \frac{\partial \phi_i}{\partial z} \frac{\partial \phi_j}{\partial x} + \mu \frac{\partial \phi_i}{\partial x} \frac{\partial \phi_j}{\partial z} \right) (U_z)_i + \frac{\partial \phi_i}{\partial x} \phi_j p_i \right] dx$$

$$= \oint_{\partial \Omega} (n_x t_{xx} + n_z t_{zx}) \phi_j \, dx,$$

where $(U_x)_i$ and $(U_z)_i$ stand for the x- and z- components of the unknown coefficient $\mathbf{U}_i$. For the second component of Equation (6.4-7a) we get

(6.4-8b)
$$\sum_{i=0}^N \int_\Omega \left[\left(\lambda \frac{\partial \phi_i}{\partial x} \frac{\partial \phi_j}{\partial z} + \mu \frac{\partial \phi_i}{\partial z} \frac{\partial \phi_j}{\partial x} \right) (U_x)_i \right.$$

$$\left. + \left(\lambda \frac{\partial \phi_i}{\partial z} \frac{\partial \phi_j}{\partial z} + 2\mu \frac{\partial \phi_i}{\partial z} \frac{\partial \phi_j}{\partial z} + \mu \frac{\partial \phi_i}{\partial x} \frac{\partial \phi_j}{\partial x} \right) (U_z)_i + \frac{\partial \phi_i}{\partial z} \phi_j p_i \right] dx$$

$$= \oint_{\partial \Omega} (n_x t_{xz} + n_z t_{zz}) \phi_j \, dx.$$

Finally, for Equation (6.4-7b), we get

(6.4-8c)
$$\sum_{i=0}^N \int_\Omega \left[\frac{\partial \phi_i}{\partial x} \phi_j \frac{d(U_x)_i}{dt} + \frac{\partial \phi_i}{\partial z} \phi_j \frac{d(U_z)_i}{dt} \right.$$

$$\left. + \frac{K}{\rho g} \left(\frac{\partial \phi_i}{\partial x} \frac{\partial \phi_j}{\partial x} + \frac{\partial \phi_i}{\partial z} \frac{\partial \phi_j}{\partial z} \right) p_i + \phi \beta \phi_i \phi_j \frac{dp_i}{dt} \right] dx$$

$$= \oint_{\partial \Omega} \frac{K}{\rho g} \left(n_x \frac{\partial p}{\partial x} + n_z \frac{\partial p}{\partial z} \right) \phi_j \, dx.$$

The set of equations (6.4-8) can be conveniently written in matrix form. The equations for the unknown variables at a single spatial node i are as follows:

(6.4-9)
$$
\begin{bmatrix} A_{ji} & B_{ji} & C_{ji} \\ D_{ji} & E_{ji} & G_{ji} \\ 0 & 0 & L_{ji} \end{bmatrix} \begin{bmatrix} (U_x)_i \\ (U_z)_i \\ p_i \end{bmatrix}
$$

$$
+ \begin{bmatrix} 0 & 0 & 0 \\ 0 & 0 & 0 \\ H_{ji} & K_{ji} & M_{ji} \end{bmatrix} \frac{d}{dt} \begin{bmatrix} (U_x)_i \\ (U_z)_i \\ p_i \end{bmatrix} = \begin{bmatrix} (F_x)_j \\ (F_z)_j \\ (F_p)_j \end{bmatrix} .
$$

Note that this matrix equation constitutes one block in the global matrix equation for the unknown coefficients at all nodes. The functional forms of the matrix entries in Equation (6.4-9) can be determined through inspection of Equations (6.4-8); they are as follows:

$$
A_{ji} = \int_\Omega \left[(\lambda + 2\mu) \frac{\partial \phi_i}{\partial x} \frac{\partial \phi_j}{\partial x} + \mu \frac{\partial \phi_i}{\partial z} \frac{\partial \phi_j}{\partial z} \right] dx,
$$

$$
B_{ji} = \int_\Omega \left(\lambda \frac{\partial \phi_i}{\partial z} \frac{\partial \phi_j}{\partial x} + \mu \frac{\partial \phi_i}{\partial x} \frac{\partial \phi_j}{\partial z} \right) dx,
$$

$$
C_{ji} = \int_\Omega \frac{\partial \phi_i}{\partial x} \phi_j \, dx = H_{ji},
$$

$$
D_{ji} = \int_\Omega \left(\lambda \frac{\partial \phi_i}{\partial x} \frac{\partial \phi_j}{\partial z} + \mu \frac{\partial \phi_i}{\partial z} \frac{\partial \phi_j}{\partial x} \right) dx,
$$

$$
E_{ji} = \int_\Omega \left[(\lambda + 2\mu) \frac{\partial \phi_i}{\partial z} \frac{\partial \phi_j}{\partial z} + \mu \frac{\partial \phi_i}{\partial x} \frac{\partial \phi_j}{\partial x} \right] dx,
$$

$$
G_{ji} = \int_\Omega \frac{\partial \phi_i}{\partial z} \phi_j \, dx = K_{ji},
$$

$$
L_{ji} = \int_\Omega \frac{K}{\rho g} \left(\frac{\partial \phi_i}{\partial x} \frac{\partial \phi_j}{\partial x} + \frac{\partial \phi_i}{\partial z} \frac{\partial \phi_j}{\partial z} \right) dx,
$$

$$
M_{ji} = \int_\Omega \phi \beta \phi_i \phi_j \, dx.
$$

The entries of the forcing vector are just integrals of the known boundary values:

$$(F_x)_j = \oint_{\partial\Omega} (n_x t_{xx} + n_z t_{zx})\phi_j \, d\mathbf{x},$$

$$(F_z)_j = \oint_{\partial\Omega} (n_x t_{xz} + n_z t_{zz})\phi_j \, d\mathbf{x},$$

$$(F_p)_j = \oint_{\partial\Omega} \frac{K}{\rho g} \left(n_x \frac{\partial p}{\partial x} + n_z \frac{\partial p}{\partial z} \right) \phi_j \, d\mathbf{x}.$$

We can reduce the set of ordinary differential equations (6.4-9) to a set of algebraic equations using a standard finite-difference approximation of the time derivatives. Thus, using a time step of length k, we get

$$\left.\frac{d(U_x)_i}{dt}\right|^{n+\theta} = \frac{1}{k}[(U_x)_i^{n+1} - (U_x)_i^n] + \mathcal{O}(k),$$

$$\left.\frac{d(U_z)_i}{dt}\right|^{n+\theta} = \frac{1}{k}[(U_z)_i^{n+1} - (U_z)_i^n] + \mathcal{O}(k),$$

$$\left.\frac{dp_i}{dt}\right|^{n+\theta} = \frac{1}{k}(p_i^{n+1} - p_i^n) + \mathcal{O}(k).$$

(The truncation errors for the case $\theta = 1/2$ will be $\mathcal{O}(k^2)$.) Corresponding to these difference approximations we have the following approximations of the unknown nodal values in time:

$$(U_x)_i^{n+\theta} \equiv \theta(U_x)_i^{n+1} + (1 - \theta)(U_x)_i^n,$$

$$(U_z)_i^{n+\theta} \equiv \theta(U_z)_i^{n+1} + (1 - \theta)(U_z)_i^n,$$

$$p_i^{n+\theta} \equiv \theta p_i^{n+1} + (1 - \theta)p_i^n.$$

With this temporal discretization, Equations (6.4-9) can be rewritten in the matrix form

$$(6.4\text{-}10) \quad \left(\theta \begin{bmatrix} A_{ji} & B_{ji} & C_{ji} \\ D_{ji} & E_{ji} & G_{ji} \\ 0 & 0 & L_{ji} \end{bmatrix} + \frac{1}{k} \begin{bmatrix} 0 & 0 & 0 \\ 0 & 0 & 0 \\ H_{ji} & K_{ji} & M_{ji} \end{bmatrix} \right) \begin{bmatrix} (U_x)_i^{n+1} \\ (U_z)_i^{n+1} \\ p_i^{n+1} \end{bmatrix}$$

$$+ \left((1 - \theta) \begin{bmatrix} A_{ji} & B_{ji} & C_{ji} \\ D_{ji} & E_{ji} & G_{ji} \\ 0 & 0 & L_{ji} \end{bmatrix} - \frac{1}{k} \begin{bmatrix} 0 & 0 & 0 \\ 0 & 0 & 0 \\ H_{ji} & K_{ji} & M_{ji} \end{bmatrix} \right) \begin{bmatrix} (U_x)_i^n \\ (U_z)_i^n \\ p_i^n \end{bmatrix}$$

$$= \begin{bmatrix} (F_x)_j^{n+\theta} \\ (F_z)_j^{n+\theta} \\ (F_p)_j^{n+\theta} \end{bmatrix}.$$

Given initial and boundary conditions, one can solve the global matrix equations generated by the nodal equations (6.4-10) by proceeding stepwise through time. At the end of each time step, the finite-element nodes are moved according to the calculated displacements $\mathbf{U}(\mathbf{x}, t)$, and the new geometry of the region is then used to compute the next time step. For a more detailed explanation of this procedure, we refer the reader to Safai (1977).

6.5. Oil Reservoir Modeling.

One of the most active areas of current research and development in large-scale numerical modeling is the field of oil reservoir simulation. An oil reservoir is a deposit of petroleum and associated fluids, usually gas and brine, held underground in the interstices of a porous rock formation. To produce the oil, petroleum engineers must use the natural forces in the reservoir, such as pressure, together with artificial techniques such as fluid injection to overcome the actions of viscosity and capillary forces that impede the flow of oil into production wells. In devising optimal production strategies, reservoir engineers commonly use mathematical simulators to assess the effects of various fluid production and injection schemes on the flow of oil, gas, and water in the rock formation

Our purpose in this section is twofold. First, we wish to show how the basic equations of petroleum reservoir flow arise from the continuum-mechanical considerations presented in Chapter One. Second, we wish to outline some of the most common numerical schemes used to approximate these governing equations. Readers interested in more thorough introductions to oil reservoir simulation should consult monographs by Peaceman (1977) and Aziz and Settari (1979) together with the other references cited below.

Compositional flows in porous media.

Oil reservoirs are mixtures. They consist of several phases, including rock, oil, water, and gas, and therefore the theory of multiphase mixtures applies. They also consist of many molecular species. The hydrocarbon species in the oil and gas phases are especially important, since transfers of species between the oil and gas phases, including the processes of dissolution, evaporation, condensation, and gas percolation, often have profound effects on a reservoir's production. Thus the theory of multispecies mixtures is important here, too. Flows involving multiphase, multispecies mixtures are called **compositional** flows. To accommodate the presence of both phases and species, we shall begin by extending the mixture-theoretic formalism outlined in Section 1.5.

For simplicity, let us assume that there are three fluid phases in the reservoir, namely, water (W), oil (O), and gas (G), with chemical species indexed by $i = 1, \ldots, N + 1$. Let us label the rock phase by the index R. Conceivably, at least, each species can exist in any phase and can transfer between phases via dissolution, evaporation, condensation, and so forth, subject to thermodynamic constraints. We shall assume here that the rock is chemically inert and that there are no intraphase or stoichiometric chemical reactions, although in some applications to certain enhanced oil recovery technologies reactions of this kind may be important.

In this mixture, each pair (i, α), with i chosen from the species indices and α chosen from the phases, is a constituent. Thus, for example, CH_4 in the gas phase is one constituent, CH_4 in oil another, and $n\text{-}C_4H_{10}$ in oil yet another. Each constituent (i, α) has its own **intrinsic mass density** ρ_i^α, measured as mass of i per unit volume of α, and its own velocity $\mathbf{v}_i^\alpha$. To accommodate the established kinematics of phases, we shall still associate with each phase α its volume fraction ϕ_α. Moreover, letting $\phi = 1 - \phi_R$ be the porosity, we define the **saturation** of fluid phase α as $S_\alpha = \phi_\alpha/\phi$. Using these basic quantities, we then define several useful variables. The **intrinsic mass density of phase** α is

$$\rho^\alpha = \sum_{i=1}^N \rho_i^\alpha;$$

the **mass fraction of species** i **in phase** α is

$$\omega_i^\alpha = \rho_i^\alpha/\rho^\alpha;$$

the **bulk density of the fluids** is

$$\rho = \phi \sum_{\alpha \neq R} S_\alpha \rho^\alpha;$$

the **total mass fraction of species** i **in the fluids** is

$$\omega_i = (\phi/\rho) \sum_{\alpha \neq R} S_\alpha \rho^\alpha \omega_i^\alpha;$$

the **barycentric velocity of phase** α is

$$\mathbf{v}^\alpha = (1/\rho^\alpha) \sum_{i=1}^N \rho_i^\alpha \mathbf{v}_i^\alpha;$$

and the **diffusion velocity of species** i **in phase** α is

$$\mathbf{u}_i^\alpha = \mathbf{v}_i^\alpha - \mathbf{v}^\alpha.$$

Suppose the index $N + 1$ represents the species making up the inert rock phase. Then the following constraints hold:

$$\sum_{i=1}^{N} \omega_i = \sum_{i=1}^{N} \omega_i^\alpha = \sum_\alpha \phi_\alpha = \sum_{\alpha \neq R} S_\alpha = 1,$$

where the index α in the second sum can represent any fluid phase, and

$$\sum_{i=1}^{N} \rho_i^\alpha \mathbf{u}_i^\alpha = \mathbf{0}.$$

Each constituent (i, α) has its own mass balance, given by analogy with Equation (1.5-2) as

$$\frac{\partial}{\partial t}(\phi_\alpha \rho_i^\alpha) + \nabla \cdot (\phi_\alpha \rho_i^\alpha \mathbf{v}_i^\alpha) = r_i^\alpha,$$

where the exchange terms r_i^α must obey the restriction $\sum_{i=1}^{N} \sum_{\alpha \neq R} r_i^\alpha = 0$. If we impose the further constraint that there are no intraphase chemical reactions, then we have in addition $\sum_{\alpha \neq R} r_i^\alpha = 0$ for each species $i = 1, \ldots, N$. Since phase velocities are typically more accessible to measurement than species velocities, it is convenient to rewrite the constituent mass balance, following the species balance derivation in Section 1.5, as

$$\frac{\partial}{\partial t}(\phi S_\alpha \rho^\alpha \omega_i^\alpha) + \nabla \cdot (\phi S_\alpha \rho^\alpha \omega_i^\alpha \mathbf{v}^\alpha) + \nabla \cdot \mathbf{j}_i^\alpha = r_i^\alpha,$$

where $\mathbf{j}_i^\alpha = \phi S_\alpha \rho^\alpha \omega_i^\alpha \mathbf{u}_i^\alpha$ stands for the **diffusive flux** of constituent (i, α). Summing this equation over all fluid phases α and using the restrictions gives a total mass balance for each species i:

$$\frac{\partial}{\partial t}(\rho \omega_i) + \nabla \cdot [\phi(S_W \rho^W \omega_i^W \mathbf{v}^W + S_O \rho^O \omega_i^O \mathbf{v}^O + S_G \rho^G \omega_i^G \mathbf{v}^G)]$$

$$+ \nabla \cdot (\mathbf{j}_i^W + \mathbf{j}_i^O + \mathbf{j}_i^G) = 0, \qquad i = 1, \ldots, N.$$

To establish flow equations for each species, we need velocity field equations for each fluid phase together with some constitutive equations for the diffusive fluxes $\mathbf{j}_i^\alpha$. For the fluid velocities we may postulate Darcy's law, Equation (1.5-6), assuming that the porous medium is isotropic. However, when several fluid phases occupy the porous rock, we must alter Darcy's law to account for the interference to flow that each fluid feels when other

fluids occupy the same pore space. The most common way to do this is to
write, for each fluid α,

$$\mathbf{v}^\alpha = -\frac{kk_{r\alpha}}{\mu^\alpha \phi S_\alpha}(\nabla p^\alpha - \rho^\alpha g \nabla Z).$$

The factor $k_{r\alpha}$ is called the **relative permeability** of fluid α and satisfies
$0 \leq k_{r\alpha} \leq 1$. In many cases it is reasonable to assume that $k_{r\alpha}$ is a function
of fluid saturations only for a given rock-fluid system. Figure 6-5 shows a
typical pair of relative permeability functions for an oil-water-rock mixture.

  The existence of several fluid phases also allows for the existence of
several fluid pressures. This effect actually occurs in nature, owing to the
physics of fluid-fluid and fluid-rock interfaces. The difference between two
fluid pressures at a given point in the reservoir is the **capillary pressure**,
defined by $p_{C\alpha\beta} = p_\alpha - p_\beta$. Clearly, when three fluid phases exist, only two
capillary pressures can be independent. Capillary pressures, like relative
permeabilities, are typically functions of fluid saturation in a given rock-
fluid system.

  For the diffusive fluxes the appropriate assumption is not so clear. In
single-phase flows through porous media, the diffusive flux of a species with
respect to the fluid's barycentric velocity is called **hydrodynamic dispersion**. The literature on single-phase hydrodynamic dispersion is large but

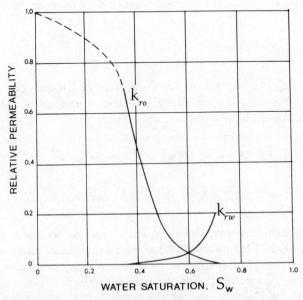

Figure 6-5. Typical relative permeability curves for water-wet rock matrix.

by no means conclusive. By contrast, the literature on hydrodynamic dispersion in multifluid flows is quite sparse. The most common approach in multiphase oil reservoir simulation is to assume that hydrodynamic dispersion is a small enough effect that the diffusive fluxes in the mass balance for each species are negligible. Provided this assumption is reasonable, we arrive at the following flow equation for species i in the fluids:

$$(6.5\text{-}1) \qquad \frac{\partial}{\partial t}[\phi(S_W \rho^W \omega_i^W + S_O \rho^O \omega_i^O + S_G \rho^G \omega_i^G)]$$

$$-\nabla \cdot \left[\frac{k k_{rW} \rho^W \omega_i^W}{\mu_W}(\nabla p_W - \rho^W g \nabla Z) + \frac{k k_{rO} \rho^O \omega_i^O}{\mu_O}(\nabla p_O - \rho^O g \nabla Z) \right.$$

$$\left. + \frac{k k_{rG} \rho^G \omega_i^G}{\mu_G}(\nabla p_G - \rho^G g \nabla Z) \right] = 0, \qquad i = 1, \ldots, N.$$

To close this set of equations, we need some supplementary constraints giving relationships among the variables. One class of supplementary constraints consists of the thermodynamic relationships giving phase densities and compositions as functions of pressure and overall fluid mixture composition. Conceptually, these relationships take the forms

$$\rho^\alpha = \rho^\alpha(\omega_1^\alpha, \ldots, \omega_{N-1}^\alpha, p_\alpha), \qquad \alpha = W, O, G,$$
$$\omega_i^\alpha = \omega_i^\alpha(\omega_1, \ldots, \omega_{N-1}, p_\alpha), \qquad \alpha = W, O, G; \quad i = 1, \ldots, N-1,$$
$$S_\alpha = S_\alpha(\omega_1, \ldots, \omega_{N-1}, p_\alpha), \qquad \alpha = W, O, G.$$

However, it is important from a computational viewpoint to observe that the actual thermodynamic statements of these relationships may yield simultaneous sets of nonlinear algebraic equations giving phase densities, compositions, and saturations implicitly. If this is the case, then the calculation of fluid-phase thermodynamics may constitute a major part of the computational effort in a simulation. We refer the interested reader to Coats (1980) or Allen (1984a) for discussions of two different approaches to modeling flows with such complicated thermodynamics.

The other class of supplementary constraints includes constitutive relationships for the particular rock-fluid system being modeled. These relationships involve the capillary pressures and relative permeabilities, typically taking the forms

$$p_O - p_W = p_{COW} = p_{COW}(S_O, S_G),$$
$$p_G - p_O = p_{CGO} = p_{CGO}(S_O, S_G),$$
$$k_{r\alpha} = k_{r\alpha}(S_O, S_G), \qquad \alpha = W, O, G.$$

In practice, petroleum engineers derive these functions through measurements of fluid and rock samples extracted from the reservoir under study.

Black-oil models.

Black-oil models are special cases of the general compositional equations that allow limited interphase mass transfer, the composition of each phase depending on pressures only. This class of models has become a standard engineering tool in the petroleum industry. We shall review the formulation of the black-oil equations and discuss a few selected aspects of their numerical solution.

The fundamental premise of the black-oil model is that a highly simplified, three-species system can often serve as an adequate model of the complex mixtures of brine and hydrocarbons found in natural petroleum reservoirs. For practical purposes, petroleum engineers define these three "pseudo-species" according to what appears at the surface, at **stock-tank conditions** (STC), after production of the reservoir fluids. Thus, we have the species o, which is stock-tank oil; g, which is stock-tank gas, and w, which is stock-tank water. Underground, at **reservoir conditions** (RC), these species may partition themselves among the three fluid phases O, G, and W in a distribution depending on the pressures in the formation.

Now we impose a set of thermodynamic constraints on this partitioning of species. First, we assume that there is no exchange of water w into the nonaqueous phases O and G, so that $\omega_w^W = 1$, and $\omega_w^O = \omega_w^G = 0$. Second, we allow no exchange of oil o into the vapor phase G or the aqueous liquid W, so that $\omega_o^O = 1$, and $\omega_o^W = \omega_o^G = 0$. Third, we prohibit the dissolution of gas g into the aqueous liquid W, so that $\omega_g^W = 0$. However, we allow the gas g to dissolve in the hydrocarbon liquid O according to a pressure-dependent relationship called the **solution gas-oil ratio**, defined by

$$R_S(p_O) = \frac{\text{volume of } g \text{ in solution at RC}}{\text{volume of } o},$$

where the volumes refer to volumes at STC.

To facilitate further reference to volumes of species at STC, we relate the phase densities ρ^α at RC to the species densities ρ_i^{STC} at STC by defining the **formation volume factors**. For W and G these definitions are fairly simple:

$$B_W(p_W) = \frac{\rho_w^{\text{STC}}}{\rho^W(p_W)}, \qquad B_G(p_W) = \frac{\rho_g^{\text{STC}}}{\rho^G(p_G)}.$$

For the hydrocarbon liquid O, however, we must also account for the mass of dissolved gas at RC:

$$B_O(p_O) = \frac{\rho_o^{\text{STC}} + R_S(p_O)\rho_g^{\text{STC}}}{\rho^O(p_O)}.$$

If we substitute these definitions into the flow equations (6.5-1) for the species o, g, w and divide through by the constants ρ_i^{STC}, we obtain the three **black-oil equations**. The flow equations for water, oil, and gas are, respectively,

(6.5-2a)
$$\frac{\partial}{\partial t}\left(\frac{\phi S_W}{B_W}\right) - \nabla \cdot [\lambda_W (\nabla p_W - \gamma_W \nabla Z)] = 0;$$

(6.5-2b)
$$\frac{\partial}{\partial t}\left(\frac{\phi S_O}{B_O}\right) - \nabla \cdot [\lambda_O (\nabla p_O - \gamma_O \nabla Z)] = 0,$$

(6.5-2c)
$$\frac{\partial}{\partial t}\left[\phi\left(\frac{S_G}{B_G} + \frac{R_S S_O}{B_O}\right)\right] - \nabla \cdot [\lambda_G (\nabla p_G - \gamma_G \nabla Z)]$$
$$-\nabla \cdot [R_S \lambda_O (\nabla p_O - \gamma_O \nabla Z)] = 0.$$

To keep the notation tractable, we have adopted the abbreviations $\lambda_\alpha = kk_{r\alpha}/(\mu_\alpha B_\alpha)$ and $\gamma_\alpha = \rho^\alpha g$.

These equations constitute a system of coupled, nonlinear, transient PDEs. As we have argued throughout this book, the numerical solution methodology appropriate for a given system depends to a great extent on the classification of the governing PDEs. Although each of the black-oil equations is formally parabolic in appearance, the system can often exhibit behavior more typical of an elliptic-hyperbolic set if capillary influences are small. To see this, consider the two-phase version of Equations (6.5-2) in which gas is absent, porosity is constant, and fluid compressibilities and gravity forces have no effect. The flow equations in this simple case reduce to

$$-\phi\frac{\partial S_W}{\partial t} = \nabla \cdot (\lambda_O \nabla p_O),$$
$$\phi\frac{\partial S_W}{\partial t} = \nabla \cdot (\lambda_W \nabla p_W),$$

since $S_O + S_W = 1$. Adding these two equations gives a total flow equation $\nabla \cdot \mathbf{Q} = 0$, where $\mathbf{Q} = -\lambda_O \nabla p_O - \lambda_W \nabla p_W$ represents the total rate of fluid flow. Calling $\lambda = \lambda_O + \lambda_W$ and $p = (p_O + p_W)/2$, we can rewrite this total flow equation as follows:

$$\nabla \cdot \left[\lambda \nabla p - \left(\frac{\lambda_W - \lambda_O}{2}\right)\nabla p_{COW}\right] = 0.$$

If we examine the case when $\nabla p_{COW} \simeq 0$, the total flow equation reduces to an elliptic **pressure equation**,

$$\nabla \cdot (\lambda \nabla p) = 0.$$

We discussed the solution of such equations in depth in Chapter Three. Next, defining the **fractional flow function** $f_W = \lambda_W/(\lambda_O + \lambda_W)$, we can rewrite the water flow equation as a hyperbolic PDE,

$$\phi\frac{\partial S_W}{\partial t} + \nabla \cdot [\mathbf{Q} f_W (S_W)] = 0.$$

This is the three-dimensional **Buckley-Leverett equation** (Buckley and

Leverett, 1942). We discussed the solution of such nonlinear hyperbolic conservation laws in some detail in Chapter Five. The Buckley-Leverett equation demonstrates that, in an oil reservoir where capillary effects are small, fluid saturations can flow according to essentially hyperbolic PDEs in response to an elliptic pressure field.

Several approaches to solving the general system (6.5-2) numerically have appeared in the petroleum engineering literature. Much of the classic work on black-oil models has focused on schemes for advancing the system of coupled PDEs in time. We shall review two of the most popular methods: the **simultaneous solution** (SS) method and the **implicit pressure–explicit saturation** (IMPES) method.

Simultaneous solution (SS).

The SS method, introduced by Douglas, Peaceman, and Rachford (1959) and further developed by Coats et al. (1967), treats the flow equations (6.5-2) as simultaneous equations for the fluid pressures p_O, p_G, and p_W. Inverting the capillarity relationships and imposing the restriction on fluid saturations then yields the saturations S_O, S_G, and S_W. For ease of presentation, let us examine the two-phase case, assuming that the vapor phase G does not appear and that the porosity ϕ is constant.

The first step in the formulation is to rewrite the flow equations so that the pressures p_O and p_W appear as explicit unknowns. To do this, we apply the chain rule to the accumulation terms, giving

$$\frac{\partial}{\partial t}\left(\frac{\phi S_W}{B_W}\right) = \phi S_W b_W \frac{\partial p_W}{\partial t} + \frac{\phi S_W'}{B_W}\left(\frac{\partial p_O}{\partial t} - \frac{\partial p_W}{\partial t}\right),$$

$$\frac{\partial}{\partial t}\left(\frac{\phi S_O}{B_O}\right) = \phi S_O b_O \frac{\partial p_O}{\partial t} - \frac{\phi S_W'}{B_O}\left(\frac{\partial p_O}{\partial t} - \frac{\partial p_W}{\partial t}\right),$$

where $b_\alpha = d(1/B_\alpha)/dp_\alpha$ and S_W' signifies the derivative of the inverted capillarity relationship $S_W(p_{COW})$. This device allows us to write the system (6.5-2) as follows:

$$\phi \begin{bmatrix} (S_W b_W - S_W'/B_W) & (S_W'/B_O) \\ (S_W'/B_O) & (S_O b_O - S_W'/B_O) \end{bmatrix} \frac{\partial}{\partial t}\begin{bmatrix} p_W \\ p_O \end{bmatrix}$$

$$-\nabla \cdot \begin{bmatrix} (\Lambda_W/B_W)\nabla & 0 \\ 0 & (\Lambda_O/B_O)\nabla \end{bmatrix}\begin{bmatrix} p_W \\ p_O \end{bmatrix}$$

$$+ \begin{bmatrix} \rho^W g\nabla Z \\ \rho^O g\nabla Z \end{bmatrix} = \begin{bmatrix} 0 \\ 0 \end{bmatrix}.$$

The terms involving time derivatives in this equation represent the accumulation of fluid, while the terms involving the pressure gradients govern the Darcy flux of fluids.

Now we can employ some finite-difference or finite-element method to approximate the spatial derivative, getting a system of evolution equations having the matrix form

$$\mathbf{M}\frac{d\mathbf{p}}{dt} + \mathbf{S}\mathbf{p} = \mathbf{f}.$$

Here $\mathbf{M}$ is the matrix of coefficients arising from the spatial discretization of the accumulation terms, $\mathbf{S}$ is the matrix of coefficients arising from the discretization of fluid fluxes, $\mathbf{p}$ represents the vector of unknown nodal values of oil and water pressure, and $\mathbf{f}$ is a vector containing information from the discretized boundary conditions. Since the entries of $\mathbf{M}$ and $\mathbf{S}$ exhibit pronounced functional dependence on the unknown pressures, this system is strongly nonlinear. Therefore, as discussed in Section 6.3, our time-stepping approximation must be iterative. As an example, given a time step k_t, we might use the Newton-like procedure introduced at the end of Section 6.3 to advance from $t = nk_t$ to $t = (n+1)k_t$, yielding

$$\left(\frac{1}{k_t}\mathbf{M}^{n+1,m} + \mathbf{S}^{n+1,m}\right)\delta\mathbf{p}^{n+1,m+1}$$

$$= -\left[\frac{1}{k_t}\mathbf{M}^{n+1,m}\left(\mathbf{p}^{n+1,m} - \mathbf{p}^n\right) - \mathbf{S}^{n+1,m}\mathbf{p}^{n+1,m} + \mathbf{f}^{n+1,m}\right]$$

$$= -\mathbf{R}^{n+1,m}.$$

Here, $\delta\mathbf{p}^{n+1,m+1}$ is a vector of iterative increments, so that, starting from iteration level m, $\mathbf{p}^{n+1,m+1} = \mathbf{p}^{n+1,m} + \delta\mathbf{p}^{n+1,m+1}$ gives the vector of nodal pressures at the next iteration for the unknown time level $n+1$. In this scheme the notation $\mathbf{R}^{n+1,m}$ suggests that we regard the right side as a residual, iterating at each time step until $\|\mathbf{R}^{n+1,m}\|$ is small enough in some norm.

The formulation presented above is not unique. In fact, several variants of the SS method have appeared, including formulations treating different sets of variables as principal unknowns. Aziz and Settari (1979) provide a survey of these alternative approaches.

Implicit pressure-explicit saturation (IMPES).

In the IMPES formulation, the basic idea is to combine the flow equations (6.5-2) to get an equation for one of the fluid pressures (Breitenbach, Thurnau, and van Poolen (1969)). Solving this equation implicitly provides the information necessary to update the saturations explicitly at each time step, using an independent set of flow equations and the restriction that saturations sum to unity. Sheldon, Zondek, and Cardwell (1959) and Stone and Garder (1961) introduced this method.

We begin, as in the SS method, by expanding the accumulation terms, this time leaving saturations and pressures as principal unknowns. For the three-phase system, this leads to the following finite-difference approximations to the time derivatives:

$$\phi \frac{\partial}{\partial t}\left(\frac{S_W}{B_W}\right) = \frac{1}{k_t}(C_1 \Delta_t S_W + C_2 \Delta_t p_w) + \mathcal{O}(k_t),$$

$$\phi \frac{\partial}{\partial t}\left(\frac{S_O}{B_O}\right) = \frac{1}{k_t}(C_3 \Delta_t S_O + C_4 \Delta_t p_O) + \mathcal{O}(k_t),$$

$$\phi \frac{\partial}{\partial t}\left(\frac{S_G}{B_G} + \frac{R_S S_O}{B_O}\right) = \frac{1}{k_t}(C_5 \Delta_t S_G + C_6 \Delta_t p_G$$
$$+ C_7 \Delta_t S_O + C_8 \Delta_t p_O) + \mathcal{O}(k_t).$$

The coefficients $C_1, \ldots, C_8$ appearing here stand for the appropriate derivatives extracted using the chain rule. The reader should derive expressions for these coefficients. The notation $\Delta_t u = u^{n+1} - u^n$ defines the time-difference operator.

The next step involves the crucial assumption that the capillary pressures p_{COW} and p_{CGO} change negligibly over a time step. This assumption implies that $\Delta_t p_O = \Delta_t p_W = \Delta_t p_G$ and, furthermore, that we can treat the capillary contributions to the flux terms explicitly. Thus, our implicit, temporally discrete approximations to Equations (6.5-2) become

(6.5-3a) $C_1 \Delta_t S_W + C_2 \Delta_t p_O$
$$= k_t \nabla \cdot [\lambda_W^{n+1}(\nabla p_O^{n+1} - \nabla p_{COW}^n - \gamma_W^{n+1} \nabla Z)],$$

(6.5-3b) $C_3 \Delta_t S_O + C_4 \Delta_t p_O = k_t \nabla \cdot [\lambda_O^{n+1}(\nabla p_O^{n+1} - \gamma_O^{n+1} \nabla Z)],$

(6.5-3c) $C_5 \Delta_t S_G + C_7 \Delta_t S_O + (C_6 + C_8)\Delta_t p_O$
$$= k_t \nabla \cdot [\lambda_G^{n+1}(\nabla p_O^{n+1} + \nabla p_{CGO}^{n+1} - \gamma_G^{n+1} \nabla Z)$$
$$+ R_S^{n+1} \lambda_O^{n+1}(\nabla p_O^{n+1} - \gamma_O^{n+1} \nabla Z)].$$

To get a single pressure equation from this set, we multiply Equation (6.5-3c) by the coefficient $B = C_3/(C_5 - C_7)$ and multiply Equation (6.5-3a) by $A = BC_5/C_1$. Now add Equations (6.5-3a–c). Observe that the saturation differences in the accumulation terms arising from Equations (6.5-3) in this way now sum to an expression proportional to $\Delta_t(S_W + S_O + S_G) = 0$. Therefore our weighted sum of the time-differenced flow equations yields

(6.5-4) $C^{n+1}\Delta_t p_O = k_t \{ A^{n+1}\nabla \cdot [\lambda_W^{n+1}(\nabla p_O^{n+1} - \nabla p_{COW}^n)]$
$$+ \nabla \cdot (\lambda_O^{n+1}\nabla p_O^{n+1})$$
$$+ B^{n+1}\nabla \cdot [(\lambda_G^{n+1} + R_S^{n+1}\lambda_O^{n+1})\nabla p_O^{n+1}$$
$$+ \lambda_G^{n+1}\nabla p_{CGO}^n] - \Gamma^{n+1}\}.$$

The new parameter Γ is shorthand for the weighted sum of the gravity terms, and $C = AC_2 + C_4 + B(C_6 + C_8)$. Equation (6.5-4) is the desired pressure equation.

Now, provided we have an appropriate technique for producing discrete approximations to the spatial derivatives appearing in these equations, we can implement the following time-stepping procedure.

(i) Solve the nonlinear pressure equation (6.5-4) implicitly, using some iterative scheme such as those introduced in Section 6.3.

(ii) Solve Equation (6.5-3a) explicitly for $\Delta_t S_W$ and update the water saturation; solve (6.5-3b) for $\Delta_t S_O$ and update the oil saturation; then set $S_G^{n+1} = 1 - S_W^{n+1} - S_O^{n+1}$.

(iii) Compute p_{COW}^{n+1} and p_{CGO}^{n+1} using the new saturations; then use these to update p_W and p_G.

(iv) Begin the next time step.

As with the SS methods, variants on this development have appeared; see Aziz and Settari (1979) for a survey.

Notice that, in contrast to the SS formulation, the IMPES approach requires the implicit solution of only one flow equation at each time step. Thus IMPES schemes offer the obvious advantage that, with only one implicit equation to solve per time step, the algorithm requires smaller matrix inversions at each iteration. The resulting computational savings can be significant in problems involving large numbers of grid points. On the other hand, because it treats capillary pressures explicitly, the IMPES method suffers from instability when the time step k_t exceeds a critical value. This limitation can be inconvenient if the critical value of k_t is unknown or small compared with the life of a field project. The SS method, while requiring more computation per time step, boasts greater stability. This can prove to be a decided advantage when the problem to be solved exhibits strongly nonlinear phenomena such as gas percolation, when the pressures of liquid hydrocarbons pass through bubble points and enter the gas phase.

The temporal weighting of the flux coefficients λ_α also affects the stability of discrete solutions to the black-oil equations. It is a fairly common practice to treat these coefficients explicitly. However, this tactic leads to limits on time steps allowable for stable solutions. The limitation is especially severe in problems with gas percolation. The implicit treatment of the flux coefficients partially alleviates this stability problem.

Matrix computations.

One of the most important problems in black-oil simulation is the computational inefficiency associated with the solution of large systems of linear algebraic equations. In either the SS or the IMPES approach, the iterative time-stepping scheme calls for the solution of matrix equations at each iteration of each time step. For simulations at practical scales these calculations alone can tax the storage and CPU-time resources of the largest machines currently available. A great deal of research has focused on the development of fast iterative techniques for the solution of the large matrix systems arising in applications. Early investigations along this line explored the use of various forms of relaxation, as introduced in Section 3.13, to solve the matrix equations. The block-iterative techniques, also discussed in Section 3.13, have been especially attractive in this regard. Alternating-direction iterative techniques, as reviewed in Section 4.3, have also attracted interest. More recently, algorithms of the preconditioned conjugate-gradient type, as outlined in Section 3.15, have proven to be effective.

Some considerations for spatial discretization.

So far we have left hanging the discretization of the spatial derivatives that appear in the black-oil equations. Historically, the method of finite differences has dominated the petroleum engineering literature. The approximation of spatial derivatives using difference analogs in Equations (6.5-2) is relatively straightforward except for one special consideration: Most black-oil simulators employ upstream-weighted approximations to the flux coefficients λ_α. In the simple case of one-dimensional flow, for example, the discretization of a term like $\partial(\lambda \partial p/\partial x)/\partial x$ leads to a difference analog of the form

$$(6.5\text{-}5) \qquad \frac{1}{h}\left[\lambda_{i+\frac{1}{2}}\left(\frac{p_{i+1}-p_i}{h}\right) - \lambda_{i-\frac{1}{2}}\left(\frac{p_i-p_{i-1}}{h}\right)\right],$$

where h is the grid spacing. In its most primitive form, upstream weighting calls for evaluating $\lambda_{i\pm\frac{1}{2}}$ at integer node numbers $i-1$, i, or $i+1$ as follows:

$$\lambda_{i\pm\frac{1}{2}} = \begin{cases} \lambda_{i\pm\frac{1}{2}-\frac{1}{2}}, & \text{if flow is from } i-1 \text{ to } i+1, \\ \lambda_{i\pm\frac{1}{2}+\frac{1}{2}}, & \text{if flow is from } i+1 \text{ to } i-1. \end{cases}$$

The motivation for upstream weighting arises from the partly hyperbolic character of the flow equations, as discussed previously. In Chapter Five we mentioned the lack of uniqueness associated with nonlinear hyperbolic conservation laws and the need to guarantee that solutions to such equations reflect the limiting behavior of their dissipative counterparts. When capillary effects are globally small compared with pressure forces, high-order spatial discretizations can fail to capture the physically important dissipation and therefore converge to incorrect solutions. Upstream weighting introduces an $\mathcal{O}(h)$ error term that restores the necessary dissipation in a numerically consistent fashion. We refer the reader to Allen (1984b) for a more detailed analysis.

The reader should beware that, in highly complex systems of PDEs, upstream weighting has its own perils. In particular, simple extensions of the scheme given in Equation (6.5-5) to several spatial variables can yield numerical schemes that suffer from physically unrealistic bias in favor of flow in coordinate directions. In fact, with such coordinate-direction upstream weighting schemes it is possible for the same numerical method to produce results converging to *different* solutions, depending on the orientation of the finite-difference grid. A considerable amount of current research aims at the construction of upstream weighting methods that exhibit less dependence on the orientation of the coordinate axes. The interested reader may consult Koebbe et al. (1986) for an example of such a method.

To date, finite-element methods for discretizing the oil-reservoir equations have only started to make inroads in industrial simulators. Rather than attempting to survey these wide-ranging results, we refer the reader to a volume in the SIAM series on Frontiers in Applied Mathematics (Ewing, 1983) for a fairly recent overview.

6.6. Problems for Chapter Six.

1. Given the PDE

$$\frac{\partial^2 u}{\partial t^2} + E\frac{\partial^4 u}{\partial x^4} = 0,$$

derive the finite-difference approximation

$$u_i^{n+1} - 2u_i^n + u_i^{n-1}$$
$$= -\left(\frac{Ek^2}{h^4}\right)\left(u_{i+2}^n - 4u_{i+1}^n + 6u_i^n - 4u_{i-1}^n + u_{i-2}^n\right).$$

Here, k denotes the time step, and h stands for the spatial grid mesh. Use von Neumann stability analysis to derive the stability criterion $Ek^2/h^4 < \frac{1}{2}$.

2. The aim of this problem is to solve the fourth-order boundary-value problem

$$\nabla^4 u + au = f(x, y) \quad \text{on} \quad \Omega = (0, 1) \times (0, 1),$$
$$u(x, y) = \alpha \quad \text{on} \quad \partial\Omega,$$
$$\frac{\partial u}{\partial n}(x, y) = \beta \quad \text{on} \quad \partial\Omega,$$

where $a > 0$, α, and β are all constants. One finite-difference approximation for the PDE is

$$\left(\frac{20}{h^4} + a\right)u_{i,j} = f(x_i, y_j)$$
$$- \frac{1}{h^4}\big[u_{i+2,j} + u_{i-2,j} + u_{i,j+2} + u_{i,j-2}$$
$$+ 2(u_{i+1,j+1} + u_{i+1,j-1} + u_{i-1,j+1} + u_{i-1,j-1})$$
$$- 8(u_{i+1,j} + u_{i-1,j} + u_{i,j+1} + u_{i,j-1})\big].$$

Here h signifies the grid mesh, assumed uniform. Clearly, whenever (x_i, y_j) is a boundary node, we can fix $u_{i,j} = \alpha$. To handle the normal-derivative condition, we establish a layer of "fictitious" nodes outside the true boundary $\partial \Omega$. Along the left boundary $(x = 0)$, for example, Taylor's theorem suggests

$$u_{-1,j} = u_{0,j} - h \frac{\partial u}{\partial n}(x_0, y_j) + \mathcal{O}(h^2),$$

$$u_{1,j} = u_{0,j} + h \frac{\partial u}{\partial n}(x_0, y_j) + \mathcal{O}(h^2),$$

so that, to within $\mathcal{O}(h^2)$,

$$u_{-1,j} = u_{1,j} - 2h \frac{\partial u}{\partial n}(x_0, y_j) = u_{1,j} - 2h\beta.$$

Use this idea to develop a closed system of algebraic equations approximating the original PDE, then write a computer program to solve these equations using the Gauss-Seidel iterative procedure.

3. Consider the ordinary differential equation $du/dt = f(u, t)$, where f is a differentiable function. One implicit finite-difference scheme for this equation is $u_{n+1} - kf(u_{n+1}, t_{n+1}) = u_n$, where subscripts denote time levels and k is the time step. Develop a Newton method for solving this scheme.

4. Give heuristic reasoning showing that, if an iterative scheme for a nonlinear system has asymptotic convergence rate α, then a plot of $\log \|e^{(m+1)}\|$ versus $\log \|e^{(m)}\|$, where $e^{(m)}$ is the error at the m-th iteration, should yield a line with slope α. Use this approach to check the convergence rates of successive substitution and Newton's method for the following equation sets:

(a) $x - \cos x = 0$,

(b) $\dfrac{x^2}{16} - 1 = 0$,

(c) $x^2 + y^2 - x = 0$ and $x^2 - y^2 - y = 0$.

5. For the special case of one nonlinear equation in one unknown, say $f(u) = 0$, the method of successive substitution reduces to $u^{(m+1)} = g(u^{(m)})$, where $g(u) = u - f(u)$. Assuming f (and hence g) is differentiable, show that the Lipschitz condition $|g(w) - g(v)| \leq L|w - v|$ holds with $L = \sup_u |g'(u)|$, that is, L can be taken as the least upper bound of $|g'|$. Hint: The mean value theorem states that, if g is continuous on a closed interval $[v, w]$ and differentiable on (v, w), then there is a point $\varsigma \in (v, w)$ such that $g'(\varsigma) = [g(w) - g(v)]/(w - v)$; in other words, $g'(\varsigma)$ equals the average slope of g over the interval.

6. As an alternative to the procedures discussed in Section 6.3, one may consider **predictor-corrector** methods for nonlinear problems. For example, suppose the PDE has the form

$$\frac{\partial^2 u}{\partial x^2} = \mathcal{F}\left(x, t, u, \frac{\partial u}{\partial x}, \frac{\partial u}{\partial t}\right),$$

where $\mathcal{F}$ is linear in its fifth argument $\partial u/\partial t$. Given a uniform spatial grid of mesh h and a uniform temporal grid of mesh k, a suitable predictor step using finite differences would be

$$\delta_x^2\left(u_i^{n+1/2}\right) = h^2 \mathcal{F}\left(ih, (n+\tfrac{1}{2})k, u_i^n, \frac{u_{i+1}^n - u_{i-1}^n}{2h}, \frac{u_i^{n+1/2} - u_i^n}{k/2}\right).$$

Here, δ_x denotes the central difference operator in the x-direction; see Equation (2.6-12). This step gives a linear algebraic system for the unknown values $u_i^{n+1/2}$ associated with the $n + \frac{1}{2}$ time level. The appropriate corrector step is then

$$\delta_x^2\left(\frac{u_i^{n+1} - u_i^n}{2}\right)$$

$$= h^2 \mathcal{F}\left(ih, (n+\tfrac{1}{2})k, u_i^{n+1/2}, \frac{u_{i+1}^{n+1/2} - u_{i-1}^{n+1/2}}{2h}, \frac{u_i^{n+1} - u_i^n}{k}\right),$$

which gives a linear problem for the values u_i^{n+1} at the new time level. Construct such a predictor-corrector scheme for the nonlinear heat equation,

$$\frac{\partial}{\partial x}\left[K(u)\frac{\partial u}{\partial x}\right] = \frac{\partial u}{\partial t}.$$

7. Recall that successive overrelaxation (SOR) offers an attractive technique for solving the linear algebraic systems arising from linear elliptic PDEs. When the PDE is nonlinear, its algebraic analog will be, too. **Nonlinear overrelaxation** furnishes an extension of the standard SOR procedure based on Newton's method. Given a system of nonlinear equations $f_i(u_1, \ldots, u_n) = 0$, $i = 1, \ldots, n$, we perform a set of iterations, each involving a sequential pass through the list $u_1, \ldots, u_n$ of unknowns:

$$u_i^{(m+1)} = u_i^{(m)} - \omega \frac{f_i\left(u_1^{(m+1)}, \ldots, u_{i-1}^{(m+1)}, u_i^{(m)}, \ldots, u_n^{(m)}\right)}{\dfrac{\partial f_i}{\partial x_i}\left(u_1^{(m+1)}, \ldots, u_{i-1}^{(m+1)}, u_i^{(m)}, \ldots, u_n^{(m)}\right)}.$$

Here, $\omega \in [1, 2]$ is an overrelaxation parameter. Formulate a nonlinear overrelaxation scheme for the nonlinear Poisson problem

$$\nabla^2 u = u^2 \quad \text{on} \quad \Omega = (0, 1) \times (0, 1),$$

$$u(x, 0) = u(x, 1) = 1, \quad u(0, y) = u(1, y) = 0.$$

Write a computer program implementing the scheme.

8. For oil reservoirs in which only oil and water are present, the average pressure $\bar{p} = (p_O + p_W)/2$ and the capillary pressure p_{COW} furnish alternative independent unknowns to the actual fluid pressures p_O and p_W. Reformulate the SS procedure for the black-oil equations, outlined in Section 6.5, in terms of $\bar{p}$ and p_{COW}.

9. The PDE

$$\frac{\partial}{\partial x} \left(K \frac{\partial u}{\partial x} \right) = 0, \quad K > 0,$$

admits a "factored" form,

$$K^{-1} v + \frac{\partial u}{\partial x} = 0,$$

$$\frac{\partial v}{\partial x} = 0,$$

that serves as a starting point for mixed finite-element schemes like those mentioned for fourth-order equations in Section 6.2. Given a uniform grid $x_0 < \ldots < x_N$ of mesh h, suppose we adopt a piecewise constant trial function $\hat{u} = \sum_{i=1}^{N} u_i c_i(x)$ for u. Here,

$$c_i(x) = \begin{cases} 1, & \text{if } x_{i-1} \le x \le x_i, \\ 0, & \text{otherwise.} \end{cases}$$

Let us also adopt a piecewise linear trial function $\hat{v} = \sum_{i=1}^{N} v_i \ell_i(x)$ for v, where $\ell_i(x)$ is the usual Lagrange piecewise linear basis function. The appropriate Galerkin equations have the forms

$$\int \left(K^{-1} \hat{v} + \frac{\partial \hat{u}}{\partial x} \right) \ell_j dx = 0,$$

$$\int \frac{\partial \hat{v}}{\partial x} c_j dx = 0.$$

Develop the matrix equation for this system with, say, $N = 4$. For simplicity, assume the Neumann boundary conditions $v(x_0) = v(x_N) = 0$, although this leads to a singular equation set with no unique solution. Hint: Use integration by parts where appropriate. For an extension of this approach to two space dimensions, see Allen et al. (1985).

10. Using the factoring described in Problem 9, the PDE

$$\frac{\partial u}{\partial t} = \frac{\partial}{\partial x}\left(K\frac{\partial u}{\partial x}\right), \quad K > 0,$$

can be written as a first-order system,

$$v + K\frac{\partial u}{\partial x} = 0,$$

$$\frac{\partial u}{\partial t} + \frac{\partial v}{\partial x} = 0.$$

assign unknown approximate values of v to the nodes, as in $v_i^n \simeq v(ih, nk)$, and we associate approximate values of u with the element midpoints, as in $u_i^n \simeq u((i + \frac{1}{2})h, nk)$. Thus we can derive the discrete analogs

$$u_i^{n+1} - u_i^n = -\frac{k}{h}(v_i^n - v_{i-1}^n),$$

$$v_i^{n+1} = -\frac{K}{h}(u_{i+1}^{n+1} - u_i^{n+1}).$$

Use von Neumann stability analysis to derive a stability criterion for this scheme. Hint: Let u_i^n, v_i^n have typical Fourier components $\xi^n \exp(\hat{\imath}jh)$, $\eta^n \exp(\hat{\imath}\ell h)$, respectively, and find an expression for η^n in terms of ξ^n.

6.7. References.

Allen, M.B., *Collocation Techniques for Modeling Compositional Flows in Oil Reservoirs*, Berlin: Springer-Verlag, 1984.

Allen, M.B., "Why upwinding is reasonable," in J.P. Laible et al., Eds., *Proceedings of the Fifth International Conference on Finite Elements in Water Resources*, June 18-22, 1984, Burlington, Vermont, Berlin: Springer-Verlag, 1984, pp. 13-23.

Allen, M.B., Ewing, R.E., and Koebbe, J.V., "Mixed finite-element methods for computing groundwater velocities," *Num. Methods Partial Differential Equations* 3(1985), 195-207.

Allen, M.B. and Murphy, C.L., "A finite-element collocation model for variably saturated flows in porous media," *Num. Methods Partial Differential Equations, 3* (1985), 229-239.

Allen, M.B. and Murphy, C.L., "A finite-element collocation method for variably saturated flow in two space dimensions," *Water Resour. Res.,* *22:11* (1986), 1537-1542.

Aziz, K. and Settari, A., *Petroleum Reservoir Simulation*, London: Applied Science Publishers, 1979.

Birkhoff, G. and Lynch, R.E., *Numerical Solution of Elliptic Problems*, Philadelphia: SIAM, 1984.

Breitenbach, E.A., Thurnau, D.H., and van Poolen, H.K., "Solution of the immiscible fluid flow simulation equation," *Soc. Pet. Eng. J.* (1969), 155-169.

Buckley, S.E. and Leverett, M.C., "Mechanism of fluid displacement in sands," *Trans. AIME.*, *146* (1942), 107-116.

Carey, G.F. and Oden, J.T., *Finite Elements: A Second Course,* Englewood Cliffs, New Jersey: Prentice-Hall, 1983.

Clough, R.W. and Tocher, J.L., "Finite element stiffness matrices for analysis of plate bending," in *Proceedings, Conference on Matrix Methods in Structural Mechanics*, AFFDL, TR-66-80, Wright-Patterson Air Force Base, Ohio, 1966, 15-26.

Coats, K., "An equation-of-state compositional model," *Soc. Pet. Eng. J.* (1980), 363-376.

Coats, K., Nielsen, R.L., Terhune, M.H., and Weber, A.G., "Simulation of three-dimensional, two-phase flow in oil and gas reservoirs," *Soc. Pet. Eng. J.* (1967), 377-388.

Douglas, J., Peaceman, D.W, and Rachford, H.H., "A method for calculating multidimensional immiscible displacement," *Trans. AIME, 216* (1959), 297-306.

Ewing, R.E., Ed., *The Mathematics of Reservoir Simulation,* Philadelphia: SIAM, 1983.

Griffiths, D.F. and Mitchell, A.R., "Nonconforming elements," in D.F. Griffiths, ed., *The Mathematical Basis of Finite Element Methods*, Oxford: Clarendon Press, 1984, pp. 41-70.

Koebbe, J.V., Ewing, R.E., and Lagnado, R.P., "Accurate velocity weighting techniques," presented at the Second Wyoming Symposium on Enhanced Oil Recovery, Casper, Wyoming, May 15-16, 1986.

Lapidus, L. and Pinder, G.F., *Numerical Methods for Partial Differential Equations in Science and Engineering*, New York: John Wiley and Sons, 1982.

Ortega, J. and Rheinboldt, W.C., *Iterative Solution of Nonlinear Equations in Several Variables*, New York: Academic Press, 1970.

Peaceman, D.W., *Fundamentals of Numerical Reservoir Simulation*, Amsterdam: Elsevier, 1977.

Safai, N.M., "Simulation of Saturated and Unsaturated Deformable Porous Media," Ph.D. Thesis, Department of Civil Engineering, Princeton University, Princeton, New Jersey, 1977.

Sheldon, J.W., Zondek, B., and Cardwell, W.T., "One-dimensional, incompressible, non-capillary two-phase flow in a porous medium," *Trans. AIME, 216* (1959), 290-296.

Stone, H.L. and Garder, A.O., "Analysis of gas-cap or dissolved-gas reservoirs," *Trans. AIME, 222* (1961), 92-104.

Strang, G. and Fix, G.F., *An Analysis of the Finite Element Method*, Englewood Cliffs, New Jersey: Prentice-Hall, 1973.

APPENDIX
SUMMARY OF VECTOR
AND TENSOR ANALYSIS

This Appendix reviews some of the basic facts about vectors and tensors used in our discussions of mechanics. We begin by defining notation and reviewing the algebraic operations frequently encountered in the text. Next, we briefly treat the differentiation of vectors and tensors. Finally, we review some integral theorems used throughout the text.

Notation.

We shall restrict our attention to three-dimensional Euclidean space, and in what follows we assume a Cartesian coordinate system with axes labeled $1, 2, 3$. In this framework, a **quantity of rank** n is a collection of 3^n real numbers or variables, called **entries**, each of which is labeled by n indices. The cases encountered in most treatments of mechanics are $n = 0, 1, 2$. When $n = 0$, the quantity in question is a single real number or variable s, that is, a **scalar**. Pressure, for example, is a scalar. When $n = 1$, the quantity in question is a **vector**, $\mathbf{V}$. We may write a vector explicitly as a column array,

$$\mathbf{V} = \begin{bmatrix} V_1 \\ V_2 \\ V_3 \end{bmatrix}.$$

Velocity, for example, is an ordered triple of components along the three coordinate axes and hence is a vector. Informally, a vector is a quantity having both direction and magnitude. When $n = 2$, the quantity in question is a **tensor**, $\mathbf{T}$. We can list the entries of a tensor in array form as follows:

$$\mathbf{T} = \begin{bmatrix} T_{11} & T_{12} & T_{13} \\ T_{21} & T_{22} & T_{23} \\ T_{31} & T_{32} & T_{33} \end{bmatrix}.$$

Stress is perhaps the prototypical example of a tensor. Its entries represent the three components of momentum flux across planes perpendicular to each of the three coordinate axes, as explained in Chapter One.

We use two notations to signify these different types of quantities. One is **direct** notation, in which we write scalars in ordinary type, s; vectors in boldface, $\mathbf{V}$, and tensors in boldface sans-serif type, T. This notation is the one we emphasize in the main text of the book. The other notation is **index** notation, in which we write quantities of all ranks in ordinary type, specifying the rank of a quantity through the number of free (unrepeated, nonnumerical) indices. Thus a scalar appears as s, a vector as V_i, and a tensor as T_{ij}. Observe that the variable indices i, j, and so forth, appearing in the index notation stand, not for particular numerical values, but for generic values. Thus in V_i the index i represents at once all of its possible values $1, 2$, and 3, so that V_i is actually shorthand for the ordered triple whose entries are V_1, V_2, and V_3. Compared with index notation, direct notation seems better to facilitate physical intuition, and therefore the text exhibits a bias in its favor. However, index notation has greater flexibility and is less vulnerable to ambiguity. In what follows we shall outline the basic algebraic and differential operations using both notations.

Basic algebra.

Addition of two scalars r and s is the usual operation, $r + s$. To compute the sum of two vectors $\mathbf{V}$ and $\mathbf{W}$, we add entries having the same index:

$$
\begin{bmatrix} V_1 \\ V_2 \\ V_3 \end{bmatrix} + \begin{bmatrix} W_1 \\ W_2 \\ W_3 \end{bmatrix} = \begin{bmatrix} V_1 + W_1 \\ V_2 + W_2 \\ V_3 + W_3 \end{bmatrix} = \begin{cases} \mathbf{V} + \mathbf{W} & \text{(direct notation)} \\ V_i + W_i & \text{(index notation)}. \end{cases}
$$

Addition of two tensors T and U is similar:

$$
\begin{bmatrix} T_{11} & T_{12} & T_{13} \\ T_{21} & T_{22} & T_{23} \\ T_{31} & T_{32} & T_{33} \end{bmatrix} + \begin{bmatrix} U_{11} & U_{12} & U_{13} \\ U_{21} & U_{22} & U_{23} \\ U_{31} & U_{32} & U_{33} \end{bmatrix}
$$

$$
= \begin{bmatrix} T_{11} + U_{11} & T_{12} + U_{12} & T_{13} + U_{13} \\ T_{21} + U_{21} & T_{22} + U_{22} & T_{23} + U_{23} \\ T_{31} + U_{31} & T_{32} + U_{32} & T_{33} + U_{33} \end{bmatrix}
$$

$$
= \begin{cases} \mathsf{T} + \mathsf{U} & \text{(direct notation)} \\ T_{ij} + U_{ij} & \text{(index notation)}. \end{cases}
$$

One can easily see that $\mathbf{V} + \mathbf{W} = \mathbf{W} + \mathbf{V}$, so that vector addition is commutative, and that $(\mathbf{V} + \mathbf{W}) + \mathbf{Y} = \mathbf{V} + (\mathbf{W} + \mathbf{Y})$, so that it is associative as well. Corresponding properties hold for tensor addition.

To **multiply** a vector $\mathbf{V}$ by a scalar s, simply multiply the entries of the vector by s:

$$s \begin{bmatrix} V_1 \\ V_2 \\ V_3 \end{bmatrix} = \begin{bmatrix} sV_1 \\ sV_2 \\ sV_3 \end{bmatrix} = \begin{cases} s\mathbf{V} & \text{(direct notation)} \\ sV_i & \text{(index notation)}. \end{cases}$$

The rule for tensors is similar:

$$s = \begin{bmatrix} T_{11} & T_{12} & T_{13} \\ T_{21} & T_{22} & T_{23} \\ T_{31} & T_{32} & T_{33} \end{bmatrix} = \begin{bmatrix} sT_{11} & sT_{12} & sT_{13} \\ sT_{21} & sT_{22} & sT_{23} \\ sT_{31} & sT_{32} & sT_{33} \end{bmatrix}$$

$$= \begin{cases} s\mathbf{T} & \text{(direct notation)} \\ sT_{ij} & \text{(index notation)}. \end{cases}$$

Multiplication by scalars is **commutative** in the sense that $s\mathbf{V} = \mathbf{V}s$, and it is **distributive** in the following two senses:

$$(r + s)\mathbf{V} = r\mathbf{V} + s\mathbf{V},$$

$$r(\mathbf{V} + \mathbf{W}) = r\mathbf{V} + r\mathbf{W}.$$

Similar properties hold for tensors. We can use addition and multiplication by scalars to decompose vectors into their components along the three Cartesian axes. Let us denote the **usual basis** for three-dimensional space by

$$\mathbf{e}_1 = \begin{bmatrix} 1 \\ 0 \\ 0 \end{bmatrix}, \quad \mathbf{e}_2 = \begin{bmatrix} 0 \\ 1 \\ 0 \end{bmatrix}, \quad \mathbf{e}_3 = \begin{bmatrix} 0 \\ 0 \\ 1 \end{bmatrix}.$$

Then a vector $\mathbf{V}$ decomposes as follows:

$$\begin{bmatrix} V_1 \\ V_2 \\ V_3 \end{bmatrix} = \sum_{i=1}^{3} V_i \mathbf{e}_i.$$

Transposition.

If we consider vectors to be column arrays, then the **transpose** of a vector $\mathbf{V}$ is, formally, a row array:

$$\begin{bmatrix} V_1 \\ V_2 \\ V_3 \end{bmatrix}^\top = [V_1, V_2, V_3].$$

For tensors, transposition amounts to a flip across the diagonal:

$$\begin{bmatrix} T_{11} & T_{12} & T_{13} \\ T_{21} & T_{22} & T_{23} \\ T_{31} & T_{32} & T_{33} \end{bmatrix}^{\mathsf{T}} = \begin{bmatrix} T_{11} & T_{21} & T_{31} \\ T_{12} & T_{22} & T_{32} \\ T_{13} & T_{23} & T_{33} \end{bmatrix}$$

$$= \begin{cases} \mathbf{T}^{\mathsf{T}} & \text{(direct notation)} \\ T_{ij}^{\mathsf{T}} = T_{ji} & \text{(index notation).} \end{cases}$$

A tensor $\mathbf{T}$ is **symmetric** if $\mathbf{T} = \mathbf{T}^{\mathsf{T}}$; it is **antisymmetric** if $\mathbf{T} = -\mathbf{T}^{\mathsf{T}}$. Notice that the diagonal entries of an antisymmetric tensor must all be zero.

Dot products.

The **dot product** of two vectors $\mathbf{V}$ and $\mathbf{W}$ is the following:

$$\begin{bmatrix} V_1 \\ V_2 \\ V_3 \end{bmatrix}^{\mathsf{T}} \cdot \begin{bmatrix} W_1 \\ W_2 \\ W_3 \end{bmatrix} = [V_1, V_2, V_3] \begin{bmatrix} W_1 \\ W_2 \\ W_3 \end{bmatrix} = \sum_{i=1}^{3} V_i W_i.$$

By extension of the Pythagorean theorem, we can use the dot product to define the **Euclidean length** of any vector $\mathbf{V}$:

$$\|\mathbf{V}\|_2 = (\mathbf{V} \cdot \mathbf{V})^{1/2} = \sqrt{\sum_{i=1}^{3} V_i V_i}.$$

While it may not be immediately apparent, in the last two sums the notation "$\sum_{i=1}^{3}$" is actually superfluous. Since sums of this sort appear quite frequently in vector and tensor analysis, it is common to adopt a convention that the repetition of a variable index such as i or j in a single term implies summation over the values $1, 2, 3$ for that index. This rule is called the **Einstein summation convention**, and it affords great notational convenience with very little risk of ambiguity. Using this convention, we may write the dot product of two vectors as follows: $\mathbf{V}$ and $\mathbf{W}$ in index notation as $V_i W_i$, the symbol $\sum_{i=1}^{3}$ being understood since the index i appears repeated. Notice, however, that while $V_i W_i$ is a quantity of rank 0, $V_i W_j$ has no repeated indices and is therefore a quantity of rank 2, sometimes called the **dyadic** (or **tensor**) **product** of $\mathbf{V}$ and $\mathbf{W}$:

$$\begin{bmatrix} V_1 W_1 & V_1 W_2 & V_1 W_3 \\ V_2 W_1 & V_2 W_2 & V_2 W_3 \\ V_3 W_1 & V_3 W_2 & V_3 W_3 \end{bmatrix} = \begin{cases} \mathbf{V}\mathbf{W} & \text{(direct notation)} \\ V_i W_j & \text{(index notation).} \end{cases}$$

Using the Einstein summation convention, the dot product of vectors **V** and **W** is as follows:

$$\sum_{i=1}^{3} V_i W_i = \begin{cases} \mathbf{V} \cdot \mathbf{W} & \text{(index notation)} \\ V_i W_i & \text{(direct notation)}. \end{cases}$$

This operation is commutative, and it is distributive:

$$\mathbf{V} \cdot (\mathbf{W} + \mathbf{Y}) = \mathbf{V} \cdot \mathbf{W} + \mathbf{V} \cdot \mathbf{Y}.$$

If **V** and **W** are vectors whose magnitudes are $\|\mathbf{V}\|_2$ and $\|\mathbf{W}\|_2$, respectively, and whose directions in three-space differ by an angle θ, then $\mathbf{V} \cdot \mathbf{W}$ is a scalar whose magnitude is $\|\mathbf{V}\|_2 \|\mathbf{W}\|_2 \cos \theta$. This formula implies that two vectors **V** and **W** are orthogonal precisely when $\mathbf{V} \cdot \mathbf{W} = 0$.

The dot product of a vector and a tensor is analogous to matrix multiplication:

$$\begin{bmatrix} T_{11} & T_{12} & T_{13} \\ T_{21} & T_{22} & T_{23} \\ T_{31} & T_{32} & T_{33} \end{bmatrix} \begin{bmatrix} V_1 \\ V_2 \\ V_3 \end{bmatrix} = \sum_{j=1}^{3} \begin{bmatrix} T_{1j} V_j \\ T_{2j} V_j \\ T_{3j} V_j \end{bmatrix}$$

$$= \begin{cases} \mathbf{T} \cdot \mathbf{V} & \text{(direct notation)} \\ T_{ij} V_j & \text{(index notation)}. \end{cases}$$

Alternatively, we can compute

$$\begin{bmatrix} V_1 \\ V_2 \\ V_3 \end{bmatrix}^{\mathsf{T}} \begin{bmatrix} T_{11} & T_{12} & T_{13} \\ T_{21} & T_{22} & T_{23} \\ T_{31} & T_{32} & T_{33} \end{bmatrix} = \sum_{i=1}^{3} \begin{bmatrix} V_i T_{i1} \\ V_i T_{i2} \\ V_i T_{i3} \end{bmatrix}$$

$$= \begin{cases} \mathbf{V}^{\mathsf{T}} \cdot \mathbf{T} & \text{(direct notation)} \\ V_i T_{ij} & \text{(index notation)}. \end{cases}$$

Notice that $\mathbf{T} \cdot \mathbf{V} = \mathbf{V}^{\mathsf{T}} \cdot \mathbf{T}$ for general **V** only if **T** is symmetric.

Finally, given two tensors **T** and **U**, we can from the **double-dot product**, or **contraction**, as follows:

$$\begin{bmatrix} T_{11} & T_{12} & T_{13} \\ T_{21} & T_{22} & T_{23} \\ T_{31} & T_{32} & T_{33} \end{bmatrix} : \begin{bmatrix} U_{11} & U_{12} & U_{13} \\ U_{21} & U_{22} & U_{23} \\ U_{31} & U_{32} & U_{33} \end{bmatrix}$$

$$= \sum_{i=1}^{3} \sum_{j=1}^{3} T_{ij} U_{ij} = \begin{cases} \mathbf{T} : \mathbf{U} & \text{(direct notation)} \\ T_{ij} U_{ij} & \text{(index notation)}. \end{cases}$$

It is easy to show that the double-dot product is commutative: $\mathbf{T} : \mathbf{U} = \mathbf{U} : \mathbf{T}$.

The Kronecker and Levi-Civita symbols.

Two notational tools are quite useful in defining higher algebraic operations among vectors and tensors. The first is the **Kronecker symbol**, defined as follows:

$$\delta_{ij} = \begin{cases} 1 & \text{if } i = j \\ 0 & \text{if } i \neq j. \end{cases}$$

This symbol, especially useful in index notation, corresponds in direct notation to the **identity tensor**

$$\mathbf{1} = \begin{bmatrix} 1 & 0 & 0 \\ 0 & 1 & 0 \\ 0 & 0 & 1 \end{bmatrix}.$$

The second notational tool is the **Levi-Civita symbol**. This symbol has rank 3, and its value depends on whether the indices (i, j, k) form an even permutation of $(1, 2, 3)$, an odd permutation of $(1, 2, 3)$, or a combination with one index repeated:

$$\epsilon_{ijk} = \begin{cases} 1 & \text{if } (i, j, k) = (1, 2, 3), (2, 3, 1), \quad \text{or} \quad (3, 1, 2), \\ -1 & \text{if } (i, j, k) = (1, 3, 2), (3, 2, 1), \quad \text{or} \quad (2, 1, 3), \\ 0 & \text{if any index is repeated.} \end{cases}$$

Trace and determinant.

The **trace** of a tensor is the sum of its diagonal entries:

$$\text{tr} \begin{bmatrix} T_{11} & T_{12} & T_{13} \\ T_{21} & T_{22} & T_{23} \\ T_{31} & T_{32} & T_{33} \end{bmatrix} = T_{11} + T_{22} + T_{33}$$

$$= \begin{cases} \text{tr}\,\mathbf{T} & \text{(direct notation)} \\ \delta_{ij} T_{ij} = T_{ii} & \text{(index notation)}. \end{cases}$$

The **determinant** of a tensor is defined as the determinant of the matrix having the same entries:

$$\det \begin{bmatrix} T_{11} & T_{12} & T_{13} \\ T_{21} & T_{22} & T_{23} \\ T_{31} & T_{32} & T_{33} \end{bmatrix} = T_{11}(T_{22}T_{33} - T_{23}T_{32}) - T_{12}(T_{21}T_{33} - T_{23}T_{31})$$

$$+ T_{13}(T_{21}T_{32} - T_{22}T_{31}) = \begin{cases} \det \mathbf{T} & \text{(direct notation)} \\ \epsilon_{ijk} T_{1i} T_{2j} T_{3k} & \text{(index notation)}. \end{cases}$$

One can show that both the trace and the determinant of a tensor remain invariant under transformations to new orthogonal coordinate systems.

Cross products.

The most familiar cross product is that defined for two vectors:

$$\begin{bmatrix} V_1 \\ V_2 \\ V_3 \end{bmatrix} \times \begin{bmatrix} W_1 \\ W_2 \\ W_3 \end{bmatrix} = \begin{bmatrix} V_2 W_3 - V_3 W_2 \\ V_3 W_1 - V_1 W_3 \\ V_1 W_2 - V_2 W_1 \end{bmatrix}$$

$$= \begin{cases} \mathbf{V} \times \mathbf{W} & \text{(direct notation)} \\ \epsilon_{ijk} V_j W_k & \text{(index notation)}. \end{cases}$$

It is easy to prove, using the Levi-Civita symbol, that the cross product is **anticommutative**: $\mathbf{V} \times \mathbf{W} = -\mathbf{W} \times \mathbf{V}$. Geometrically, $\mathbf{V} \times \mathbf{W}$ is a vector perpendicular to both $\mathbf{V}$ and $\mathbf{W}$. If $\mathbf{V}$ and $\mathbf{W}$ stand at an angle θ to each other, the magnitude of $\mathbf{V} \times \mathbf{W}$ is $\|\mathbf{V}\|_2 \|\mathbf{W}\|_2 \sin\theta$, which is twice the area of the parallelogram formed by $\mathbf{V}, \mathbf{W}$, and $\mathbf{V} - \mathbf{W}$.

It is also possible, though somewhat uncommon, to define cross products using tensors. For example, we can form the cross product of a vector and a tensor, yielding another tensor:

$$\begin{bmatrix} V_1 \\ V_2 \\ V_3 \end{bmatrix} \times \begin{bmatrix} T_{11} & T_{12} & T_{13} \\ T_{21} & T_{22} & T_{23} \\ T_{31} & T_{32} & T_{33} \end{bmatrix}$$

$$= \begin{bmatrix} V_2 T_{13} - V_3 T_{12} & V_2 T_{23} - V_3 T_{22} & V_2 T_{33} - V_3 T_{32} \\ V_3 T_{11} - V_1 T_{13} & V_3 T_{21} - V_1 T_{23} & V_3 T_{31} - V_1 T_{33} \\ V_1 T_{12} - V_2 T_{11} & V_1 T_{22} - V_2 T_{21} & V_1 T_{32} - V_2 T_{31} \end{bmatrix}$$

$$= \begin{cases} \mathbf{V} \times \mathbf{T} & \text{(direct notation)} \\ \epsilon_{ijk} V_j T_{mk} & \text{(index notation)}. \end{cases}$$

Similarly, we can define the cross product of two tensors to be the following scalar:

$$\begin{bmatrix} T_{11} & T_{12} & T_{13} \\ T_{21} & T_{22} & T_{23} \\ T_{31} & T_{32} & T_{33} \end{bmatrix} \times \begin{bmatrix} U_{11} & U_{12} & U_{13} \\ U_{21} & U_{22} & U_{23} \\ U_{31} & U_{32} & U_{33} \end{bmatrix}$$

$$= \sum_{m=1}^{3} \begin{bmatrix} T_{2m} U_{m3} - T_{3m} U_{m2} \\ T_{3m} U_{m1} - T_{1m} U_{m3} \\ T_{1m} U_{m2} - T_{2m} U_{m1} \end{bmatrix} = \begin{cases} \mathbf{T} \times \mathbf{U} & \text{(direct notation)} \\ \epsilon_{ijk} T_{jm} U_{mk} & \text{(index notation)}. \end{cases}$$

TABLE A-1. Algebraic identities for vectors and tensors.

Dot Products

$$\mathbf{V}_1 \cdot \mathbf{V}_2 = \mathbf{V}_2 \cdot \mathbf{V}_1$$

$$\mathbf{V}_1 \cdot (\mathbf{V}_2 + \mathbf{V}_3) = \mathbf{V}_1 \cdot \mathbf{V}_2 + \mathbf{V}_1 \cdot \mathbf{V}_3$$

$$\mathbf{T} : \mathbf{U} = \mathbf{U} : \mathbf{T}$$

Cross products

$$\mathbf{V}_1 \times \mathbf{V}_2 = -\mathbf{V}_2 \times \mathbf{V}_1$$

$$\mathbf{V}_1 \times (\mathbf{V}_2 + \mathbf{V}_3) = \mathbf{V}_1 \times \mathbf{V}_2 + \mathbf{V}_1 \times \mathbf{V}_3$$

$$\mathbf{V}_1 \times (\mathbf{V}_2 \times \mathbf{V}_3) = (\mathbf{V}_3 \cdot \mathbf{V}_1)\mathbf{V}_2 - (\mathbf{V}_1 \cdot \mathbf{V}_2)\mathbf{V}_3 \quad \text{(vector triple product)}$$

$$\mathbf{V} \times \mathbf{V} = 0$$

$$\mathbf{V}_1 \cdot (\mathbf{V}_2 \times \mathbf{V}_3) = \mathbf{V}_2 \cdot (\mathbf{V}_3 \times \mathbf{V}_1) = \mathbf{V}_3 \cdot (\mathbf{V}_1 \times \mathbf{V}_2)$$

$$(\mathbf{V}_1 \times \mathbf{V}_2) \cdot (\mathbf{V}_3 \times \mathbf{V}_4) = (\mathbf{V}_1 \cdot \mathbf{V}_3)(\mathbf{V}_2 \cdot \mathbf{V}_4) - (\mathbf{V}_1 \cdot \mathbf{V}_4)(\mathbf{V}_2 \cdot \mathbf{V}_3)$$

$$(\mathbf{V}_1 \times \mathbf{V}_2) \times (\mathbf{V}_3 \times \mathbf{V}_4) = [(\mathbf{V}_1 \times \mathbf{V}_2) \cdot \mathbf{V}_4]\mathbf{V}_3 - [(\mathbf{V}_1 \times \mathbf{V}_2) \cdot \mathbf{V}_3]\mathbf{V}_4$$

Table A-1 lists some useful algebraic identities for scalars, vectors, and tensors.

Partial differentiation.

The fundamental differential operator in vector and tensor analysis is the **del** or **nabla** operator, denoted in Cartesian coordinates by

$$\nabla = \begin{bmatrix} \partial/\partial x_1 \\ \partial/\partial x_2 \\ \partial/\partial x_3 \end{bmatrix}.$$

The vector notation used here, known as Gibbs' notation, is less appropriate for use with non-Cartesian coordinates; however, it is useful mnemonically.

The operator ∇ can act on scalar functions, vector functions, and tensor functions, provided their component functions are differentiable. In the case of a scalar function s, we can form the **gradient** ∇s, which is a function of rank 1:

$$\begin{bmatrix} \partial s/\partial x_1 \\ \partial s/\partial x_2 \\ \partial s/\partial x_3 \end{bmatrix} = \begin{cases} \nabla s & \text{(direct notation)} \\ \partial s/\partial x_i & \text{(index notation)}. \end{cases}$$

For a vector function $\mathbf{V}$ there are more possibilities. Formally taking the dot product of ∇ with $\mathbf{V}$ gives the **divergence** of the vector function,

$$\frac{\partial V_1}{\partial x_1} + \frac{\partial V_2}{\partial x_2} + \frac{\partial V_3}{\partial x_3} = \begin{cases} \nabla \cdot \mathbf{V} & \text{(direct notation)} \\ \partial V_i/\partial x_i & \text{(index notation)}. \end{cases}$$

This function is a quantity of rank 0. Similarly, it is possible to take the formal cross product of ∇ and $\mathbf{V}$, producing the **curl** of $\mathbf{V}$:

$$\begin{bmatrix} \partial V_3/\partial x_2 - \partial V_2/\partial x_3 \\ \partial V_1/\partial x_3 - \partial V_3/\partial x_1 \\ \partial V_2/\partial x_1 - \partial V_1/\partial x_2 \end{bmatrix} = \begin{cases} \nabla \times \mathbf{V} & \text{(direct notation)} \\ \epsilon_{ijk}\partial V_k/\partial x_j & \text{(index notation)}. \end{cases}$$

The curl is clearly a function of rank 1. To get a function of rank 2 we can form the **vector gradient** of $\mathbf{V}$:

$$\begin{bmatrix} \partial V_1/\partial x_1 & \partial V_1/\partial x_2 & \partial V_1/\partial x_3 \\ \partial V_2/\partial x_1 & \partial V_2/\partial x_2 & \partial V_2/\partial x_3 \\ \partial V_3/\partial x_1 & \partial V_3/\partial x_2 & \partial V_3/\partial x_3 \end{bmatrix} = \begin{cases} \nabla \mathbf{V} & \text{(direct notation)} \\ \partial V_j/\partial x_i & \text{(index notation)}. \end{cases}$$

In the special case when the vector function $\mathbf{V}$ is the gradient of a scalar, say ∇s, taking the divergence of $\mathbf{V}$ yields

$$\frac{\partial^2 s}{\partial x_1^2} + \frac{\partial^2 s}{\partial x_2^2} + \frac{\partial^2 s}{\partial x_3^2} = \begin{cases} \nabla \cdot (\nabla s) = \nabla^2 s & \text{(direct notation)} \\ \partial^2 s/\partial x_i \partial x_i & \text{(index notation)}. \end{cases}$$

TABLE A-2. Differential identities for vectors and tensors.

Gradients

$$\nabla(r+s) = \nabla r + \nabla s$$

$$\nabla(rs) = r\nabla s + s\nabla r$$

$$\nabla(\mathbf{V} \cdot \mathbf{W}) = \mathbf{V} \cdot \nabla\mathbf{W} + \mathbf{W} \cdot \nabla\mathbf{V} + \mathbf{V} \times \nabla \times \mathbf{W} + \mathbf{W} \times \nabla \times \mathbf{V}$$

Divergences

$$\nabla \cdot (\mathbf{V} + \mathbf{W}) = \nabla \cdot \mathbf{V} + \nabla \cdot \mathbf{W}$$

$$\nabla \cdot (s\mathbf{V}) = \mathbf{V} \cdot \nabla s + s\nabla \cdot \mathbf{V}$$

$$\nabla \cdot (\mathbf{V} \times \mathbf{W}) = \mathbf{W} \cdot \nabla \times \mathbf{V} - \mathbf{V} \cdot \nabla \times \mathbf{W}$$

$$\nabla \cdot (\nabla \times \mathbf{V}) = 0$$

$$\nabla \cdot (s\mathbf{1}) = \nabla s \quad (\mathbf{1} = \text{identity tensor})$$

$$\nabla \cdot (\mathbf{T} \cdot \mathbf{V}) = \mathbf{T} : \nabla\mathbf{V} + \mathbf{V} \cdot (\nabla \cdot \mathbf{T})$$

$$\nabla \cdot (\mathbf{V} \times \mathbf{T}) = \mathbf{V} \times (\nabla \cdot \mathbf{T}) + (\nabla\mathbf{V}) \times \mathbf{T}$$

Curls

$$\nabla \times (\mathbf{V} + \mathbf{W}) = \nabla \times \mathbf{V} + \nabla \times \mathbf{W}$$

$$\nabla \times (s\mathbf{V}) = s\nabla \times \mathbf{V} - \mathbf{V} \times \nabla s$$

$$\nabla \times (\mathbf{V} \times \mathbf{W}) = (\nabla \cdot \mathbf{W})\mathbf{V} - (\nabla \cdot \mathbf{V})\mathbf{W} + \mathbf{W} \cdot \nabla\mathbf{V} - \mathbf{V} \cdot \nabla\mathbf{W}$$

$$\nabla \times \nabla \times \mathbf{V} = \nabla(\nabla \cdot \mathbf{V}) - \nabla^2 V$$

$$\nabla \times \nabla s = 0$$

The operator ∇^2 appears frequently in mechanics; it is called the **Laplace operator**.

Finally, given a tensor function **T** we can form its divergence:

$$\sum_{i=1}^{3} \begin{bmatrix} \partial T_{i1}/\partial x_i \\ \partial T_{i2}/\partial x_i \\ \partial T_{i3}/\partial x_i \end{bmatrix} = \begin{cases} \nabla \cdot \mathbf{T} & \text{(direct notation)} \\ \partial T_{ij}/\partial x_i & \text{(index notation)}. \end{cases}$$

This derivative is a function of rank 1.

Table A-2 contains some useful differential identities using the operator ∇.

Integral theorems.

There are three important integral theorems that find application in various parts of the book. While we review the basic statements of these theorems here, a formal treatment of them would require attention to quite a few technicalities that lie beyond the scope of this text. We refer the interested reader to Crowell, Williamson, and Trotter, *Calculus of Vector Functions* (1972), cited in the references to Chapter One.

The first of the theorems, the **Gauss** or **divergence theorem**, is essentially a generalization of the fundamental theorem of calculus to volume integrals. Suppose $\mathcal{V}$ is a connected region in three-dimensional Euclidean space having a smooth, closed boundary $\partial\mathcal{V}$. Suppose further that $\partial\mathcal{V}$ is orientable, so that it is possible to choose an unambiguous outward direction on $\partial\mathcal{V}$. Let **n** stand for a vector function defined on the points **x** of $\partial\mathcal{V}$ such that $\mathbf{n}(\mathbf{x})$ has unit length, is normal to the plane tangent to $\partial\mathcal{V}$ at **x**, and points in the outward direction as drawn in Figure A-1. Then if **V** is a continuously differentiable vector-valued function defined on $\mathcal{V}$,

$$\int_{\mathcal{V}} \nabla \cdot \mathbf{V} \, dx = \oint_{\partial\mathcal{V}} \mathbf{V} \cdot \mathbf{n} \, dx.$$

The integral on the left in this equation is a volume integral over $\mathcal{V}$; the integral on the right is a surface integral over the closed surface $\partial\mathcal{V}$.

The second integral theorem is sometimes called **Green's theorem**. This theorem is a three-dimensional analog of integration by parts in that it allows one to shift differential operations in a volume integral while incurring a contribution from the boundary of the volume. Let $\mathcal{V}$, $\partial\mathcal{V}$, and **n** be as above, and suppose f and g are two scalar functions that are sufficiently smooth to permit all of the differential operations called for below. Then

$$\int_{\mathcal{V}} f\nabla^2 g \, dx = -\int_{\mathcal{V}} (\nabla f) \cdot (\nabla g) \, dx + \oint_{\partial\mathcal{V}} f \nabla g \cdot \mathbf{n} \, dx.$$

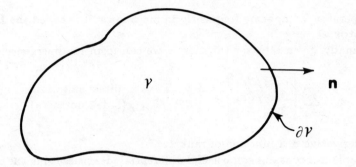

Figure A-1. A connected region $\mathcal{V}$ with smooth, orientable boundary $\partial \mathcal{V}$ and unit outward normal vector **n**.

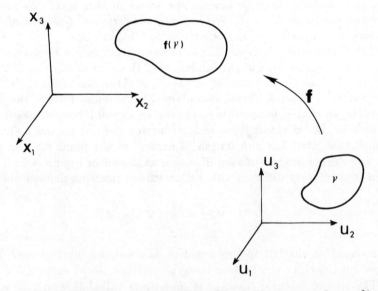

Figure A-2. Schematic of the relationship between a volume $\mathcal{V}$ in u-coordinates and the same volume as it appears under the change of variables $\mathbf{x} = \mathbf{f(u)}$.

The last integral theorem we shall review tells how to change variables in a volume integral. Suppose ϕ is an integrable scalar function defined on three-dimensional Euclidean space, and consider a continuously differentiable change of variables $\mathbf{x} = \mathbf{f}(\mathbf{u})$ mapping points in a region $\mathcal{V}$ in $\mathbf{u}$-coordinates to points in a region $f(\mathcal{V})$ in $\mathbf{x}$-coordinates, as drawn in Figure A-2. Denote by $\mathbf{J}$ the Jacobian matrix of the mapping $\mathbf{f}$, namely,

$$
\mathbf{J} = \begin{bmatrix} \partial x_1/\partial u_1 & \partial x_1/\partial u_2 & \partial x_1/\partial u_3 \\ \partial x_2/\partial u_1 & \partial x_2/\partial u_2 & \partial x_2/\partial u_3 \\ \partial x_3/\partial u_1 & \partial x_3/\partial u_2 & \partial x_3/\partial u_3 \end{bmatrix}.
$$

Then the change of variables theorem relates the integral of ϕ over volumes in $\mathbf{x}$-coordinates to integrals over volumes in $\mathbf{u}$-coordinates as follows:

$$
\int_{\mathbf{f}(\mathcal{V})} \phi(\mathbf{x})d\mathbf{x} = \int_{\mathcal{V}} \phi(\mathbf{f}(\mathbf{u}))\,|\det(\mathbf{J})|d\mathbf{x}.
$$

INDEX